THE PRESSUREMETER AND ITS MARINE APPLICATIONS: SECOND INTERNATIONAL SYMPOSIUM

A symposium sponsored by
ASTM Committee D-18 on
Soil and Rock,
Minerals Management
Service/Technology Assessment and
Research Program, and
U.S. Army Engineer Waterways
Experiment Station
Texas A&M University
College Station, Texas, 2–3 May 1986

ASTM SPECIAL TECHNICAL PUBLICATION 950

Jean-Louis Briaud, Texas A&M University, and
Jean M. E. Audibert, Earth Technology
Corporation, editors

ASTM Publication Code Number (PCN)
04-950000-38

ASTM 1916 Race Street, Philadelphia, PA 19103

Library of Congress Cataloging-in-Publication Data

The Pressuremeter and its marine applications.

(ASTM special technical publication; 950)
"The Second International Symposium on the Pressuremeter and Its Marine Applications"—Foreword.
"A symposium sponsored by ASTM Committee D-18 on Soil and Rock [and] Minerals Management Service/ Technology Assessment and Research Program and U.S. Army Engineer Waterways Experiment Station, [held at] Texas A&M University, College Station, Texas, 2–3 May 1986."
"ASTM publication code number (PCN) 04-95000-38."
Includes bibliographies and index.
1. Soil penetration test—Congresses. 2. Marine geotechnique—Congresses. 3. Soil mechanics—Congresses. I. Briaud, J.-L. II. Audibert, Jean M. E. III. International Symposium on the Pressuremeter and Its Marine Applications (2nd : 1986 : Texas A&M University, College Station) IV. ASTM Committee D-18 on Soil and Rock. V. United States. Minerals Management Service. Technology Assessment and Research Program. VI. U.S. Army Engineer Waterways Experiment Station. VII. Series.
TA710.5.P74 1986 624.1′5136 86-3337
ISBN 0-8031-0454-5

Library of Congress Catalog Card Number: 86-3337

NOTE

The Society is not responsible, as a body, for the statements and opinions advanced in this publication.

Printed in Baltimore, MD
April 1986

Foreword

The Second International Symposium on the Pressuremeter and Its Marine Applications was presented at Texas A&M University, College Station, Texas, on 2–3 May 1986. The symposium was sponsored by ASTM Committee D-18 on Soil and Rock, Minerals Management Service/Technology Assessment and Research Program, and the U.S. Army Engineer Waterways Experiment Station. Jean-Louis Briaud, Texas A&M University and Jean M. E. Audibert, Earth Technology Corporation, served as chairmen of the symposium and as editors of this publication.

Related ASTM Publications

Strength Testing of Marine Sediments: Laboratory and In-Situ Measurements, STP 883 (1985), 04-883000-38

Hydraulic Barriers in Soil and Rock, STP 874 (1985), 04-874000-38

Measurements of Rock Properties at Elevated Pressures and Temperatures, STP 869 (1985), 04-869000-38

Laterally Loaded Deep Foundations: Analysis and Performance, STP 835 (1984), 04-835000-38

Behavior of Deep Foundations, STP 670 (1979), 04-670000-38

A Note of Appreciation to Reviewers

The quality of the papers that appear in this publication reflects not only the obvious efforts of the authors but also the unheralded, though essential, work of the reviewers. On behalf of ASTM we acknowledge with appreciation their dedication to high professional standards and their sacrifice of time and effort.

ASTM Committee on Publications

ASTM Editorial Staff

Bill Benzing
Janet R. Schroeder
Kathleen A. Greene

Contents

SOIL PROPERTIES FROM PRESSUREMETER DATA

FOUNDATION DESIGN USING PRESSUREMETER DATA

PRESSUREMETER AND LATERALLY LOADED PILES

Introduction

The first Symposium on the Pressuremeter and Its Marine Applications was held in Paris on April 19 and 20, 1982. It was organized by the Institut Francais du Petrole and the Laboratoires des Ponts et Chaussees. The proceedings are available from Editions Technip, 27 rue Ginoux, 75737, Paris, France. This Special Technical Publication (STP) is based on the Second International Symposium on the Pressuremeter and Its Marine Applications which took place at Texas A&M University on May 2 and 3, 1986. This Second International Symposium was sponsored by the American Society of Testing and Materials(ASTM), the Minerals Management Service/Technology Assessment and Research Program, and the U.S. Army Engineer Waterways Experiment Station. The coeditors suggest that this type of symposium become a regular event under the possible acronym of ISPA: International Symposium on the Pressuremeter and Its Applications.

The need for having a second symposium and for having it in the United States stemmed from the effort undertaken by ASTM Committee D.18 on Soil and Rock to prepare a standard on pressuremeter testing. The effort started in 1980 when Section D.18.02.07 on Pressuremeter Testing was created as a subdivision of Subcommittee D.18.02 on Sampling and Related Field Testing for Soil Investigations. Section D.18.02.07 focussed on developing a standard for the onshore pressuremeter test. Great progress has been made, but a consensus has not yet been reached. In 1983, ASTM Subcommittee D18.13 on Marine Geotechnics decided to start work on guidelines for the offshore pressuremeter test. Soon after, they realized the drastic difficulties that one would face in establishing such guidelines. It was then decided to organize this symposium in order to gather the latest information on the pressuremeter test. This information and the discussions which took place during the symposium will greatly help the work of Subcommittee D18.13 and Section D18.02.07.

Of the 47 abstracts received,   papers from the USA, France, the United Kingdom, the Netherlands, Italy, Canada, and Thailand were finally accepted for publication in STP 950. They cover 5 different topics. The first topic is pressuremeter testing: these papers describe how the pressuremeter is deployed offshore and to some extent onshore, and how the test is run. The second topic is theory and the pressuremeter: these papers describe recent advances in the theoretical interpretation of the pressuremeter test. The third topic is soil properties from pressuremeter data: these papers describe what soil parameters can be obtained from pressuremeter data and how to obtain them. The fourth topic is foundation design using pressuremeter data: these papers describe using pressuremeter parameters for the design of deep and shallow foundations subjected to vertical loading. The fifth topic is pressuremeter and laterally loaded piles: these papers describe using the pressuremeter curve in the design of piles subjected to horizontal loading. Although many questions remain unanswered, this STP is a valuable tool for the engineers and researchers who seek current knowledge on the pressuremeter and its application.

Finally, the coeditors wish to sincerely thank the ASTM staff in Philadelphia for the flawless preparation of this symposium and for the timely completion of this publication.

Jean-Louis Briaud
Civil Engineering Department
Texas A&M University
College Station, TX 77843
Symposium cochairman and editor

Jean M.E. Audibert
Earth Technology Corporation
7020 Portwest Drive, Suite 150
Houston, TX 77024
Symposium cochairman and editor

Pressuremeter Testing

Françoise Brucy and Pierre LeTirant

USE OF PAM AND PRESSUREMETERS IN OFFSHORE FOUNDATION DESIGN

REFERENCE: Brucy, F. and Le Tirant, P., "Use of PAM and Pressuremeters in Offshore Foundation Design," The Pressuremeter and Its Marine Applications: Second International Symposium, ASTM STP 950, J.L. Briaud and J.M.E. Audibert, Eds. American Society for Testing and Materials, 1986.

ABSTRACT: The pressuremeter is an in situ investigation method and also describes the stress-strain behavior of the soil. The Self-Boring technique is now available for use in water depths up to 1000 m as the result of the development of the "PAM." It consists principally of a seabed frame from which a measuring probe is made to penetrate into the seabed. Pressuremeter data and self-boring parameter logs are evaluated on the basis of the results acquired to date from different sites. Some comparative data are presented. Then all different ways to use pressuremeter data results in the design of foundations are reviewed. Fundamental analyses still have to face many problems. Correlations with other soil parameters are needed but must be used judiciously. Finally, direct design methods appear to be the most promising ones.

KEYWORDS: pressuremeter, offshore, self-boring, correlation, design method.

INTEREST OF THE PRESSUREMETER AND SELF-BORING PRESSUREMETER FOR OFFSHORE GEOTECHNICS

Pressuremeter testing is increasingly being used in offshore geotechnical engineering. It has the advantage of

Mrs. Brucy and Mr. Le Tirant are respectively geotechnical engineer and head of Marine Geotechnics Department at Institut Français du Pétrole, B.P. 311, 92506 Rueil-Malmaison, France.

being an in situ investigation method from which the stress-strain behavior of the soil can be assessed.

Emphasis has long been placed in Europe on the in situ determination of soil properties. Offshore environmental conditions are likely to give more and more importance to in situ measuring techniques. Certainly, drilling and coring methods are the most widely used. There are obvious advantages in obtaining a soil sample that can be observed and tested in the laboratory. But coring techniques now have to cope with the difficulty of operating in deeper and deeper water. More or less small-diameter core samples taken at great water depths are necessarily subjected to strain relief. This is especially the case when sediments contain gas. Taking samples of very soft soils may also be very difficult under great water depths. In this context, in situ measurements and particularly pressuremeter measuring have an obvious role to play. By getting free of sample-taking problems, they provide direct access to the soil properties in situ and are therefore already a way of assessing and completing the results of conventional laboratory tests on samples.

The pressuremeter test, like the cone penetration test, provides an in situ measurement of the failure or limit strength characteristics of the soil. But it also has the unique advantage of providing a stress-strain description of the soil. The standard test makes a complete loading curve available as a matter of simple routine. Deformation moduli can then be obtained. Moreover, the principle of pressuremeter loading can be used in a great many ways to modulate a wide variety of soil tests. Relaxation tests and the infinite range of cyclic loading tests can be performed with the pressuremeter, so as to obtain answers to specific problems engineers have to cope with.

The self-boring process developed over the last fifteen years jointly by the Laboratoires des Ponts et Chaussées (France) and the University of Cambridge (U.K.) has led to very fruitful applications of the pressuremeter method. It offers the attractive feature of minimum disturbance of borehole walls and thus of the soil to be tested.

It is well known in geotechnics that testing conditions are of great importance for the results obtained. The self-boring technique provides an opportunity for careful control of the way the probe is inserted in the soil. At any rate it ensures the uniformity of the conditions under which the probe is positioned at the test location, which is not the case for other processes of inserting probes. Good repeatability of the test can thus be obtained.

From a practical point of view, the self-boring technique has the subsequent advantage of combining boring and the insertion of the testing probe, resulting in a more efficient operation. It may be pointed out here that, for offshore applications, this condition has to be fulfilled by all conventional type pressuremeters. Solutions have been developed involving either the vibrodriving of a testing probe like the Pressio-Penetrometer or the thrusting in of the probe like the Push-In-Pressuremeter (1).

The pressuremeter and the self-boring pressuremeter technique has not, up to now, played a very important part in marine geotechnical investigations, but it appears at present to be emerging as an important new offshore in situ technology. The self-boring technique is now available for use under great water depths as the result of the development of the PAM device (from the French "Pressiomètre Autoforeur Marin"). The PAM can make pressuremeter measurements in water depths up to 1000 m (3300 ft) with a penetration of 60 m into the seabed soil.

DESCRIPTION OF THE PAM DEVICE AND OPERATION

The PAM consists principally of a seabed frame from which a measuring probe is made to penetrate into the seabed. Detailed descriptions were given at the 1982 Symposium on the Pressuremeter and its Marine Applications (2)(3). Thus only a brief review will be given hereunder.

The PAM Probe and Seabed Frame

The PAM measuring probe is 6 m (20 ft) long and 160 mm (6 1/2 in) in diameter. It has been derived from the PAF-76 probe developed in France by the Laboratoires des Ponts et Chaussées. Figure 1 shows the PAM probe, with its self-boring module, pressuremeter module and driving module.

The self-boring module makes the PAM penetrate into the seabed soil without rotating. The soil is first cut through by a cutting ring then pulverized by an inside grinder and washed up inside the probe by water circulation. The soil from the borehole wall is almost certainly not subjected to any compression or displacement during penetration by the probe. The pressuremeter module consists of a cylindrical cell 160 mm in diameter, with a length-to-diameter ratio of 2. Dilation may attain 25% of the initial volume. The driving module is usually a drill collar weighing 8 kN (1500 lb), sometimes a packer and jack module for step-by-step penetration.

The PAM probe is connected to the seabed frame by an

electric and hydraulic umbilical 70 m (230 ft) long and 125 mm (5 in) in diameter.

The seabed frame supports all the equipment that is required to operate the PAM probe (Fig. 2). It has the following main characteristics :

. height	8.5 m (28 ft)
. weight in air	16 t
. width for transportation	2.5 m (8 ft)
. overall dimensions when operating	6 x 7 m (20 x 23 ft)
. area of footpads	20 m^2 (215 sq ft)

It is equipped with a winch for the storage of the probe umbilical and also with a seawater circulation pump for self-boring operations, a "pressure-volume" controller (called CPV) for pressuremeter testing, and two main electrohydraulic plants in equipressure.

The PAM is operated from the surface via a single electric umbilical which supplies power to the seabed equipment and transmits remote controls and data by multiplex. On the surface the monitoring and data-acquisition equipment is centralized in a control cabin on the support vessel.

Operating the PAM

The PAM is operated from a ship equipped with an A-frame or crane and a winch. Figure 3 shows a DP support vessel which has already been used to operate the PAM. The PAM frame is lowered to the seabed by the hoisting cable, with the electric umbilical secured to it. It is generally lowered at a rate of about 10 meters per minute. This frame is equipped with two remote-controlled propellers as well as a compass so that its orientation in the water can be monitored from the control cabin. The orientation of the PAM on the seabed can also be controlled to compensate for slopes.

Boring can then begin. The probe umbilical is paid out from its winch, while the probe leaves the frame and self-bores its way into the seabed. The rate of penetration is chosen between 5 and 35 cm/min, and is generally set at 12 cm/min (6 in/min). When the probe with the pressuremeter cell has reached the required depth in the seabed, it is stopped and a pressuremeter test is performed.

Due to offshore operating criteria, the following procedures are mostly recommended: (1) use of a 1% over-sized cutting edge; (2) limiting the relaxation time of the pressuremeter cell to one or two minutes before its expan-

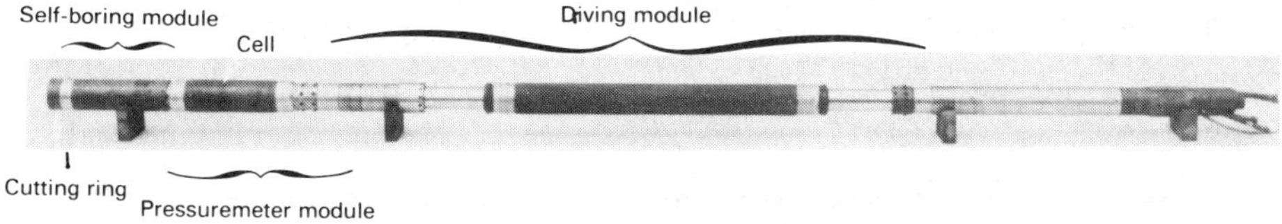

FIG. 1. PAM Self-Boring Probe.

FIG. 2. PAM Seabed Frame.

FIG. 3. Operating Support Vessel.

sion; (3) setting the rate of expansion at 5% per minute.

Experience

The PAM was systematically tested on onshore sites in 1979 and 1980. Since 1981, it has been tested at sea, in water depths ranging from several meters to a record depth of 625 m (4).

A typical 50 m boring operation with pressuremeter tests every one or two meters takes about 24 hours to perform.

The PAM has now acquired thorough offshore operating experience. Altogether, it has been used for investigations in a wide range of soil types.

EVALUATION OF PAM SELF-BORING PRESSUREMETER MEASUREMENTS

The PAM self-boring pressuremeter measurements will be evaluated hereunder on the basis of the results acquired to date from different sites.

Self-boring parameters are recorded every 10 centimeters while the probe is penetrating into the seabed. Pressuremeter diagrams are recorded while the pressuremeter is inflated at a given level.

Self-Boring Control Parameter Logs

The self-boring control parameters consist in the pressure inside the undilated pressuremeter p_i, the torque on the grinder C_r and the water circulation pressure p_w. These parameters enable the operator to be constantly aware of the operating conditions during the advancement of the borehole and thus of the pressuremeter testing conditions.

The water circulation pressure measures the efficiency of sediment evacuating inside the probe. Deficient evacuation may result in both the formation of a sediment plug at the end of the probe and the consequent increase in the tip force as well as the displacement of sediments outside the cutting edge which causes compression and remolding of the soil. A piston effect on the tip may result. Fracturing of the formation has even been observed by Clough and Denby (5). Early PAM experience in some highly plastic hard clay led to the development of the evacuating system with particular care. The water circulation flow rate, the passage of the sediment inside the probe, and sediment grinding above the cutting ring were adjusted. No main problems were encountered afterwards, while circulation pressures always remain stable, at about 0.1 MPa.

Pressure p_i was found to provide a significant reflection of the quality of the self-boring penetration because it depends directly on the state of the borehole walls. An accidental displacement of the sediment outside the cutting edge produces a pressure rise inside the pressuremeter. Under regular operating conditions, the use of an oversized cutting edge produces p_i logs very similar to the hydrostatic pressure because of the lateral decompression of the soil around the pressuremeter.

Soil Reaction Logs

The log of the probe's driving force describes the reaction of the soil to the penetration of the probe. It is used as a control parameter since its value governs the feasibility of probe penetration, and it may also be used for geotechnical interpreting since the soil reaction varies according to the nature and stiffness of the formations crossed through. Before interpreting, the following elements also have to be taken into consideration. First, it must be borne in mind that penetration by gravity will produce a very smooth profile, while step-by-step penetration will produce typical sawtooth profiles. The rate of penetration may also theoretically have an effect on soil reaction values, but since the rate is commonly set at the same values in most soils, no differences in the PAM soil reaction profiles have to be considered for this parameter. Temporary increases in the soil reaction measurement have also been observed when probe penetration was resumed after pressuremeter testing at a given depth. This is attributed to some sticking of the soil on the inflatable cell. Moreover, temporary refusals have even been encountered, and the probe has had to be lifted slightly in its hole before penetration could be resumed. Then the driving force observed is low. These interpretations are illustrated in Figure 4 where a corrected soil reaction log is also shown. It can be seen there that temporary refusals were obtained below 43 m as the soil reaction exceeded the total weight of the probe and drill collar. However, boring could proceed and pressuremeter tests be performed, down to a depth of 50 m below the mudline, thanks to the probe lifting procedure.

Figure 5 plots the reaction averages Q versus the corresponding p_{20} dilation pressure (see next section) obtained for three sites. Both parameters are believed to be indicative of the limit property of the soil. The still limited available data for Q do not enable any simple correlations to be established. Q may also be compared to

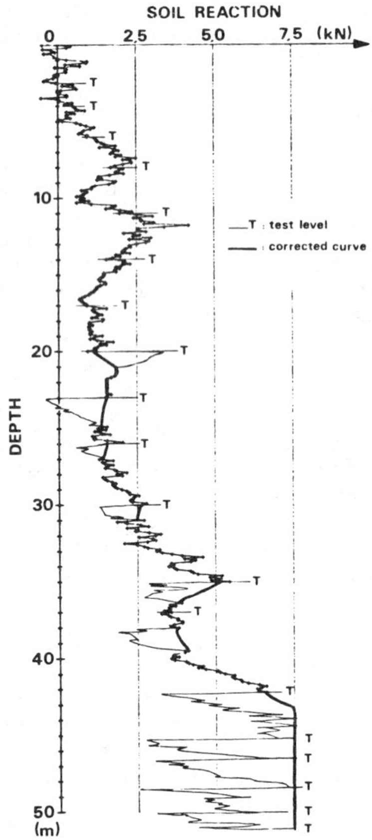

FIG. 4 - Example of an Interpreted Soil Reaction Log

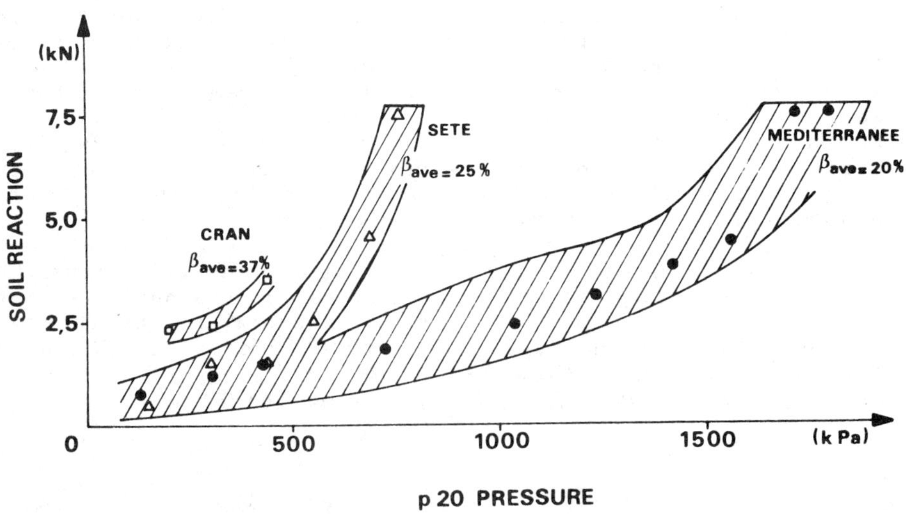

FIG. 5 - Soil Reactions versus p_{20} Dilation Pressures

cone penetration resistance q_c. Actually it was found that Q was proportional to q_c but again at this stage, more data would be necessary to determine a truly reliable relationship.

Pressuremeter Diagrams

Pressuremeter diagrams and the shear curve diagrams derived from them have been obtained in various types of soils such as clays, silts, and sands, including one calcareous sand. According to the nature of the soil, the shape of the diagrams is more or less rounded and may be characterized by a coefficient called the soil coefficient which measures a sort of curvature rate $\beta = (p_{20} - p_5)/(p_{20} - p_0)$.

As stated in a companion paper (6), experimental evidence with the PAM and PAF-76 apparatus has shown that soils can be identified through self-boring pressuremeter measurements using coefficient β conjugated with values of p_{20}.

Figure 6 shows typical diagrams obtained in a carbonate sand at the Plouasne site. It is emphasized here that carbonate sands show specific features compared to siliceous sands. The initial part of the pressure/volume curve looks like a sand curve, while the end part is similar to a clay curve, and the derived shear curve features a plateau like a clay. Coefficient β typically lies in the range of 30% to 35% like for clays and silty clays. The shape of unload-reload loops confirms the importance of compressibility for such calcareous sands, but work remains to be done to characterize and quantify it from pressuremeter results.

Pressuremeter Data Logs

Pressure p_0: The pressure p_0 existing inside the pressuremeter cell when it just comes into contact with the soil is an important parameter as the starting point for the interpretation of pressuremeter measurements and because of its potential for estimating the total horizontal stress in situ. It should be remembered that only two minutes are routinely allowed for relaxation in offshore operations. The use of an oversized cutting ring produces an inflection on the pressure/volume curve whose abcissa is assumed to measure p_0. A comparison of the volume ordinate at p_0 to the theoretical annulus created by the ring serves to evaluate the reliability of the p_0 estimate. Actually, experience in soft to medium clay or silty clay shows that p_0 is a good estimate of the horizontal stress of the soil (Fig. 7). In sands and silts, p_0 measures a certain decompression of the soil

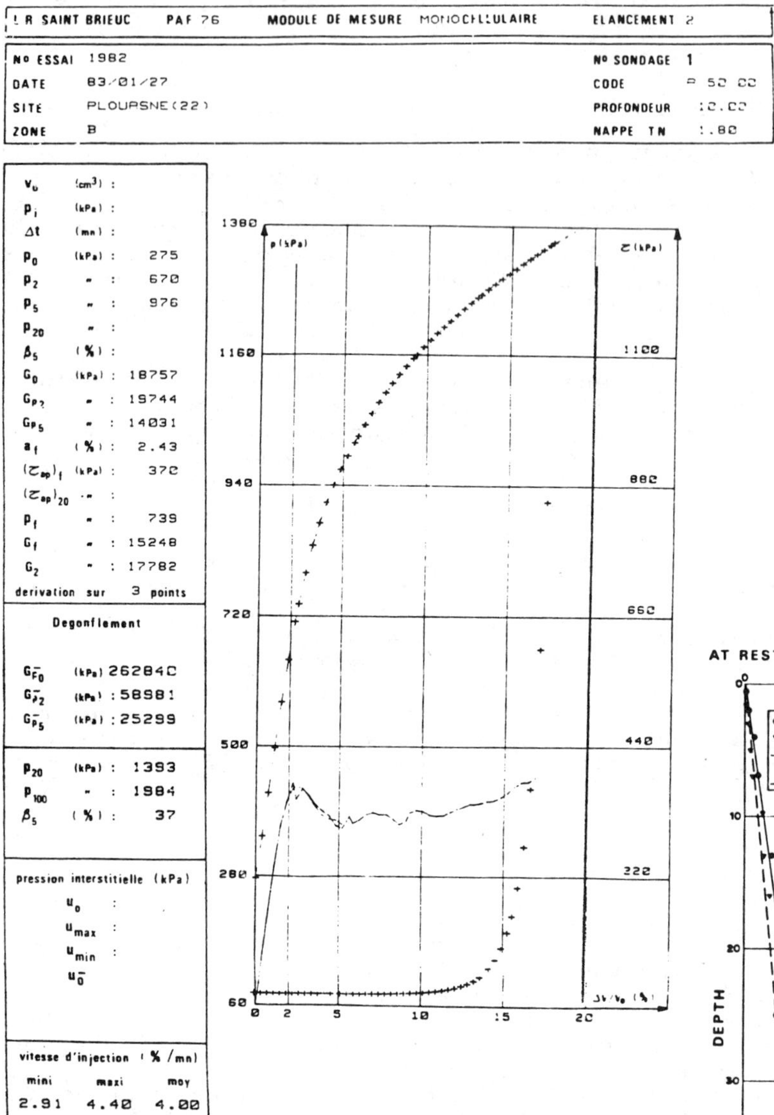

FIG. 6 - Typical Pressuremeter Diagrams in Calcareous Soil

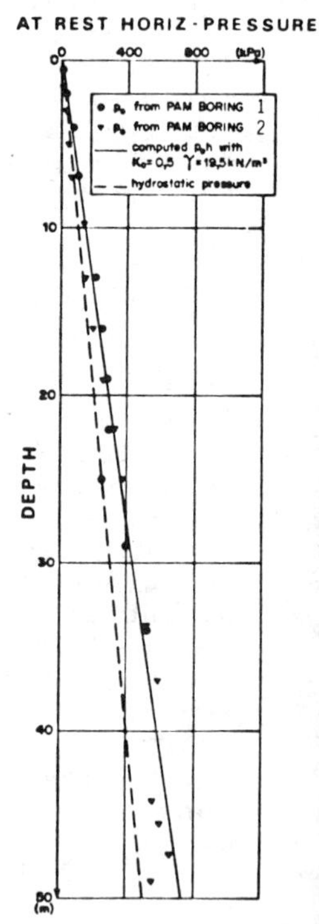

FIG. 7 - At Rest Pressure Log

because the borehole tends to close in.

Pressure p_{20}: The dilation pressure p_{20} applied to the soil when the cell has reached a volumetric strain of 20% characterizes the limit property of the soil and is widely used as input for foundation designing (6). The initial conditions of the pressuremeter test may have an effect on the p_{20} values measured as was observed on two tests off Sète, France (Fig. 8), but in all other cases, the values measured show practically no dispersion (Fig. 9). The p_{20} pressures can be related to cone resistance values which are basically a measurement of the soil property at failure. Available data from different sites where both Cone Penetration Tests and Self-Boring Pressuremeter Tests were performed (4)(7)(8) are given in Figure 10 for clays and silty clays. The correlation between the two variables appears quite simple and can be taken as $1 \leqslant q_c/p_{20} \leqslant 3$.

for clayey deposits. Figure 11 shows similar data obtained in Plouasne calcareous formations where it is seen that the p_{20} profile follows the cone resistance profile quite closely and gives a good representation of the different cemented seams.

Shear strength : Shear strength profiles analytically derived from self-boring pressuremeter test diagrams are known to be much higher than other reference data. Some authors have discussed whether the difference could be partly due to the presence of a remolded soil annulus, to the use of an incorrect value of p_0 (9), and also to the simplifying plane-strain assumption (10) which can induce false peaks. It is also argued that the criteria used in the analysis of results from different types of tests may be essentially different with respect for example to the strain rate or to the stress paths followed (11) so that a single parameter value cannot be defined. Figure 12 shows a comparison of relevant data obtained in a Mediterranean Sea site.

Moduli: Moduli are obtained from self-boring pressuremeter tests. Initial tangent moduli are very sensitive to even slight borehole wall disturbances and show much dispersion, while secant moduli (typically Gp_2) provide reliable data that can be correlated to the usual undrained shear modulus (10). Also, as stated above, the investigation of the unload-reload modulus may prove useful, especially with respect to calcareous formations.

USE OF PAM AND PRESSUREMETERS IN OFFSHORE FOUNDATION DESIGN

The results of PAM or other pressuremeter tests may be

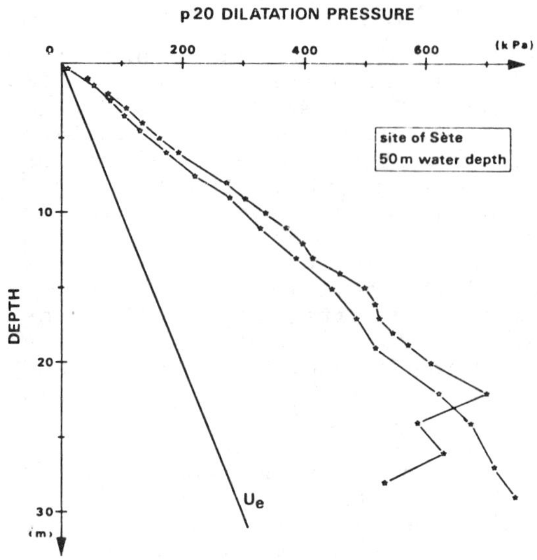

FIG. 8 - P_{20} Pressure Log.
2 Borings off Sète

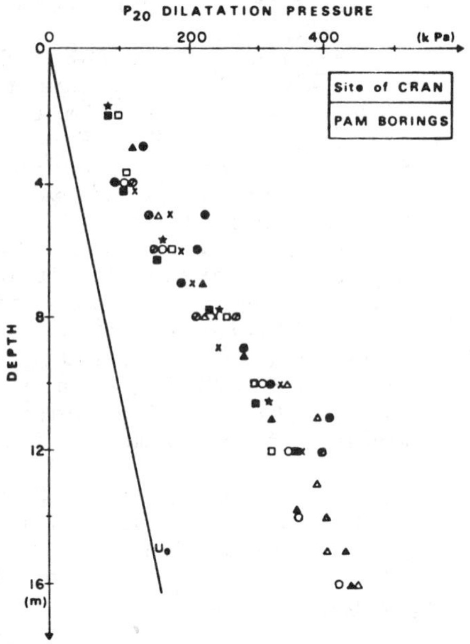

FIG. 9 - P_{20} Pressure Log.
8 Borings at Cran

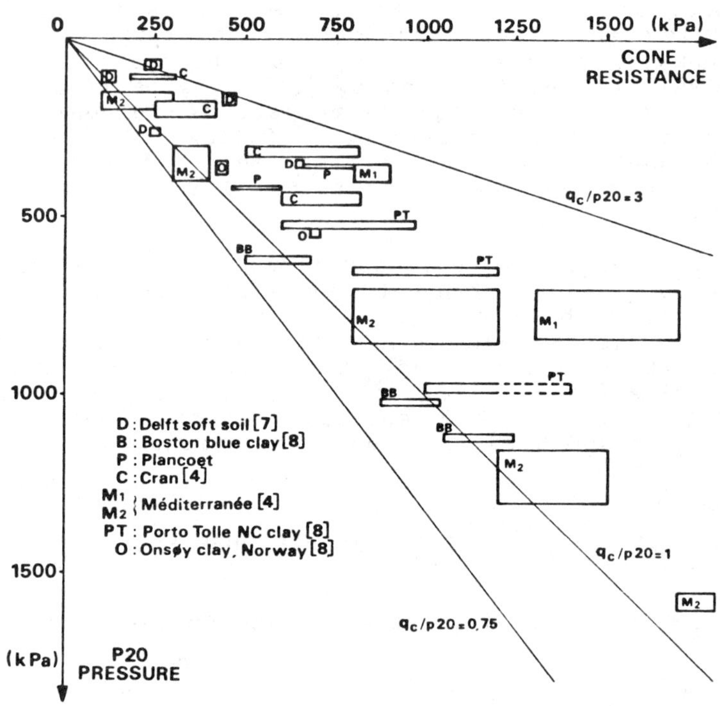

FIG. 10 - Correlation p_{20} Pressure to q_c Cone Resistance

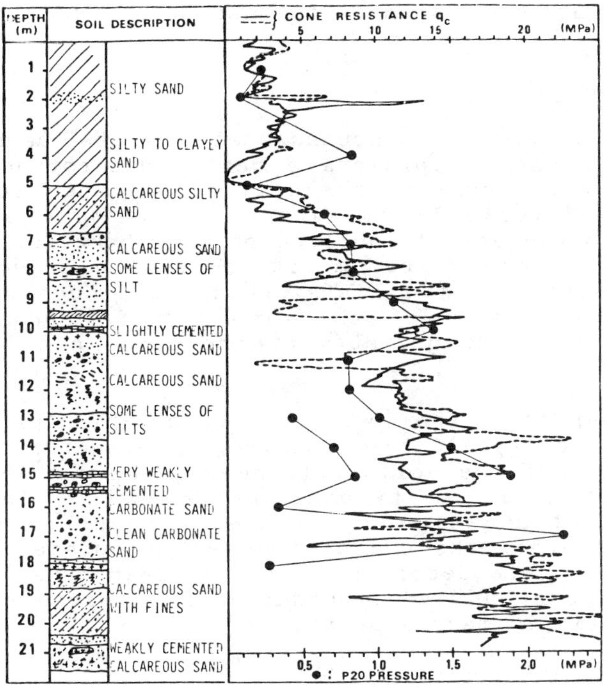

FIG. 11 - Cone Resistance and p_{20} Logs at Plouasne Calcareous Sand Site.

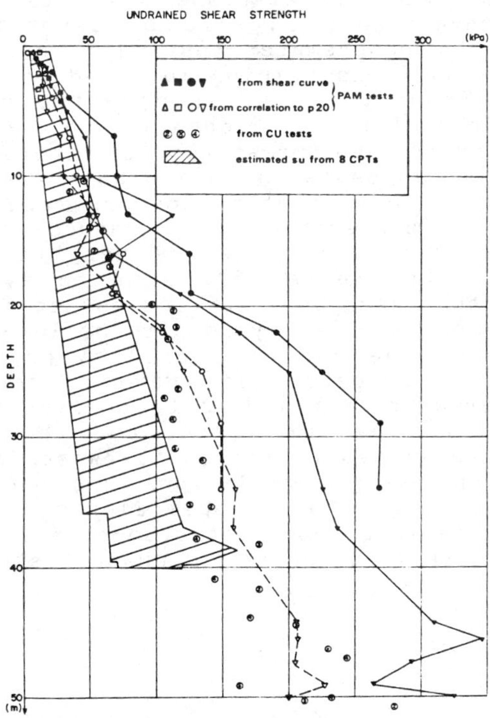

FIG. 12 - Undrained Shear Strength Logs. Mediterranean Site

used in three different ways in the design of foundations.

Fundamental Analysis

Numerical analyses of pressuremeter tests have been used by many authors to promote a better understanding of the response of the bulk of the soil around the pressuremeter probe. Numerical simulations using more or less sophisticated mathematical modeling of soil behavior (10) (12) have helped assess the importance of the assumptions made for pressuremeter test interpretations, such as, for example, the observance of the plane strain conditions. The role played by the geometry of the probe can then be fully understood.

On the other hand, the conditions under which pressuremeter test data can yield the fundamental parameters of the constitutive law of the soil still need to be examined and checked against good quality experimental data. The next stage which is foundation designing is thus not yet at hand. In this area, the use of such an integrated procedure for the complete description of the phenomena governing the behavior of a given foundation in a given soil is still problematic in many aspects.

Use of Correlations

Another practical approach to using pressuremeter test results is by means of correlations with soil parameters obtained from conventional in situ or laboratory tests. The application of pressuremeter testing is not widespread in the world, though it is gaining more and more appreciation outside of France. This explains the need, among practicing engineers, for a link between their routine soil investigation results and pressuremeter characteristics. Correlations may be of assistance when some transpositions from one characteristic to another one are required in the use, for example, of a specific design method more familiar to the engineer. Moreover, they are very useful in the search for better knowledge of a site to evaluate the results of tests that have been carried out at this particular site. The results of every in situ or laboratory test may be enhanced by the use of correlations. However, correlations may be dangerous if the user forgets their origin, the different criteria used to analyse the results from the different types of tests, and the dispersion parameters that are linked to them. Likewise, one must be very cautious when using correlations between parameters having different natures, for example linking a mechanical characteristic at failure (cone resistance, shear strength, limit pressure) and a deformation charactistic (Young's modulus, etc.).

A theoretical basis for the use of correlations between parameters has been proposed by Tavenas and Leroueil (11) who show that all aspects of the behavior of clays are strongly linked to the stress paths followed by the soil element and proceed from a unique limit state principle. Since any soil test has its own stress path, sometimes a complicated one, it is impossible to define single parameters values (such as one shear strength value, Fig. 13), and therefore the use of correlations is necessary and justified. All the same, an empirical use of test results in the design of foundations is induced by the fact that all problems corespond to different, variable and complicated stress paths that no simple test can usually recreate.

The establishment of correlations proceeds from a statistical analysis of data from different soil tests and/or from different sites (Figs. 10, 11, 12). The study can be more or less extensive by taking into account all possible cases (7)(8)(9), but it must be clear that it is generally impossible to integrate all specific site conditions.

Another approach was used by Lassoudière and Zanier (10) where the simple elasto-plastic (G, Cu) parameters were fitted, through a finite element method, to simulate a series of true pressuremeter tests performed at given sites. The comparison of the results of these tests to the data standardly obtained from the same pressuremeter diagrams was then made and proved to be reliable. The parameters determined in this way could then theoretically be used for the prediction of other foundation or geotechnical work behavior, thus leading to direct design methods.

Direct Design Methods

Precisely, the most promising feature of the pressuremeter lies in the possibility of inputting the test parameters directly into the design method. In this area, the definite analogy between the cylindrical expansion of a pressuremeter and the lateral movement of a pile subjected to horizontal loads is a must. As a consequence, procedures have been proposed using pressuremeter curves to generate the (p,y) curve elementary model used in the design of laterally loaded piles (6), (13). Many aspects still have to be solved, but these methods are very interesting tools.

The first design equations and charts relating pressuremeter test results directly to foundation design were published in 1963 (1). They are based on a number of tests and observations on actual structures and have been applied successfully for office buildings, port facilities and bridge designing, allowing reliable evaluation of the bearing capacity and settlement of shallow or deep foundations.

For specific offshore applications, a general project, based on the state of the art, has been undertaken at the Institut Français du Pétrole, aimed at determining generalized procedures for soil classification and routine analysis of offshore pile and shallow foundation design, making direct use of self-boring pressuremeter test results. The project is described in a companion paper at this symposium (6).

CONCLUSIONS

1. The Pressuremeter appears at present to be emerging as an important new offshore in situ investigation method in many respects.

2. The self-boring technique is now available for use in water depths up to 1000 m as the result of the development of the "PAM." Self-boring parameters are used as control parameters, and pressuremeter measurements give reliable data for geotechnical interpreting.

3. The results of PAM or other pressuremeter tests may be used in three different ways in the design of foundations.

3a. Fundamental analyses still have to face many problems.

3b. Correlations with other soil parameters are needed but must be used judiciously.

3c. Direct design methods appear to be the most promising ones.

REFERENCES

(1) Reid, W.M., St. John H.D., Fyffe S. and Rigden W.J., "The Push-In Pressuremeter", Symposium on the Pressuremeter and its Marine Applications, Ed. Technip, Paris, 1982.

(2) Brucy, F., Faÿ, J-B., and Marignier, J., "Offshore Self-Boring Pressuremeter: the LCP/IFP Probe", Symposium on the Pressuremeter and its Marine Applications, Paris 1982.

(3) Faÿ, J-B. and Le Tirant, P., "Offshore Self-Boring Pressuremeter for Deep Water, Symposium on the Pressuremeter and its Marine Applications, Paris.

1982.

(4) Brucy, F., Faÿ, J.B. and Le Tirant, P., "Three Years Experience with the Self-Boring Pressuremeter 'PAM'," Paper 4677, Offshore Technology Conference, Houston, 1984.

(5) Clough, G. W. and Denby, G. M., "Self-Boring Pressuremeter Study of San Francisco Bay Mud," Journal of the Geotechnical Foundation Engineering Division, ASCE, vol. 106, paper 15148, pp. 45-63, January 1980.

(6) Bécue, J.P., Brucy, F., and LeTirant, P., "Application of Pressuremeter Test Results to Designing of Offshore Foundations," The Pressuremeter and Its Marine Applications, ASTM STP 950, 1986.

(7) Kruizinga, J., "PAF tests compared with classic tests in Delft soft soil," Proceedings of the Second Symposium on Penetration Testing, Amsterdam, 1982, pp. 655-659.

(8) Ghionna, V., Jamiolkowski, M., Lacasse, S., Ladd, C. C., Lancellotta, R. and Lunne, T. "Evaluation of Self-Boring Pressuremeter," International Symposium on Soil and Rock Investigations by In Situ Testing, Paris, 1983, vol. 2, pp. 293-301.

(9) Powell, J. J. M., Marsland, A. and Al-Khafagi, A. N., "Pressuremeter Testing of Glacial Clay Tills," International Symposium on Soil and Rock Investigations by In Situ Testing, Paris, 1983, vol. 2, pp. 373-378.

(10) Lassoudière, F.M. and Zanier, F.B. "Numerical Analysis of Pressuremeter Tests by the Finite Element Method," The Pressuremeter and Its Marine Applications, ASTM STP 950, 1986.

(11) Tavenas, F. and Leroueil, S., "Clay Behaviour and the Selection of Design Parameters," Seventh European Conference on Soil Mechanics and Foundation Engineering, Brighton, 1979, vol. 1, pp. 281-991.

(12) Borsetto, M., Imperato, L., Nova, R. and Peano, A., "Effects of Pressuremeters of Finite Length in Soft Clay," International Symposium on Soil and Rock Investigations by In Situ Testing, Paris, 1983, vol. 2, pp. 211-215.

(13) Briaud, J.L., Smith, J. and Meyer, B. "Laterally Loaded Piles and the Pressuremeter: Comparison of Existing Methods", Design and Performance of Laterally Loaded Piles and Pile Groups, ASTM STP 835, Kansas City, 1983.

Stanley Fyffe, William M Reid, and James B Summers

THE PUSH-IN PRESSUREMETER: 5 YEARS OFFSHORE EXPERIENCE

REFERENCE: Fyffe, S., Reid, W.M., and Summers, J.B., "The Push-In Pressuremeter: 5 Years Offshore Experience," The Pressuremeter and Its Marine Applications, Second International Symposium: ASTM STP 950, J.-L. Briaud and J. M.E. Audibert, Eds., American Society for Testing and Materials, 1986.

ABSTRACT: The importance of in-situ testing of soils has been recognised for some time. To extend the equipment available for in-situ testing in an offshore environment, the British Government financed the development of a wireline operated push-in pressuremeter and since 1980 the Stressprobe has been used extensively offshore. This paper updates the record of tests carried out with the Stressprobe equipment since 1982 and describes the software used for test evaluation. Examples of test data are given and alternative forms of analysis are compared.

KEYWORDS: Push-in pressuremeter, in-situ testing, soil testing, offshore testing, stress/strain relationship.

BACKGROUND

In 1975, the Department of Energy; on behalf of the British Government financed the development of a push-in pressuremeter. This pressuremeter is compatible with standard wireline equipment, thus allowing pressuremeter tests to be carried out in the same borehole as cone tests, vane tests and soil sampling.

Mr Fyffe is a lecturer in soil mechanics at Dundee College of Technology, Bell Street, Dundee, DD1 1HG, Scotland; Messrs Reid and Summers are a director and development engineer respectively of Stressprobe Ltd., 145 West Regent Street, Glasgow G2 4SA.

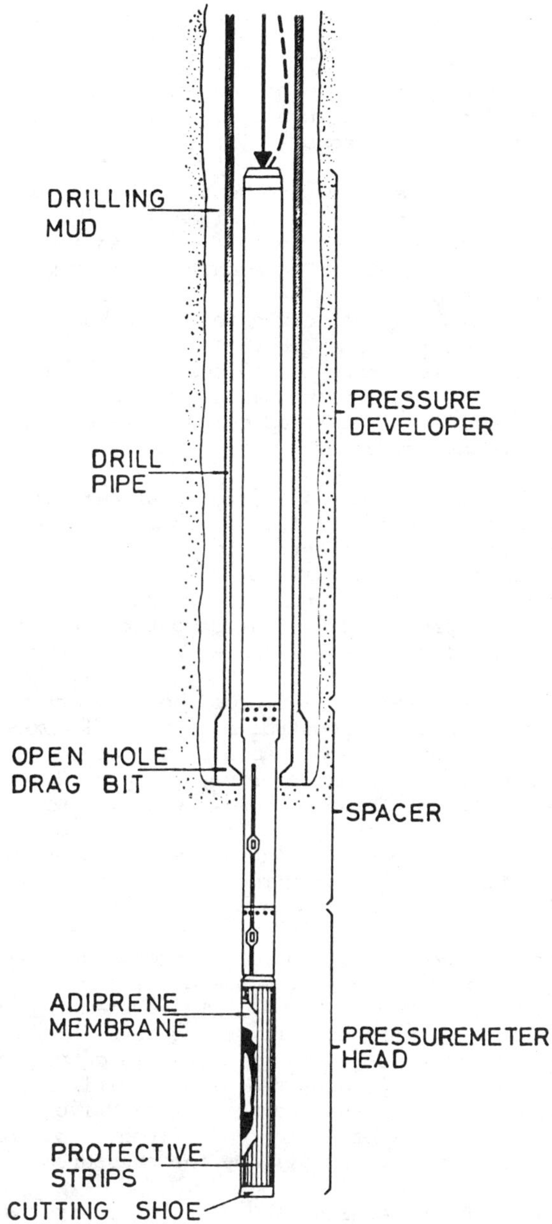

FIG. 1 -- The Stessprobe Pressuremeter

Since its commercial launch in 1980, the push-in pressuremeter (Stressprobe) has completed some 225 tests with 138 of these being offshore.

The form of equipment can be seen in Figure 1 This illustrates the Stressprobe located within a typical wireline operated drill string. The downhole elements consist of:

1) a hollow pressuremeter head with a protected adiprene membrane on its outside
2) a spacer section up which soil can pass
3) a pressure developer from which oil is pumped to inflate the pressuremeter
4) an electro-mechanical cable to the surface.

The wireline is connected on deck to a control box and recording system. In the latest update of the equipment, the data recording is carried out simultaneously by a chart recorder and by a micro computer.

Papers by Henderson et al, Fyffe et al and Reid et al, [1-3], have described in detail the development and early experience with the Stressprobe pressuremeter and reference should be made to these papers for specific details of the equipment. One element of the current Stressprobe configuration which has been updated since these papers were presented is the computer data logging system which records and processes the signals from the pressure developer.

The two signals sent from the downhole electronics of the push-in pressuremeter to the control box are cell pressure, in analogue form, and volume change as a series of pulses derived from the rotation of the screw which drives the piston. In earlier versions of the equipment these pulses were summed up digitally and then output, along with pressure, in analogue form, to a two pen chart recorder. This gave a continuous record of the test as shown in Fig.2. The system was reliable, but processing of the test data thereafter was time-consuming.

In order, therefore, to provide offshore analysis of the test a data logging and processing system was developed, the main elements of which are described in Fig 3. The system is based on the Hewelett Packard HP 85B micro-computer using IEEE-488 interfacing. The HP 85B computer is widely used offshore, is compact, robust, includes a small monitor and printer and allows fairly rapid data storage and retrieval via an internal magnetic tape drive. It also supports the important feature of interrupt driven processing, triggered by internal timers or externally by the pressuremeter control box.

When a reading is required the computer instructs the data logger to take an analogue reading of pressure and a digital reading of volume change, it also records the time elapsed since the beginning of the test.

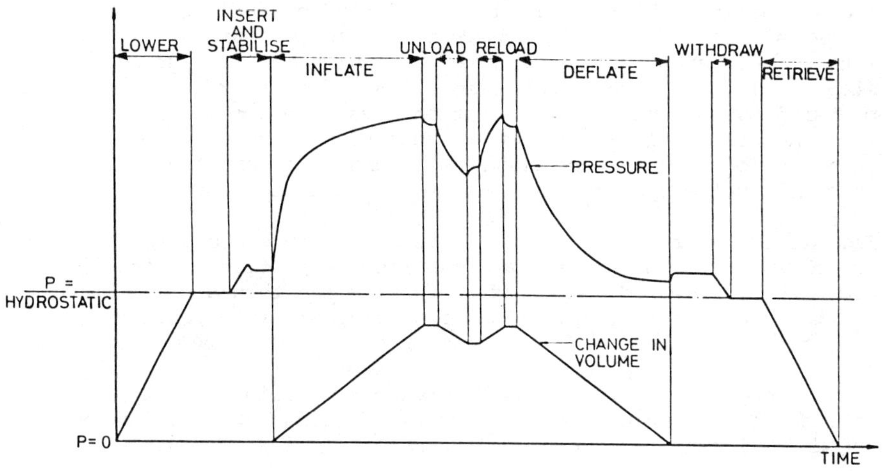

FIG 2--Typical Output from Stressprobe

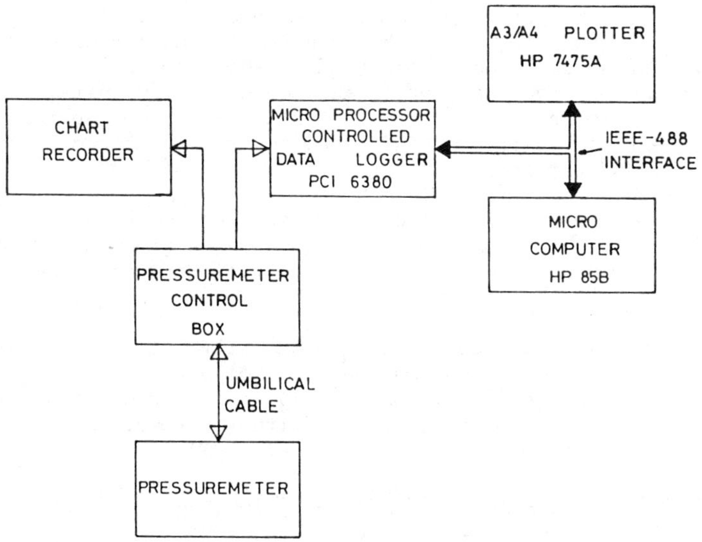

FIG. 3--The Stressprobe Control and Data Logging Equipment

During the main part of the test (inflation/deflation cycles) the readings are triggered automatically by the volume pulses from the pressuremeter. This yields almost 200 readings during the inflation stage alone. Throughout the rest of the test timed readings are taken at operator selected intervals which may be varied at any stage. The operator may also trigger a reading at any time by pressing one of the special function keys on the computer. Logging is therefore fully automatic but can respond to operator intervention, when necessary. A chart recorder is still used as this allows the operator to monitor the test as it proceeds.

Immediately after a test has finished the recorded data is post-processed by the computer and the results plotted on an HP 7475 plotter. Processed and plotted results are then available within minutes of test completion with minimal intervention of the operator.

Since computer test records have become available new numerical routines have been developed for curve fitting the test data and for the determination of the shear strength of the soil.

OFFSHORE EXPERIENCE

In July 1980, the first offshore tests using the Stressprobe were carried out in Block 7/12 of the Norwegian Sector of the North Sea, at depths of up to 20m below the seabed, in 70m of water.

Since 1980 the Stresprobe pressuremeter has been used on 23 contracts performing a total of 225 tests, of which 138 have been carried out offshore. The majority of offshore tests have been undertaken in the Norwegian Sector of the North Sea (90 tests), with the balance in the British Sector (34 tests) and the Australian Continental Shelf (14 tests). Land based tests have been performed both in Norway and Britain.

To date, the deepest water where tests have been carried out was 330m, in Block 31/2 of the Norwegian Sector. The greatest overall depth of a test was 385m, in Block 31/6 of the Norwegian Sector, where the recorded water depth was 300m.

Of the 225 tests, 173 have been in clays and silty clays, yielding interpreted shear strength values ranging from 9 kPa to 470 kPa. 16 tests have been carried out in silts, 14 in calcareous sands in Australian waters, 14 in other sands and with the remaining 8 tests being in varied fill material.

A summary of the offshore tests are given in Table 1.

DEVELOPMENTS IN INTERPRETATION

As standard, soil strength and stiffness parameters are determined from pressuremeter tests by the method due to Gibson and Anderson, [4] and Hughes, Wroth and Windle,[5] for cohesive and non cohesive soils respectively.

TABLE 1 - OFFSHORE STRESSPROBE TESTS

Year	Location	Country	No of Tests	Test Depth	Soil Type	Water Depth
80	Ula Field Block 7/12	Norway	4	5-20m	Silty clay	70m
81	North Rankin Field	Australia	10	3-110m	Silty calcareous sands	125m
81	Block 16/16	Britain	4	4-15m	Clay	135m
81	Magnus Field, block 20/12	Britain	5	4-70m	Stiff boulder clay	185m
81	Andrew Field, Block 16/28	Britain	16	4-75m	Very soft to very stiff silty clay	120m
81	Block 34/10	Norway	10	5-30m	Sandy clay	135m
82	Ula;Field, Block 7/12	Norway	17	6-25m	Soft to stiff silty clay	75m
83	Sleipner Field, Block 15/9	Norway	11	28-35m	Silty sandy clay	80m
83	Troll Field, Block 31/2	Norway	18	6-42m	Silty sandy clay	330m
83	Oseberg Field	Norway	4	11-30m	Very stiff to hard silty clay	110m
84	Tern Field, Block 210/25	Britain	9	4-35m	Slightly sandy silty clay	170m
84	Troll Field, Block 31/6	Norway	16	5-85m	Very soft to very stiff silty clay	300m
84	Gullfaks Field, Block 34/10	Norway	10	4-65m	Silty sandy clay	220m
84	Bass Strait	Australia	4	26-127m	Silty calcareous sands	75m

In addition, however, a method has been developed to enable assessment of the stress/strain response of cohesive soils during a pressuremeter test. The procedure is computer based, developed from a method due to Palmer, [6].

Palmer demonstrates that for an undrained soil during a pressuremeter test the principal stress difference is equal to twice the slope of the graph of applied pressure (p) against the natural logarithm of volumetric strain, as shown in eq. 1.

$$(\sigma_r - \sigma_\theta) = 2 \frac{dp}{d[\ln(\Delta V/V)]} \qquad (1)$$

where

σ_r = radial stress (compression positive)
σ_θ = circumferential stress (compression positive)
p = cell inflation pressure corrected for internally generated pressures
ΔV = change in volume of pressuremeter cell from initial reference state
V = volume of pressuremeter cell in deformed state

He further derives an equation for principal extension, eq 2 and a relationship between radial and circumferential strain, eq. 3.

$$\varepsilon_\theta = (1 - \Delta V/V)^{-1/2} - 1 \qquad (2)$$

$$(1 + \varepsilon_\theta)(1 + \varepsilon_r) = 1 \qquad (3)$$

where

ε_θ = circumferential strain (extension positive)
ε_r = radial strain (extension positive)

Equations 2 and 3 may be re-arranged to give radial compression as follows.

$$\varepsilon_r = \frac{1}{1 + \varepsilon_\theta} - 1$$

$$\varepsilon_r = (1 - \Delta V/V)^{1/2} - 1$$

substituting $e_r = -\varepsilon_r$ gives eq 4, which expresses this extension in the conventional soil mechanics terms as.

$$e_r = 1 - (1 - \Delta V/V)^{1/2} \qquad (4)$$

where

e_r = radial strain (compression positive)

The stress/strain relationship for plane strain is then obtained by plotting (σ_r - σ_θ) against e_r as in fig 4(c).

In order to obtain a smooth stress strain relationship it is necessary to curve fit the pressuremeter inflation data. Experience has shown that an efficient procedure is to fit a polynomial curve to the relationship between inflation pressure and the natural logarithm of volumetric strain. This is done by a specially developed computer program which uses a least squares method of polynomial regression to fit the curve. The polynomial obtained is then differentiated and the stress/strain relationship plotted by computer. The program uses data logged automatically during the pressuremeter test and typical plots obtained are shown in figs 4(a), (b) and (c).

When curve-fitting it is important to ensure that the most appropriate degree of polynomial is chosen. This may be done by a combination of visual examination of the fit to the data and by plotting the coefficient of correlation against the degree of polynomial. Figs 5(a) and (b) show the same pressuremeter data fitted by 2nd and 3rd degree polynomials respectively, and it can be seen that the 2nd degree polynomial is a poor fit. This is borne out by the plot of coefficient of correlation against degree of polynomial shown in fig. 6 which also shows that no appreciable improvement in fit is obtained by using any polynomial higher than the 3rd degree. Stress/strain relationships obtained from the 2nd, 3rd and 4th degree polynomial are shown in fig 7.

At present it is necessary to adopt a trial and error procedure for each test although it has generally been observed that a polynomial of a degree in the range of 3 to 5 gives the best fit. The practice of choosing the lowest polynomial which gives an acceptable fit is advocated as higher degree polynomials tends to oscillate to fit the data points too closely. This may lead to problems, particularly as the stress/strain relationship is obtained from the derivative of the fitted curve.

Typical stress/strain relationships obtained, by the above method, from pressuremeter tests in silty clay in the same area, are shown in fig 8. The values of principal stress difference corresponding to the undrained shear strengths previously obtained by Gibson and Anderson analyses are also shown. A plot of undrained shear strength with depth comparing the above values with conventional test results (average of Torvane, Fall Cone and Pocket Penetroneter results) is given in fig 9.

It is considered that this curve-fitting method is to be preferred to the Gibson and Anderson method because it gives an idea of the soil behaviour rather than simply yielding a value for undrained shear strength assuming the soil behaviour to be perfectly elastic/plastic. In particular, for a soil exhibiting peak and residual strengths the designer may take heed of this in his calculations, whereas previously he had only one value from the test.

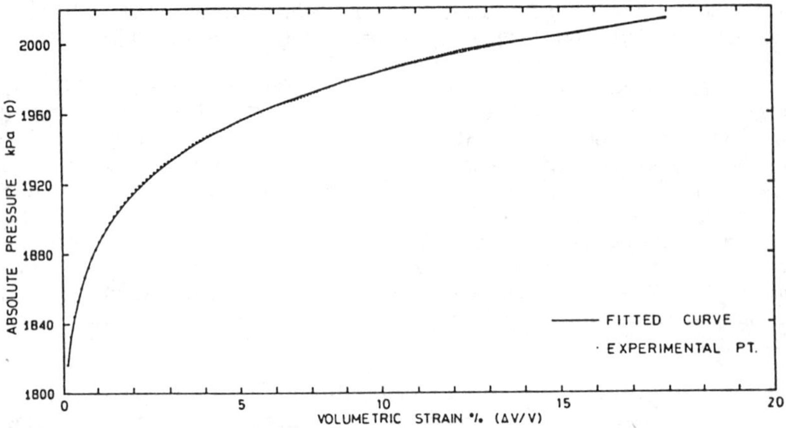

FIG. 4(a)--Typical Inflation Curve

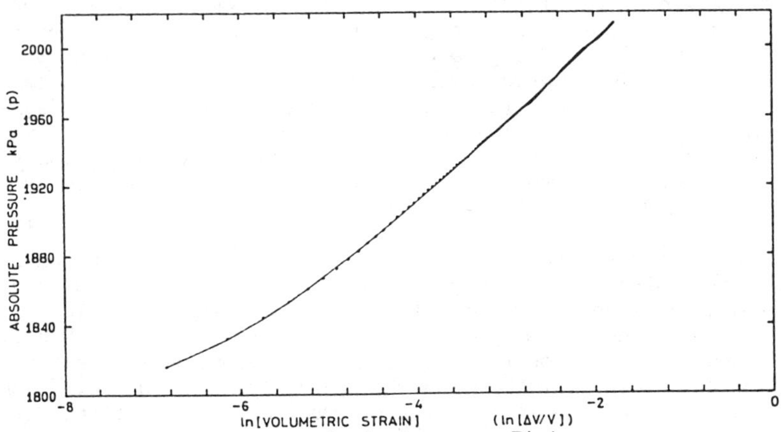

FIG. 4(b)--Typical Gibson and Anderson Plot

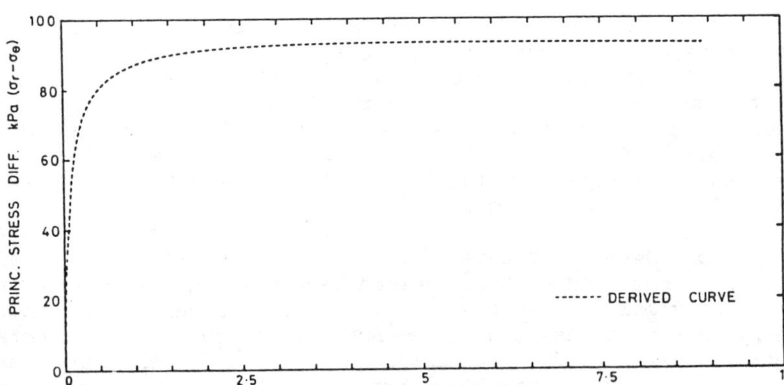

FIG. 4(c)--Derived Stress/Strain Curve

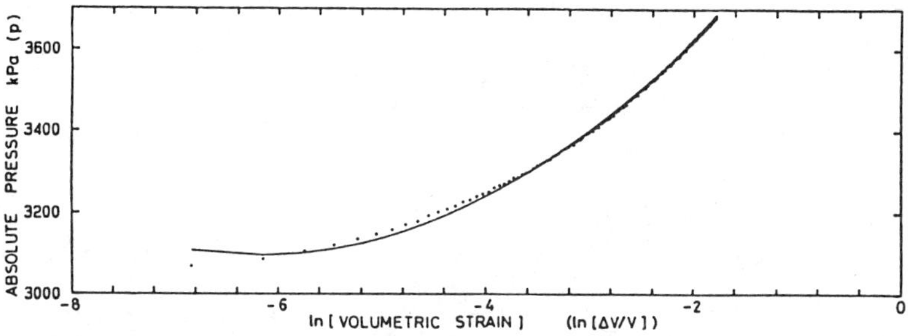

FIG. 5(a)--2nd Degree Polynomial

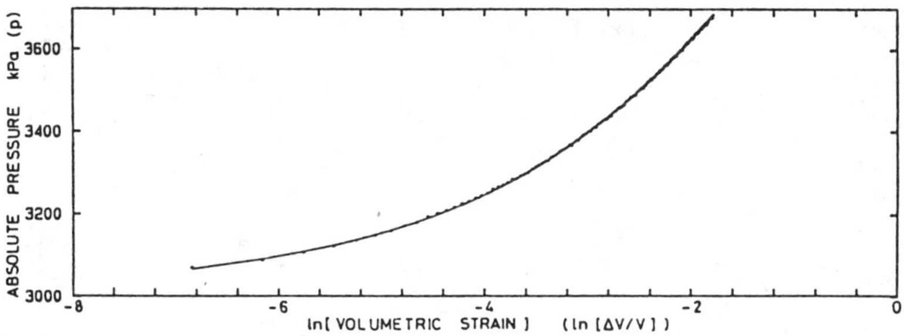

FIG. 5(b) --3rd Degree Polynomial

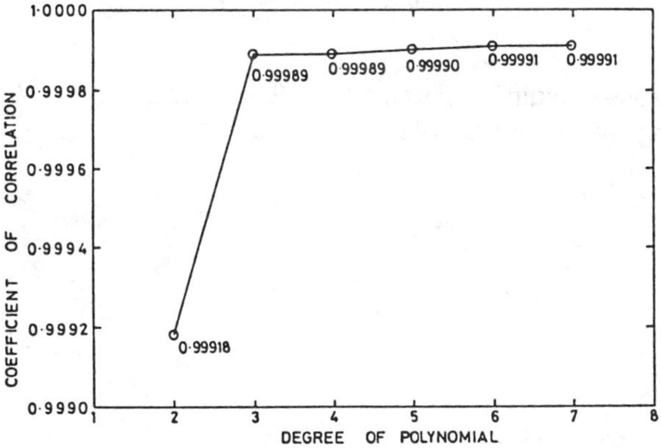

FIG. 6--Degree of Polynomial versus Coefficient of Correlation

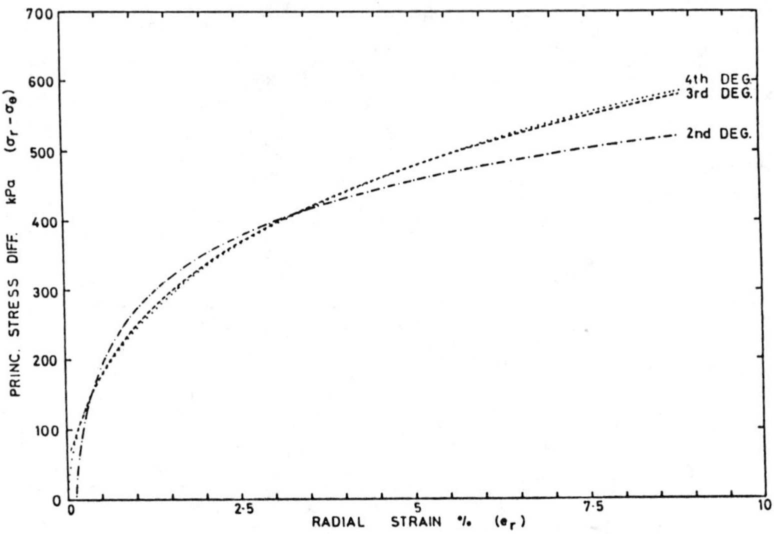

FIG. 7 -- Stress/Strain Curves for 2nd, 3rd, and 4th Degree Polynomials

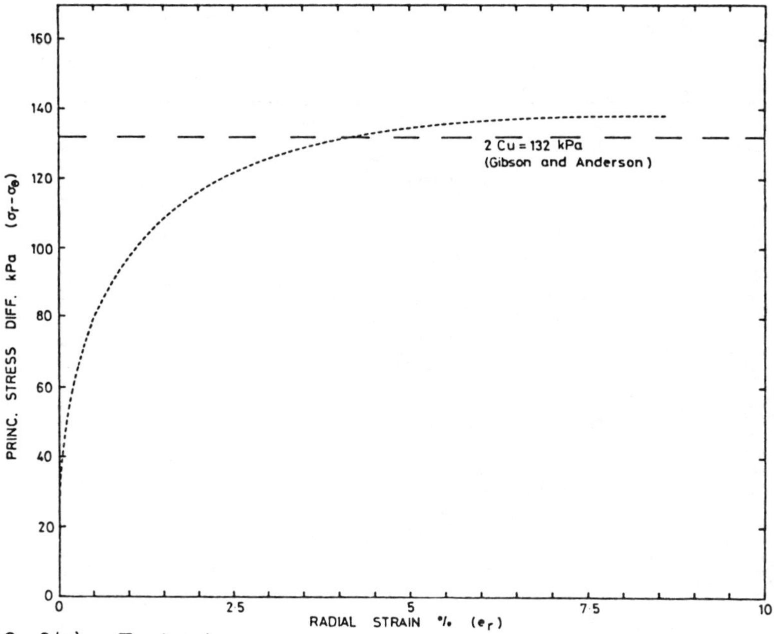

FIG. 8(a) -- Test A

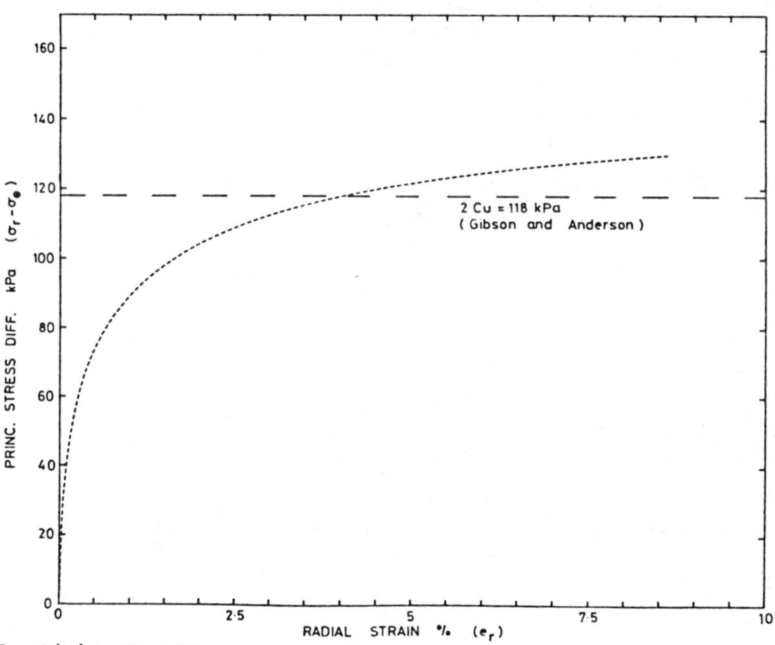

FIG. 8(b)--Test B

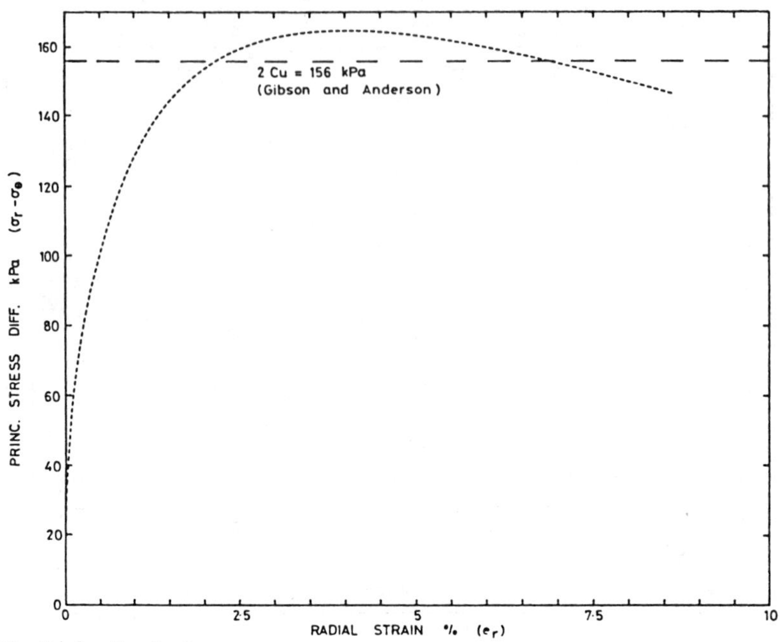

FIG. 8(c)--Test C

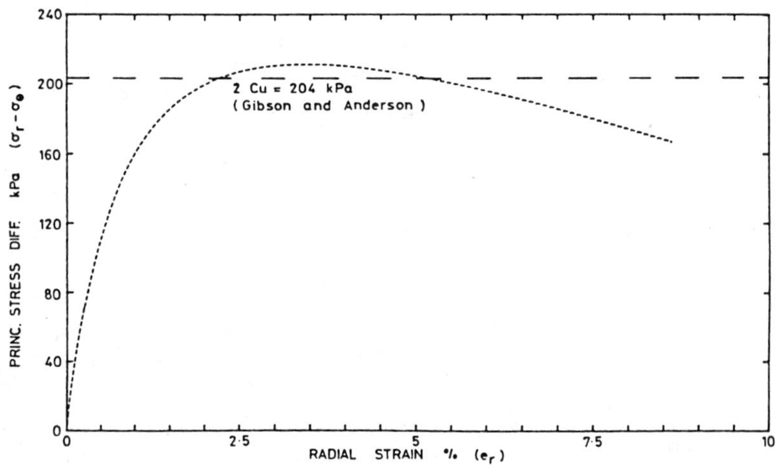

FIG. 8(d)--Test D

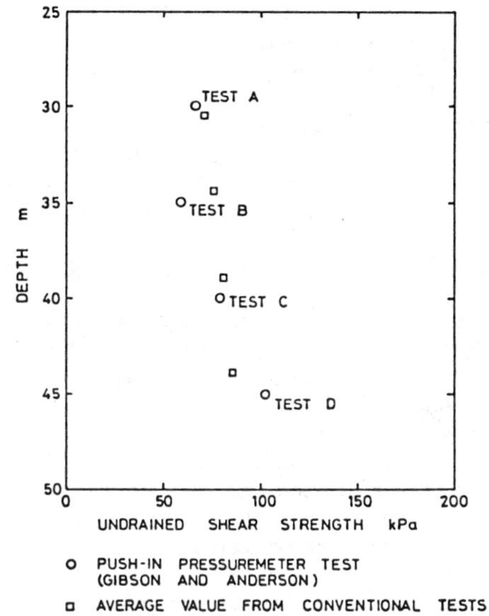

FIG. 9--Comparison of Results from Pressuremeter Tests with those from Conventional Tests

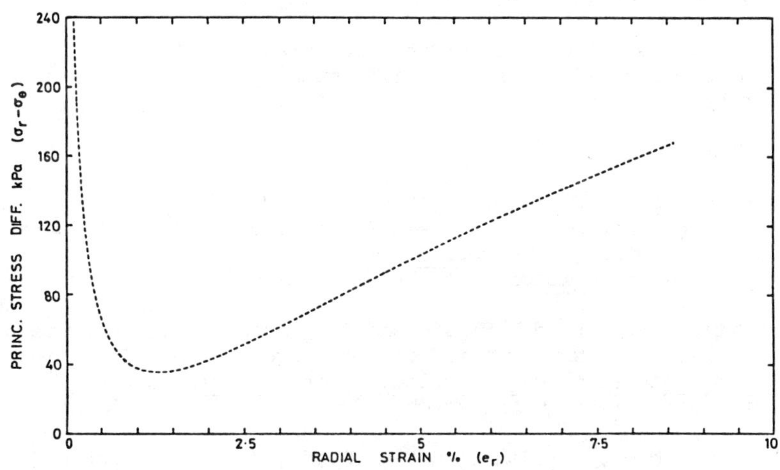

FIG. 10--Stress/Strain Curve for Test on Fill Material

The curve fitting method may also be used to detect a suspect test or one which has not exhibited substantially undrained behaviour. For example, fig 10 shows the strange stress/strain relationships derived from a pressuremeter test on fill material. From the plot it is clear that it would have been unwise to report a value of undrained shear strength for such a material, however, this may not have been obvious from a Gibson and Anderson plot.

The axes used for the stress strain plot can be varied but the choice of deviator stress, ($\sigma_r - \sigma_\theta$) against radial strain, (e_r) is preferred by the Authors as this is considered to be more familiar to the designer.

CONCLUDING REMARKS

1) The Stressprobe push-in pressuremeter has completed, in the five year period from 1980, a total of 225 successful tests, 138 of which have been in conjunction with wireline drill string equipment offshore.

2) A sophisticated computer based data logging and post processor facility is available as standard on all Stressprobe equipment. This allows the results of tests to be evaluated within a short space of time after completion of the tests.

3) Using data logged results in conjunction with curve fitting routines it has been possible to produce stress strain relationships from the pressuremeter inflation cycle. The information available from this form of test analysis is considered to be superior to that obtained by other more simple forms of evaluation.

4) The shear strengths derived from pressuremeter tests offshore have yielded results of similar magnitude to those obtained by other more conventional techniques.

REFERENCES

[1] Henderson, G., Smith, P.D.K., and St. John, H.D., "The Development of the Push-in Pressuremeter for Offshore Site Investigation," Proceedings of the S.U.T. Conference, Offshore Site Investigation, London, March 1979.

[2] Fyffe, S., Reid, W.M., and St. John, H.D., "The use of the Push-in Pressuremeter in Offshore Site Investigation," Proceedings of the Oceanology International Conference, Brighton, March 1982.

[3] Reid, W.M., St.John, H.D., Fyffe, S., and Rigden,W.J., "The Push-in Pressuremeter," Symposium Sur La Pressiometrie et Ses Applications en Mer, Paris, 19-20 April 1982.

[4] Gibson, R.E., and Anderson, W.F., "Insitu Measurement of Soil Properties with the Pressuremeter," Civil Engineering and Public Works Review, Vol.56, No. 658, 1961, pp 615-618.

[5] Hughes, J.M.O., Wroth, C.P., and Windle, D., "Pressuremeter Tests in Sands" Geotechnique, Vol. 27, No.4, 1977, pp 455-477

[6] Palmer, A.C., "Undrained Plane-Strain Expansion of a Cylindrical Cavity in Clay; A Simple interpretation of the Pressuremeter Test, " Geotechnique, Vol. 22, No. 3, 1972, pp 451-457.

Nick J. Withers, Leo H.J. Schaap, and Clive P. Dalton

THE DEVELOPMENT OF A FULL DISPLACEMENT PRESSUREMETER

REFERENCE: Withers, Nick J., Schaap, Leo H.J., Dalton, Clive P., "The Development of a Full Displacement Pressuremeter." The Pressuremeter and Its Marine Applications: Second International Symposium. ASTM STP 950, J.L. Briaud and J.M.E. Audibert Eds., American Society for Testing and Materials, 1986.

ABSTRACT: The Fugro Full Displacement Pressuremeter (FDPM) has been developed to make it easier to obtain offshore in-situ measurements of stress-strain properties which can be combined with the continuous profiling of electric piezocone penetration testing. The instrument is designed to follow behind a standard Fugro piezocone with a diameter of 43.7 mm.
The prototype has a differential pressure capability of 10 MPa and a radial expansion limit of 50%.
Results of some preliminary field trials on land are reported and its future application offshore is discussed.
Continuing research with the prototype FDPM and conventional pressuremeters will improve both empirical and theoretical interpretation and extend the application of the data.

KEYWORDS: Soils, pressuremeter, cone penetrometer test, offshore.

Nick J. Withers and Leo H.J. Schaap are respectively Senior Soils Engineer and Manager Electronics Division at Fugro Geotechnical Engineers B.V., P.O. Box 63, 2260 AB Leidschendam, Holland. Clive P. Dalton is Managing Director of Cambridge Insitu, Little Eversden, Cambridge CB37HE, United Kingdom.

INTRODUCTION

Soil investigation for major offshore structures mainly consists of soil sampling by stationary push sampling or piston sampling and of (piezo) cone penetration testing [1]. Both techniques have advanced to a degree that in combination with modern soil investigation vessels, offshore investigations have become standard and routine practice [2].

Soil engineering parameters are basically derived from laboratory tests on the samples. However, the samples are always disturbed to a certain degree which causes uncertainty about several parameters, specifically the stress-strain properties. Cone penetration tests provide a detailed insight to the stratification and through previously established empirical correlations, assist the derivation of engineering parameters.

In situ measurement of stress strain properties has been promoted for some time as a way of reducing the uncertainties. Tests used for this purpose have included plate load tests of various sorts, dilatometer and pressuremeter tests.

Of these, pressuremeter testing provides the most satisfactory way to measure insitu soil strength and stiffness offshore, because it relies on no correlations like the dilatometer. It can cover a reasonably wide strain range and the equipment is in principle easier to deploy and operate by remote control in conventional offshore drilling operations, than plate load tests.

Nevertheless pressuremeter testing offshore is not yet used routinely. One of the reasons presumably is that the offshore versions of the pressuremeter currently on the market are perceived not to be optimally cost effective on the grounds that:

1. These tools are suitable only for restricted ranges of soil types and strengths.
2. The tools are deployed and operated as separate operations within the site investigation and involve significant extra costs and time offshore. Furthermore due to the robustness of the tool necessary for offshore application, embedment forces are high and sometimes require precoring.

To overcome these practical problems Fugro has investigated the feasibility of developing a small size pressuremeter mounted behind a standard 15 cm^2 electric cone penetrometer.

THE FULL DISPLACEMENT PRESSUREMETER

The concept of the Full Displacement Pressuremeter (FDPM) is to enable tests to be performed as part of the cone penetration test operations. The pressuremeter section is included in the string of cone rods and is of the same diameter. Cone penetration is halted when the centre of the membrane reaches the desired test depth and a pressuremeter test is performed. Cone penetration can be resumed as soon as the pressuremeter membrane has deflated. The data and control equipment are built into the routine CPT site investigation package.

Apart from the ease of operation the other advantages of the FDPM are:

1. The large soil disturbance caused by insertion is repeatable and operator independant.
2. Measurement of stress strain properties at a depth known precisely in relation to the cone resistance profile. This opens up possibilities for reducing scatter and improving correlation of soil strength and stiffness with cone resistance.
3. Direct measurement of the lateral load deformation response of the soil [3, 4] have shown the value of such data for the design of laterally loaded piles.

In-house studies by Fugro, later confirmed by theoretical studies and experimental evidence by others [5], indicated that in situ shear modulus values obtained from FDPM unload-reload cycles, would be comparable with those obtained from other types of pressuremeter (Selfboring, Pushed in Selfboring Type, Menard, P18). Furthermore, theoretical studies indicated that very likely reliable analyses to derive soil strength from a full unload curve could be developed, provided that the unloading was commenced after very large expansion strains ($\Delta R/R_o$ > 20 to 30%) had been imposed on the soil.

These geotechnical and practical considerations gave confidence in the viability of this in situ tool, and in its potential for offshore applications.

This paper describes the development since 1983 and initial field trials of Fugro's prototype Full Displacement Pressuremeter. The likely methods of interpretation and applications of the data are also briefly outlined.

OTHER PRESSUREMETERS

There are three other types of pressuremeter. The first to be developed was by L. Menard [6]. This type of pressuremeter is lowered into a predrilled hole. Measurements of pressure and volumetric strain are made. The results are highly sensitive to soil disturbance from drilling and stress relief at the borehole wall, neither of which can be measured. With careful control of procedures, repeatable results are obtained but application of the data remains empirical and based on correlations with measured foundation performance. Difficulties of measuring foundation performance offshore reduce its potential for this environment.

The Self Boring Pressuremeter was developed [7, 8] to minimise soil disturbance so that interpretation of the data could be based on more fundamental theories. This instrument is slowly pushed into the ground. At its lower end a sharp cutter shoe and rotating cutter remove sufficient soil to accomodate the volume of the pressuremeter. Measurements are made of total pressure, pore pressure, and circumferential strains at three positions at the mid height of the expanding length. These data can be interpreted using cavity expansion theories to provide fundamental stress-strain properties, as well as information about in situ stress conditions. However, skill and experience are required to control the push rate, the rate of removal of soil and the position of the cutter with respect to the cutter shoe so that the disturbance is in fact minimised during installation in different soils.

Neither of these two types of pressuremeter lends itself easily to deep water offshore operation unless a fully dedicated installation system is constructed.

The Push In Pressuremeter (PIP) was developed [9] primarily for offshore use. This instrument is a thick walled tube of area ratio 40%. Soil displaced as the instrument is pushed into the ground enters the tube and is withdrawn with it after the test. Inflation is by oil and the pressure and volumetric strain are measured. Because it is impossible to design the cutting shoe so as to minimise the disturbance in all types of soil the boundary conditions at the start of the test can vary. The large cross-sectional area of the instrument means that in dense sands, cemented soils and hard clays large reaction forces may be required. Moreover measurement of volumetric strain is, in principle, less accurate than direct radial or circumferential strain measurement.

DESCRIPTION OF THE PROTOTYPE FUGRO FDPM

Figure 1 shows the general arrangement of the FDPM in a cut away view of the pressuremeter sections. Figure 2 shows the pressuremeter inflated in air. The overall

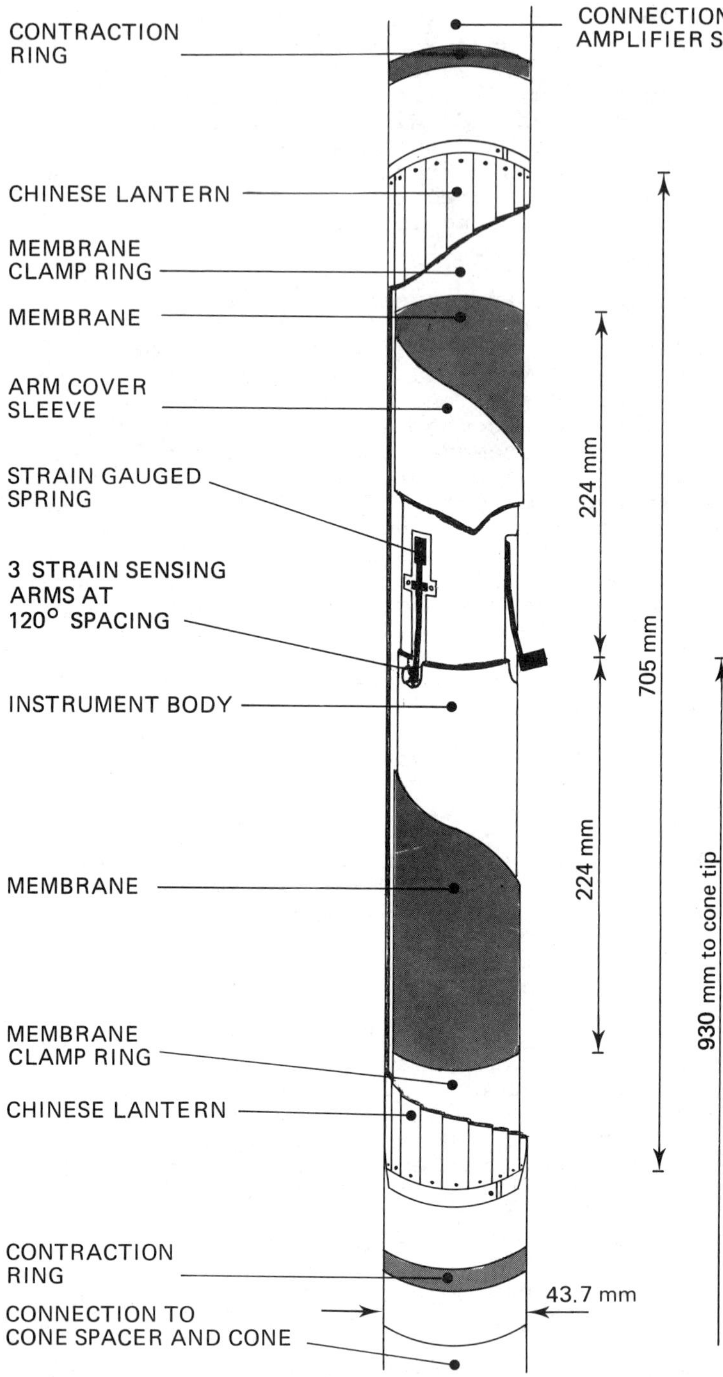

FIG. 1 Details of Prototype FDPM

FIG. 2 Inflated FDPM

length of the membrane is 450 mm. The outside diameter is the same as that of the 15 cm^2 cone (43.7 mm). The L/D ratio is 10.3. The minimum distance between the centre of the membrane and the conical tip of a 15 cm^2 Fugro piezocone is 930 mm. By means of spacers this distance may be enlarged.

Above the pressuremeter is a module which amplifies the pressuremeter signals. The cone signals are passed independently through the pressuremeter modules to the surface on separate wires.

The pressuremeter measures inflation pressure and, at three locations 120° apart, the circumferential strain.

The pressure capability of the equipment is 10 MPa. The radial strain capacity is 50%, which is more than double the capacity of existing pressuremeters.

Inflation of the prototype is achieved with Nitrogen gas. Oil will be used in offshore operations.

The detailed design of the prototype was made by Cambridge Insitu to specifications prepared by Fugro after joint discussions.

The general arrangement and form of construction is modelled on the proven self boring pressuremeter construction. A number of new features have been developed specifically to meet the design specifications and the constraints imposed by the small diameter.

Some of the notable features are:

1. The connectors for the inflation fluid and the electrical cable, which must seal 10 MPa yet pass through a succession of standard 20 tonne cone rods of 16 mm bore.
2. The diameter limitations in (1) above mean the electrical cable cannot be too large, and this restricts the number of cores if it is to be strong enough. This problem requires there to be an electronic circuit at the pressuremeter and that its components shall be subject to the full working pressure of the instrument.
3. The instrument is much smaller in diameter than any other pressuremeter yet must withstand pushing and extraction forces up to 20 tonnes. Careful attention had to be paid to material cross sections and strut buckling loads at certain critical points, especially where the strain measuring feelers operate and where the various parts screw together.
4. The instrument expands by 50% in diameter which, for 43.7 mm diameter, is 11 mm. Once allowance has been made for the thickness of the membrane and of the chinese lantern protective sheath on the O.D. and for the cable duct up the centre, very little space remains to accomodate the strain feelers with their supporting arms. Nevertheless the well tried system of pivoted arm and strain gauged spring that is used in the Cambridge self boring pressuremeters has been fitted in and works well.

5. A better system than screws for securing the ends of the chinese lantern was called for and a collet lock has been employed. This is stronger and does not disturb the smooth outside surface of the instrument.

6. The chinese lantern will not withstand compression and must therefore be drawn in by its lower end and extracted by its top end. It must also expand so a degree of end float is also required and this must be prevented from filling with dirt at any time.
Great care has been taken to provide correctly positioned stops at both ends, taking account of both manufacturing and fitting tolerances.

7. A piezocone is required for which eight wires have been provided. This cone must be capable of fitting on the end of a spacer which may vary considerably in length. It is not yet clear whether the distance between the cone and the centre of the expanding length is, or is not, important.

8. Finally, a considerably altered Strain Control Unit has been made for carrying out tests at constant, and high rates of strain using much higher than usual gas pressures. Appropriate attention has been paid to hydraulic testing, to the provision of safety valves and to safety generally, as well as to the much altered control characteristics required.

The unit provides as output:

1. The cone signals which have come independently via separate cores from the cone

2. Readings of total pressure, radial displacement for each of the strain arms, average (or total) radial displacement and pore pressure (when fitted to the FDPM).

PRELIMINARY FIELD TRIALS

The primary objective of the preliminary field trials was to check that the operation of the prototype met the specifications. Bench testing had already established that the design pressure rating of 10 MPa could be achieved.

Further research testing is currently (1985) in progress to provide data bases for different soil types. These data bases will help the development of methods of interpretation based on theoretical analyses and on empirical correlations.

Some results are demonstrated by two soundings, with FDPM tests, which were made at Fugro's test site at the Middenweg. Measurements of pressure and three radial strains were made, recorded on a chart recorder, an X-Y plotter and an HP85 cassette. For the purpose of these tests the FDPM was connected to a dummy cone. A separate 15 cm^2 CPT was also performed. Fig. 3 shows the set up of the prototype testing.

Fig. 4 compares the total thrust to push the FDPM and a standard Fugro 15 cm^2 cone to a depth of 17 m. This upper 10 m of the soil profile consists of peat and very soft clay in which penetration resistance is negligible. For the sand below this, the figure shows that only a small amount of additional thrust was required, despite the large difference in the length of the 15 cm^2 size section. Fig. 5 shows the cone resistance profile at this Middenweg site.

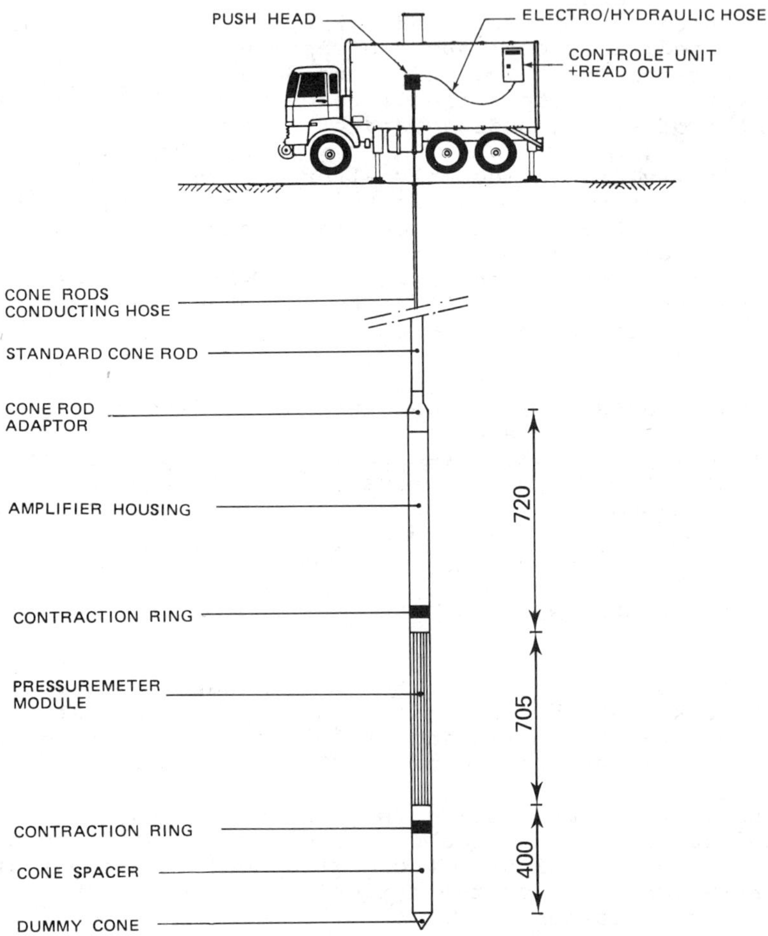

FIG. 3 Set-up for Prototype Testing

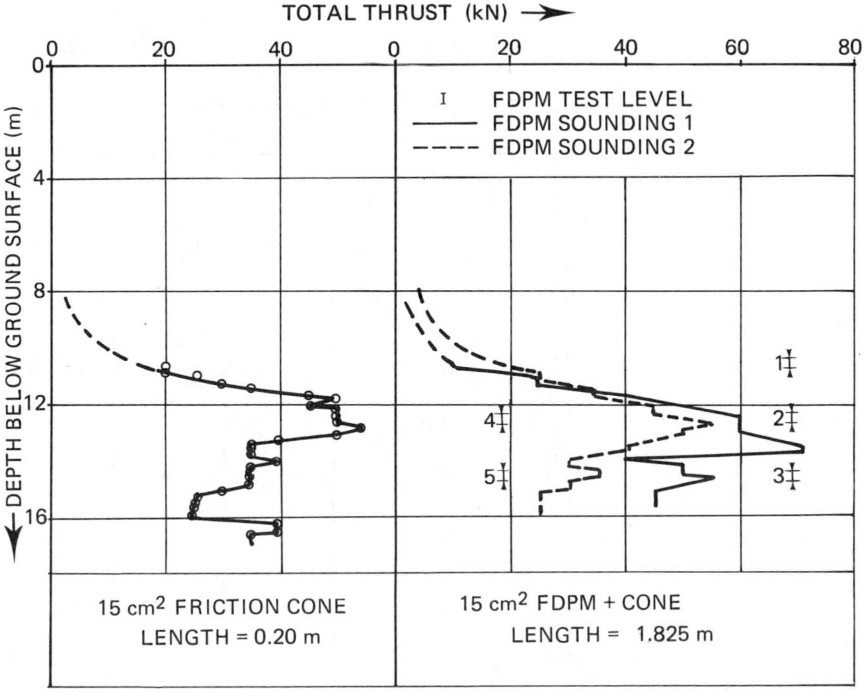

FIG. 4 Total Thrust Comparisons

Fig. 6 shows two FDPM tests carried out at a depth of 12.6 m. The following conclusions are drawn:

1. The lift off pressures are similar, varying between 300 and 360 kPa.

2. The difference in limit pressure at 40% radial displacement is about 24%. This is similar to the difference in the total thrust of 20%.

3. The shear modulus values measured in the two tests are listed in Table 1.

TABLE 1--Measured Shear Moduli 12.6 m Depth.

Test No.	Approximate Radial Displacement		
	10%	20%	45%
FDPM 2	71 MPa	83 MPa	53 MPa
FDPM 4	63 MPa	63 MPa	56 MPa

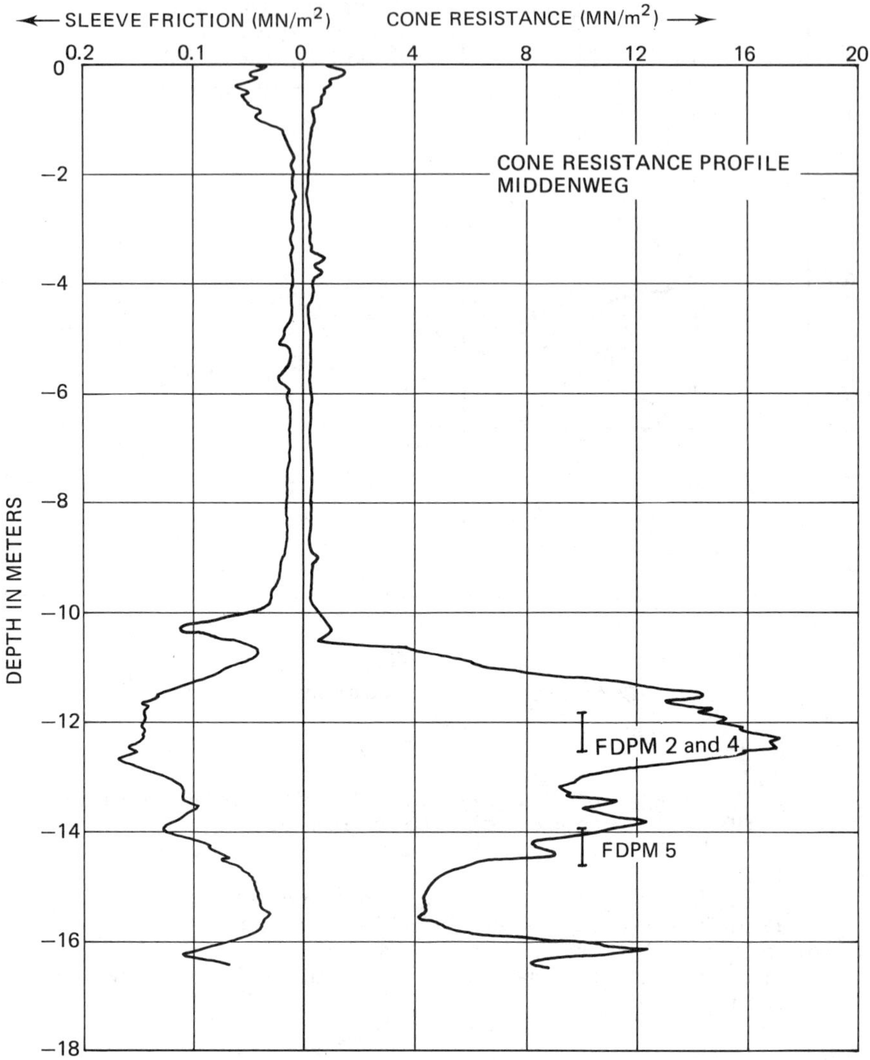

FIG. 5 15 cm^2 Cone Penetration Test

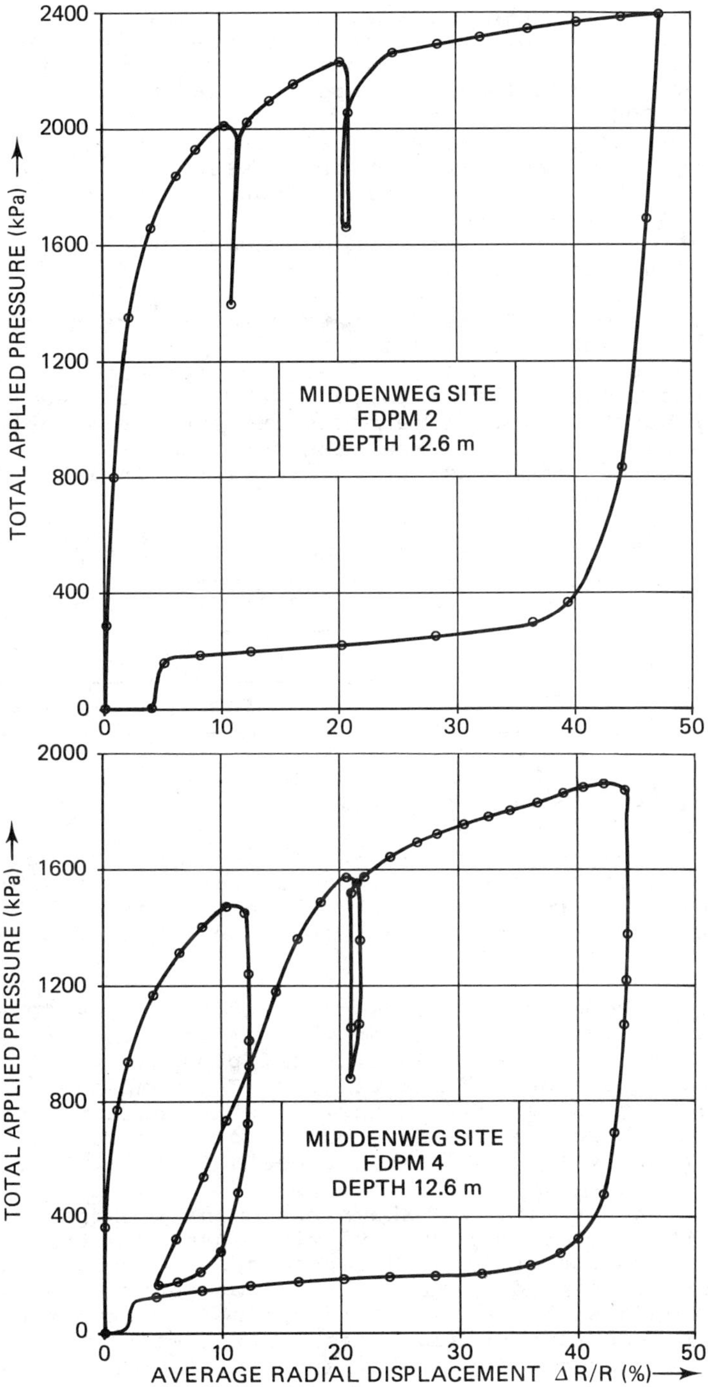

FIG. 6 FDPM Test 2 and 4

The agreement between the values for FDPM 4 are good and better than that for FDPM 2 in which some initial problems were experienced in controlling the rate of unloading as precisely as possible. These problems had been solved by the time FDPM 4 was performed.

4. The tests suggest a radial displacement capacity exceeding 30% is required for the pressure-expansion curve to define a limiting slope with adequate certainty. The form of the curves suggest that the effects of installation have not yet been overcome for radial displacement less than 20%.

Fig. 7 shows FDPM test 5 performed at a depth of 14.6 m. The following conclusions are drawn:

1. The pressure at 40% radial displacement is 47% to 70% of the limit pressure found in FDPM tests 2 and 4. This reflects similar differences in cone resistance and total thrust pressure between the depths.

2. The shear modulus values are listed in Table 2.

TABLE 2--Measured Shear Moduli 14.6 m Depth

Test No.	Approximate Radial Displacement		
	12%	25%	45%
FDPM 5	16 MPa	17 MPa	17 MPa

The consistency of the values is good.

One of the sets of design criteria was that the membrane corrections needed to be as small as possible, repeatable, and comparable with those of existing types of pressuremeters because it is a major determinant of data accuracy.

Fig. 8 shows the membrane correction for the FDPM. This was obtained by inflating the membrane in air, with chinese lantern fitted. The lift off pressure was 27 kPa. Hysteresis was about 10 kPa maximum. The membrane correction at 50% expansion was 138 kPa. These values are similar to those for the self boring pressuremeter. Further tests will confirm the repeatability of this data.

It is instructive to make a preliminary comparison between the prototype FDPM data with that from the self boring pressuremeter. Fig. 9 reproduces two published curves by Hughes and Robertson [5] from tests in sand. One test was made with a normal self boring pressuremeter, the other with the same pressuremeter fitted with a 60° solid cone tip to convert it into a larger version of the FDPM. The two curves in Fig. 9 have been plotted on the same scale as the prototype FDPM curves (Figs. 6 and 7).

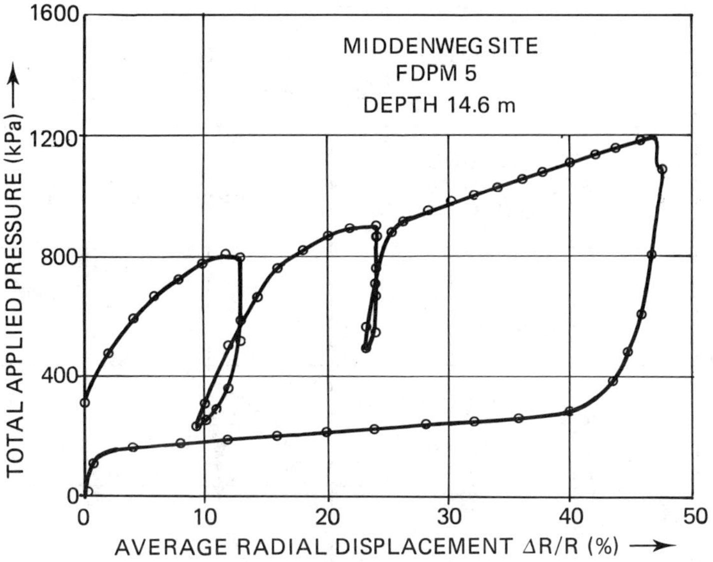

FIG. 7 FDPM Test 5

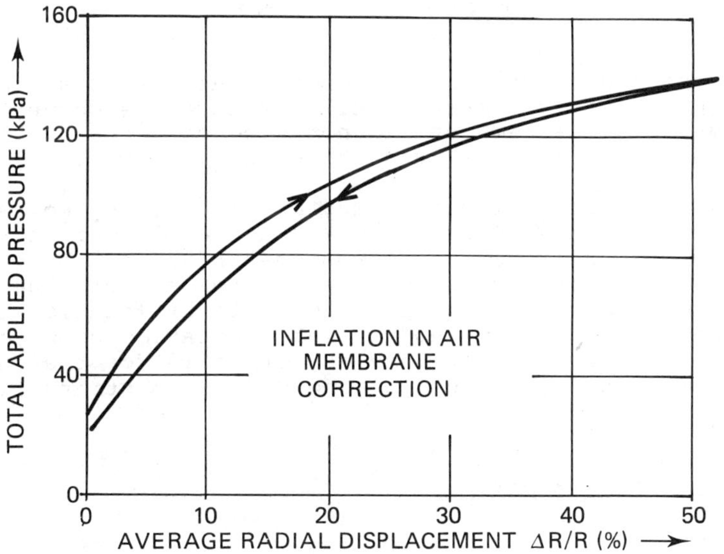

FIG. 8 Membrane Correction

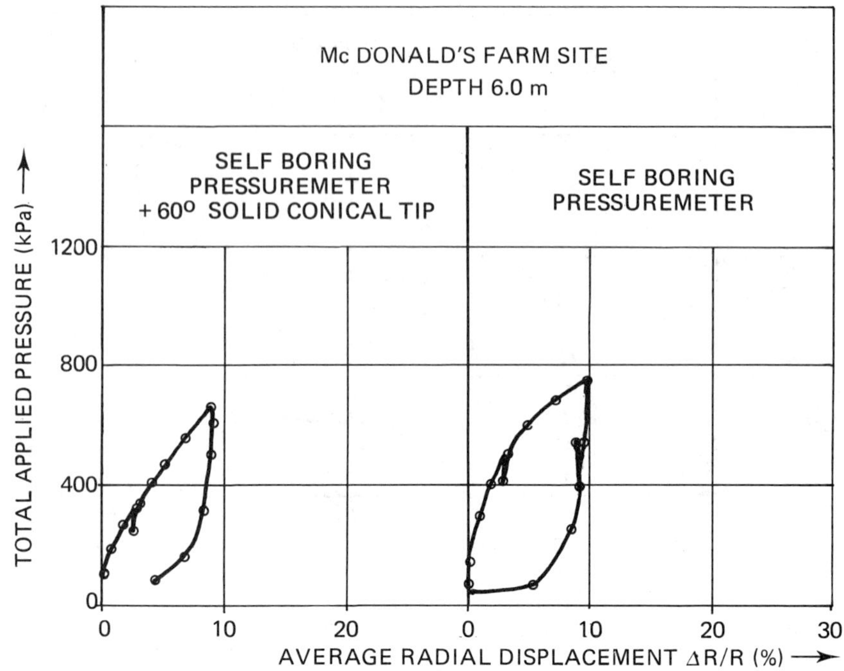

FIG. 9- Pressuremeter Tests adapted from Hughes & Robertson, 1985

Hughes and Robertson [5] report that the shear moduli from the two tests compared well but that the converted self boring pressuremeter did not produce a reasonable friction angle, when calculated in the same way.

One possible explanation may be provided by comparison of the curves in Figures 7 and 9. This would suggest that the converted self boring pressuremeter did not have sufficient radial expansion capacity to overcome the effects of insertion. A second possible explanation is that the theoretical method of interpretation used for the selfboring pressuremeter data is not valid for the full displacement type, because of the different stress fields.

INTERPRETATION OF DATA

A research programme is being executed by Fugro, the Universities of Oxford and British Columbia and Dr. J.M.O. Hughes. Test data is being collected eith the prototype full displacement pressuremeter, with one created by attaching a 60° solid cone to a self boring pressuremeter, and with a self boring pressuremeter.

Shear Modulus

The prototype FDPM (and the converted self boring pressuremeter) show that shear modulus can be measured repeatably, by making unload-reload loops for small stress increments. The shear modulus is then given by

$$G = 1/2\ \Delta p/(\Delta R/R_0)$$

where Δp is the increment in total applied pressure
$\Delta R/R_0$ is the radial displacement ratio.

It would appear in sands [5] that the insertion method does not appreciably affect the value of shear modulus measured from small unload-reload loops. Also the FDPM appears to provide reliable and representative in situ measured data, which is comparable to that obtained by other pressuremeters.

With respect to fine grained soils, preliminary indications are that the full displacement shear modulus is comparable with that obtained from orhter pressuremeters. The repatability of the values is improved if pore pressures are allowed to dissipate before the small unload-reload loops are made.

Soil Strength

Work is proceeding to establish how cavity expansion theory can be adapted and used to calculate undrained shear strength and friction angle from the expansion and unloading curve. Preliminary indications [5] would suggest that the FDPM curve can be correlated to the self boring pressuremeter curve and to the cone resistance. Therefore, until analyses based on fundamental principles have been validated, interpretation for strength will remain largely empirical.

Friction angle in sands:

Insertion of the pressuremeter in sands causes an initial increase in radial efective stresses sufficient to yield the soil out to some radius around the cone tip. The boundary condition imposed by the cylindrical surface behind the cone mean that the radial stress reduces so that the soil in effect arches around the pressuremeter. The yield stress of the soil now varies with radius. Subsequent expansion of the pressuremeter membrane reloads the soil. As the radial strain increases the stresses at different radii successively approach their yield stresses. At large enough strains the outer limit of the yield surface created by the cone will be further expanded.

This is one reason for building a large strain capacity into the FDPM design.

The interpretation of self boring pressuremeter tests for friction angle uses the gradient, s, of the asymptotic straight line through a plot of log radial effective stress (σ_r-u_o) against log radial displacement ($\Delta R/R$) [11]. This is based on a closed form cavity expansion model for a frictional soil with dilation.

One of the objectives of the ongoing theoretical research and testing is to confirm whether or not a similar type of analysis can be used for the large strains imposed by the FDPM. It is expected that the complex stress distribution in the soil will necessitate the use of a numerical rather than a closed form solution.

Undrained shear strength of clay:
Conventional analysis of pressuremeter data are based on an initial horizontal effective stress which is near the in situ condition. However, the stresses after insertion of the FDPM are expected to be significantly different. This is likely to complicate derivation of undrained shear strength values from the loading curve.

However at large strains, the effects of cone insertion have been overcome, and in priciple the stress field in the soil around the pressuremeter can be determined. The unloading curve after large expansion strains starts from a known state of stress and thus can in principle be analysed to yield the undrained shear strength of the soil. Research is currently in progress to confirm this hypothesis.

This research includes:
a) Theoretical analyses of the unloading curve
b) Comparisons of self boring and FDPM curves
c) Comparison of interpreted strengths from FDPM tests with results of other tests at well documented sites.

In situ Horizontal Stress

Early indications are that the drained lift off pressure (σ_{ro}- u_o) for the FDPM in sands is similar to the in situ effective horizontal stress. The testing programme in progress is evaluating this more thoroughly, but there would appear to be the intriguing possibility to use the FDPM to estimate approximately the in situ horizontal stresses in sands.

In clays the lift of pressure is complicated by pore pressure dissipation effects. Estimation of the insitu horizontal effectives stress would appear to depend on an analysis of the limit pressure achieved and the unloading curve since the limit pressure must be related to σ_{no}, G, and C_u, and the latter two parameters can be obtained from the unloading curve. This approach is complicated by the sensitivity of the result to assumed strain patterns imposed by the membrane (cylindrical or spherical).

Modelling

The solid FDPM is a model of a solid or plugged pile. The pressure expansion curve therefore lends itself to the possibility of direct application as a p-y curve for laterally loaded pile design [3], [4].

FUTURE EQUIPMENT DEVELOPMENT

For offshore use the FDPM will be adapted for application with standard offshore carrier tools such as Wison and Seacalf [1]. Inflation will be achieved by oil pressurized by a piston driven by a rotating lead screw.

Furthermore it is planned that digital telemetry will be used to monitor the measurements and to control the strain and pressure rates.

The application of the oil inflation system and digital telemetry will be a further development of existing technology.

The current pressure capacity of the pressuremeter will not need to be increased because a 10 MPa differential capacity across the membrane should be sufficient for most soils.

The present design is such that the prototype FDPM can be adapted easily but its introduction offshore will depend on further onshore testing.

The prototype FDPM is therefore being put through a comprehensive programme of testing with the following objectives:

1. Establish the accuracy and repeatibility of the test.
2. Compare FDPM data with other pressuremeter data, and insitu test data, particularly the estimates of shear modulus and undrained strength of clays and silts.
3. Optimise the design for offshore deployment.
4. Establish a data base for interpretation and understanding the test results.

Field testing work is being carried out by Fugro Geotechnical Engineers B.V. and the University of British Columbia with the assistance of Cambridge Insitu and Dr J.M.O. Hughes. Oxford University are assisting with the development of theoretical analyses.

REFERENCES

[1] Toolan, F.E., "Recent Improvements in Soil Investigation Techniques," Design in Offshore Structures, Thomas Telford Ltd., London, 1983, pp. 29-36.

[2] Zuidberg, H.M., Richards, A.F. and Tsuzuki, M. "Vessel Outfit for Advanced Geotechnical Site Investigations in Deep Water". Ocean Space Utilization '85 Proceedings, Nihon University, Tokyo, Tokyo, Springer-Verlag, 1985, pp. 433-440.

[3] Robertson, P.K., Hughes, J.M.O., Campanella, R.G. and Sy, A., "Design of Laterally Loaded Displacement Piles Using a Driven Pressuremeter", Laterally Loaded Deep Foundations: Analysis and Performance, ASTM STP 835, J.A. Langer, E.T. Mosley and C.D. Thompson, Eds., American Society for Testing and Materials, 1984, pp. 229-238.

[4] Kay, S., Griffiths, D.V. and Kolk, H.J., "Application of Pressuremeter Testing to Assess Lateral Pile Response in Clays," The Pressuremeter and Its Marine Applications, ASTM STP 950, J.L.Briaud and J.M.E. Audibert, Eds. American Society for Testing and Materials, 1986.

[5] Hughes, J.M.O., and Robertson, P.K., "Full Displacement Pressuremeter Testing in Sand" Canadian Geotechnical Journal, August, 1985.

[6] Ménard, L., "Mesures In Situ des Propriétés Physique des Sols", Annales des Ponts et Chaussées, Paris, no. 14, mai-juin, 1957, pp.357-377.

[7] Baguelin, F.J., and Jézéquel, J.F., "Le Pressiomètre autoforeur", Annales de l'Istitut Technique du Bâtiment et des Travaux Publics, Supplement no. 307-308, Série Sols et Fondations no. 97, juillet-août, 1973, pp. 133-160.

[8] Baguelin, F.J., Jézéquel, J.F., and Shields, D.H., "Self Boring Pressuremeters", The Pressuremeter and Foundation Engineering, Transtech Publications, Aedermannsdorf, Switzerland, 1978, pp.506-573.

[9] Reid, W.M., St. John, H.D., Fyffe, S. and Rigden, W.J., "The Push-In Pressuremeter", Proc. Symp. on the Pressuremeter and Its Marine Applications, Paris, April 1982, Edtions Technip, Paris, 1982, pp. 247-261.

[10] Hughes, J.M.O. "Push in Pressuremeter Tests on Soft Clays in Langley B.C.", Private Communications, 1983.

[11] Hughes, J.M.O., Wroth, C.P., and Windle, D. "Pressuremeter Tests in Sands", Géotechnique, Vol. 27, No. 4, 1977, pp. 455-477.

Domenico Bruzzi[(1)], Vidio Ghionna[(2)], Michele Jamiolkowski[(2)], Renato Lancellotta[(2)], and Giovanni Manfredini[(3)].

SELF-BORING PRESSUREMETER IN PO RIVER SAND

REFERENCE: Bruzzi D., Ghionna V., Jamiolkowski M., Lancellotta R., and Manfredini G., "Self-Boring Pressuremeter In Po River Sand", The Pressuremeter and Its Marine Applications: Second International Symposium; ASTM STP 950, J.-L. Briaud and J.M.E. Audibert, Eds., American Society for Testing and Materials, 1986.

ABSTRACT: The results of a series of self-boring pressuremeter tests (SBPT) carried out in the quaternary Po river sand are presented. Assessment of total horizontal in situ stress, shear modulus and shear strength, of the tested sand is discussed. Comments on the use of the above mentioned parameters in practice are advanced.

KEYWORDS: in-situ tests, self-boring pressuremeter, sands, in-situ horizontal stress, shear modulus, shear strength.

INTRODUCTION

This paper presents the results of a series of self-boring pressuremeter tests (SBPT's) performed in 1984 on the recent alluvial deposits of the Po river valley. The tests were part of a much more extensive geotechnical investigation undertaken by the National Electricity Board of Italy (ENEL) to select and qualify a site for a possible Nuclear Plant. The test site was also subjected to a detailed geological investigation. In this paper the results of 53 SBPT's, using the English Camkometer MK VIII and French PAF 76, in a well documented geotechnical deposit are presented and discussed.

TEST SITE

The test site is located in the central part of the Po river valley, close to the ancient town of Mantova.

(1) ISMES, Bergamo (Italy)
(2) Technological University of Torino (Italy)
(3) ENEL (National Electricity Board), Rome (Italy)

From a geological point of view, the site is characterized by the presence of 1500 to 2000 m thick quaternary deposits. The upper 15 m consisting of a thin clayey silt stratum covering a layer of clean or only slightly silty sands that are very recent holocen deposits less than 10,000 years old. For this deposit any mechanical consolidation caused by erosion may be virtually eliminated. The underlying clean or slightly silty sand with embedded thin lenses and layers of cohesive materials represents the most recent glacial deposit (Würm, 3rd), with thickness ranging between 30 to 40 m. For this deposit the geological information suggests the possibility of limited mechanical overconsolidation caused by erosion of few meters of soil before the holocen deposition.

The geotechnical properties of the above mentioned soil deposits have been investigated by numerous borings in which frequent sampling has been performed using the 100 mm OD Osterberg piston sampler. Also a significant number of in situ tests consisting of SPT's, electrical CPT's, Piezocone tests (CPTU) have been carried out at the site. A representative soil profile and the results of the above mentioned in-situ tests are shown in Figs. 1 to 3. At one location the shear wave velocity has also been measured by means of the cross hole techniques and the results are shown in Fig.10.

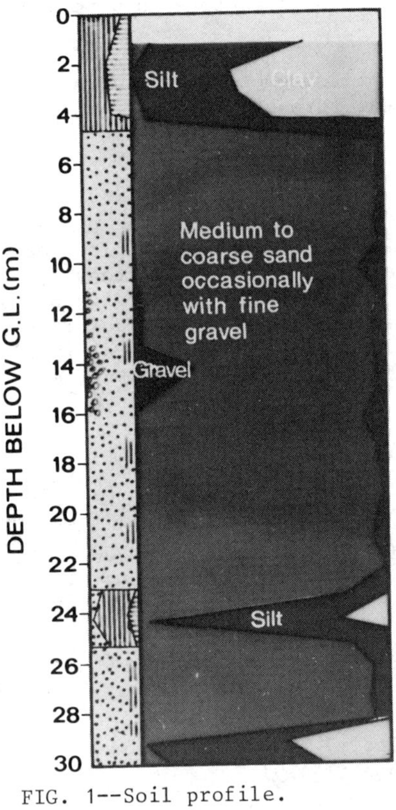

FIG. 1--Soil profile.

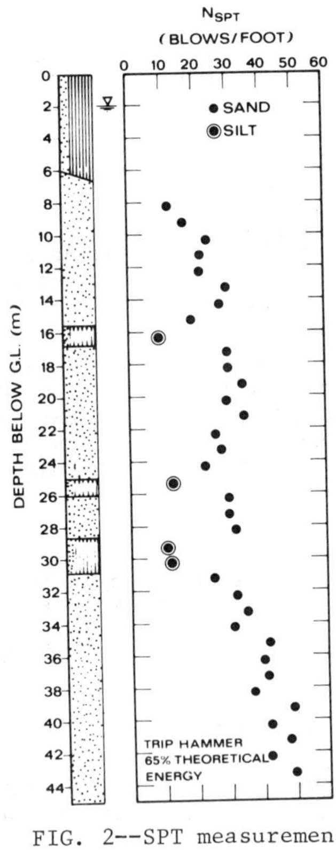

FIG. 2--SPT measurements.

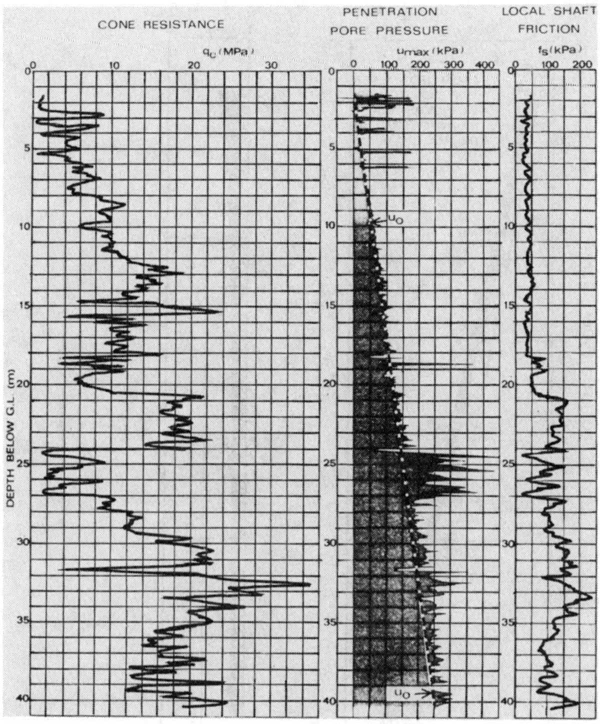

FIG. 3--Piezocone measurements.

The position of the static G.W.L. at the site has been ascertained and monitored by piezometer observations for a period of two years. From these observations it was found that an unique aquifer exists at the site at least within the explored depth of 150 m. The static G.W.L. corresponds to the depth normally ranging between 1 and 4 below the existing G.W.L. and it may be subjected to seasonal oscillations which should not exceed 3 to 5 m in magnitude.

Based on the above information and on the results of oedometer tests run on samples of cohesive materials embedded within the predominantly sand layers, it is possible to establish the preconsolidation pressure (σ'_p) and overconsolidation ratio (OCR) profiles given in Fig. 4. This tentative assessment of the stress history leads to the conclusion that the deposits under consideration are slightly OC, partially due to aging and partially due to small erosion and limited G.W.L. oscillations.
Here only the results of tests performed in sand are reported neglecting those run in cohesive and silty sand materials for which the expansion under fully drained conditions has not been achieved.
The index properties of the tested sand are given in Table 1.

This is predominantly quartz sand with subangular to subrounded particles. Based on the results of calibration tests [1] it may be defined as only moderately curshable material.

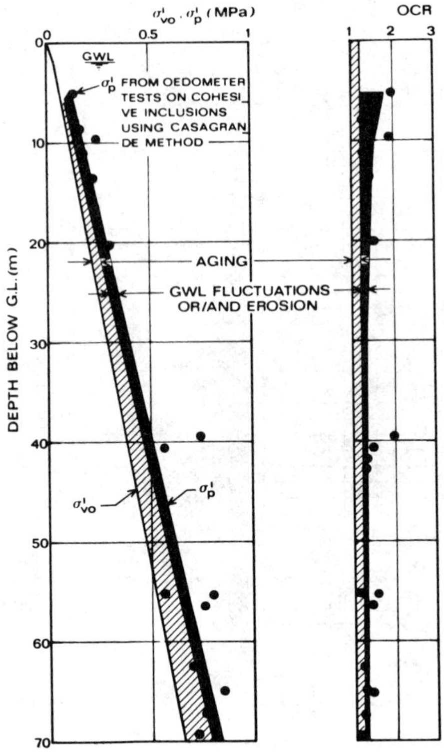

FIG. 4--Stress history.

TABLE 1--Index properties of Po river sand.

Bulk density γ:	18.1 to 19.0 kN/m^3
Fraction passing sieve No.200 ASTM:	0 to 12% *
d_{50}:	0.20 to 0.40 mm
d_{60}/d_{10}:	1.5 to 3.0

*Low plasticity and non plastic silt particles.

EQUIPMENT AND TEST PROCEDURE

The commercially available self-boring pressuremeters shown in Fig.5 have been used. The Camkometer device has been modified in order to allow independent reading of the three strain arm feelers.

The characteristics of the insertion and test procedure for the Camkometer may be summarized as follows:

- cutter speed, rev./min: 40 to 70
- water pressure, kPa: 150 to 250
- water discharge, lt/min: 10 to 16

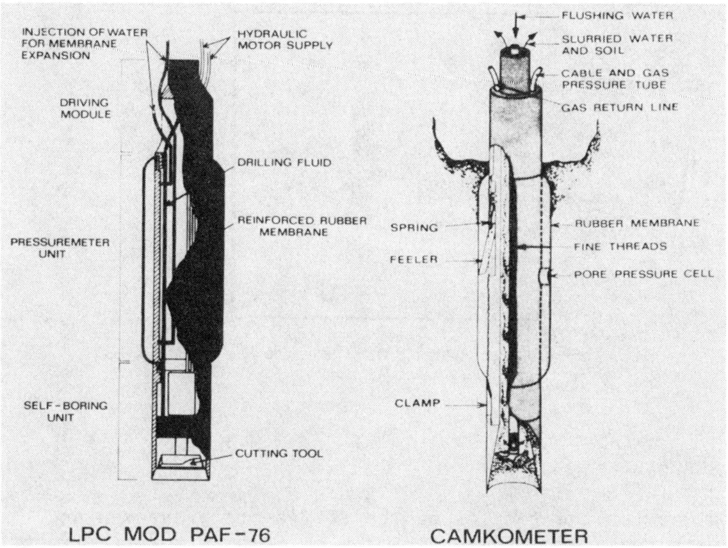

FIG. 5--Self-boring pressuremeter devices.

- cutter setting*, mm: 15 to 47
- penetration rate, mm/min: 20 to 35
- relaxation time, hours: 1 to 2 reaching for few tests 10 to 15.

The Camkometer expansion tests were intended to be run at a constant rate of cavity strain ε_o equal to 1%/min. In stiff soils however, like the tested sand, the existing equipment was not able to maintain such a rate of strain. Thus at least the first part of the expansion curve has been obtained under approximately constant rate of load conditions with the cavity pressure rate $\dot{p} \cong$ 30 to 60 kPa/min. A new type of cutter with a helical shape was used to insert the Camkometer probe, as shown in Fig.6. To minimize the scour effect in the sand, the cutter has a dished water deflector which turns the discharge to spray sideways.

The insertion of the PAF-76 probe has been performed under the following average conditions:

- cutter speed, rev./min: 50 to 60
- water pressure, kPa: 800 to 1500
- water discharge, lt/min: 10 to 15
- cutter setting, mm: 60
- penetration rate, mm/min: 100 to 2500
- relaxation time, hours: 0.5 to 3

The PAF-76 expansion have been performed under constant rate of cavity strain $\varepsilon_o \cong$ 0.8% / min.

*distance between the cutting edge and the cutter apex.

FIG. 6--Cutter used for Camkometer tests.

TEST RESULTS

In this chapter the results of the SBPT's are presented and discussed with regard to: the evaluation of in situ total horizontal stress σ_{ho}, determination of the "elastic" shear modulus yielded from unload-reload (G_{UR}) and reload-unload (G_{RU}) loops performed during the tests, and assessment of shear strength. The analysis of experimental data is restricted mainly to the Camkometer test results, while data for the PAF-76 are only examined for the determination of the total horizontal stress.

Horizontal Stress

It is generally postulated that if self-boring devices are inserted into the ground with minimum disturbance then it is possible to measure the total horizontal stress σ_{ho} existing prior to insertion. Among the different procedures suggested for evaluating σ_{ho} from SBPT's [2], the most straightforward is that which assumes that the corrected cavity pressure causes, after a sufficiently long relaxation period, the early movement of the pressuremeter membrane "lift-off pressure" p_o to be equal to σ_{ho}. This approach yields reasonable results in soft clays [3], [4], [5], [6], and encouraging but still problematic results in stiff clays [7], [5], [8], [9]. The results in sand are so far largely contradictory [7], [10], [11], and it is questioned [12], [13], [19], whether existing self-boring devices, even after an ideal insertion, are able to correctly measure σ_{ho}.

Table 2 shows the values of the lift-off pressure p_o^* measured in the Po river sand at four different locations using the Camkometer probe. These values are compared with the best estimate of σ_{ho}, computed assuming an average coefficient of earth pressure at rest K_o=0.55 based on OCR** values obtained from measured pressure σ'_p, knowing hydrostatic pore pressure u_o and by assuming an average bulk density γ = 19.4 kN/m^2.

The following observations are made on the above data.

- On average the measured p_o values are appreciably higher than the best estimate of in-situ total horizontal stress σ_{ho}.

* Average of the three strain arms readings

** Postulating K_o=function of mechanical over consolidation only

TABLE 2-- Total horizontal stress from Camkometer tests.

	Depth, m	u_o, kPa	σ_{ho}, kPa	p_o, kPa	CS, mm	K_o,-	Notes
Borehole 4003	6.2	44.9	86.3	91.7	20	0.62	
	7.7	60.2	109.1	59.8	20	≅ 0	TOM
	9.2	75.5	132.0	74.9	26	≅ 0	TOM
	10.7	90.8	154.9	183.0	26	0.79	
	12.2	106.1	176.9	195.1	26	0.69	
	15.2	136.7	222.6	257.0	35	0.77	TG
	16.7	152.0	245.5	299.9	30	0.86	MS
	18.2	165.2	268.4	323.9	26	0.85	
	19.7	180.5	291.2	270.2	26	0.45	
	21.2	195.8	314.1	502.6	26	1.43	TG
	23.2	216.2	344.6	336.5	26	0.52	
Borehole 4017	7.4	64.3	107.8	123.3	26	0.75	
	8.9	78.5	130.7	156.1	26	0.82	
	10.4	94.9	153.5	168.2	26	0.69	
	11.9	109.1	176.0	226.0	26	0.96	
	13.4	126.5	198.8	341.4	26	1.60	TG
	14.8	138.7	220.2	239.7	26	0.68	TOM
	15.9	149.9	236.9	323.1	26	1.09	TOM
	17.9	170.3	267.4	254.2	20	0.47	TOM
	19.4	185.6	290.3	279.3	20	0.49	
	20.9	200.9	313.2	286.1	20	0.42	
	22.8	220.3	342.2	361.0	20	0.63	TG
	24.9	244.8	375.6	380.0	20	0.57	MS
Borehole 5050/1	6.3	19.4	75.8	91.7	35	0.71	
	7.3	29.6	91.0	82.6	35	0.47	
	8.8	44.9	114.0	109.3	35	0.51	
	10.3	60.1	136.8	143.5	35	0.60	
	11.8	75.4	159.6	123.8	35	0.32	
	13.3	90.7	182.5	106.2	35	0.09	TOM
	16.3	121.3	228.2	145.3	47	0.12	TOM
	17.8	136.6	251.1	216.5	47	0.38	TOM
	19.3	151.9	273.9	219.6	15	0.31	TG
	20.8	167.2	296.8	220.3	15	0.23	TOM
	22.3	182.5	319.7	358.8	23	0.71	
	23.8	197.8	342.5	521.9	23	1.23	TG
Borehole 5050/2	6.3	19.4	75.8	98.5	23	0.77	
	7.3	29.6	91.0	74.9	23	0.41	MS
	8.8	44.9	114.0	113.7	23	0.55	
	10.3	60.1	136.8	60.5	15	≅ 0	TG

u_o = Hydrostatic pore pressure during SBPT
σ_{ho} = Best estimate of the total in-situ horizontal stress
p_o = Lift-off pressure, average of the three strain arms
CS = Cutter setting
K_o = Coefficient of earth pressure at rest from p_o
TOM = Traces of organic matter
TG = Traces of gravel
MS = More silt

- There is no direct correlation between yielded values of K_o and special soil features like traces of organic matter and presence of more silt.
- In contrast, the presence of small amounts of gravel seems to be linked with the manifestly unreliable values of K_o which are either too high or low.
- In some cases a cutter setting [5], equal or less than 20 mm leads to apparently disturbed tests, yielding K_o values that are too low.

Excluding the tests used to calibrate the self-boring technique in which the measured p_o values are manifestly too low (yielding $K_o < 0.35$) and those where a small amount of gravel was encountered, the average results of Camkometer's tests performed in the Po river sand may be summarized as follows:

$$\frac{\sigma_{ho}}{p_o} = 0.95 \pm 0.13; \qquad K_o = 0.65 \pm 0.18$$

This data indicates an overestimate of the in situ lateral stress and consequently a significant overestimate of K_o in comparison with what would be considered reasonable for the assumed stress history of the deposit.

The opposite result is obtained from tests with the PAF-76 probe. This yielded Po values that are consistently too low, leading to K_o values between 0 and 0.30. A comparison of p_o and K_o measurements from the Camkometer and PAF-76 tests in Po river sand is given in Fig. 7.

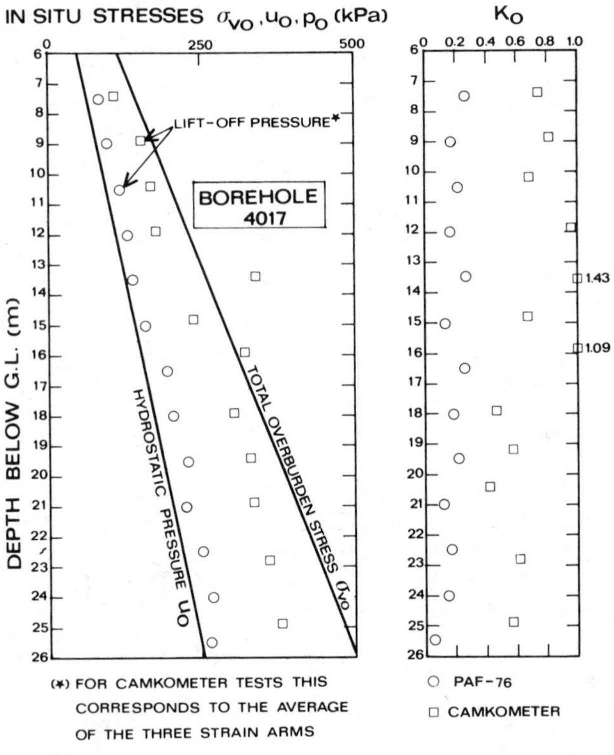

FIG. 7-Comparison of p_o and K_o measurements from Camkometer and PAF-76 devices.

Shear Modulus

Fig. 8 shows a typical expansion curve for Camkometer test in Po river sand. Three unload-reload loops (1 to 3) and one reload-unload loop (4) have been performed. If a sufficiently long pressuremeter is performed in an elastic medium, the surrounding soil is only subjected to shear, thus the test only measures the shear modulus G. The slope of previously executed loops correspond approximately to twice the unload-reload G_{UR} or reload-unload G_{RU} shear moduli of the tested sand. In a similar manner the slope of the secant to any point on the expansion curve is usually considered [14], as twice the secant shear modulus G_s.

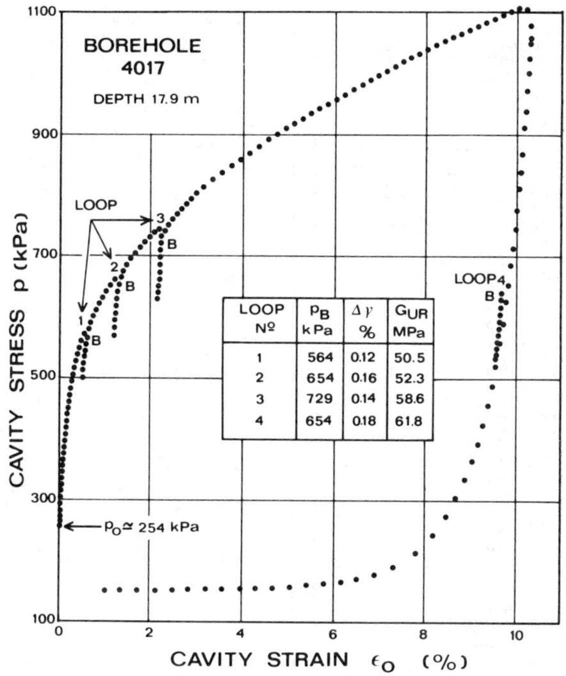

LOOP Nº	p_B kPa	$\Delta\gamma$ %	G_{UR} MPa
1	564	0.12	50.5
2	654	0.16	52.3
3	729	0.14	58.6
4	654	0.18	61.8

FIG. 8--A typical expansion curve.

Table 3 presents values of G_{UR} and G_{RU} obtained from the Camkometer tests run in Po river sand at three different locations.

At present it is well recognized that if the amplitude of the loops is properly limited, i.e. so that the surrounding soil does not fail in extension [10], the values obtained for G_{UR} and G_{RU} represent a reliable measure of the so called "elastic" shear stiffness of the tested sand. This statement applies to the drained expansion test. For which it may be postulated, within the framework of the theory elasto-plasticity, that any unload of the expanding cavity wall will bring the surrounding soil below the currently expanded yield surface in the domain where strains are small and to a large extent reversible [15], [16]. This concept is illustrated in Fig.9, which shows a typical segment of the expansion curve AB with an unload-reload loop BCD and the related effective stress path plotted in the MIT q-p' plane.

TABLE 3--Shear moduli of Po river sand from Camkometer tests.

Borehole	Depth, m	G_{UR} MPa	$\Delta\gamma$ %	p_B kPa	K_G	G^c_{ur}	σ_{ho}/p_o
4003	6.2	22.3	0.32	190.6	227.6	15.5	0.94
		27.0	0.28	258.9	239.4	16.3	
		25.7*	0.32	239.6	235.9	16.0	
	7.7	12.8	0.38	165.1	143.9	10.5	1.83
		11.3*	0.46	137.6	137.6	10.1	
	9.2	30.9	0.32	274.2	273.8	21.4	1.76
		32.4	0.32	370.0	249.0	19.5	
		35.3	0.30	453.6	246.1	19.2	
		41.2	0.18	602.5	251.7	19.7	
	10.7	34.5	0.34	373.1	266.3	22.0	0.85
		34.1	0.32	469.9	236.2	19.5	
		38.5*	0.30	456.7	270.2	22.3	
	12.2	47.4	0.24	384.3	363.2	31.8	0.69
		51.8	0.28	519.9	342.8	30.0	
		49.2	0.24	487.3	336.2	29.4	
	15.2	44.6	0.22	407.8	338.7	32.3	0.87
		45.2	0.24	489.3	312.5	29.8	
		33.5*	0.26	370.0	268.9	25.6	
	16.7	47.1	0.26	515.8	319.9	31.7	0.82
		50.5	0.22	642.2	335.3	33.2	
		56.4	0.26	751.3	318.1	31.5	
		58.8*	0.26	637.1	359.7	35.6	
	18.2	40.0	0.22	192.4	280.1	28.7	0.83
		43.9	0.22	592.3	280.4	28.7	
		52.8*	0.22	581.0	340.2	34.8	
	19.7	49.3	0.20	560.7	325.9	34.5	1.08
		46.9	0.34	646.3	288.2	30.5	
		56.8	0.24	1026.5	277.8	29.4	
		50.3*	0.20	542.3	337.7	35.1	
	21.2	77.3	0.10	709.5	455.6	49.6	0.63
		70.8	0.18	754.3	405.0	44.1	
		70.2	0.22	784.9	393.7	42.9	
		69.3*	0.30	840.0	377.3	41.1	
	23.2	24.7	0.10	412.8	196.7	22.2	1.02
		23.3	0.21	433.	180.6	20.4	
		26.6	0.34	450.6	201.4	22.8	
4017	7.4	26.0	0.24	276.3	230.4	16.0	0.87
		28.0	0.20	324.2	229.9	16.0	
		27.9*	0.26	313.0	233.0	16.2	
	8.9	37.8	0.18	324.2	305.4	22.8	0.84
		36.0	0.20	383.3	275.1	20.5	
		38.1	0.18	446.5	270.7	20.2	
		34.6*	0.26	374.1	268.1	20.0	
4017	10.4	29.1	0.18	376.2	226.4	18.0	0.91
		31.4	0.16	425.1	230.3	18.3	
		30.9*	0.22	358.8	246.4	19.6	
	11.9	34.3	0.20	449.5	246.8	20.8	0.78
		37.3	0.20	498.5	254.7	21.5	
		40.0	0.18	543.3	261.6	22.0	
		40.4*	0.20	482.2	280.2	23.6	
	13.4	58.4	0.18	733.9	332.4	29.4	0.58
		56.5	0.18	745.2	319.0	28.2	
		58.5*	0.20	722.7	335.4	29.7	
	14.8	42.4	0.24	651.4	257.1	23.7	0.92
		47.4	0.12	742.1	269.6	24.8	
		45.4	0.20	816.5	246.3	22.7	
	15.9	50.3	0.16	586.1	322.6	30.6	0.73
		51.2	0.18	675.8	305.9	29.0	
		51.9	0.18	736.0	297.4	28.2	
	17.9	50.0	0.12	563.7	330.2	33.0	1.05
		52.3	0.16	654.4	320.2	31.9	
		58.6	0.14	728.9	339.9	33.9	
		61.8*	0.18	653.4	378.5	37.7	
	19.4	49.8	0.14	598.4	321.6	33.2	1.04
		50.0	0.16	673.8	303.1	31.3	
		60.0	0.16	682.0	362.1	37.3	
	20.9	61.2	0.12	675.8	372.7	39.7	1.09
		63.1	0.14	761.5	361.7	38.5	
		65.3	0.16	845.1	355.3	37.8	
		72.9*	0.18	769.6	415.5	44.2	
	22.8	62.4	0.10	732.9	367.2	40.5	0.95
		67.1	0.14	777.8	382.8	42.3	
		61.8	0.12	820.6	343.0	37.9	
		68.3*	0.16	772.7	391.1	43.2	
	24.9	50.9	0.14	615.7	332.7	37.9	0.99
		50.5	0.16	657.5	318.4	36.3	
		46.6	0.16	700.3	284.1	32.4	
		41.9	0.26	950.1	317.8	24.8	
		54.6*	0.16	658.5	344.8	39.0	
5050/1	6.3	34.9	0.20	201.8	319.4	24.9	0.83
		38.0	0.24	618.8	220.5	17.2	
	7.3	31.6	0.30	396.5	221.9	18.0	1.10
		36.0	0.24	556.6	200.0	16.2	
		48.7*	0.14	525.0	304.1	24.7	
		39.3*	0.22	380.2	281.2	22.8	
	8.8	45.4*	0.12	553.5	278.5	23.8	1.04
		35.5*	0.32	374.1	275.4	22.0	

TABLE 3--Continued

Depth, m	G_{UR} MPa	$\Delta\gamma$ %	p_B kPA	K_G	G^c_{ur}	σ_{ho}/p_o	Borehole
10.3	33.3	0.26	399.6	236.5	21.2	0.95	
	39.9*	0.22	431.2	274.4	24.6		
11.8	28.5	0.32	338.4	220.4	20.6	1.29	
13.3	36.5	0.24	345.6	281.9	27.4	1.72	5050/1
	39.8*	0.26	447.5	273.5	26.6		
16.3	34.7	0.30	425.1	248.1	25.8	1.57	
	61.0*	0.18	710.5	344.1	35.8		
17.8	44.2	0.28	414.9	323.1	34.7	1.16	
	79.2*	0.22	1146.8	360.3	38.7		
19.3	41.2	0.28	415.9	303.1	33.5	1.25	
	31.3*	0.42	403.2	234.0	25.9		
20.8	77.1	0.20	492.4	520.6	59.1	1.35	
22.3	48.8	0.26	664.6	290.5	33.9	0.89	
	48.5*	0.28	658.5	290.0	33.8		
23.8	71.9	0.16	872.6	377.0	45.0	0.66	
6.3	27.0	0.32	239.6	231.0	18.0	0.77	5050/2
	25.1*	0.40	226.3	219.2	17.1		
7.3	17.7	0.38	235.5	154.0	12.5	1.22	
	20.4*	0.48	240.6	175.5	14.3		
8.8	33.1	0.24	287.5	268.5	23.0	1.00	
	25.0*	0.58	303.8	197.8	16.9		
10.3	27.3*	0.44	198.8	261.5	23.4	2.26	
	27.3	0.40	317.0	214.8	19.3		
11.8	29.0	0.28	291.5	238.6	22.3	1.65	
	27.3	0.40	301.7	221.8	20.7		

G_{UR} = Measured unload-reload shear modulus
p_b = Stress from the loop starts
$\Delta\gamma$ = Shear strain amplitude
G^c_{ur} = Shear modulus refered to the in situ σ'_{ho}
K_G = Modulus number obtained with n = 0.45
* = Realod-unload modulus

Despite this qualification of the G_{UR} and G_{RU}, there is still the problem of how to use these moduli in design. This requires assessing the average stress and shear strain levels which correspond to these moduli. As for all the boundary value problems this is difficult to define. Therefore to use the SBP moduli for design it is necessary to adopt a number of simplified hypotheses.

As the relevant stress level is concerned the existing trend is to refer to some kind of average stress existing around the expanding pressuremeter probe. This average stress may be either octohedral effective stress [19] or the mean value of the plane strain effective stresses [12].

When a value for reference stress is selected, it is possible to adopt the following procedure to relate the measured G_{UR} and G_{RU} values to the level of effective stress:

a. Firstly the value of G_{UR} is considered, which corresponds to a given level of shear strain and the cavity stress p_B from which the loop starts (see Fig. 8).

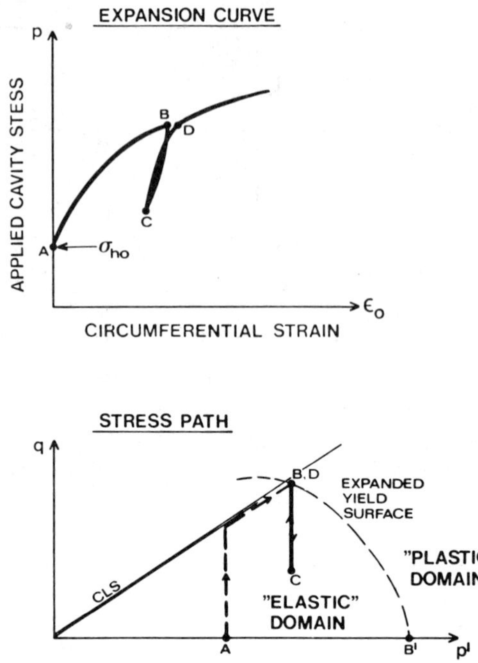

FIG. 9--Concept of "elastic" shear stiffness as obtained from unload-reload loops.

b. The current weighted average of the stress around the probe is computed using an appropriate constitutive equation. For perfect elasto-plasticity, this is approximately the initial stress plus 0.4 to 0.5 times the net cavity stress increase.

c. With this value of average stress, and assuming an appropriate modulus exponent n, the modulus K_G in the Janbu empirical formula as given below can be computed:

$$G = K_G \cdot p_a \left(\frac{\sigma'_o}{p_a} \right)^n$$

where:

K_G = modulus number (-)
n = modulus exponent (-)
p_a = reference stress (FL^{-2})
σ'_o = average effective stress around the probe (FL^{-2})

d. With this value of K_G, the shear modulus may be assessed for the desired level of average stress.

The above outlined procedure has been applied to the G_{RU} and G_{UR} values yielded by Camkometer tests in Po sand, with reference to the value of average effective stress acting on the horizontal plane and assuming n = 0.45*. The resulting values of K_G are shown on Table 3. They have been used to assess the values of "elastic" shear modulus (G^c_{UR}, G^c_{RU}) corresponding to the initial in situ effective horizontal stress σ'_{ho}. The values so obtained for G^c_{UR} and G^c_{RU} are reported on Table 3.

*A conventional assumption, taking into account n dependence on γ [26] it might be higher.

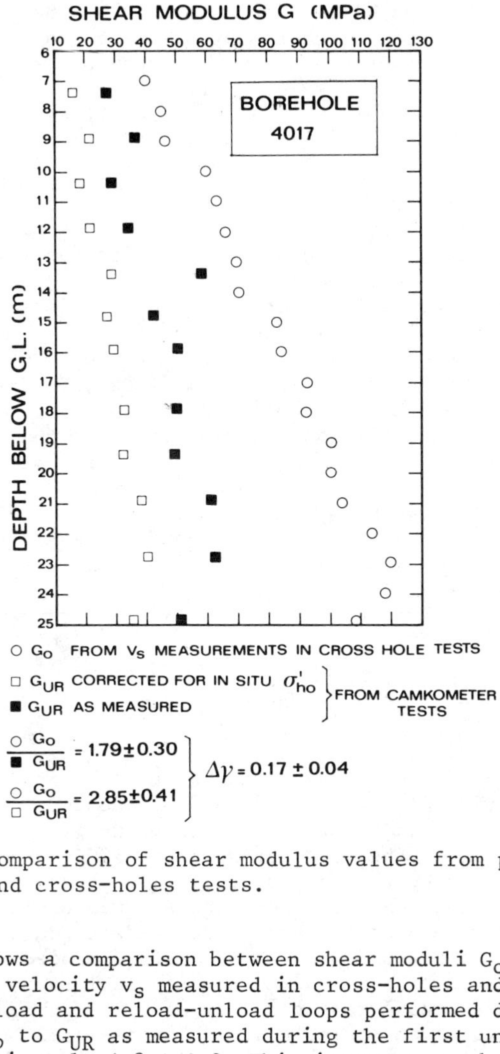

FIG. 10--Comparison of shear modulus values from pressuremeter and cross-holes tests.

Fig.10 shows a comparison between shear moduli G_o computed from the shear wave velocity v_s measured in cross-holes and those obtained from unload-reload and reload-unload loops performed during SBPT's. The ratio of G_o to G_{UR} as measured during the first unload-reload cycle is approximately 1.8 ± 0.3. This incorporates both the effects of different strain and stress levels.

After the reduction of G_{UR} to the stress equal of σ'_{ho}, at which the G_o is measured, the ratio of G_o to the corrected nodulus G^c_{UR} becomes ≅ 2.8 ± 0.4. This ratio primary reflects the different shear strain levels at which G_o and G_{UR} are assessed.

Other comments that may be worthwhile for the data shown on Table 3 are as follows.

- Despite the unavoidable scatter of values obtained for K_G due to the natural variability of the tested sand and the "a priori" assumed value of n = 0.45, a quite well defined trend is observed for K_G to decrease with increasing shear strain amplitude γ of the loop.

- For an equal level of $\Delta\gamma$, the values of K_G from the reload-unload loops tend to be 15 to 20% higher than those corresponding to the unload-reload loops. The writers are not aware if this is due to the strain history experienced by the sand surrounding the probe during the previous phases of the test or caused by other reasons.

- The measured values of G_{UR} and G_{RU} do not appear to be influenced by the current ratio of σ_{ho}/p_o which represents a qualitative indicator of the soil disturbance caused by the probe insertion.

Shear Strength

Hughes et al. [16] proposed a procedure for assessing the effective angle of shearing resistance ϕ'_{PS} and angle of dilation ν'_{PS} for plane strain condition from drained expansion tests in sands. These involve the following assumptions:

a. The sand behaves as an elasto-plastic frictional material, obeying Mohr-Coulomb failure criterion.

b. The relationship between volumetric strain rates $\dot{v}$ and shear strain rates is defined by a straight line. A zero intercept is suggested when these strains are not measured.

c. Once the failure wall has been initiated at the cavity, the relationship between log $(p-u_o)$ and log ε becomes a straight line with slope S.

d. The relation between ϕ'_{PS} and ν'_{PS} is derived from Rowe's [17, 18] stress dilatancy theory. The following equations result:

$$\sin \phi'_{PS} = \frac{(K+1)\ S}{(K-1)\ S+2} \qquad \sin \nu'_{PS} = \frac{2KS - (K-1)}{K+1}$$

where:

K $= \tan^2 (45 + \frac{\phi'_{cv}}{2})$; ϕ'_{cv} = constant volume angle of shearing at large strain

Values of ϕ'_{cv} for the Po river sand were determined using a ring shear apparatus giving $\phi'_{cv} = 34° \pm 1°$; 34° is adopted for computations.

Fig. 11 shows typical log $(p-u_o)$ vs log ε plots. Values of S from Camkometer tests in the Po river sand are given in Table 4, along with ϕ'_{PS1} and ν'_{PS1} obtained by the Hughes et al. [16] procedure.

Robertson [19] and Robertson and Hughes [20] conclude that assumption (b) for Hughes et al. procedure leads to an underestimate of both ϕ'_{PS} and ν'_{PS}. The authors propose a modified procedure at this symposium, which gives ϕ'_{PS2} and ν'_{PS2} shown in Table 4. ϕ' values computed from measured cone resistance q_c using Durgunoglu and Mitchell [22] theory are also given in Table 4. As demonstrated by Bellotti et al. [23] these ϕ' values are on average 1° to 2° lower than the peak angle from triaxial compression tests ϕ'_{TX}. The following comments are made based on the data in Table 4:

- the Hughes et al. [16] procedure leads to highly scattered values of ϕ'_{PS} and ν'_{PS} that are generally too low but sometimes too high;
- the modified procedure [19, 20] works in the right direction but still did not yield consistent values of ϕ'_{PS} and ν'_{PS};
- difficulties encountered when assessing ϕ'_{PS} and ν'_{PS} may be linked to: inadequacy of the incorporated constitutive relationship, curvilinear strength envelope [21, 23], lack of specific experimental data relating $\dot{v}$ to $\dot{\gamma}$, finite length to diameter ratio of the Camkometer probe [24], etc. On the contrary this situation should not be attributed to disturbance caused by insertion of the SBP probe, since almost all of the tests do not exhibit signs of "large initial disturbance" as intended by Hughes et al. [16], Fahey [25] and Robertson and Hughes [20].

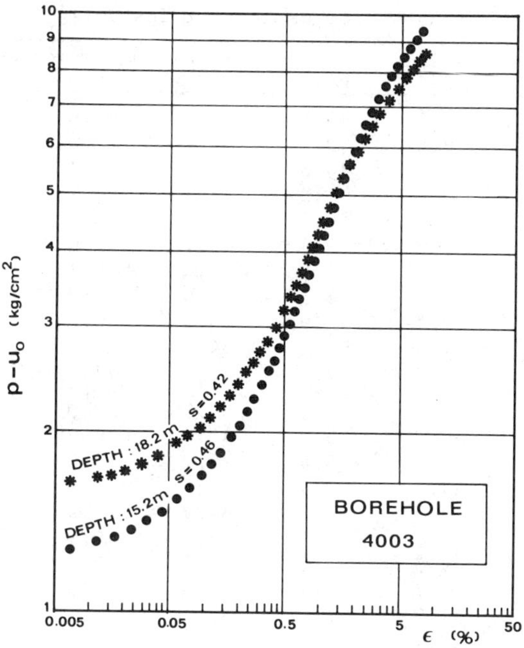

FIG. 11--Typical log $(p-u_o)$ log ε plots.

SUMMARY AND CONCLUSIONS

The results of 53 SBPT's run in the geotechnically well documented quaternary Po river sand are presented. The discussion is restricted to the following aspects: assessment of the total horizontal stress, practical significance of the shear modulus obtainable from unload-reload loops and evaluation of shear strength.

The experimental data presented and experience gained so far with SBP's in sand allow the following comment:

a. The corrected Camkometer cavity stress corresponding to the lift-off pressure p_o, which is assumed equal to σ_h, produced at the Po river site highly scattered values of derived K_o. These values are 20 to 50% higher than the authors' best estimate of K_o in situ.

b. The values of p_o obtained from tests using PAF-76 yielded both horizontal total stress and K_o values that are consistently too low.

c. The shear moduli G_{UR} and G_{RU} represent a reliable measure of the horizontal "elastic" shear stiffness of the tested sand. For engineers to use the moduli in design, it is necessary to relate them to the relevant mean stress and shear strain levels. The G_{UR} and G_{RU} measured at the Po River site are representative for the shear strain level of ≅ 0.15% to 0.30%.

d. Assessment of the friction angle following Hughes et al. [16] approach yielded highly scattered and frequently unrealistically low values of ϕ'. The modified procedure suggested by Robertson [19] and Robertson and Hughes [20] seems to work in the right direction but it still does not eliminate all the inconsistencies in ϕ' values obtained

TABLE 4--Shear Strength from Camkometer Tests (ϕ'_{cv} = 34°).

	Depth, m	S	ϕ'_{PS1}(°)	ν'_{PS1}(°)	ϕ'_{PS2}(°)	ν'_{PS2}(°)	$\phi'(q_c)$,(°)	NOTES
Borehole 4003	6.2	0.53	46.0	15.5	47.5	17.6	39	
	7.7	0.57	48.6	19.2	50.5	21.0	38	
	9.2	0.52	45.3	14.6	47.1	17.1	39	
	10.7	0.44	39.8	7.3	42.7	11.2	40	
	12.2	0.54	46.6	16.4	48.4	18.6	41	
	15.2	0.46	41.2	9.1	43.7	12.6	-	
	16.7	0.41	37.7	4.6	41.0	9.1	-	
	18.2	0.42	38.4	5.5	41.5	9.7	40	
	19.7	0.27	27.1	-7.9	34.5	0.9	37	NR
	21.2	0.28	27.9	-7.0	34.9	1.2	39	NR
	23.2	0.44	39.8	7.3	42.7	11.2	39	
Borehole 4017	7.4	0.31	30.3	-4.3	36.3	3.0	38	NR
	8.9	0.43	39.1	6.4	42.1	10.2	40	
	10.4	0.32	31.1	-3.4	36.7	3.6	38	NR
	11.9	0.32	31.1	-3.4	36.7	3.6	39	NR
	13.4	0.28	28.0	-7.0	34.9	1.2	39	NR
	15.9	0.27	27.1	-7.9	34.5	0.9	37	NR
	17.9	0.30	29.5	-5.2	35.8	2.5	38	NR
	19.4	0.29	28.7	-6.1	35.5	2.0	36	NR
	20.9	0.33	31.9	-2.6	37.1	4.1	-	NR
	22.8	0.27	27.1	-7.9	34.5	0.9	36	NR
Borehole 5050/1	6.3	0.53	46.0	15.5	47.5	17.6	39	
	7.3	0.58	49.3	20.2	51.2	21.6	38	
	8.8	0.62	51.9	24.0	53.9	24.6	38	NR
	10.3	0.56	48.0	18.3	49.7	20.3	36	
	11.8	0.49	43.3	11.8	45.5	14.8	39	
	13.3	0.58	49.3	20.2	51.2	21.6	37	
	16.3	0.52	45.3	14.6	47.1	17.1	38	
	17.8	0.65	53.9	27.0	55.5	27.2	38	NR
	19.3	0.58	49.3	20.2	51.2	21.6	38	
	20.8	0.78	62.8	41.1	62.8	41.1	41	NR
	22.3	0.37	34.8	1.0	39.0	6.6	35	
	23.8	0.22	23.0	-12.5	32.5	-1.5	36	NR
Borehole 5050/2	6.3	0.51	44.6	13.6	46.5	16.3	39	
	7.3	0.34	32.6	-16.7	37.6	4.9	38	NR
	8.8	0.45	40.5	8.2	43.3	11.7	38	
	10.3	0.55	47.3	17.4	49.1	19.3	36	
	11.8	0.54	46.6	16.4	48.4	18.6	39	

S = Slope of the log $(p-u_o)$ vs log ε plot

ϕ'_{PS1} = Plane strain angle of shearing resistance according Hughes et al. [16]

ϕ'_{PS2} = As above according Robertson and Hughes [20]

ν'_{PS1} = Plane strain angle of dilation according Hughes et al. [16]

ν'_{PS2} = As above according Robertson and Hughes [20]

$\phi'(q_c)$ = Angle of shearing resistance computed on the basis of the measured cone resistance using Durgunoglu and Mitchell [22] theory

NR = ϕ'_{PS} not reliable

from SBPT's. This might be due to: (i) the lack of specific experimental data relating volumetric strain to shear strain of the Po river sand; (ii) the discussed theory not incorporating the effect of curvilinear strength envelope [21, 23]; (iii) the sand and (iv) finite length to diameter ratio of the probe.

ACKNOWLEDGMENTS

National Electricity Board of Italy (ENEL) for providing unpublished experimental data.

Mr. Maniscalco R. and Mr. Impavido M. for the execution of the SBPT's.

The National Research Council (CNR) for partial economical support.

REFERENCES

[1] Baldi G., Bellotti R., Crippa V., Fretti C., Ghionna V., Jamiolkowski M., Ostricati D., Pasqualini E., Pedroni S., "Laboratory Validation of In-Situ Tests", Jubilee Volume, XI ICSMFE, S. Francisco, 1985.

[2] Lacasse S. and Lunne T., "In-Situ Horizontal Stress from Pressuremeter Tests," Proc. of the Symp. on the Pressuremeters and its Marine Applications, Paris, 1982.

[3] Ghionna V.N., Jamiolkowski M. and Lancellotta R., "Characteristics of Saturated Clays as Obtained from SBP Tests", Proc. Symp. The Pressuremeter and its Marine Applications, Paris, 1982.

[4] Clough G.W. and Denby G.M., "Self-Boring Pressuremeter Study of San Francisco Mud", JGED, ASCE, No.GT1, 1980.

[5] Clarke B.G., "In Situ Testing of Clay Using Cambridge Self-Boring Pressuremeter", Ph.D. Thesis, Cambridge Univ., 1981.

[6] Benoit J., "Analysis of Self-Boring Pressuremeter Tests in Soft Clay", Ph.D. Thesis, Stanford Univ., Calif., 1983.

[7] Windle D., "In-Situ Testing of Soils with a Self-Boring Pressuremeter", Ph.D. Thesis, University of Cambridge, U.K., 1976.

[8] Dalton J.C.P., Hawkins P.G., "Some Measurements of the In-Situ Stress in a Meadow in Cambridge Shire Country Side", Ground Engineering, No.5, 1982.

[9] Jamiolkowski M., Ladd C.C., Germaine J.T., Lancellotta R., "New Developments in Field and Laboratory Testing of Soils", Theme Lecture, XI ICSMFE, S. Francisco, 1985.

[10] Fahey M., Randolph M.F., "Effect of Disturbance on Parameters Derived from Self-Boring Pressuremeter Tests in Sand", Géotechnique 34, No.1, 81-97, 1984.

[11] Jewell R.J., Fahey M. & Wroth C.P., "Laboratory Studies of the Pressuremeter Test in Sand", Géotechnique 30, No.4, 507-531, 1980.

[12] Fahey M., Randolph M.F., "Effect of Disturbance on Parameters Derived from Self-Boring Pressuremeter Tests in Sand", Discussion Closure. Geotechnique, 35, No.2, 219-222, 1985.

[13] Gudehus G., Personal Communication, 1985.

[14] Baguelin F., Jézéquel J.F. and Shields D.H., "The Pressuremeter and Foundation Engineering", Trans. Tech. Publications, Clausthall, 1978.

[15] Hughes J.M.O., "Interpretation of Pressuremeter Tests for the Determination of the Elastic Shear Modulus. Proc. of the Conf. on Updating Subsurface Sampling of Soils and Rocks and Their In-Situ Testing, Santa Barbara, Cal, 1982.

[16] Hughes J.M.O., Wroth C.P. & Windle D., "Pressuremeter tests in Sands", Géotechnique 27, No.4, 455-477, 1977.

[17] Rowe P.W., "The Stress-Dilatancy Relation for Static Equilibrium of an Assembly of Particles in Contact", Proc. Royal, Society. A., 1962.

[18] Rowe P.W., "Theoretical Meaning and Observed Values of Deformation Parameters for Soils", Proc. of the Roscoe Memorial Symposium, Cambridge, U.K., 1971.

[19] Robertson P.K., "In-Situ Testing of Soil with Emphasis on Its Application to Liquefaction Assessment", Ph.D. Thesis, Univ. of British Columbia, Vancouver, 1982.

[20] Robertson P.K. and Hughes J.M.O., "Determination of Properties of Sand from Self-Boring Pressuremeter Tests", Proc. Symposium on The Pressuremeter and Its Marine Applications , ASTM STP 950, Texas A & M., 1986.

[21] Baligh M.M., "Theory of Deep Site Static Cone Penetration Resistance," Res. Report R75-56, No.517, Dept. of Civil Eng., Mass. Inst. of Technology, Cambridge, Mass., 1975.

[22] Durgunoglu H.T. and Mitchell J.K., "Static Penetration Resistance of Soils", Research Report Prepared for NASA Headquarters, Washington, D.C., Univ. of California, Berkeley, 1973.

[23] Bellotti R., Ghionna V., Jamiolkowski M., Manassero M. and Pasqualini E., "Evaluation of Sand Strength from CPT", Symposium on Soil and Rock Investigations by In-Situ Testing, Paris, Vol.1, 1983.

[24] Laier J.E., "Effects of Pressuremeter Probe to Length Diameter Ratio and Borehole Disturbance on Pressuremeter Test Results in Dry Sand", Ph.D. Thesis, Univ. of Florida, 1973.

[25] Fahey M., "A Study of the Pressuremeter Test in Dense Sand", Ph.D. Thesis, University of Cambridge, 1980.

[26] Wroth C.P., Randolph M.F., Houlsby G.T. and Fahey M., "A Review of the Engineering Properties of Soils with Particular Reference to the Shear Modulus". Dept. of Engineering, University of Cambridge, CUED/D Soils TR75, 1979.

Robert E. Bacciarelli

THE CALIBRATION AND USE OF A HIGH CAPACITY PRESSUREMETER TO DETERMINE ROCK STIFFNESS

REFERENCE: Bacciarelli, R.E, "The Use of a High Capacity Pressuremeter to Determine Rock Stiffness", The Pressuremeter and Its Marine Applications: Second International Symposium, ASTM STP950 J.L. Briaud and J.M.E. Audibert, Eds., American Society for Testing and Materials, 1986.

ABSTRACT: Rubber membrane type pressuremeters are in common use worldwide for the determination of the in situ deformation characteristics of rock. Standard correction values and data reduction methods quoted by the equipment manufacturer are conventionally used to process pressuremeter test data. A basic assumption of the data reduction procedure is that the rubber membrane behaves elastically. However problems experienced with the non-elastic behaviour of a thick rubber membrane during an investigation in the UK resulted in a complete reassessment of calibration techniques and data reduction procedures. Revised on site calibration and test procedures are proposed, together with a simple method for data reduction and modulus determination. It is apparent that frequent and stringent calibration of the pressuremeter test equipment is essential to avoid errors in both data reduction and modulus determination.

KEYWORDS: Rubber-Membrane Type Pressuremeter, Pressuremeter Use in Rock, Calibrations, Test Procedure, Data Reduction, Modulus Determination.

INTRODUCTION

Current design practice in both rock and soil mechanics has become increasingly complex and sophisticated, especially with the advent of routine computer use in analytical modelling for the prediction of foundation and excavation performance.

Robert Bacciarelli is the Senior Engineering Geologist at Rendel Palmer and Tritton Limited, Consulting Engineers, 61 Southwark Street, London SE1 1SA, United Kingdom.

The use of such sophisticated and precise analytical models requires the input of both realistic and accurate engineering parameters to allow realistic prediction of foundation behaviour. The engineering parameters necessary for foundation design in rock rely on an assessment of the "rockmass" strength and deformation characteristics in which allowance is made for the effect of discontinuities.

Laboratory testing of rock is inevitably biased towards the testing of small "intact" rock material and does not take into account the effect of discontinuities within the rockmass. Thus, in situ testing of the rockmass which to some extent measures the influence of discontinuities on deformation behaviour is preferred for a realistic assessment of the "rockmass" properties. In situ testing in rock such as plate load testing can be expensive, time consuming and often difficult to carry out. However, the pressuremeter provides a relatively rapid and cost effective method for the determination of the in situ horizontal deformation characteristics of the rockmass, especially if correlated with laboratory testing and other in situ testing, and is now in common world wide use for the in situ testing of rock.

All too often the designer accepts a quoted pressuremeter test result as a distinct engineering parameter without fully understanding the practicalities of how the initial field results were obtained or the subsequent processing of the data carried out. It is considered essential that the designer is aware of both the test procedure and methods employed in the processing of the data in order to appreciate the inherent limitations of the pressuremeter and its results. Otherwise the indiscriminate and inappropriate use of such pressuremeter test results could lead to erroneous and unrealistic foundation predictions.

This paper outlines the use of a thick rubber membrane type pressuremeter, in particular the OYO Elastmeter 100 (uprated to 200), in pre-drilled boreholes to determine the deformation characteristics of rocks and suggests a modified method for calibration and test procedure, data reduction and modulus determination. The OYO pressuremeter is capable of applying a pressure of up to 20MPa to the borehole wall by inflation of a hard, thick neoprene rubber membrane. Borehole wall movement is monitored by a two-arm caliper system mounted within the pressuremeter probe.

Pressuremeters of similar type to the OYO Elastmeter 100/200 are currently in use worldwide such that the observations and conclusions from this case study are directly applicable to other similar membrane type pressuremeters.

As part of a site investigation in Somerset, UK undertaken in 1983 and 1984, over 100 pressuremeter tests were performed in pre-drilled holes to determine the engineering properties of the underlying rocks, which consisted of interbedded mudstones, limestones and gypsiferous siltstones.

This particular pressuremeter was quoted as being capable of measuring deformations in rocks with an unconfined compressive strength

TABLE 1—DATA REDUCTION PROCEDURES RECOMMENDED BY OYO

OYO DATA REDUCTION PROCEDURE - 1980

A. PRE-OPERATIONAL SYSTEM COMPLIANCE CHECK

B. PRESSURE - pressure corrections not required.

C. DISPLACEMENT

(i) Display value Rn at 60 seconds on digital indicator obtained and then processed as follows:
$R_I = Rn + 23.5$
where, R_I = Inner radius of tube.
Rn = Digital indicator reading as displayed after 60 seconds.
23.5 = Radius of steel cylinder within which calipers are mounted (mm).

(ii) Tube correction applied from correction given by manufacturer from single graph which is supposed to be applicable to all supplied rubber membranes, using graph for initial borehole diameter.
$Rs = R_I - Pg$
where, Pg = tube thickness correction.
Rs = standard tube radius

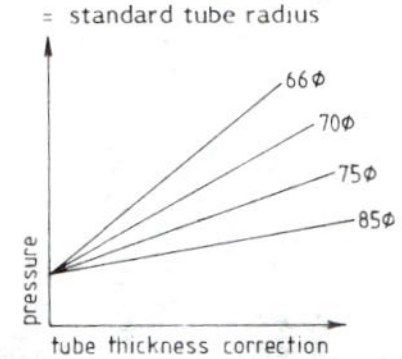

(iii) Outer tube radius (Ro) obtained from single graph supplied by manufacturer and applicable to all membranes supplied from standard tube radius (Rs).

Notes: (a) System compliance check recommended as a calibration.
(b) Final results plotted as a pressure/borehole radius deformation curve (P and Ro).

MODIFIED OYO DATA REDUCTION PROCEDURE - 1982

A. PRE-OPERATIONAL SYSTEM COMPLIANCE CHECK

B. PRESSURE

(i) Hydrostatic head correction in hydraulic system - recommended by personal communication by manufacturer.

(ii) Rubber Reaction Correction - from calibration of membrane by inflation at 1 atmosphere to 0.8 MPa.
$P' = P - Rg$
where, P = gauge pressure.
Rg = rubber reaction correction $= x\, Rn^y$ from cablibration.
P' = effective applied pressure.

This correction quoted as "not important for high pressuremeter".

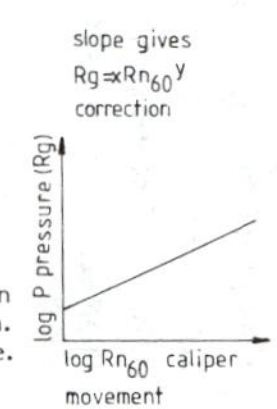

C. DISPLACEMENT

(i) Display value Rn at 60 seconds on digital indicator obtained and then processed as follows:
$R_I = \frac{Rn + 23.5}{10}$
where, R_I = Inner tube radius (cm).
Rn = Digital indicator reading as displayed after 60 seconds (mm).
23.5 = Radius of steel cylinder within which calipers are mounted (mm).

(ii) Membrane thickness correction from a calibration.
$Pg = (P - Po) \times z$
where, Pg = membrane thickness correction.
P = gauge pressure.
Po = pressure corresponding to start of membrane compression.

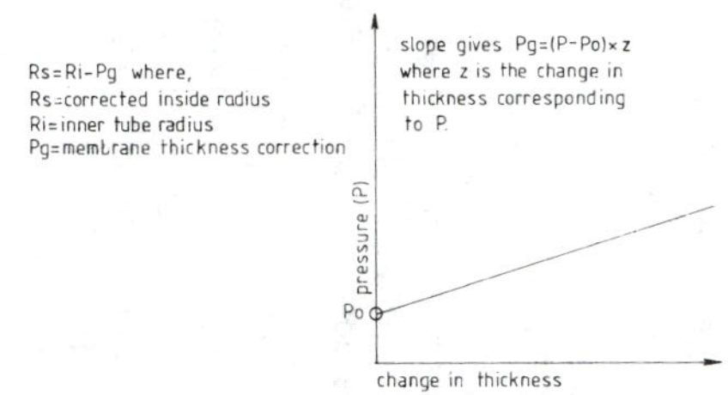

The manufacturer quoted that "the indication of the radius measuring system is slightly influenced by the applying of pressure".

(iii) Standard Cross-Sectional Area Correction is obtained by measurement of both internal and external diameter of the membrane; and dividing by 2 to obtain the radius.
e.g. R_{IO} = approx. 2.1 cm
R_{OO} = approx. 3.5 cm

where, R_{IO} = internal radius of membrane.
R_{OO} = external radius of membrane.

$S = \pi.(R_{OO}^2 - R_{IO}^2)$
where, S = standard cross-sectional area of membrane.

(iv) The outside membrane radius (Ro) is obtained, assuming that volume change is negligible, as follows:

$Ro = \sqrt{R_S^2 + S/\pi}$

Notes (a) System compliance check, cross-sectional area and rubber reaction calibrations quoted as being recommendable.
(b) Membrane thickness correction referred to and value quoted.
(c) Frequency of calibration not stated.
(d) Final results plotted as pressure/borehole radius deformation curve (P' and Ro).

of between 0.5 MPa and 100 MPa. The unconfined compressive strength of the rocks encountered at the Somerset site ranged from 2 MPa to 80 MPa, with the majority being between 13 MPa to 35 MPa. The testing was biased towards the weakest rocks which were the weak to moderately weak mudstones.

Pressuremeter field test data is generally processed and corrected by the application of "standard" correction values and procedures recommended by equipment manufacturers (See Table 1)(1)(4). These procedures involve the assumption that the rubber membrane behaves in a perfectly elastic manner.

Problems with analysis of the field test data from the Somerset site were encountered when using manufacturer recommended correction values and data reduction procedures. In certain extreme instances anomalous results were obtained where the borehole appeared to contract on the application of pressure to the borehole wall and vice versa.

It was established that the hard neoprene rubber membrane of the pressuremeter in fact behaved in a non-elastic manner. The results of the pressuremeter testing also proved to be extremely sensitive to rubber membrane behaviour, with "corrections" for rubber membrane behaviour being of a similar order of magnitude as the borehole wall deformations actually being measured.

Research and development during the site work resulted in considerable changes being made to the operational procedure for use with this particular pressuremeter. A reassessment was made of both the methods used to establish "correction values" for rubber membranes in general and use of these correction values in the reduction of data to determine rock stiffness. However, the most important conclusion was the necessity for thorough regular calibration throughout testing together with the observation that thick neoprene rubber membranes can behave in a non-elastic manner.

Experience with the Oyo pressuremeter has brought to light "problems" which are likely to be encountered by all other similar rubber membrane type pressuremeters which make use of thick, hard neoprene rubber membranes for testing in rock. In conclusion it is considered essential that regular on-site calibration be carried out during pressuremeter testing in order to avoid erroneous results and subsequent modulus determination.

DESCRIPTION OF EQUIPMENT

The OYO Elastmeter 100 (uprated to 200: Model 413B) pressuremeter equipment used in this investigation consisted of a probe, digital indicator unit and a hydraulic pump. This pressuremeter equipment is briefly described below and illustrated in Fig. 1.

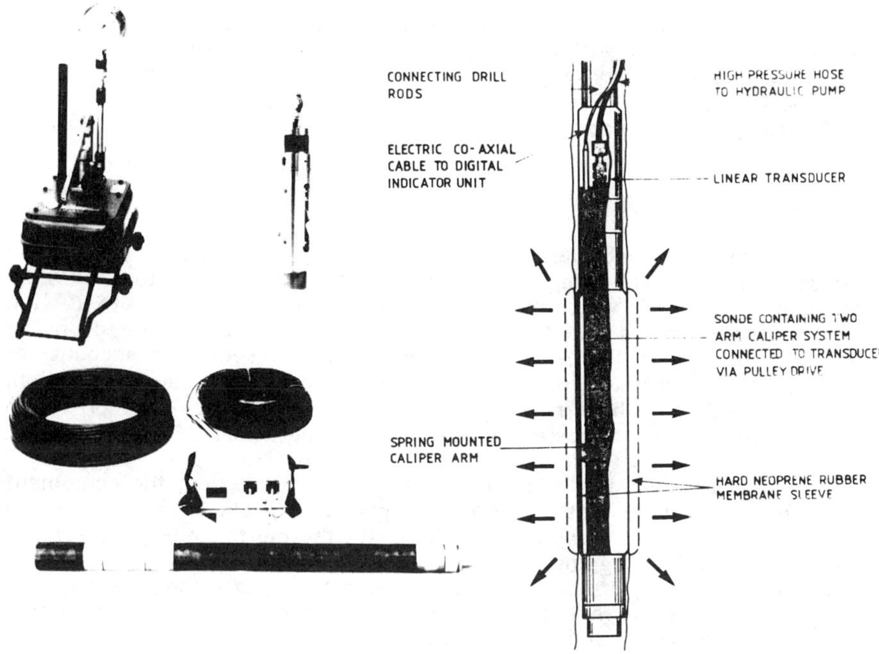

FIG. 1 -- OYO Elastmeter 100 (Uprated 200) Pressuremeter

The Probe

The probe consisted of a heavy duty steel cylinder within which were two spring-mounted calipers. The calipers were mounted diametrically opposite each other midway along the length of the probe. A thick hard neoprene rubber membrane sleeve, 520mm long, was fitted onto the steel cylinder over the two calipers. The rubber membrane had an outside diameter of 70mm and was about 13mm thick. During testing, the rubber membrane was expanded under hydraulic pressure to apply pressure to the borehole wall. The caliper movement, after correction for rubber membrane behaviour, represented the displacement of the borehole wall under loading at various pressure increments. The pressure and caliper displacement, after correction, was plotted in the form of a stress/displacement curve for each test and the deformation modulus calculated. Corrections for rubber membrane behaviour were obtained from instrument calibrations and these are described in section 4 on calibrations.

Digital Indicator Unit

Caliper movement was measured by an electrical displacement transducer mounted within the sonde and relayed as an average radial movement, via a heavy duty electric co-axial cable, to a digital indicator unit at the surface. The digital indicator unit was powered by a 12 volt DC battery power source.

The Hydraulic Pump

The rubber membrane sleeve of the probe was expanded by hydraulic pressure applied, via a high pressure hose, by a hand-operated hydraulic pump at the surface which was capable of applying pressures up to 20MPa. The pressure applied was measured by a Bourdon gauge mounted on the pump. The recorded pressure was subsequently corrected to account for the hydrostatic pressure in the system and the borehole at the test section, together with a rubber reaction correction as required. The correction procedures are fully discussed in section 4 on calibration.

Although not used during this particular investigation, the equipment can be modified so that the pressure gauge is replaced by a pressure transducer in the sonde which together with the digital indicator unit can be linked to a data logging facility, field computer and printer. A further modification can incorporate a solenoid valve to regulate the pressure from the pressure source in a pre-programmed pattern. This facility would enable the carrying out of both stress and strain controlled tests with the raw field data being automatically corrected using on site calibration data previously inputted into the computer. The corrected results can then be studied on site prior to interpretation and modulus determination.

TEST PROCEDURE

Pressuremeter testholes were required at three locations at the site and were drilled in progressive stages in order to carry out pressuremeter testing from ground surface down to a total depth of 75 metres, at approximately 2.5 metre intervals. Rotary cored boreholes put down within 5 metres of each testhole prior to testing provided information on the detailed geology and enabled the scheduling of a comprehensive pressuremeter test programme.

Subsequently an NX sized testhole, with an outside diameter of 76mm, was drilled nearby in progressive stages by rotary core drilling techniques using water flush. A diamond impregnated bit was used to provide a smoothsided testhole. The core obtained from each test location was examined and the precise test depth finalised on the basis of the actual geology encountered within the test section.

Pressuremeter testing was carried out on the same day as each test section was drilled in order to minimise disturbance to the borehole wall due to stress relief, softening or "caving" of material from above and within the test section. Generally three test sections were drilled and tested per day. Calibrations and pre-cycling of the rubber membrane were carried out during drilling operations and at suitable intervals during the testing programme.

The water level within the borehole was determined prior to testing. The pressuremeter probe was then inserted into the borehole to the required test depth and orientated to measure displacements in a specific direction.

The rubber membrane was inflated and deflated in increments by hydraulic pressure. The pressure was measured by a surface mounted Bourdon pressure gauge. Caliper movements, averaged as radial displacement, were displayed on the digital indicator unit at surface for each corresponding pressure increment. The actual radial displacement of the borehole wall was obtained at a later stage by the application of corrections which allowed for compression of the rubber membrane. The effective applied pressure was obtained after correction for hydrostatic head in both the hydraulic system and the borehole.

Corrections for the stretching of the rubber membrane on expansion and rubber reaction were not applied during the investigation as these were found to be relatively small, generally within $\pm 5\%$ of the actual final results. However, the relevant calibration data was recorded.

Measurements of radial caliper movement were recorded at 15, 30 and 60 seconds after application of each pressure increment. This procedure was generally found to be adequate as negligible displacement occurred after 60 seconds except during the initial loading stage and after onset of apparent 'plastic' deformation where displacements continued for a period greater than 60 seconds. To observe creep effects particular pressures were maintained for longer periods and displacement monitored.

A variety of loading patterns were adopted in order to investigate the rock behaviour under different loading conditions. These are briefly described below and illustrated in Fig. 2.

Small Unload/Reload Cycles

The loading procedure generally consisted of six small unload-reload cycles, with unloading to within 20% to 50% of the maximum pressure achieved for that cycle. Each cycle included approximately five equal increments of pressure applied in increasing steps upto the maximum for the cycle and five increments of decreasing pressure. (2)

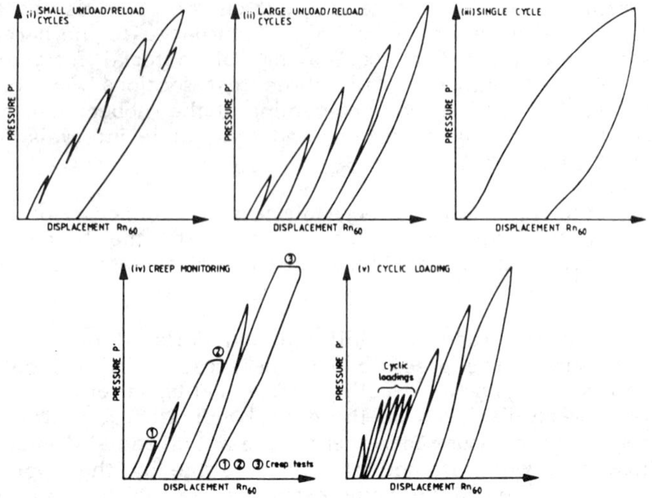

FIG. 2 --Example of Pressuremeter Loading Patterns Used

Large Unload/Reload Cycles

Five loading cycles, each consisting of approximately five equal increments of pressure were applied in increasing steps up to the maximum for the cycle and five increments of decreasing pressure down to a nominal seating pressure, which was generally about 0.5 MPa. Maximum test pressure achieved was 20 MPa. The large unload-reload cycles were often subject to plastic reverse deformation and bedding movements on unloading to the nominal seating pressure, such that the preferred loading pattern consisted of small unload-reload cycles. (3)

Single Loading Cycle

The test consisted of a single loading cycle upto 20 MPa with numerous pressure increments on the loading and unloading stage in order to obtain a detailed deformation curve for comparison with the small and large unload/reload cycle loading patterns previously described.

Cyclic Loading

The influence of cyclic loading on the behaviour of the rock was investigated by modifying the loading pattern to enable a specific cycle to be repeated, in this case up to five times.

Creep Monitoring

An attempt was made to assess whether the rockmass being tested was susceptible to creep and to observe the behaviour of the rock under a sustained load at various stress levels. The loading procedure was amended to maintain a pressure increment constant at a particular stage during the test, generally at the maximum test pressure of the first, third and fifth cycles. This pressure was maintained constant for upto two hours.

CALIBRATION

Calibration of the pressuremeter equipment is an essential part of the pressuremeter test procedure in order to obtain meaningful results for data reduction and subsequent calculation of deformation modulus.

Initially the manufacturer's test procedures and recommended calibration values were adopted (1)(4) as shown in Table 1. However, it became increasingly apparent from anomalous test results as the test programme progressed that these values were inadequate. In extreme cases using the quoted calibration and correction values the processed test results seemed to indicate borehole contraction on application of pressure to the borehole wall and vice versa. It was discovered that membrane behaviour during testing varied from membrane to membrane, as well as with use and time. Regular onsite calibration was adopted to obtain specific correction values for each individual membrane used and also to account for time dependent behaviour. Thus, it became necessary to calibrate the equipment using modified calibration procedures and at greater frequencies than originally indicated by the equipment manufacturer. The calibration procedure and frequency was amended progressively throughout the investigation. Calibration frequency was gradually increased from one calibration per membrane to calibration of the membrane every ten tests, until eventually calibration was carried out at the start and end of each day's testing.

After extensive research during the course of the project modified operational and calibration procedures were finally adopted and these are described below and illustrated in Table 2.

Pre-Operational System Compliance Check

Deformations in rock during testing tend to be relatively small, often less than 1mm, which is often close to or at the stated operating limits for both the mechanical and electrical components of the displacement sensing system such that a regular check on the accuracy of the system is considered essential.

The interaction of the caliper-arm system with the associated electronics of the digital indicator unit was checked as a pre-operational

TABLE 2—PROPOSED DATA REDUCTION PROCEDURE

A. OPERATIONAL PROCEDURE

(i) SYSTEM COMPLIANCE CHECK

(ii) PRECYCLING

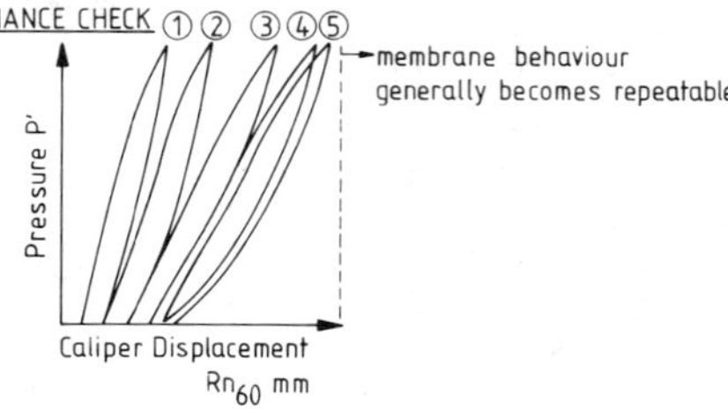

B. PRESSURE

(i) HYDROSTATIC HEAD CORRECTION

$$Po = P + [Hs - Hw]$$

P = gauge pressure
Hs = hydrostatic head in hydraulic system
Hw = hydrostatic head in borehole

(ii) RUBBER REACTION

$Rg = x Rn_{60}{}^{y}$

Data plotted from rubber reaction calibration

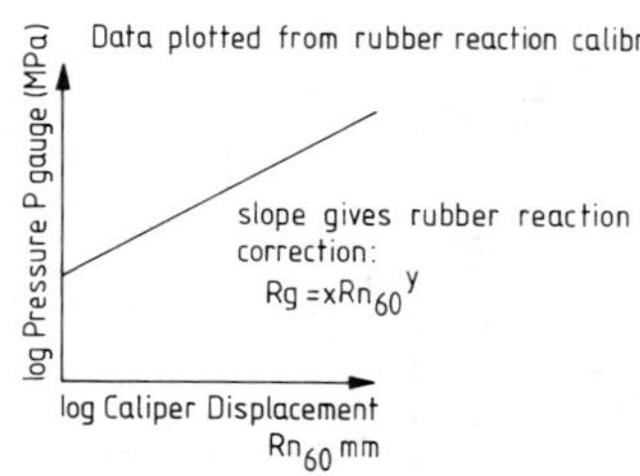

(iii) EFFECTIVE APPLIED PRESSURE

$P' = Po - Rg$

Po = pressure corrected for hydrostatic head
Rg = rubber reaction

C. DISPLACEMENT

(i) RUBBER COMPRESSION

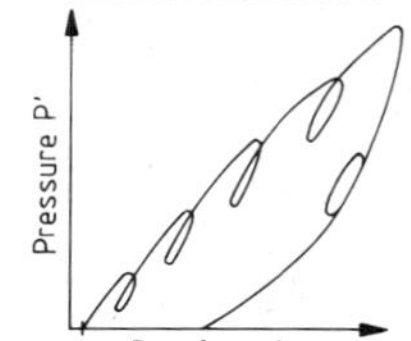

Radial Caliper Displacement
[Borehole wall movement + rubber compression]

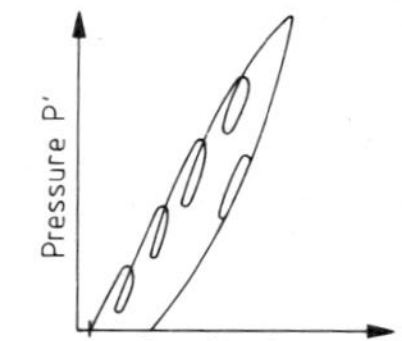

Radial Caliper Displacement
[Rubber compression against rigid steel tube]

$\Delta r^{*} = \Delta r_m - \Delta r_c$

Δr_c = caliper displacement from rubber compression calibration
Δr_m = caliper displacement from test

(ii) REDUCTION OF MEMBRANE THICKNESS WITH INCREASING HOLE SIZE

$\Delta r = \Delta r^{*} - \Delta t$ Δt = change in thickness due to stretching

Change in thickness due to stretching at different hole sizes from calibration data

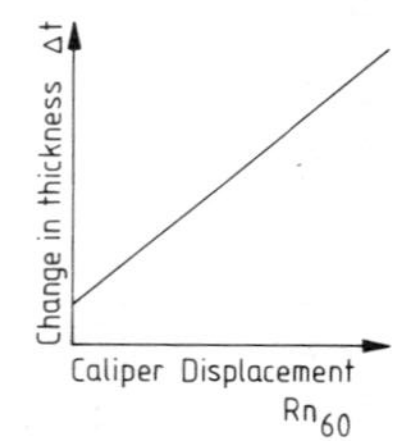

N.B.
Test results are plotted as effective applied pressure (P') versus radial caliper movement (Δr) after 60 seconds.

procedure. This check was carried out by sliding a specially machined calibration ring with two different internal diameters, which corresponded to 0.0 and 10.0 millimeters radial caliper movement on the display of the digital indicator unit, over the sonde onto the calipers. Adjustments were made to the unit until the displayed values corresponded directly to the respective ring diameters. The system was set at the start of the day's testing and then checked at the end of the day's testing to assess whether drift had occurred with time.

Head of Water in the System and Borehole

In order to obtain the effective applied pressure at the specific test depth the gauge pressure has to be corrected to take into account the hydrostatic head of water in the test equipment system above the test section, as well as the hydrostatic pressure within the borehole acting directly on the probe at the test depth.

Rubber Reaction

On expansion of the rubber membrane during testing there is a corresponding small decrease in applied pressure which can be accounted for during data reduction by the use of a rubber reaction "correction" to the recorded gauge pressure obtained from calibration. The pressuremeter can be calibrated by inflation in air at ground level, at approximately one atmosphere pressure, in the vertical position in incremental pressure stages up to 0.8 MPa to obtain the rubber membrane resistance to unrestrained expansion. Calibration was carried out vertically in order to model the actual pressuremeter orientation during testing. The effect of the correction on the final test results was consistent and small, with the recorded gauge pressure being generally within 3% of the actual effective applied pressure.

Rubber Compression

There is a decrease in membrane thickness as the membrane is expanded and compressed against the borehole wall, which due to the high applied pressures used when testing in rocks can be of a similar magnitude as the actual deformation of the rock itself during testing.

It is necessary to apply a correction, obtained from a calibration, to the corresponding pressuremeter test results in order to obtain the actual response of the rock which takes into account the compression of the rubber membrane during testing. This is the major correction used in the data reduction procedure.

A calibration to obtain information on the compressibility of the rubber membrane over the pressure range used for testing was carried out regularly in a vertically standing, rigid steel tube with the same diameter as the testhole. The calibration tube was grouted vertically into the ground. This procedure was adopted in order to model as closely as possible the conditions during actual testing.

The pressuremeter test data is highly sensitive to the rubber compression correction, especially as this correction can be of the same order of magnitude as the deformation of the rock being tested. It should be noted that compression of the rubber membrane was observed during the investigation to be as much as 5 to 10 times greater than the actual deformation of the rock tested. However, anomalous test results were observed even after application of the rubber compression corrections obtained from onsite calibration. This was finally attributed to the inelastic behaviour of the thick rubber membrane, especially over the first few inflation/deflation cycles of membrane use, during both testing and calibration.

This situation was remedied by pre-cycling the rubber membrane prior to use in order to obtain both a repeatable and an improved elastic response during use.

Research during the investigation indicated that after about five or more inflation/deflation cycles the rubber membrane behaviour and response to further expansion became relatively consistent and repeatable. A typical example of a precycling record is shown in Fig. 3. Thus precycling of the rubber membrane for a minimum of five equal cycles up to the maximum test pressure of 20 MPa, or until repeatability to within 0.02mm was achieved at the maximum test pressure of 20 Mpa, was carried out as a pre-operational procedure prior to any pressuremeter use on site.

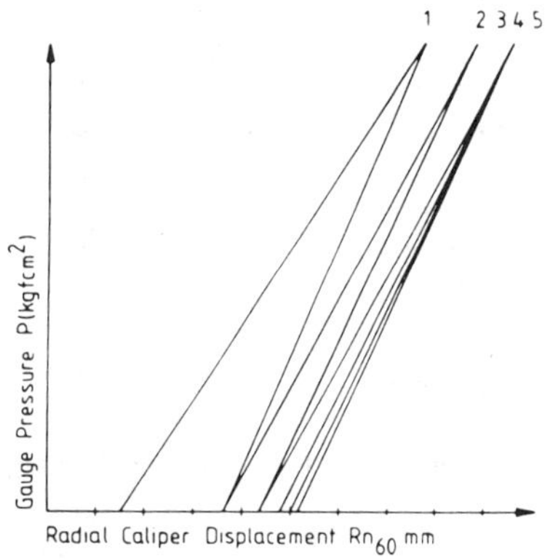

FIG. 3 --Example of Pre-cycling of Pressuremeter Equipment

Reduction of Membrane Thickness with Increasing Hole Size

There is a corresponding reduction in thickness of the rubber membrane on expansion and increasing borehole size during testing which also has to be accounted for during data reduction in order to obtain actual caliper movements during testing.

Processing of the test results according to the manufacturer's method involved the assumption that the membrane was perfectly elastic and there was no volume change during testing such that the cross-sectional area of the membrane remained constant as the membrane expanded and contracted during testing. Initially data regarding the internal and external diameter of the membrane prior to use was recorded in order to obtain the cross-sectional area of the membrane and apply this correction to the recorded caliper movement data to account for the change in membrane thickness due to stretching on expansion.

However on site experience indicated that the membrane did not behave in a perfectly elastic manner such that the cross-sectional area of the membrane did not remain constant on expansion.

In order to assess the effect of membrane stretching it is considered that prior to the use of a new membrane, a series of initial tests should be undertaken in rigid, calibration tubes of different diameters in order to determine the correction required to account for the reduction in membrane thickness with increasing borehole size. This correction is generally linear and can be accurately defined. In weak rocks borehole wall deformations are likely to be small such that the effect of the non-application of this correction on the test results will also be relatively small with uncorrected deformations generally within ±10% of the actual corrected deformations.

DATA REDUCTION

The raw field data is obtained as measurements of applied pressure and radial caliper movement. These values require the application of various correction factors in order to obtain a deformation curve suitable for interpretation and derivation of rock deformation parameters. A suggested procedure for data reduction is outlined below and illustrated in Table 2.

The "effective applied pressure" is obtained after correction of the applied field pressure for the head of water in the borehole above the test depth and rubber reaction. A further correction is required if a hydraulic pressure system is used to correct for the head of liquid medium in the system above the test depth.

The true "radial caliper movement" after 60 seconds for each individual pressure increment is obtained after correction for rubber compression and reduction in membrane thickness with increasing hole size.

The corrected field data is subsequently plotted as the "effective applied pressure" against the true "radial caliper movement" after 60 seconds, as in Fig. 4. The difference between caliper displacement after 30 and 60 seconds, is also plotted to facilitate interpretation of the test deformation curve, and is often misleadingly referred to as the "Creep Index".

DETERMINATION OF DEFORMATION MODULUS

Initial Shear Modulus

The shear modulus (Gi) may be evaluated from the initial linear tangent of the corrected deformation curve as indicated in Fig. 4. (2)

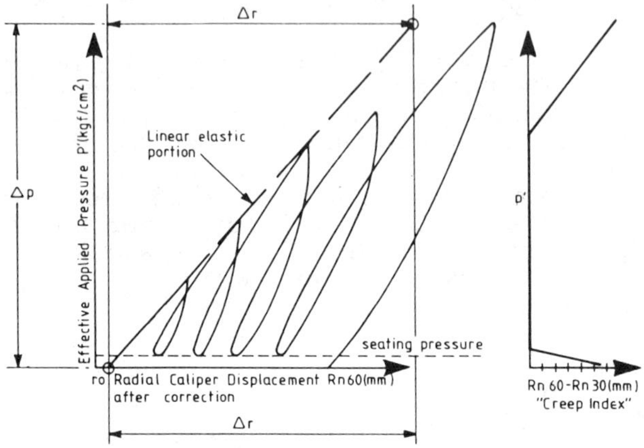

FIG. 4 --Derivation of Initial Shear Modulus

$$\text{Shear Modulus Gi (Initial)} = \frac{1}{2} \cdot ro \cdot \frac{\Delta p}{\Delta r} \qquad (1)$$

Where, r_o = original borehole radius.
Δr = change in radial caliper movement.
Δp = pressure range over linear elastic portion corresponding to Δr.

The test section was drilled at a nominal diameter of 76 millimetres, however in practice this diameter may be larger or smaller by a few millimetres. An assumed original borehole radius of 38mm was adopted because small deviations from the assumed borehole radius had a negligible effect on the calculated shear modulus in the rocks encountered, being generally with $\pm5\%$ of the actual value. However, the original borehole radius (r_o) can be determined by caliper logging of the borehole test section prior to insertion of the pressuremeter for testing.

Unload/Reload Shear Modulus

The Unload/Reload Shear Modulus (Gur) can be obtained from the gradient of a line which bisects the hysteresis loop of each unload-reload cycle as shown in Fig. 5.

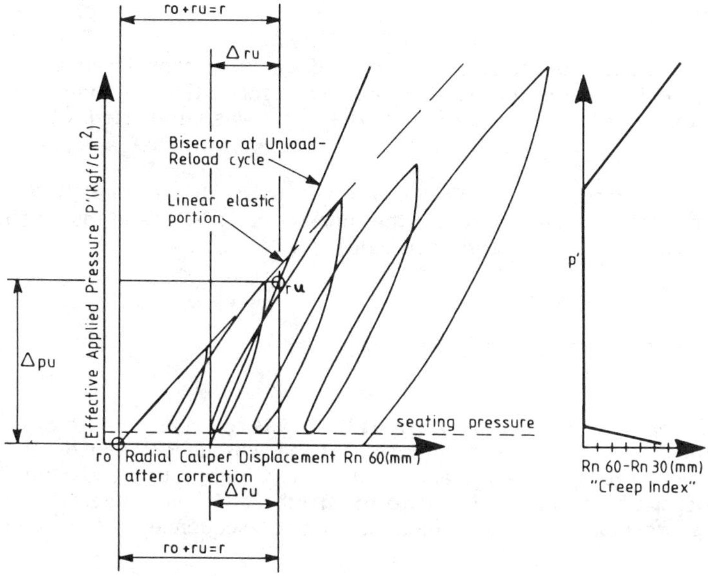

FIG. 5 --Derivation of Unload Reload Shear Modulus

$$\text{Shear Modulus Gur (Unload/Reload)} = \tfrac{1}{2} \cdot r \cdot \frac{\Delta pu}{\Delta ru} \qquad (2)$$

Where, r =original borehole radius plus the change in caliper displacement reflecting rockmass movement for a particular unload/reload cycle.

Δ ru =change in radial caliper movement for a particular unload/reload cycle.

Δ pu =pressure range over the bisected portion of the hysteresis loop of the unload/reload cycle corresponding to Δru.

Elastic Modulus

To obtain the Elastic or Young's Modulus an appropriate value of Poisson's Ratio should be chosen for both the drained or undrained case from either laboratory and insitu tests or published data.

$$\text{Elastic Modulus } E = 2(1 + \mu)\, G \tag{3}$$

Where, G = Shear Modulus
μ = Poisson's Ratio

Shear Modulus, as opposed to Elastic Modulus, was considered to be more appropriate for the purposes of reporting deformation behaviour in that the need for selecting a value of Poisson's Ratio was eliminated. (2)

It should be stressed that the above shear and elastic moduli are obtained from tests carried out in the horizontal plane and as such represent horizontal deformation moduli values.

BRIEF DESCRIPTION OF THE SITE IN SOMERSET

The site investigated in north Somerset U.K. was relatively flat though modified in the area where pressuremeter testing was undertaken by the presence of two large spoil heaps about 10 metres high. The underlying Lower Lias rocks, which gently dipped to the north at 10° were overlain by a thin cover of residual soil. A typical geological sequence is shown in Figs. 7 and 8.

The Lower Lias consisted of an alternating sequence of laminated or thinly bedded grey mudstones and calcareous mudstones, generally between 0.3 and 5.0 metres thick, interspersed with thin bands of limestone, generally less than 0.5 metres in thickness. The frequency of the limestone beds was varied but the sequence could be further subdivided into an upper interbedded mudstone/limestone unit (20% limestone) of which the top 5-10 metres is weathered, a middle mudstone unit (5% limestone) and a lower interbedded mudstone/limestone unit (30% limestone). The limestones were generally strong with compressive strengths between 60 to 80 MPa, whilst the mudstones were weak to moderately strong with strengths ranging from 13 to 35 MPa.

These Lower Lias rocks were underlain by the Cotham Limestone consisting of a 1.5 metre thick laminated muddy limestone beneath which lay a dark grey/black fissile, weak to moderately weak mudstone known as the Westbury Mudstone unit. The black mudstones were further underlain by moderately strong grey, green and red-brown gypsiferous siltstones and mudstones of the Mercia Mudstone Group.

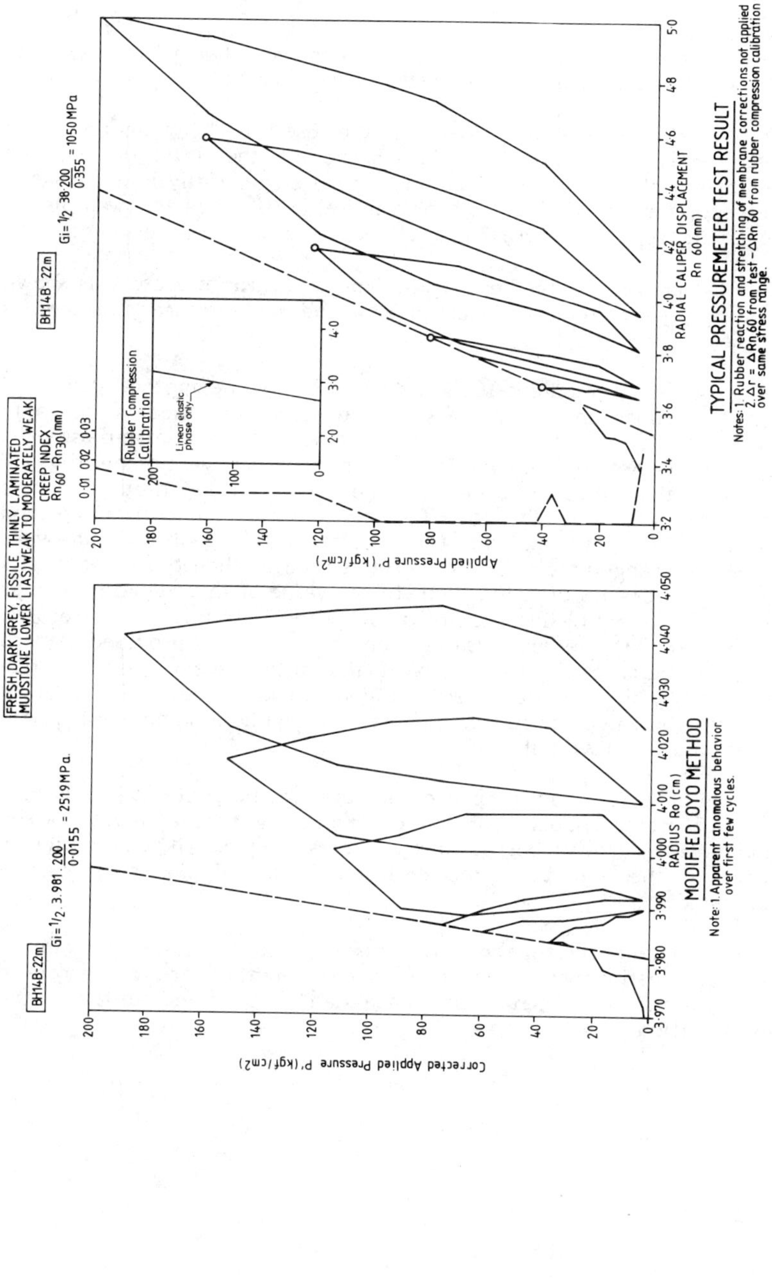

FIG. 6 - - Typical Pressuremeter Test Result.

These rocks were strongly anisotropic in terms of both geology and deformation behaviour.

DISCUSSION OF THE TEST RESULTS

In situ testing of the "rock mass" consisted of both pressuremeter and plate load testing supplemented by comprehensive laboratory testing on "intact" rock samples obtained from rotary cored boreholes.

After initial "problems" posed by the use of the manufacturer's procedures and "corrections", as discussed in the introduction, the pressuremeter test data was reduced and processed generally in accordance with the recommended procedure shown in Table 2. Emphasis was placed on thorough and regular calibration throughout the testing programme.

A typical pressuremeter test result from a Liassic mudstone is shown in Fig. 6, together with data processed using the manufacturer's method as shown in Table 1.

The initial shear modulus (Gi) was obtained by the method discussed earlier in this paper and a summary of the values obtained is shown in Fig.7. The moduli derived from test curves obtained by the manufacturer's modified method are also shown for comparison and were generally 1.4 times greater than those obtained by the recommended method. It is interesting to note that the manufacturer's initial unmodified method recommended in their manual (1) resulted in moduli which were inconsistent and ranged from 0.9 to 2.5 times greater than those eventually obtained. These results clearly illustrate the value of thorough and regular calibration during testing in order to avoid potentially misleading results. It is considered that the test results eventually obtained provided a more accurate and realistic determination of the actual shear modulus than those obtained by the manufacturer's method which was based on a factory quoted correction value for rubber compression and an assumption that the membrane behaved elastically.

It should be noted that the processed data was not, infact, in this case corrected for either rubber reaction or stretching of the membrane on expansion however these corrections are relatively small, as previously discussed, and the effect on the determined moduli was subsequently relatively minor.

Finally, for comparison, the initial shear moduli (Gi) obtained from the pressuremeter testing using the recommended method are compared in Fig.8 with those from laboratory and plate load tests carried out during the same investigation.

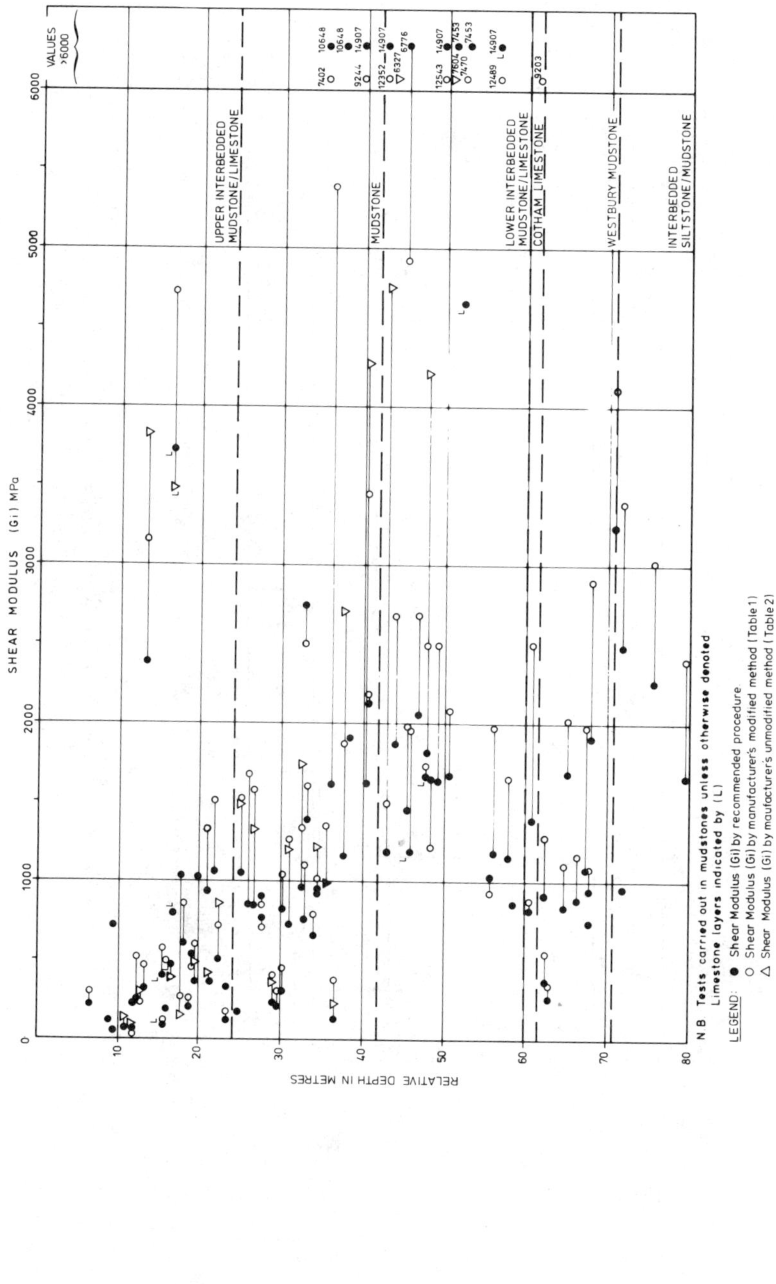

FIG. 7 - - Summary of Shear Moduli (Gi) from Pressuremeter Testing.

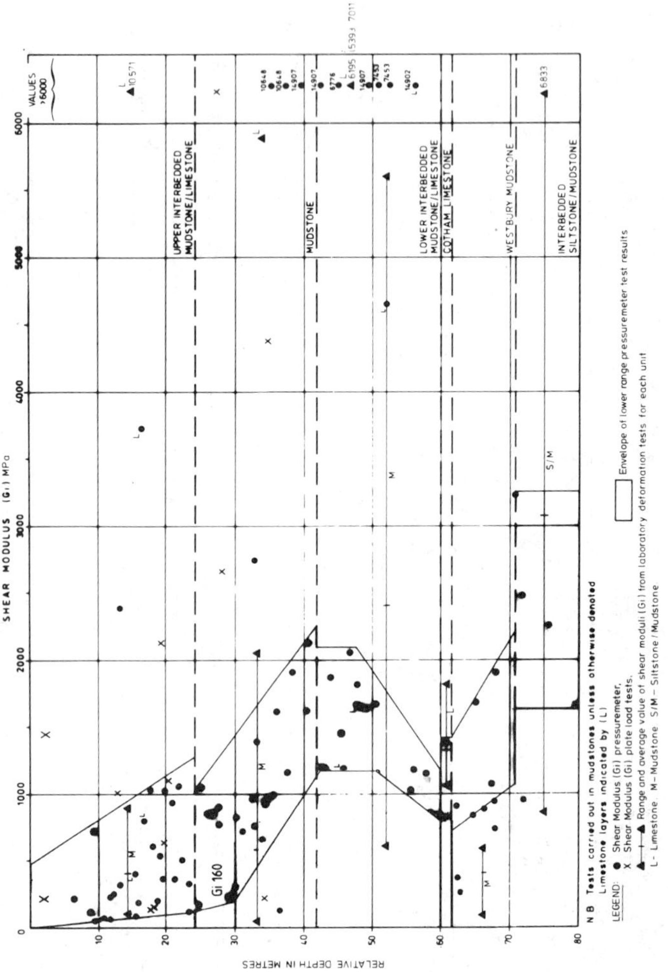

FIG. 8 - - Summary of Shear Moduli (Gi) from Pressuremeter, Plate Load and Laboratory Tests.

CONCLUSION

In order to derive realistic and meaningful in situ horizontal deformation modulus values from pressuremeter tests in rocks it is suggested that the following be incorporated into pressuremeter testing procedure:

Test Procedure

- Pre-cycling of rubber membrane prior to any calibration or test.
- Electronic and Mechanical System Compliance Check.*
- Caliper Logging to determine original borehole size.

Calibration

(i) Pressure

- Record the Hydrostatic Head in the System and the Borehole.
- Rubber Reaction Calibration.*

(ii) Displacement

- Rubber Compression Calibration.*
- Reduction in membrane thickness with increasing borehole size calibration.

All calibrations marked with an asterix should be carried out at the start and end of each day's testing. All calibration tubes should be rigid and stiff so that there is no deflection of the tube at the maximum pressure achievable by the pressuremeter.

A reasonable assessment of in situ horizontal deformation modulus can be obtained using the high capacity rubber membrane type pressuremeter in weak to moderately weak rocks if the above items are incorporated into the pressuremeter test programme. It is evident from the experience described that the equipment was working at the limit of its capacity, particularly where the rubber membrane experienced deformations of the same order of magnitude or greater than the rock tested.

The large number of corrections required for rubber membrane type pressuremeters is obviously a limitation of this particular type of pressuremeter. This limitation could usefully form the basis of future research together with the development of an appropriate deformation measuring system which would do away with the present rather tedious calibration corrections required for pressuremeter data reduction.

ACKNOWLEDGEMENTS

The author wishes to acknowledge the co-operation of the Central Electricity Generating Board of the United Kingdom in giving their permission to publish this paper. The author is also greatly indebted to the members of the Geotechnical Department of Rendel Palmer and Tritton Limited and Dr. R.J. Mair for their assistance during the preparation of this paper.

REFERENCES

(1) Oyo Corporation, "Operation Manual for Model 4149 Elastmeter 100", 1980.

(2) Mair, R.J. and Wood, D.M. "A Review of the Use of Pressuremeter for In Situ Testing" CIRIA Report Project 335, 1985.

(3) Wroth, C.P. "British Experience with the Self-Boring Presuremeter". Symposium on the Pressuremeter and Its Marine Applications", Paris, April 1982.

(4) Oyo Corporation, "Untitled Report for a Pressuremeter Investigation in Texas", 1982.

Jean Canou and Mehmet T. Tumay

FIELD EVALUATION OF FRENCH SELF-BORING PRESSUREMETER PAF 76 IN A SOFT DELTAIC LOUISIANA CLAY

REFERENCE: Canou, J., and Tumay, M. T., "Field Evaluation of French Self-Boring Pressuremeter PAF 76 in a Soft Deltaic Louisiana Clay," The Pressuremeter and Its Marine Applications: Second International Symposium, ASTM STP 950, J. L. Briaud and J. M. E. Audibert, Eds., American Society for Testing and Materials, 1986.

ABSTRACT: The results of a field investigation carried out with the French self-boring pressuremeter PAF 76 in a soft deltaic clay deposit located in south Louisiana are presented. The in-situ test is performed using an electronic control and data acquisition system, monitored by an HP85 microcomputer. A brief description of the equipment used, the general in-situ test procedure, and the approach used for the interpretation of a standard expansion test are presented. General characteristics of the deposit as well as an outline of the testing program are summarized. The standard pressuremeter characteristics are then plotted versus depth and compared with results obtained using other testing methods at the same site (cone penetrometer, dilatometer, triaxial test). This comparison shows that the self-boring pressuremeter yields shear strength and elastic modulus values much higher than the other classical methods, which agrees with previously reported observations.

KEYWORDS: in-situ testing, soft deltaic deposits, self-boring pressuremeter, pore pressure, shear strength, elastic modulus, computer monitoring, data acquisition and processing

INTRODUCTION

In-situ geotechnical evaluations have gained increasing interest during the last decade in particular. Among these methods, the

Mr. Canou is a research engineer at CERMES, Paris, France (formerly Graduate Research Assistant, Department of Civil Engineering, LSU); Professor Tumay is Coordinator of Geotechnical Engineering, Department of Civil Engineering, Louisiana State University, Baton Rouge, LA 70803.

self-boring pressuremeter (PAF 76), developed and widely used in France, is one of the most recent and sophisticated, and allows the acquisition of comprehensive and reliable information about engineering properties of soils. A field experiment with a piezometric PAF 76 system (LA-PPMS), modified to conduct tests on well-documented soft deltaic deposits of south Louisiana is described herein. A brief description of the rather complex equipment and the general procedural outlines are given. The results obtained with LA-PPMS tests are presented, and the geotechnical soil characteristics calculated are compared with the values obtained by other in-situ and conventional testing methods.

DESCRIPTION OF THE EQUIPMENT, DATA ACQUISITION AND CONTROL SYSTEM

Basic Equipment

The self-boring pressuremeter with piezometric attachment used in this study (LA-PPMS) is a modified version of the system developed and widely used in France by the "Laboratoires des Ponts et Chaussees." The model of the basic probe used is called "PAF 76" [1]. A schematic of the LA-PPMS setup is depicted in Fig. 1. Detailed operational and theoretical information is available elsewhere [2].

The equipment is mainly composed of two parts: the probe, driven into the soil; and the control unit, CPV 1000, which controls the test. The probe itself is comprised of three adjacently connected hollow modules which are the driving module, the measuring module (expanding part of the probe), and the self-boring module.

The driving module is the connection between the drill rods and the probe, and provides support and housing for various monitoring systems (i.e., transducers, electrovalves, etc.). The self-boring module is composed of a cutting shoe equipped with a blade type grinder, driven by a hydraulic motor. This motor is hydraulically powered by a unit located at the ground surface.

The measuring module is composed of a rubber membrane tightened at both ends on the body of the probe. An annular cavity expanded by injection of water can thus be created. The rubber membrane is protected with metallic strips vertically attached on the membrane. A small element of porous bronze, located in the middle of the measuring module and flush with it, is connected to a pressure transducer through a thin copper tubing, and allows the simultaneous measurement of the pore water pressure during driving, relaxation and expansion modes. When properly gauged, the measuring cell is 269 mm long and has a diameter of 132 mm.

Auxiliary Equipment

Running the self-boring pressuremeter test in the field requires substantial auxiliary equipment, namely: a drill rig to drive the

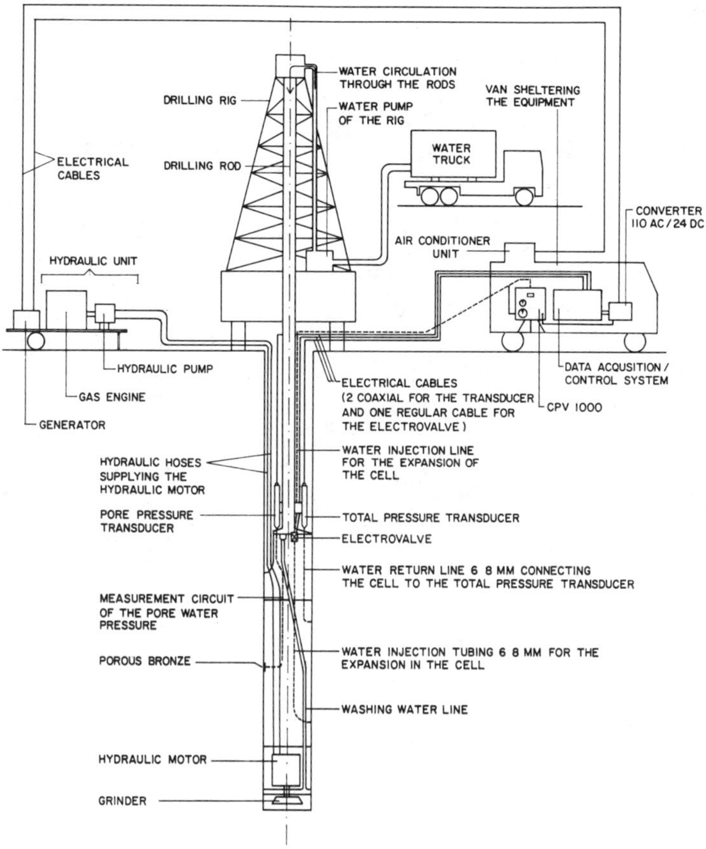

FIG. 1--Schematic of Louisiana piezometric pressuremeter system (LA-PPMS).

probe; a water supply used as washing water for self-boring process; a sheltering vehicle for the electronic equipment, and the control unit, etc.; a hydraulic unit to supply the power to the grinder incorporated to the probe; and a power supply for the control unit and the electronic equipment (Fig. 1).

Data Acquisition and Control System

LA-PPMS is entirely computer controlled through an HP data acquisition and control system (Fig. 2). Three measuring channels are presently utilized: two pressure transducers protected in their housings and mounted on the driving module measure respectively the total pressure in the cell and the pore water pressure at the cell-soil interface; and a potentiometer placed inside the control unit CPV 1000 allows the measurement of the volume of water injected into the measuring cell.

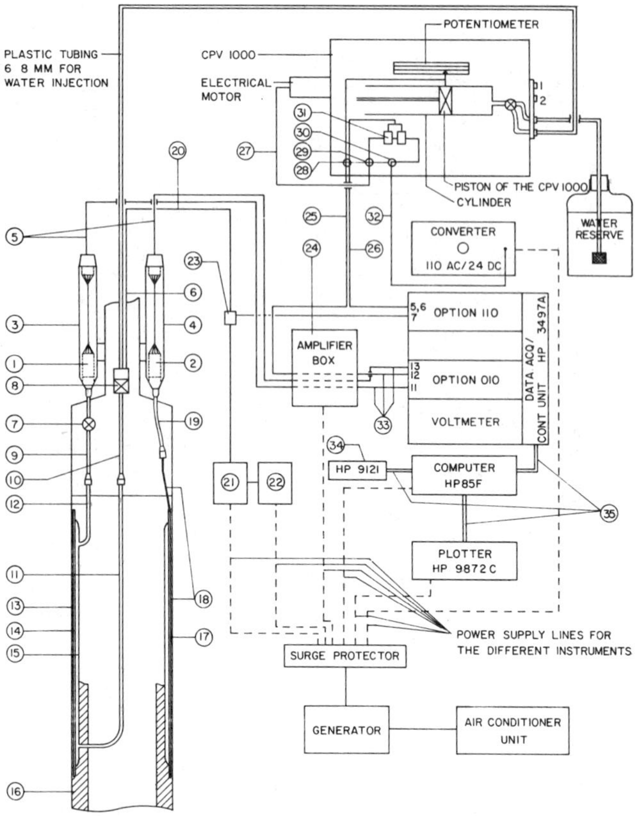

1 Total pressure transducer KISTLER
2 Pore water pressure transducer KISTLER
3,4 Transducer housing
5 Electrical coaxial cables linking the transducers to the data acquisition system
6 Electrical cable of the electrovalve
7 Manual valve on the return water line of the cell
8 Electrovalve
9,10 Plastic tubing 6.8 mm
11 Metallic tube linking the feeding water line to the measuring cell
12 Metallic tube of the return line of the cell
13 Rubber sheath equipped with metallic strips
14 Rubber membrane under the sheath
15 Expandable cavity between the body of the probe and the membrane
16 Body ofthe probe
17 Porous bronze for pore pressure measurement
18 Copper tube 1-2 mm linking the porous bronze to a piece of 4.6 plastic tubing
19 4.6 mm plastic tubing linking the copper tube to the transducer
20 Electrical cable oft he electrovalve
21,22 12DC volts power supply for the electrovalve
23 Switch box for the electrovalve
24 Cable linking the electrovalve to the controlling relay 7 in option 110 of HP3497A
25 Cable connecting the potentiometer to the data acquisition system
26 Cable linking relays 31 to relays 5 and 6 of option 110 of HP3497A
27 Supplying cable of the electrical motor
28,29,30 Measurement, motor supply and general supply
31 Relays controlling the electrical motor
32 CPV 1000 power supply line
33 Connections from the amplifier box to the HP3497A
34 Disc drive HP9121D
35 Interface connections between the different components of the HP system

FIG. 2--Schematic of computer based test monitoring/control and data acquisition system of LA-PPMS.

The electrical motor driving the piston inside the CPV 1000 and controlling water injection direction and rate for the expansion/deflation is monitored by the HP 85 computer through electrical relays of the data acquisition/control system, HP 3497A. An electrovalve, located on the driving module of the probe, can also be automatically opened or closed, thus allowing the isolation of the probe from the rest of the system or to let the water flow in/out of the cell at preprogrammed times.

Calibration and acquisition computer programs are necessary to run standard as well as cyclic pressuremeter tests. Data reduction programs subsequently process the data to evaluate the soil characteristics (elastic moduli, shear strength, failure characteristic, limit pressure, etc.). This data reduction can be done either in the field or later in the office.

GENERAL SET-UP AND PROCEDURE FOR A STANDARD TEST

After completion of prescribed operations [1,2] (i.e., saturation of the CPV 1000 circuits, calibration of the measuring cell, saturation of the pore water pressure measurement circuit), running a classical self-boring pressuremeter profile requires the completion of three successive operations:

1. Self-boring (Fig. 3). This phase consists of driving the probe down to the depth at which an expansion test will be performed. The self-boring process is the association of three different operations carried out simultaneously: pushing of the probe at a constant rate (about 30 cm per minute) with the rig; rotation of the grinder, breaking up the core of soil penetrating the cutting shoe; and injection of washing water under pressure at the grinder's level to carry up the sediments to ground surface through the body of the probe. The self-boring technique thus allows the soil to be smoothly replaced by the probe with the least possible amount of disturbance.

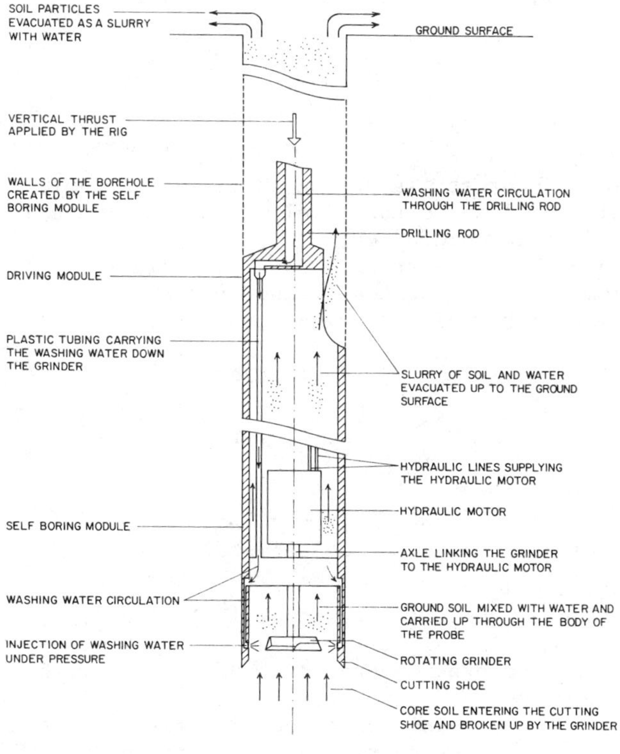

FIG. 3--Self-boring process.

2. Relaxation phase. During this period of rest (usually about 40 minutes long), the measuring cell is kept at a constant volume. The soil around the cell thus resumes a condition similar to the "at rest, K_o" state of stress. During this phase, the total pressure and the pore water pressure are monitored. Generally, these values should decrease until they reach a stable condition, close to the initial K_o state.

3. Expansion-deflation of the measuring cell. The expansion can be carried out up to any maximum deformation desired (usually 10 or 20 percent of original probe volume) at a constant rate of deformation. The standard volumetric deformation rate $\Delta V/V_o$ normally used is 2 percent per minute.

TEST INTERPRETATION

The theoretical interpretation of the pressuremeter test is based on the expansion of an infinitely long cylindrical cavity. Plane strain conditions are assumed. A treatise has been developed [3,4,5,6], to calculate the maximum shear stress for the soil. For expansion in undrained conditions (i.e., no volume change in the soil mass during the test), which is assumed valid for saturated clays:

$$\tau_m = \frac{\sigma_r - \sigma_\theta}{2} = f(g) = f\left(\frac{\Delta V}{V_o}\right)$$

with

τ_m = maximum shear stress acting on a soil element,
σ_r, σ_θ = total radial and orthoradial principal stresses in cylindrical coordinates,
g = Green orthoradial strain, equal to $(1/2)(\Delta V/V_o)$,
ΔV = volume variation of the cylindrical cavity,
V_o = initial volume of the cylindrical cavity.

The relation between τ_m and $\Delta V/V_o$ is given by:

$$\tau_m = \frac{dp}{d(\Delta V/V_o)} \frac{\Delta V}{V_o} \left(1 + \frac{\Delta V}{V_o}\right) \tag{1}$$

where

$\frac{dp}{d(\Delta V/V_o)}$ is the slope of the pressuremeter curve,

$(p, \Delta V/V_o)$, p being the total pressure in the cell.

The treatment of the experimental data developed by the "Laboratoire des Ponts et Chaussees", and used in this study is based on this procedure. The data reduction computer program first makes different pressure and volume corrections and the "corrected" pressuremeter curve is plotted. The elastic shear moduli G_o (tangent), Gp2, Gp5 (secant) are then calculated on the expansion curve as well as on the deflation curve (G_o*, Gp2*, etc.). Detailed procedural outlines are given elsewhere [1,2,6].

The shear curve is then derived using (1). Two fitting methods are used to calculate the slope $dp/d(\Delta V/V_o)$. The first one uses a regression on 3 or 5 experimental points, and provides a derivation of the whole expansion curve up to the maximum deformation reached during the test. The second method is based on a mathematical derivation of a polynomial curve fitting of the corrected pressuremeter curve, and provides a smooth shear curve calculated on the first 15 experimental points.

The failure characteristics (failure deformation a_f, creep pressure p_f, shear strength $(\tau ap)_f$, second elastic modulus G_f) are determined on this shear curve.

Finally, the program performs a general curve fitting of the expansion curve and extrapolates P20 and P100 on this fitting. It also plots the pore water pressures obtained during the test (expansion-deflation).

FIELD INVESTIGATION

The LA-PPMS investigation reported in this paper was carried out on the Borgne Canal site, in the vicinity of New Orleans [7,8,9]. This site is mostly composed of soft grey clay, with presence of silty layers down to about 22 meters. Hard sandy layers are found below. An electric piezo-cone penetrometer log is given in Fig. 4 showing the profile of the site and the two principal clay layers, between 4.50 m and 7 meters and between 11 and 22 meters. Typical soil properties are also given in Table 1. Two pressuremeter soundings, PR01 and PR02, were performed with LA-PPMS at a distance of meters from each other. PR01 was carried out down to 18.84 meters and PR02 down to 15 meters. Thirteen standard expansion tests were made for PR01 and 5 for PR02. Table 2 gives a list of these tests.

General Comments on the Field Investigation

General procedure: The self-boring process must be carried out with extreme care to obtain meaningful results. The pressure of the washing water has to be properly maintained to provide adequate flow in order to carry up the sediments. In soft clay, the driving process is relatively easy, no refusal being noticed, and the grinder works properly. However, the driving rate of the probe has to be closely monitored because abrupt changes in this rate results in the clogging of the circulation passages of the probe around the hydraulic motor (see Fig. 3), especially in sticky clay. It was also observed that the presence of pieces of gravel could substantially hamper the self-boring operation.

The saturation of the high volume pore water pressure measurement circuit is very delicate, and saturation is hard to maintain in the field with the system used so far. Fig. 5 depicts the deairing procedure presently utilized (i.e., Fig. 5B: flushing the lines; Fig. 5A: final connection and calibration). This procedure is followed to

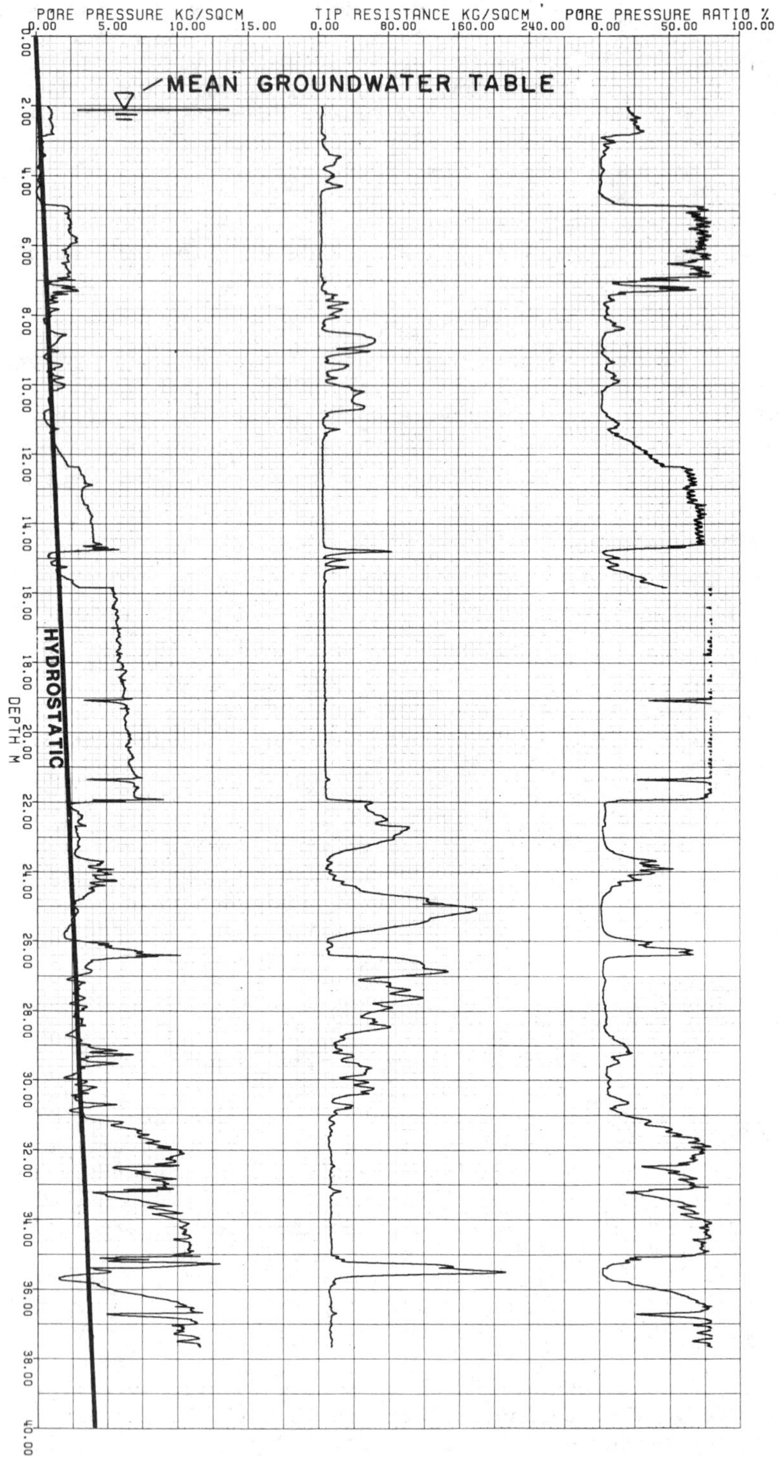

FIG. 4--Piezocone penetration test (PCPT) log of Borgne Canal Site [7,8,9].

TABLE 1--Typical soil properties of Borgne Canal Site [7].

Depth ft	Depth (m)	Stratum	Consistency, % W_M	W_L	W_P	I_P	I_L	γ Unit Weight (lb/ft^3)
0								
3								
6								
9		Medium gray silty clay	36.2	40.4	20.2	20.2	0.79	114.6
12		Medium gray silt and clayey silt	29.6	20.8	N.P.	N.P.	N.P.	115.8
15	(4.60)		41.5	42.8	21.5	21.3	0.94	112.?
18		Soft gray silty clay with silt lenses and traces	74.7	73.0	27.1	45.9	1.04	97.?
21		of organic material	60.9	75.1	29.0	46.1	0.69	92.8
24			69.8	79.0	27.7	51.3	0.82	99.0
27		Medium gray silt, silty clay with clay lenses	27.7	23.6	N.P.	N.P.	N.P.	122.7
30	(9.15)	and traces of organic material	35.6	28.2	N.P.	N.P.	N.P.	110.2
33			31.9	22.0	N.P.	N.P.	N.P.	112.7
36			40.0	40.1	21.8	18.3	1.00	112.1
39		Medium and soft gray silty clay	45.3	49.8	22.6	27.2	0.84	110.2
42			47.3	61.1	22.1	39.0	0.65	109.4
45	(13.75)		44.0	59.0	21.5	37.5	0.60	111.5
48			62.6	72.5	25.2	47.3	0.79	100.0
51			64.4	91.5	28.0	63.5	0.57	100.2
54		Soft + medium gray clay	51.4	72.6	24.5	48.1	0.56	107.1
57			55.7	83.0	25.2	57.8	0.53	96.5
60			58.4	98.0	28.7	69.3	0.43	94.6
63	(19.20)		60.4	86.2	27.5	58.7	0.56	95.9

TABLE 2--List of the tests performed.

Profile No.	Expansion No.	Depth meters	Deform. Rate % per min.	Max. Deform. %
PRO 1	EXP1	4.34	2	20
	EXP2	4.84	2	20
	EXP3	5.84	2	20
	EXP4	6.84	2	20
	EXP5	9.24	2	20
	EXP6	10.24	2	20
	EXP7	11.24	2	20
	EXP8	12.24	2	20
	EXP9	13.24	2	20
	EXP10	14.24	2	20
	EXP11	15.24	2	20
	EXP12	16.84	2	20
	EXP13	17.64	2	20
PRO 2	EXP15	4.14	2	10
	EXP16	5.14	2	10
	EXP17	5.59	2	10
	EXP18	12.24	2	10
	EXP19	14.94	2	10

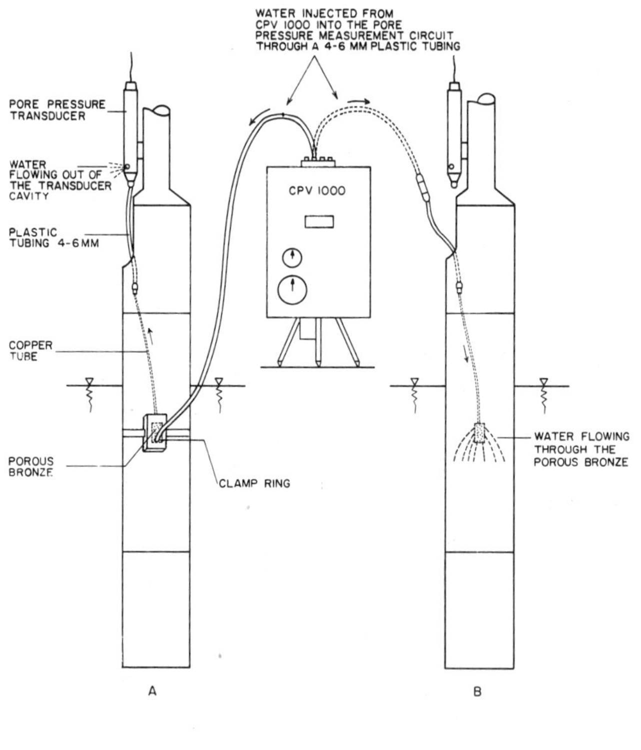

FIG. 5--Deairing of piezometric measurement system [1,2].

deair the piezometric system while the probe is at the ground surface. No method yet exists to perform deairing operations while boring and/or at depth. Clogging of the porous bronze due to fine soil particles frequently happens, resulting in prolonged time response. Further equipment development and modification to minimize volume and deformability of measurement circuit, and provision for in-place deairing is essential.

<u>Quality control of the results</u>: In the field, two curves are plotted for each test: the relaxation curve, plotted in real time, and the raw expansion curve (without corrections), plotted after the end of the test. A typical relaxation curve is shown in Fig. 6. The quality of the self-boring can be controlled by examination of the total and the pore water pressures monitored during the process itself, and subsequently during the relaxation phase. During the self-boring process, both total and pore water pressures should remain close to the "at rest" values. An important increase (i.e., 50%) of these parameters above the in-situ "at rest" state during the self-boring is characteristic of a forcing back of the soil around the probe and thus, of an unsatisfactory self-boring. Also, a total pressure very close to the pore water pressure is characteristic of decompression of the soil and the creation of a gap between the probe and the soil. This also is the result of an unsatisfactory self-boring process. When this problem occurs, a characteristic decompression phase will appear on the initial part of the expansion curve.

A poor quality self-boring process usually results in lower mechanical characteristics for the soil. The soil around the probe is more disturbed than it should be (forced back or decompressed) and

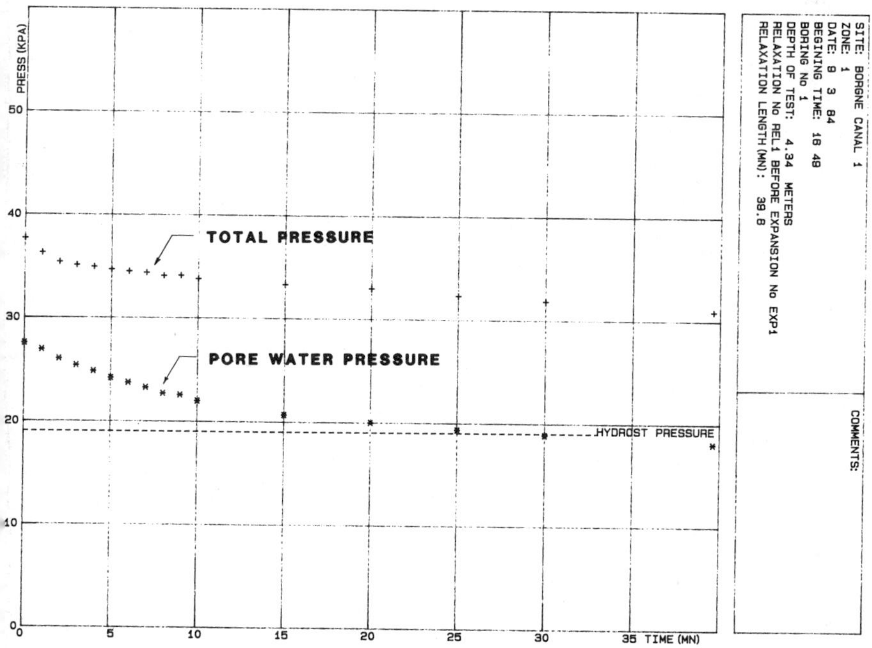

FIG. 6--Typical relaxation test (real-time) results.

thus, it appears weaker (i.e., lower elastic moduli, lower shear strength). However, the effect of disturbance on limit pressures (P20, P100) is negligible.

During the relaxation phase, both total pressure and pore water pressure should decrease to reach a stable condition after a reasonable time (circa 20-40 minutes). An increase of the pore water pressure readings after the end of the self-boring is characteristic of unsatisfactory response of the measuring system (i.e., inadequate saturation and/or loss of saturation, clogging of the porous bronze, etc.).

The examination of the shape of the raw expansion curve plotted at the end of the test is indicative of the quality of the test. Final "corrected" curves are then plotted, either in the field or later in the laboratory (see Fig. 7). All the soil characteristics are then calculated by the reduction computer program and printed on the chart. The legend for all the parameters are given in the list of symbols. A print-out of the complete corrected results is finally done.

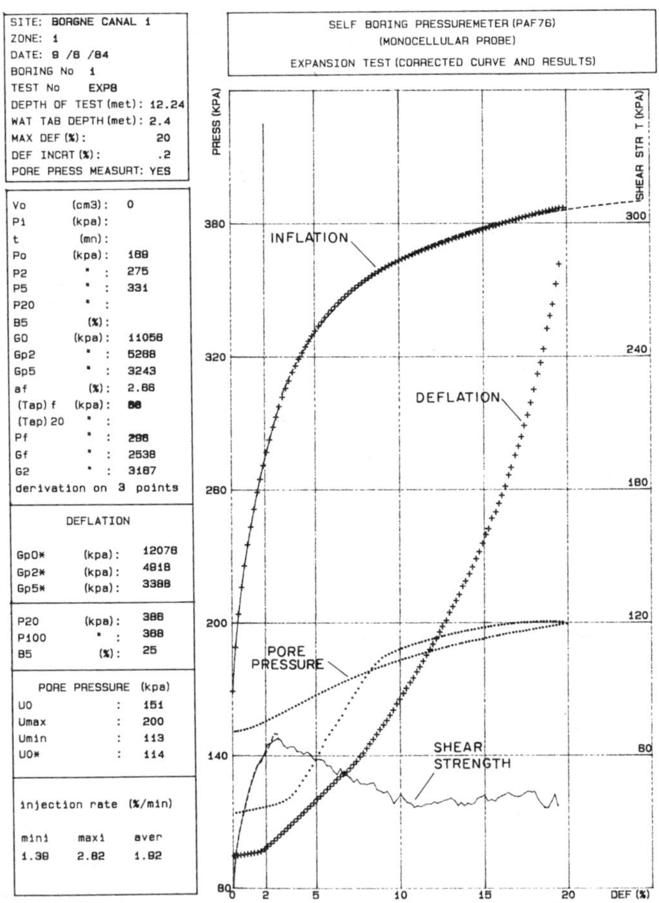

FIG. 7--"Corrected" expansion curve and results for EXP8.

Examination of the Results

Cell pressure: Figure 8 shows P0, P2, P5 and P20 for the first soil profile, PR01. There is a regular increase of P0 with depth. Three drops in P0 appear at a depth of 9.24, 13.24 and 15.24 meters, corresponding to tests EXP5, EXP9 and EXP11. For these tests, P0 was very close to the hydrostatic pressure U0. However, these pressure drops, visible for P0, P2, and Pf, disappear for P5, P20 and P100 (Fig. 9). This demonstrates that the pressuremeter curve does not seem to be very much affected by the initial conditions of the test (P0) after a certain deformation. P100, obtained as a mathematical extrapolation of the experimental curve, ranges from 260 to 580 kpa, which is fairly well representative of a soft clay. A similar range of values (70-500 kpa) was obtained by the authors training with PAF 76 in the soft clay of Cran, the testing and calibration site of the "Laboratoire des Ponts et Chaussees" of St. Brieuc, France. The P20 and P100 peak observed (P20 = 404 kpa, P100 = 738 kpa) at a depth of 9.24 meters (test EXP5), is the result of the presence of a stronger layer of silt at that depth.

Pore water pressure: The pore water pressures U0, Umax, Umin and U0* for the two soil profiles are presented in Fig. 10. For PR01 relaxation, U0 is very close to the hydrostatic pressure down to 9.24 meters, however the response of the measuring system during the

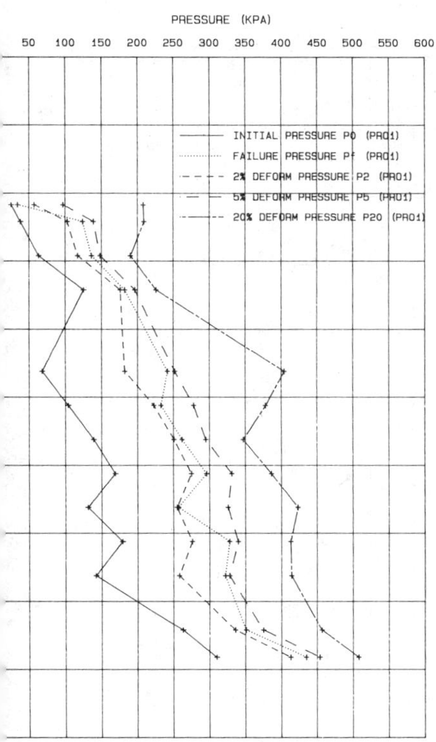

FIG. 8--Total cell pressure for different volumetric deformations (PR01).

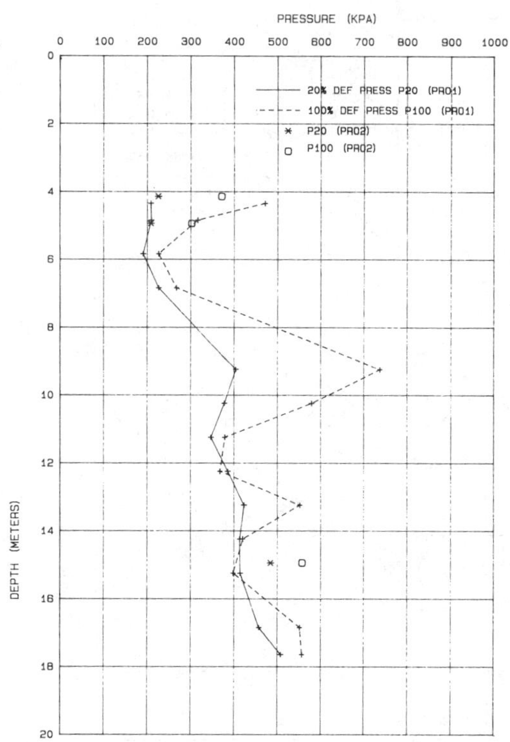

FIG. 9--P20 and P100 for PR01 and PR02.

subsequent expansion phase is not satisfactory (Umax is almost equal to UO). For the deeper tests, the response of the system becomes better, but UO increases with respect to the hydrostatic pressure. It is important to note that UO*, which is almost equal to Umin, always remains very close to the hydrostatic pressure. It must also be noted that with proper compliance of the measuring system, the maximum pore water pressure reaches 1.5 to twice the hydrostatic pressure. UO and Umax obtained for PRO2 have also been plotted in Fig. 10. UO compares fairly well with the results obtained for PRO1. Umax cannot be directly compared with the results of the first soil profile because the expansion tests were carried out up to 10 percent deformation for PRO2, instead of 20 percent for PRO1. In a normally or slightly overconsolidated clay, the maximum excess pore pressure ΔUmax obtained for an undrained test should reach at least 40 to 50 percent of the corresponding total pressure increase, ΔP_{Umax}. The ratio $\Delta Umax/\Delta P_{Umax}$ is shown in Table 3 for the site of Borgne Canal. These same ratios have been calculated for a series of tests performed on Cran site of the "Laboratoire des Ponts et Chaussees." The comparison between both results shows that the values of the ratio obtained for Borgne Canal tests are considerably lower for the type of soil considered. This indicates a rather low time response of the measuring system. It also appears that on Cran site, the maximum pore water pressure occurs for a fairly low deformation, usually 3 to 4 percent. In Borgne Canal, the pore water pressure keeps increasing

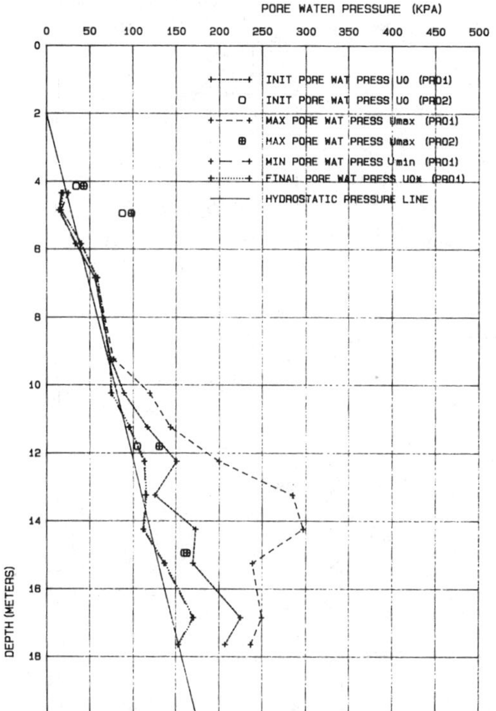

FIG. 10--Initial, final, maximum minimum pore water pressure versus depth.

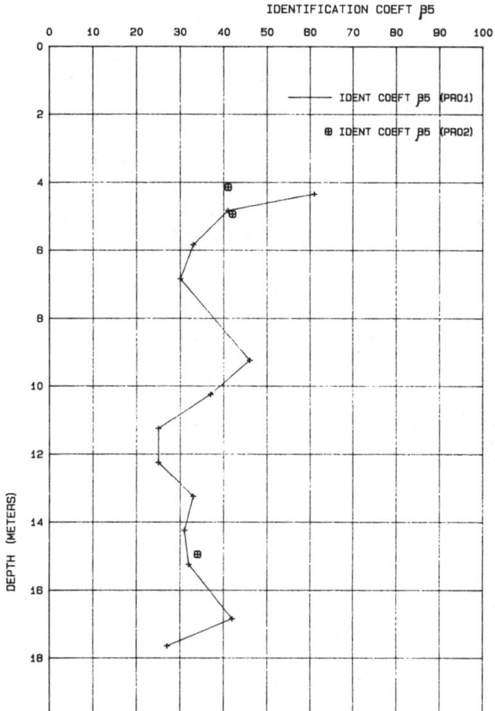

FIG. 11--Soil identification coefficient, β5, versus depth.

TABLE 3--Pore water pressure characteristics Borgne Canal and Cran sites.

Site	Profile	Test	Depth	Fail. Deform.	Def. Umax	(ΔV/ΔP) Fail.	$\Delta U_{max}/\Delta P_{Umax}$
			meters	%	%	%	%
B.C.	PRO1	EXP1	4.34	0.4		0	3
		EXP2	4.84	3.52		0	1
		EXP3	5.84	3.60		1	5
		EXP4	6.84	2.66		0	2
		EXP5	9.24	4.50		1	1
		EXP6	10.24	2.36	19.54	4	11
		EXP7	11.24	2.56	19.20	14	13
		EXP8	12.24	2.66	19.54	6	23
		EXP9	13.24	2.05	18.62	8	54
		EXP10	14.24	4.22	16.15	21	52
		EXP11	15.24	4.60	19.13	9	24
		EXP12	16.84	2.92	19.38	9	13
		EXP13	17.64	3.38	18.8	6	14
Cran	A2	1933	1	1.05	10.01	9	29
		1934	2	4.47	10.10	21	22
		1935	3	1.84	4.27	4	21
		1936	4	2.27	5.24	17	39
		1937	5	2.08	5.87	31	52
		1938	6	2.00	4.00	21	26
		1939	7	2.02	5.27	42	57
		1940	8	1.98	3.24	42	48
		1941	9	1.03	2.01	39	59
		1942	10	2.07	3.03	52	57
		1943	11	2.06	2.62	55	56
		1944	12	2.01	2.62	60	45
		1945	13	2.05	2.05	48	42
		1946	14	2.07	1.63	43	46
		1947	15	1.00	3.77	59	34

with a low rate during expansion. This time response problem of the measuring system results from the difficulty experienced in saturating the system and keeping it saturated during the test.

Identification coefficient of the soil, β5 [6]: This coefficient is defined as follows: β5 = (P20 - P5)(100)/(P20 - P0). It is characteristic of the soil type. For a soft clay, it should range from 20 to 40. For a sandy or silty material, β5 should increase above 40. Figure 11 shows that most β5 values are below 40. The value of 61 obtained for the first test results from an unsatisfactory pressuremeter curve at that depth due to CPV 1000 malfunction. The value 46 obtained at 9.24 meters for EXP5 is characteristic of a stronger type of soil at that depth (silty material). The range of β5 values calculated is thus in good agreement with the type of soil tested.

Shear strength: The method used to derive the shear curve has already been described. When a maximum is found on this curve, it is taken as the undrained shear strength and the corresponding point is the failure point. If no maximum exists, the failure point is estimated when a noticeable drop occurs in the slope of the shear curve. The PAF 76 shear strength profile for Borgne Canal site is presented in Fig. 12. Table 3 shows that the failure generally occurs for a deformation ranging from 2.5 to 4 percent. The shear strength values range from 23 to 91 kpa, which are in good agreement with comparative Cran data.

The shear strength values obtained for the second soil profile are much lower than the one obtained for the first one. This again is consistent with the shear modulus results, and confirms the fact that the quality of the self-boring has a great influence on the soil characteristics calculated. The very low shear strength values at depths of 6.84 and 16.84 meters confirm self-boring problems (forcing back of the soil due to inappropriate drill rig control) at this depth (test EXP4 and EXP12).

Elastic shear moduli: GO, Gp2 and Gp5 are represented in Fig. 13. The range of GO values (30-140 bars) is in good agreement with PAF 76 results expected for that type of soil. The very low value obtained for the very first test of PRO1 (GO = 29.7 bars at 4.24 meters) resulted from an inappropriate probe insertion. The high values of GO noticed between 10 and 11.5 meters correspond to the harder silty layer described previously. It must be noted that except for EXP15, the shear modulus values obtained for the second soil profile are systematically lower than for the first profile. This is an indication that the soil was less disturbed for the first

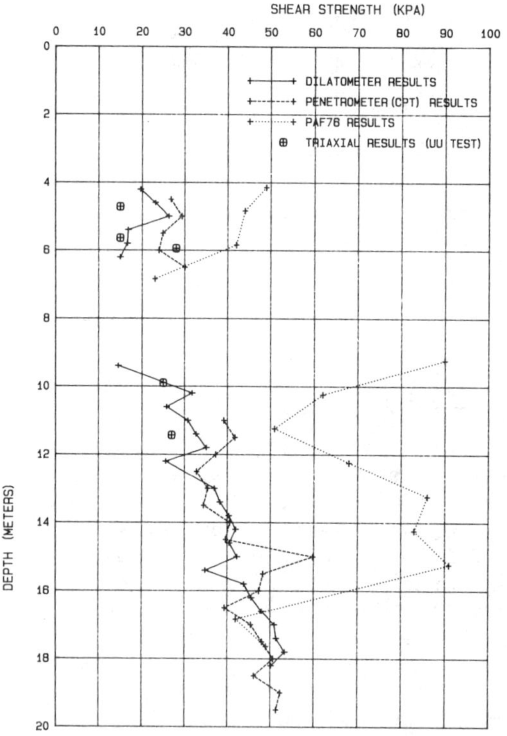

FIG. 12--Shear strength profile for Borgne Canal Site

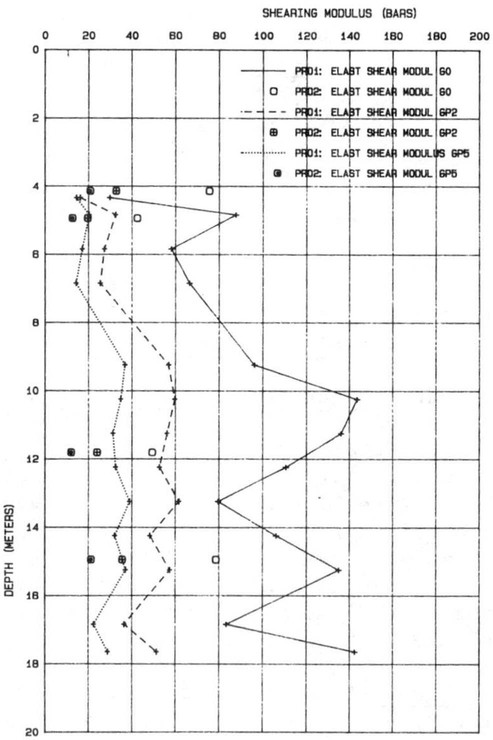

FIG. 13--Elastic shear moduli GO, GP2, GP5 versus depth.

profile than for the second one, and it gives an idea on the quality of the self-boring process for each sounding.

COMPARISON BETWEEN PAF 76 DATA AND OTHER IN-SITU AND CONVENTIONAL TESTING SYSTEMS

The site of Borgne Canal has been previously investigated with different in-situ testing devices [7,8]. In particular, piezo-cone penetrometer and dilatometer tests have been carried out at this location within context of ongoing LSU geotechnical research. In July 1984, "undisturbed" samples were obtained from the site in order to run laboratory tests and compare the results with PAF 76 data.

Piezo-cone Penetration Tests (PCPT)

An LSU/Fugro type piezo-cone penetrometer [9,10] with simultaneous measurement of tip resistance, sleeve friction and pore water pressure was used on the site. The following formula was used to evaluate the undrained shear strength S_u:

$$S_u = (q_c - q_o)/N_c$$

where q_c is the tip resistance, q_o the overburden pressure and N_c the cone factor. A cone factor of 9 was taken [7] to relate the cone tip resistance to shear strength obtained with the field vane in soft clays of Louisiana. An average specific weight of 1.7 tons/m^3 was assumed to calculate the overburden pressure. An evaluation of the elastic Young modulus E was also made based on the empirical relation $E = K\ q_c$. A value of 7, typical of a soft clay, was taken for K [7,11].

Dilatometer Tests

These tests were performed in collaboration with Schmertmann and Crapps, Inc., Gainesville, Florida. The dilatometer provides different soil characteristics, and in particular the shear strength S_u and the dilatometer modulus E_d. E_d is related to the Young modulus E by the following formula [12,13]:

$$E_d = \frac{E}{1 - \nu^2}$$

Triaxial Tests

Unconsolidated undrained triaxial tests were run on "undisturbed" samples collected on the site to obtain the undrained shear strength. An estimation of the Young modulus E was also done, based on the elasticity formula:

$$\varepsilon_1 = \frac{1}{E}(\sigma_1 - \nu(\sigma_2 + \sigma_3))$$

with ε_1 vertical deformation of the sample, σ_1 vertical principal stress and σ_2 equal to σ_3, confining pressure. ν, Poisson's ratio was taken equal to 0.5 (undrained conditions), which yielded:

$$E = \frac{\sigma_1 - \sigma_3}{\varepsilon_1}$$

E was calculated for a deformation of 1 percent ($\varepsilon_1 = 0.01$) on the experimental triaxial curves.

Comparison of the Results

Shear strength: PAF 76, PCPT, dilatometer and triaxial shear strength values are summarized in Fig. 12. The dilatometer and the PCPT values are very close to each other, regularly increasing with depth. The triaxial yields slightly lower values than the other tests, but still close to them. However, the range of PAF 76 values is very different (1.5 to 3 times higher) from the other procedures, except for few abnormally low values due to operating problems. This finding is similar to the conclusion drawn from the Young modulus results.

A comparison made between PAF 68 and field vane undrained shear values (Fig. 14) [5] shows that PAF results are systematically higher by 30 to 50 percent than vane test results. An investigation carried out at eight locations in France by the "Laboratoire des Ponts et Chaussees" has shown that PAF results are about 65 percent higher than triaxial and 30 percent higher than field vane [5].

Elastic modulus: Figure 15 shows that the dilatometer, the PCPT and the triaxial results are very close to each other. However, the range of these values is very different from PAF 76 data. PAF 76 yields values usually 10 to 20 times higher than the other techniques. This confirms the fact that PAF 76 soil characteristics are generally much higher than other in-situ or laboratory testing methods. Table 4 shows a comparison between Menard pressuremeter and PAF data [6]. Most of the time this is a ratio of 5 to 10 between Menard and PAF shear moduli.

CONCLUSIONS

In this paper, a computer-controlled set-up of the French self-boring pressure PAF 76 system, and its field evaluation in a soft deltaic Louisiana clay are presented. The following observations and conclusions are drawn from this study:

1. PAF 76 yields mechanical characteristics of soil (shear strength, elastic moduli) generally much higher than other in-situ and conventional laboratory testing methods. The shear strength

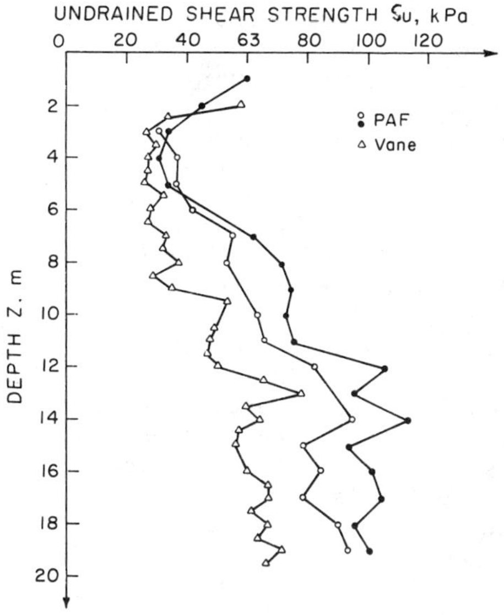

FIG. 14--Comparison between undrained shear strength from field vane and PAF 68 [3].

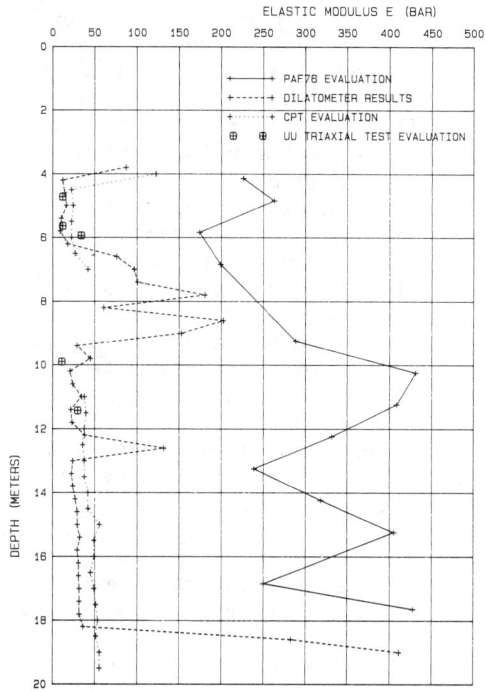

FIG. 15--Comparative Elastic modulus profile for Borgne Canal Site.

TABLE 4--Flanders clay: comparison between PAF and Menard test results.

Depth	Self-Boring Pressuremeter				Menard Pressuremeter		
	P_f	P20	E_{pf}	τ_{max}	P_ℓ	S_u	E_m
meters	kpa	kpa	kpa	kpa	kpa	kpa	kpa
1	108	163	7,500	40	240	37	4,200
2	250	411	88,100	119	340	55	6,000
3	200	431	33,800	116	500	85	6,400
4	310	442	25,300	102	460	63	5,500
5	313	519	38,200	106	560	84	6,600
6	316	554	30,800	150	520	89	5,100
7	332	589	31,500	135	600	98	7,000
8	352	634	28,700	190	700	112	9,000

values are 1.5 to twice higher, and the shear moduli are 10 to 20 times higher. These findings are consistent with results obtained by Laboratoire Central des Ponts et Chaussees, France, in similar soils. It has been observed that soil disturbance is partly responsible for this phenomenon, but other reasons (i.e., stress and strain path, specific characteristics of each test) should also be sought.

2. The dependability of the PAF 76 sounding is contingent upon the quality of the self-boring procedure. Thus, this process should be carried out with extreme care (i.e., well coordinated effort of

drill rig crew - practice oriented; and PAF 76 crew - research oriented), and proper monitoring (i.e., probe insertion rate, washing fluid pressure, washing fluid circulation rate, pore pressure, total cell pressure) should be maintained during self-boring.

3. The pore water pressure measurement procedure is very delicate to be carried out routinely because of he complicated and cumbersome saturation process of the circuit. Minimizing the volume and obtaining/maintaining full saturation of the measuring circuit is essential for a short time-response (compliance) and high quality measurements. Improvements of the procedure should be sought to render the process more adaptable to site conditions. In particular, equipment development/modification is desirable for resaturation and calibration of the measuring circuit from the ground surface.

4. Self-boring pressuremeter tests generally gave realistic results, consistent with other PAF 76 data obtained in similar soils. The range of shear strength, elastic moduli and limit pressures obtained on the site of Borgne Canal compares well with data obtained in a similar clay on Cran site in France. Likewise, the soil identification coefficient $\beta 5$, ranging from 20 to 40, is well representative of a soft clay. It appears to be an acceptable characterization index for soils.

5. It was observed that even trace amounts of gravel detrimentally impede the proper operation of the grinder. Washing fluid circulation in fat clays can also decelerate, occasionally halt, the self-boring procedure.

During the course of this study, PAF 76 equipped with a computer-based test control and data acquisition/reduction system appeared to be a very versatile in-situ testing device, capable of providing complete and reliable information in soft deltaic clayey soil. The procedure is relatively complicated to be carried out routinely. However, the electronic control/acquisition capability substantially facilitates this method to render it desirable and economically feasible to be employed in research and in high-cost foundation problems. The unique feature of PAF 76, allowing the soil to be tested in an almost undisturbed state, is especially very attractive for comparative research purposes. Future improvements of various other in-situ determination techniques to evaluate soil properties will certainly be related to further refinement of the self-boring probes, and the comprehensive theoretical treatise of soil behavior during failure.

ACKNOWLEDGEMENTS

The research described in this paper was sponsored by the National Science Foundation Grant No. CEE8113679. Messieurs J. F. Jezequel and F. Baguelin of LCPC, France; I. Juran and F. Schlosser of ENPC, France, are gratefully acknowledged for their support and advice during the course of this study. Any opinions, findings, conclusions and recommendations expressed are those of the authors.

REFERENCES

[1] Jezequel, J. F. and LeMehaute, A., "The Self Boring Pressuremeter Model 76 (PAF 76) - User's Manual," English Translation by J. Canou, M. T. Tumay (editor), Louisiana State University, Department of Civil Engineering Geotechnical Engineering Report No. GE-84/05, November 1984, 104 pp.

[2] Canou, J. and Tumay, M. T., "Calibration and Field Evaluation of French Self Boring Pressuremeter (PAF 76)," Louisiana State University, Department of Civil Engineering Geotechnical Engineering Report No. GE-84/04, November 1984, 416 pp.

[3] Baguelin, F., Jezequel, J. F., Lemee, E. and LeMehaute, A., "Expansion of Cylindrical Probes in Cohesive Soils," Journal of the Soil Mechanics and Foundation Division, American Society of Civil Engineers, Vol. 98, SM11, November 1972, pp. 1129-1142.

[4] Baguelin, F., Jezequel, J. F. and LeMehaute, A., "Etudes des Pressions Interstitielles Developpes Lors de l'Essai Pressiometrique," Proceedings of the Eighth International Conference on Soil Mechanics and Foundation Engineering, Moscow, Vol. 1.1, 1973, pp. 19-24.

[5] Baguelin, F. and Jezequel, J. F., "Further Insights on the Self Boring Technique Developed in France," Proceedings of the Conference on In-Situ Measurement of Soil Properties, American Society of Civil Engineers, Raleigh, N.C., Vol. 2, 1975, pp. 231-243.

[6] Baguelin, F., Jezequel, J. F. and Shields, D. H., "The Pressuremeter and Foundation Engineering," Series on Rock and Soil Mechanics, Trans Tech Publication, 1978.

[7] Tumay, M. T. and Yilmaz, R., "In-Situ Determination of Undrained Shear Strength of Louisiana Soils by Quasi-Static Penetration Test," Louisiana State University, Department of Civil Engineering, Geotechnical Engineering Report No. GE-82/04, December 1981, 282 pp.

[8] Tumay, M. T. and Hajibakar, I., "In-Situ Determination of Compressibility of Louisiana Soils Using Piezo Cone Penetration Tests (PCPT)," Louisiana State University, Department of Civil Engineering, Geotechnical Engineering Report No. GE-82/06, December 1982, 253 pp.

[9] Tumay, M. T., Acar, Y. and Boggess, R., "Subsurface Investigations with Piezo Cone Penetrometer," Special Publication on Cone Penetration Testing and Experience, American Society of Civil Engineers, New York, 1981, pp. 325-342.

[10] Rad, N. S. and Tumay, M. T., "A Study on Pore Pressure Response of Piezo Cone Penetration Test (PCPT)," Geotechnical Testing Journal, American Society for Testing and Materials, Philadelphia, PA, Vol. 8, September 1985, pp. 125-131.

[11] Bowles, J. E., Foundation Analysis and Design, McGraw Hill Book Company, New York, 1982.

[12] Marchetti, S., "In-situ Tests by Flat Dilatometer," Journal of the Geotechnical Engineering Division, American Society of Civil Engineers, Vol. 106, GT3, March 1980, pp. 299-321

[13] Schmertmann, J., Discussion of "In-Situ Tests by Flat Dilatometer," by S. Marchetti, Journal of the Geotechnical Engineering Division, American Society of Civil Engineers, Vol. 107, No. GT6, June 1981, pp. 831-837.

LIST OF SYMBOLS

a_f	volumetric deformation of the cylindrical cavity at failure
E	Young modulus
E_d	dilatometer modulus
g	orthoradial Green strain
g_r	radial Green strain
G	elastic shear modulus
GO	tangent shear modulus
Gpx	secant shear modulus evaluation on the expansion part of the pressuremeter curve (x% deformation)
Gpx*	secant shear modulus evaluated on the deflation part of the pressuremeter curve (x% deformation)
Gx	secant shear modulus evaluated on the shear curve (x% deformation)
Gf	secant shear modulus evaluated on the shear curve at failure
I_L	liquidity index
I_P	plasticity index
N_c	cone factor
p,P	total cell pressure
p_f	creep (failure) pressure
p_ℓ	limit pressure (Menard test)
P_o	initial horizontal soil pressure
P_x	cell pressure for x% deformation
q_c	tip resistance (PCPT test)
q_o	overburden pressure
S_u	undrained shear strength
u_o	initial pore water pressure
Umax	maximum pore water pressure
Umin	minimum pore water pressure
u_o*	final pore water pressure
V	current volume of the measuring cell
V_o	initial volume of the measuring cell
βS	identification coefficient of the soil
ΔPx	increase in total cell pressure for x% deformation
ΔU	excess pore water pressure
ΔV	volume increment
W_M	natural moisture content
W_L	liquid limit
W_P	plastic limit
γ	specific weight of soil (unit weight)
ν	Poisson's ratio
$\sigma_r, \sigma_\theta, \sigma_z$	radial, orthoradial and vertical principal stresses (cylindrical coordinates)
τ_m	deviator stress (shear stress)
$(\tau ap)_f$	shear stress at failure
$(\tau ap)_x$	shear stress for x% deformation

Theory and the Pressuremeter

François M. Lassoudière[1], François B. Zanier[2]

NUMERICAL ANALYSIS OF PRESSUREMETER TESTS BY THE FINITE ELEMENT METHOD

REFERENCE: Lassoudière, F. M., Zanier F. B., "Numerical Analysis of Pressuremeter Tests by the Finite Element Method", The Pressuremeter and its Marine Applications: Second International Symposium, ASTM STP 950, J. L. Briaud and J. M. E. Audibert, Eds., American Society for Testing and Materials, 1986.

ABSTRACT: The self–boring pressuremeter test offers an exceptional means to investigate *in situ* soil behavior. However, it is important to clarify as thoroughly as possible the conditions in which one can pass from the pressuremeter curve (probe volume–pressure relation) to a stress–strain curve characterizing the local behavior of the soil in the neighborhood of the probe.

This article presents numerical simulations of real pressuremeter tests to help assess the importance of the role played by the geometric characteristics of the probe employed, on the pressuremeter test and on its interpretation.

KEYWORDS: pressuremeter test, finite element method, elasto–plasticity, strain–hardening, cyclic loading, shear strength, shear modulus

The pressuremeter can be considered as a particular structure subject to a standardized loading test. The pressuremeter curve integrates the response of a whole volume of soil around the probe (this applies to all *in situ* tests). This curve, which represents the "rough" result of the pressuremeter test, can be interpreted by means of two types of method.

---

1 Research Engineer, Institut Français du Pétrole BP 311 92506 Rueil Malmaison Cedex, France

2 Graduate Student, Institut Français du Pétrole BP 311 92506 Rueil Malmaison Cedex, France

1. Analog methods which, through correlations, establish empirical relations between certain characteristics of the pressuremeter curve and the soil characteristics to be introduced into foundation design calculations. These include the classic correlations between undrained cohesion C_u and limit pressure P_l [1,2], or applications to p−y curves [3,4].

2. Analytical methods [5,6,7] which, through an a posteriori analysis of the pressuremeter test, serve to infer the shear strength curve of the soil in contact with the probe from the pressuremeter curve, based on a simple analytical relation of the form:

$$\tau = \frac{\Delta V}{V} \frac{dp}{d\left[\Delta V / V\right]} \qquad (1)$$

where: p = pressure in the probe

ΔV = volume change of the probe

V = probe volume

τ = shear stress in the soil.

Figure 1 gives the standard shape of the pressuremeter curve and the derived shear strength curve.

The following assumptions are necessary to obtain equation (1):

i) during expansion of the probe, the soil is deformed under undrained condition,

ii) the probe slenderness ratio is presumed to be infinite, to guarantee a plane strain field.

If the following assumptions are also made:

iii) all the points of the soil at the level of the probe obey the same stress−strain curve,

iiii) the vertical stress remains intermediate between the major stress and minor stress,

the derived curve directly gives an estimate of the shear modulus G and of the undrained cohesion C_u .

PROCEDURE

It is clear that the assumption of plane strain is essential for the analytical interpretation of the pressuremeter test. The finite element method helps to discard this assumption and to evaluate the importance of end effects in the pressuremeter test. The assumption of non–drainage of the soil is made in the following discussion. A complementary study of the one presented here can be found in [8], taking account of soil permeability.

The following procedure was adopted:

1. Representation using a mesh and boundary conditions appropriate to the actual conditions of a self–boring pressuremeter test.

2. Using a sophisticated soil model (elasto–plasticity with kinematic strain hardening), simulation of pressuremeter tests performed on the clayey Cran site [9], to investigate the shape of the effective stress paths, the stress and pore pressure distribution, and the stress–strain curves around the probe.

3. Using a simple soil model (elasto–plasticity without strain hardening), simulation of a series of pressuremeter tests performed on the sites of Cran, Sète [10] and Montanazo [11], to evaluate systematically the relations existing between:

- soil shear modulus G and secant pressuremeter modulus at 2% strain Gp_2
- undrained cohesion C_u and derived shear strength τ.

Use was made of the FONDOF code developed at IFP to design foundations of offshore structures under static and cyclic loads. The mesh is shown in Figure 2. It contains eighty isoparametric elements with eight nodes and nine integration points. The mesh radius is 1m60, or twenty times the radius of the pressuremeter probe used. Computation showed that the soil zone influenced by expansion of the probe did not exceed ten radii. The probe is represented with its real dimensions: 160mm in diameter and 320mm high. The pressure was applied directly to the soil elements at the level of the probe. The nodes corresponding to the ends of the membrane are related to highly rigid elements representing the driving module and the self boring module, so that the "football" shape of the inflated membrane is reproduced. The borehole walls are maintained by the application of a hydrostatic pressure distribution.

FIRST ANALYSIS

In this part, the soil is simulated using the CYCLADE model developed at IFP. This model cannot be discussed in detail here, but involves an elasto–plastic soil model with kinematic strain hardening, which allows consideration of the effect of the loading history (especially the effect of cyclic loadings). The formulation is three–dimensional in effective stresses, uses three yield surfaces, and incorporates the notion of critical state and the concept of plastic modulus field.

References [12, 13, 14] offer a complete presentation and examples of the application of CYCLADE to calculate the foundations of an offshore structure under cyclic loads.

The first step consists in determining the parameters of the CYCLADE model on the basis of undrained consolidated triaxial tests conducted on samples taken from the clayey Cran site (onshore France). Eleven parameters were determined:

K_a, G_a, and n : non linear elastic part of strains,

$$K = K_a \left[\frac{p}{p_a}\right]^n \qquad G = G_a \left[\frac{p}{p_a}\right]^n \qquad (2)$$

ϕ = friction angle
c' = effective cohesion
λ = isotropic compressibility index
p_{co} = initial critical pressure
a = deviatory strain hardening parameter
b = isotropic strain hardening parameter
α = dilatancy parameter
r = size of elastic domain

Figure 3 shows the response of CYCLADE with the set of parameters adopted on a CU test performed on a sample taken at level 3.7m. This set of parameters is then used to simulate two pressuremeter tests performed at depths of 5 and 9m on the Cran site. Figure 4 shows that the tests are reproduced very successfully up to a volumetric strain of about 10%.

Since the calculation was conducted in undrained conditions, with calculation of the pore pressure, the change in effective radial and tangential stresses in contact with the probe can be represented, with a comparison of the calculation results with measurements of the pore pressure [15]. Figure 5 shows the change in pore pressure (measured and calculated), and Figure 6 the effective stress paths. Note that the experimental tangential stress is not measured but determined from the radial stress by equation (1), which is based on the assumption of plane strain, undoubtedly explaining a substantial part of the deviation observed between the experimental and calculated stress paths.

It is also possible incidentally to check the main stresses during expansion of the probe. Figure 7 helps to check that the vertical stress remains the intermediate stress, except at the very start of the test.

Figure 8 gives the pore pressure distribution calculated in contact with the probe in a vertical plane for a volumetric strain of 2%. The effect of the slenderness ratio is quite clear (undoubtedly magnified by the finite element discretization) with the appearance of "suction" at the probe ends. This may indicate that prudence is necessary in interpreting pore pressure measurements performed at a single point located in the probe's median plane.

Since the computation offers a realistic description of the pressuremeter tests performed, what happens in the soil around the probe will now be examined.

Figure 9 shows the stress−strain response at two points in the soil located in the probe's median plane and at a distance of 0.5cm and 4cm from it. The curves are fairly well superimposed, except that the point furthest from the probe is obviously less stressed. On the other hand, if these curves are compared with those calculated in a plane situated near the top of the probe, significantly different responses are observed (Figure 10).

This is why Figure 11 shows a difference between the analytical derivative of the simulated pressuremeter curve and the stress−strain curve calculated in contact with the probe (point 1 on figure 9). This shows clearly the result of non−observance of the plane strain condition due to the finite slenderness ratio of the probe. It is worth noting that this results in an overestimate of the soil's shear strength.

If the analytical derivative of the calculated curve is now compared with the computed stress−strain curve at point 1 (Figure 12), wide differences can be observed, that could not be anticipated from the results given in Figure 4. This stems from the fact that, due to its inherent principle, the analytical derivation method magnifies the local changes in concavity found in the experimental curve.

Extreme prudence is therefore required in using the shear strength curve derived analytically from the pressuremeter curve, especially if one attempts to interpret the peaks obtained at low volumetric strains of the probe.

The following conclusions may be drawn from this first study:

1. The finite slenderness ratio of the probe causes a systematic overestimation of the shear strength of the soil, if the method used involves the analytical derivation of the pressuremeter curve.

2. This method magnifies the irregularities in concavity of the pressuremeter curve, and it is uncertain whether the shear strength peaks observed at small volumetric strains should be taken into account.

These conclusions agree with the analyses made by various authors, including [7, 16, 17, 18].

Since the CYCLADE model helps to account for the effect of cycling loadings, a dummy cyclic pressuremeter test was simulated. The result is given in Figure 13. The progressive decrease in secant modulus of the cycles, as well as the strain hardening effect caused by the cycles, can be observed to agree with experimental observation [19]. The simulation of cyclic pressuremeter tests, which are currently very widespread, goes beyond the objective set for this paper, but it is worth keeping in mind that suitable simulation tools are already available.

SECOND ANALYSIS

A very simple rheological model is used here: linear elastic and perfectly plastic in total stresses. Only two parameters are involved, the shear modulus G and undrained cohesion C_u, with the Poisson ratio set at 0.5 (undrained condition). The mesh used is the one shown in Figure 2.

The calculation method is different from that of the first analysis. The soil parameters (G and C_u) can now be adjusted to obtain a suitable simulation of a series of pressuremeter tests performed at a given site, in order to identify a relation between the soil characteristics G and C_u, and the pressuremeter characteristics Gp_2 (secant modulus at 2%), and the derived shear strength.
Figure 14 shows the result of the numerical simulation conducted on a series of six pressuremeter tests performed on the Cran site. The same proceedings were carried

out on reconnaissance surveys with the PAM offshore self–boring pressuremeter developed by IFP on the sites of Sète (50m water depth off the French coast) and Montanazo (625m water depth off the coast of Spain). A total of ten real pressuremeter tests were simulated, with the modulus G selected to reproduce correctly the start of the pressuremeter curve, and the cohesion C_u selected to obtain the plateau of the curve.

Figure 15 shows the shear strength profiles obtained on the Cran site. It clearly appears that τ_{max} overestimates the undrained cohesion of the soil C_u, which appears to be closer to the value derived at 20% strain τ_{20}.
Figure 16 shows that the pressuremeter modulus at 2% volume variation overestimates the soil shear modulus G. Figure 17 summarizes all the data obtained on the three sites investigated, which differ by the value of the identification coefficient β , given by equation (3):

$$\beta = \frac{p_{20} - p_5}{p_{20} - p_0} \qquad (3)$$

where: p_0 = soil pressure at rest,
p_5 = pressure in probe at 5% strain,
p_{20} = pressure in probe at 20% strain

The Montanazo site was the most "silty", and the Cran site the most "clayey".

Note that G_{p2} overestimates the shear modulus G by 20 to 30%, that the peak of the derived shear strength curve τ_{max} overestimates the undrained cohesion C_u by 30 to 50%, of which a correct estimate appears to be provided by the value at 20% strain τ_{20} . Note also that attempts to compare the soil shear modulus G with the secant pressuremeter moduli at 5 and 10% strain failed to yield satisfactory results.

Based on these data, one can simulate the response of pressuremeters with different slenderness ratios on the same site (this implies assuming that the characteristics G and C_u determined by the method discussed above are intrinsic). Figure 18 gives the result of simulated tests for slenderness ratios of 2, 4 and 6 on Cran clay.

The following remark can be made in this respect. The pressure at 20%

volumetric strain p_{20} only coincides with the theoretical limit pressure of the probe with infinite slenderness ratio for a slenderness ratio of 2. For probes with higher slenderness ratios, it is necessary to exceed 40% to reach the theoretical limit value [20].

CONCLUSIONS

It must be kept in mind that the foregoing results were obtained by using a number of assumptions including:

i) undrained behavior of the soil

ii) soil undisturbed by penetration of the pressuremeter (perfect self–boring)

iii) rather complete rheological model, but obviously imperfect

With this in mind, the following points can be made.

1. The pressuremeter test can be described realistically by using the finite element method combined with a sophisticated non–linear soil model such as CYCLADE

2. The analytically derived shear strength curve must be used with caution, because the probe's finite slenderness ratio causes a systematic overestimate of the undrained cohesion (see first analysis), and this overestimate may be aggravated up to 50% if one uses the peak of the derived curve generally obtained at approximately 2% volume variation.

3. In so far as the derivation method magnifies the concavity variations of the pressuremeter curve, by causing wide oscillations to appear in the shear strength curve, caution must be exercised in using this curve to determine the parameters of a non–linear soil model as is usually done from a triaxial test.

4. The secant modulus (G_{p2}) overestimates the soil shear modulus (by 20 to 30% in the cases investigated). However, while the difference is fairly constant for a given soil, it varies from one soil to another, especially according to the soil's plasticity index Ip. It would be interesting to vary the sites analyzed to try to determine an empirical correction equation for the shear modulus and strength values determined analytically from the pressuremeter curve as a function of the type of soil.

5. Due to the influence of probe slenderness ratio on the general shape of the pressuremeter curve, care must be taken in empirical correlations (analog methods) to avoid mixing the data obtained from pressuremeters with different slenderness ratios.

REFERENCES

[1] MENARD L., "Calcul de la force portante des fondations sur la base des résultats pressiométriques", Sols−Soils, Vol 2, No 5, pp 9−24 (1962)

[2] GIBSON R.E. and ANDERSON W.F., "In situ measurement of soil properties with the pressuremeter", Civil.Eng and Pub.Works Review, Vol 56, No 658, pp 615−618 (1961)

[3] BRIAUD J.L., SMITH T.D. and MEYER,B., "Calcul de pieux chargés latéralement à l'aide de résultats pressiométriques", Symposium on the Pressuremeter and its marine applications, Paris (1982)

[4] BAGUELIN F, FRANK R and SAID Y.H., "Theoretical study of lateral reaction mechanism of piles", Geotechnique, Vol 27, N°3 (1977)

[5] PALMER A.C., "Undrained plane strain expansion of a cylindrical cavity in clay: a simple interpretation of the pressuremeter test", Geotechnique, vol 22, No 3, pp 451−457 (1972)

[6] LADANYI B., "In−situ determination of undrained stress−strain behavior of sensitive clays with the presuremeter",Canadian Geotechnical Journal, Vol. 9, 1972, pp. 313−319.

[7] BAGUELIN F, JEZEQUEL J.F., LE MEE E. and LE MEHAUTE A., "Expansion de sondes cylindriques dans les sols cohérents", Bull.Liaison Labo.P.et Ch., 61, pp 189−202 (1972)

[8] NAHRA R., "Contributions numériques et analytiques à l'étude de la consolidation autour du pressiomètre", ENPC DDI Thesis (1985)

[9] INSTITUT FRANCAIS DU PETROLE, "Pressiomètre autoforeur marin, expérimentation à Cran et au Havre", Test Report N° 29795 (1981)

[10] INSTITUT FRANCAIS DU PETROLE, "Pressiomètre autoforeur marin, expérimentation en Méditerranée", Test Report N° 30544 (1982)

[11] INSTITUT FRANCAIS DU PETROLE, "Offshore self−boring pressuremeter trials on Casablanca field", Test Report N° 31295 (1983)

[12] LASSOUDIERE F., MEIMON Y, HUJEUX J.C. and AUBRY D., "Cyclic behaviour of soils and application to offshore foundation calculation", Int. Conf. on the Behaviour of Offshore Structures, Boston, USA (1982)

[13] AUBRY D., HUJEUX J.C, LASSOUDIERE F.and MEIMON Y., "A double memory model with multiple mechanisms for cyclic behaviour of soils", Int. Symp. on Numerical Models in Geomechanics, Zurich (1982)

[14] MEIMON Y. and LASSOUDIERE F., "Application to design of marine foundations of a complete model for cyclic behaviour of soils", 4th Int. Conf. on Behaviour of Offshore Structures (BOSS'85), Delft, The Netherlands [1985)

[15] BAGUELIN F., JEZEQUEL J.F. and LE MEHAUTE A., "Etude des pressions interstitielles développées lors de l'essai pressiométrique", ICMSFE Conference, Moscow (1973)

[16] PREVOST J.H. and HOEG K., "Analysis of pressuremeter in strain–softening soil", Journ.of the Geotech.Eng.Div, ASCE GT8, pp 717–732 (1975)

[17] BORSETTO M., IMPERATO L., NOVA R. and PEANO A., "Effects of pressuremeters of finite length in soft clay", International Symposium on Soil and Rock Investigation by In Situ Techniques, Vol II, pp 211–215 (1983)

[18] WASCHKOWSKI E., "Contribution à l'étude du comportement du sol autour d'une sonde pressiométrique et d'un pieu chargé verticalement", CNAM Thesis, Orléans (1976)

[19] JEZEQUEL J.F. and LE MEHAUTE A., "Essais cycliques au pressiomètre autoforeur", Symposium on the Pressuremeter and its marine applications, Paris (1982)

[20] SALENCON J., "Expansion quasi–statique d'une cavité à symétrie sphérique ou cylindrique", Annales des P.et C., III (1966)

FIGURES:

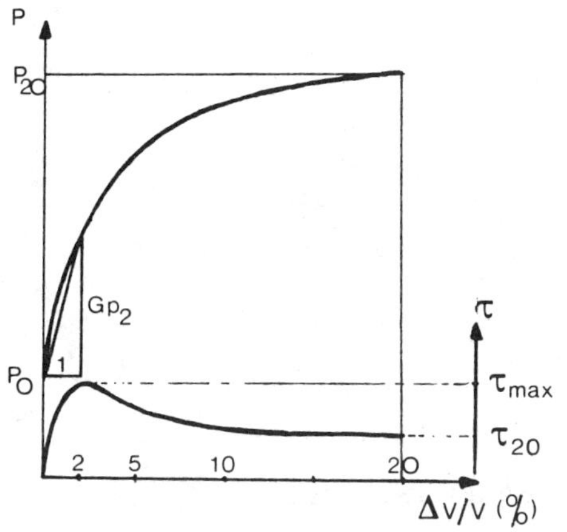

Figure 1. Pressuremeter curve and standard derived shear strength curve

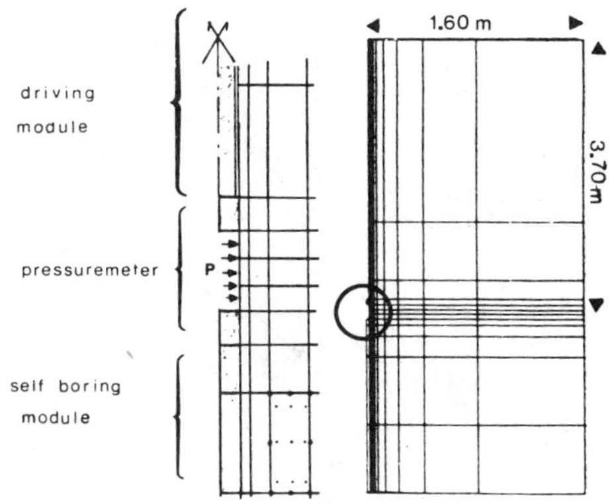

Figure 2. Finite Element mesh used to simulate probe expansion

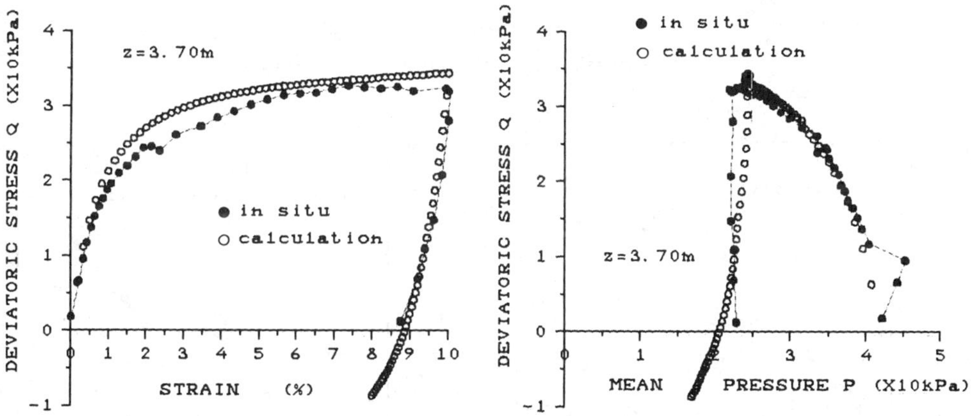

Figure 3. CYCLADE model response on a C.U triaxial test on Cran Clay. Set of parameter used: K_a = 20 MPa, G_a = 12.5 MPa, n = 0.7, ϕ = 28°, c' = 5 kPa, λ = 45, p_{co} = 0.04 MPa, a = 0.005, b = 0.7, α = 0.1

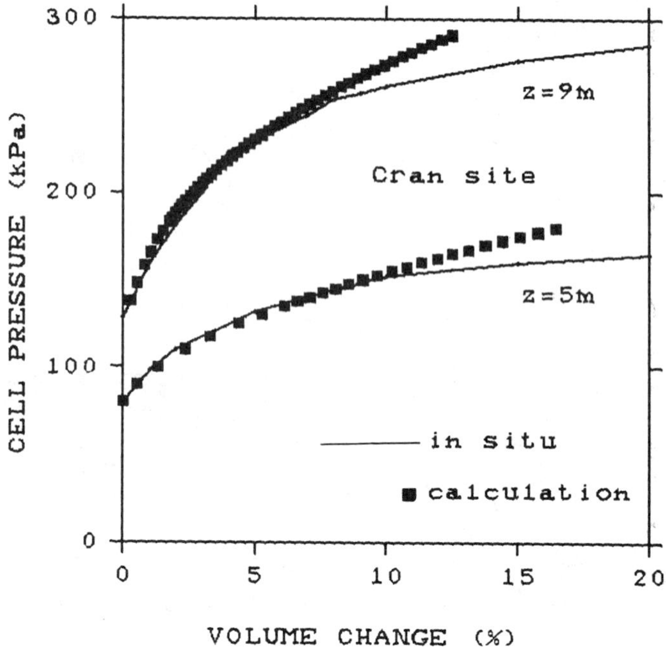

Figure 4. Pressuremeter test simulation with CYCLADE on Cran site: experiment and calculation

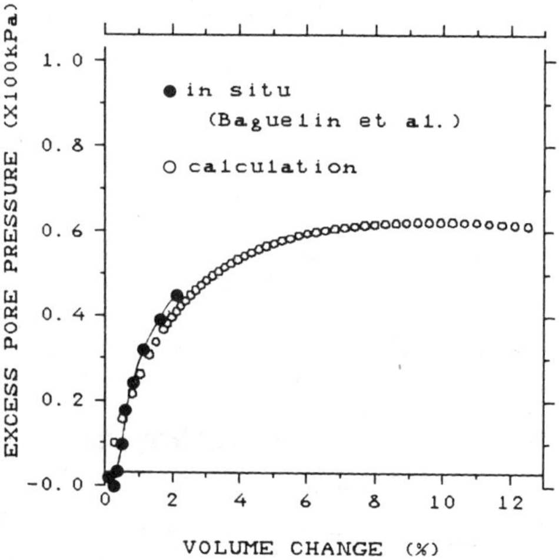

Figure 5. Pore pressure generation during probe expansion: experiment and calculation

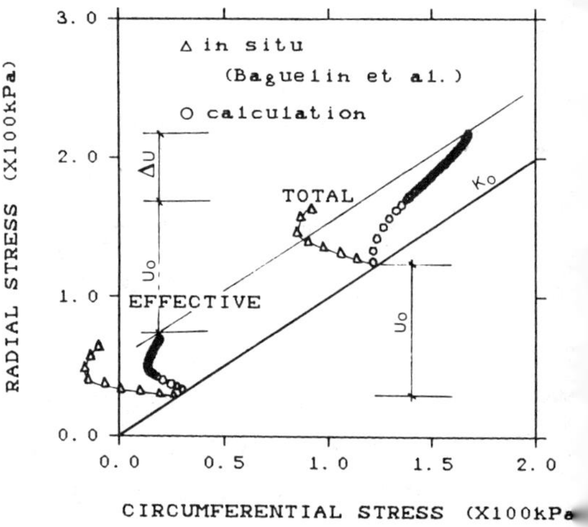

Figure 6. Effective and total stress paths during probe expansion: experiment and calculation

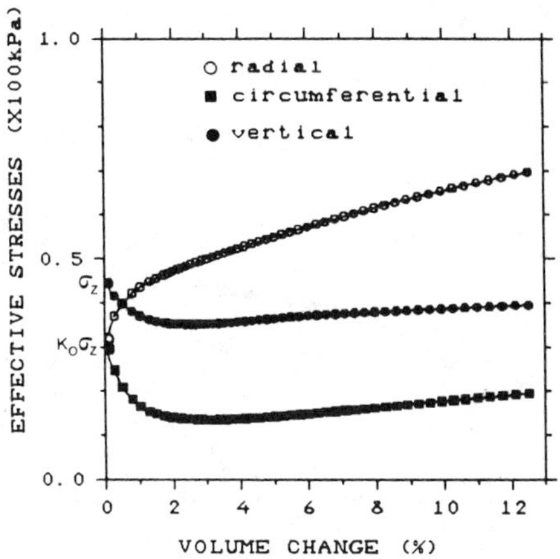

Figure 7. Effective stress variation during probe expansion (simulation)

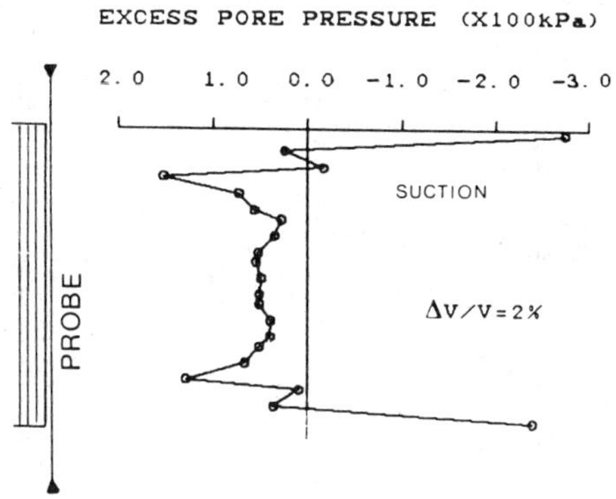

Figure 8. Pore pressure distribution along the probe at 2% volumetric strain (simulation)

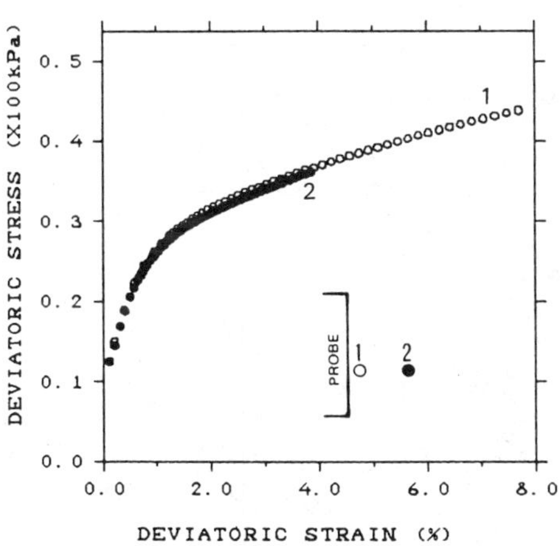

Figure 9. Stress–strain curves in the soil in the median plane of the probe: point 1 at a distance of 0.5cm, point 2 at a distance of 4cm

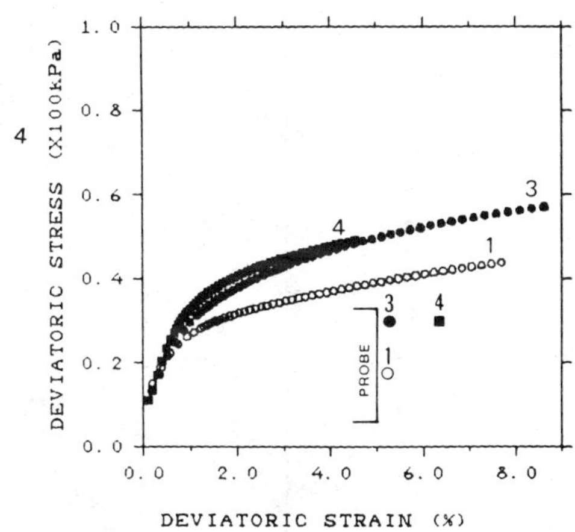

Figure 10. Comparison between stress–srain curves in the median plane of the probe and at the top: point 1 and 3 at a distance of 0.5 cm, point 4 at a distance of 4cm

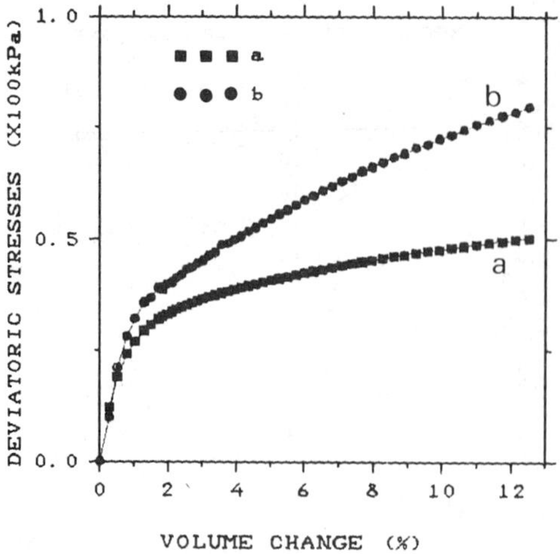

Figure 11. Comparison between the computed stress–strain curve at point 1 (a) and the derivative of the computed pressuremeter curve (b)

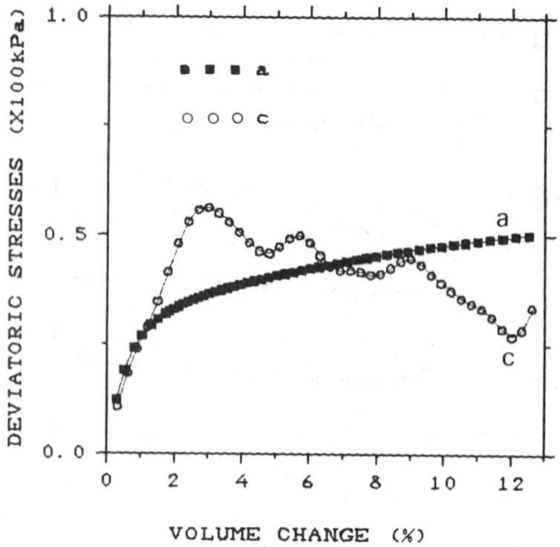

Figure 12. Comparison between the computed stress–strai curve at pint 1 (a) and the derivative of the experimental pressuremeter curve (c)

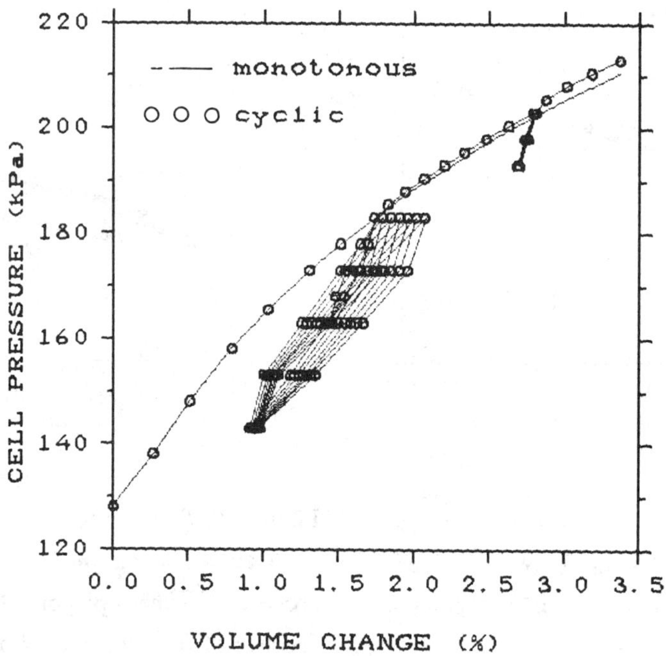

Figure 13. Simulation of a cyclic pressuremeter test with CYCLADE model showing decrease of cyclic secant modulus and hardening effect of cyclic loading

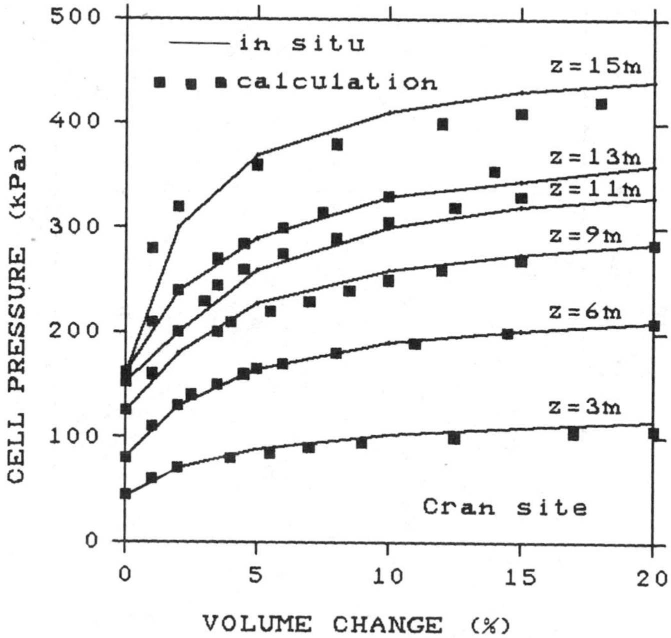

Figure 14. Simulation of six pressuremeter tests carried out on Cran Site (experiment and calculation)

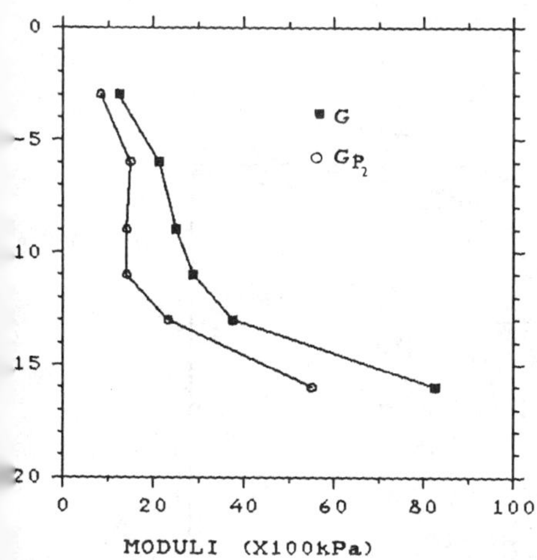

ure 15. Shear modulus log on Cran , comparison between G and ssuremeter secant modulus Gp_2

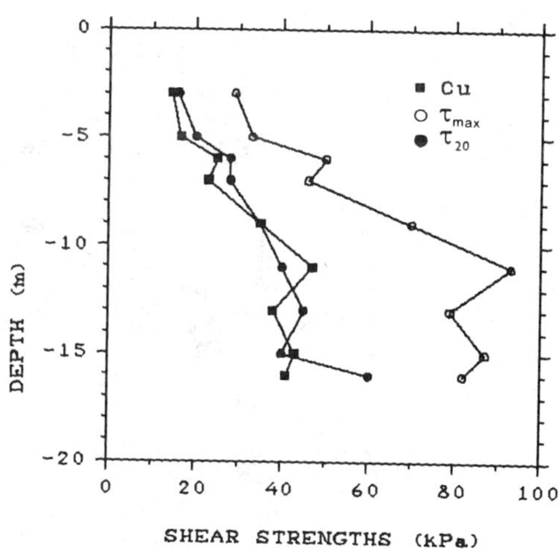

Figure 16. Shear strength log on Cran site, comparison between C_u and derived shear strength τ_{max}

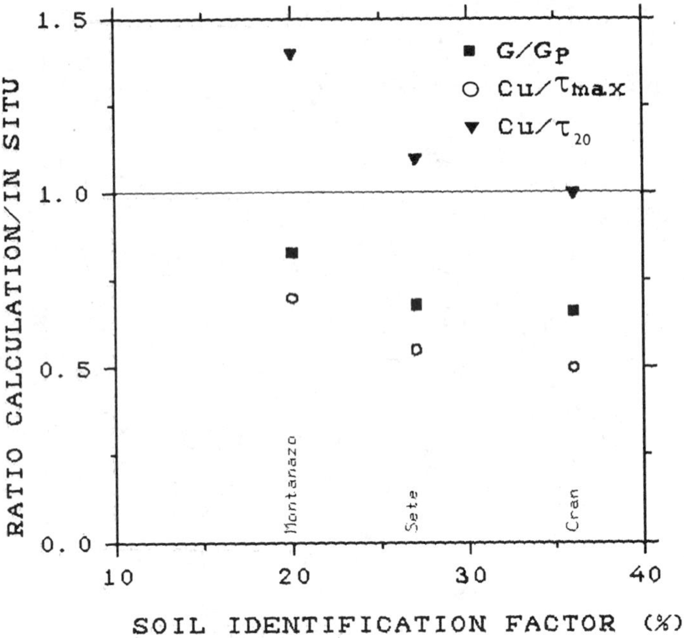

Figure 17. Comparison between soil parameters G and C_u and pressuremeter paramaters Gp_2, τ_{max} and τ_{20} carried out on three different sites

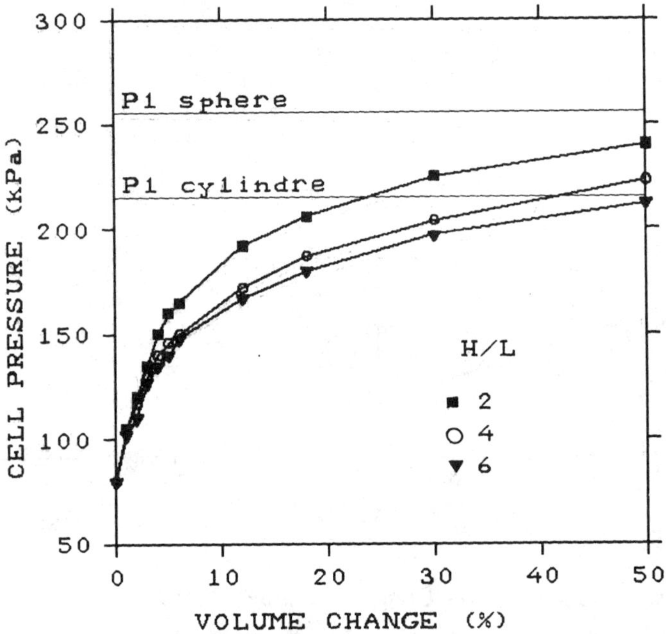

Figure 18. Influence of length/diameter ratio of the probe (H/L) on the shape of the pressuremeter curve

Jean Benoit and G. Wayne Clough

PRINCIPAL STRESSES DERIVED FROM SELF-BORING PRESSUREMETER TESTS IN SOFT CLAY

REFERENCE: Benoit, J. and Clough, G.W., "Principal Stresses Derived from Self-Boring Pressuremeter Tests in Soft Clay," The Pressuremeter and Its Marine Applications: Second International Symposium, ASTM STP 950, J.-L. Briaud and J.M.E. Audibert, Eds., American Society for Testing and Materials, 1986.

ABSTRACT: Careful evaluation of the in-situ lateral stress is important for many marine applications. Advanced versions of the self-boring pressuremeter (SBPM) allow assessment of these in-situ horizontal stresses in three different directions. Full advantage of the multiple measurements is best achieved by using a micro-computer data acquisition system. Such a system is described in this paper along with the lateral stress results obtained in a series of 32 tests in soft clay.

KEYWORDS: Self-boring pressuremeter, pressuremeter, soft clay, in-situ tests, micro-computer, data acquisition, lateral stress, anisotropy.

INTRODUCTION

Accurate determination of the in-situ lateral stress is important for many marine problems, including the analysis of load capacity of pile foundations and stability of slopes. With the increase in use of sophisticated analysis procedures involving stress path concepts and tools such as the finite element method, knowledge of the in-situ stress has become even more important. One in-situ testing tool which has shown considerable potential in this

Dr. Benoit is an Assistant Professor at the Civil Engineering Department, University of New Hampshire, Durham, NH 03824; Dr. Clough is Professor and Head of the Civil Engineering Department, Virginia Polytechnic Institute and State University, Blacksburg, VA 24061.

regard is the self-boring pressuremeter (SBPM). Recent versions of the SBPM track the movement of the membrane and measure the applied expansion pressure using electronic sensors. Typically, three points of membrane movement located at mid-probe and 120 degrees apart are monitored, allowing assessment of the in-situ stress at each of the three locations. Additionally, an average value is obtained.

Several techniques are available to evaluate the total in-situ lateral stress, and Lacasse and Lunne (1) have compared the methods commonly used. Their results suggest that most accepted procedures for determining the in-situ horizontal stress have essentially the same degree of accuracy and that none of them yield any significant improvement over the simple inspection method. The inspection method consists of visually observing the first noticeable outward movement of the membrane on the field pressure-strain curve and correcting for membrane inertia. The pressure corresponding to this initial lift-off is the total lateral stress. If the soil is isotropic, all three points tracking the membrane should indicate movement at the same time. However, Dalton and Hawkins (2) reported for the stiff Gault clay that anisotropy exists in the lateral stress field, and suggest a technique using the Mohr circle of stress representation to interpret the three independent lateral pressure values in terms of principal stresses in the horizontal plane.

In this investigation a total of 32 SBPM tests were conducted in the San Francisco Bay Mud and the influence of several key factors on measured in-situ stress were evaluated. A micro-computer data acquisition system was implemented to take full advantage of the multiple measurement capabilities of the advanced SBPM version used for this testing program. The system allowed a more careful assessment of the lateral stresses in three different directions. The results suggest that the lateral stresses can be reasonably measured using the SBPM and that, for Young Bay Mud anisotropy exists in the lateral stress field.

SITE CHARACTERISTICS

The Young Bay Mud at the Hamilton Air Force Base site extends to a depth of approximately 16.7m (55 ft). A summary of the properties and test results on the deposit is given by Bonaparte and Mitchell (3). During this investigation, the depth of the groundwater table fluctuated between 0.6 and 2.1m (2 to 7 ft). The natural water content is around 60 percent above 2m (6.6 ft) and 90 percent below this depth. Below the strongly desiccated zone at 1.5m (5 ft), the plastic and liquid limit are about 45 and 90 percent, respectively. It is classified as a CH to MH soil. The total unit weight is 15.7 kN/m^3 (100 pcf) to a depth of 2m (6.6 ft) and 14.8 kN/m^3 (94 pcf) thereafter. Sensitivities of the soil below 2m (6.6 ft) range from 6 to 8.

The effects of desiccation appear to extend to a depth of 5.5m (18 ft). The overconsolidation ratio (OCR) increases from that depth to the surface. Below 5.5m (18 ft), available data suggest that the Young Bay Mud is slightly overconsolidated at an OCR of 1.1 to 1.3. Ratios of undrained shear strengths to the vertical effective overburden pressure below 5.5m (18 ft) range from 0.29 to 0.38 depending on the type of in-situ or laboratory test conducted (3,4). The coefficients of earth pressure at rest, K_o, are reported to vary from 0.48 to 0.70.

SELF-BORING PRESSUREMETER

A photograph of the complete SBPM field testing operation is presented in Figure 1. For this testing program, the SBPM has a diameter of 8.1 cm (3.2 in.) and is approximately 104 cm (40.9 in.) long. Of the total length, a 49.5 cm (19.5 in.) section is covered by a flexible rubber membrane (L/D ≃ 6). At the end of the probe a sharp cutting shoe houses a chopping bit able to move relative to the edge of the shoe. The chopping bit is connected to a hollow drilling shaft that runs inside a EX size drilling casing onto which the SBPM is attached. As the instrument is pushed into the ground, drilling fluid is pumped down the drilling shaft, the chopping bit cuts the soil, and the material is flushed to the surface in the annulus between the EX casing and the drilling shaft.

When a testing location is reached, drilling is stopped and some time is allowed for stabilization of excess pore pressures. By applying nitrogen pressure to the inside of the instrument, the membrane is then expanded against the sides of the borehole. As the cavity is enlarged, movement of the membrane is tracked by three strain feeler arms set midway along the probe, 120 degrees apart. In addition, pore pressures are measured by two transducers, 180 degrees apart, attached to the membrane to stay in constant contact with the expanding cavity.

DATA ACQUISITION SYSTEM

In this investigation the SBPM tests were performed using an automatic loading device which controlled the membrane expansion rate as set by the operator. A micro-computer data acquisition system was developed to record the results, and allows one to obtain an average of 6000 data points per channel for each test. The system consists of a standard Apple II Plus micro-computer with a 48K memory and an analog to digital (A to D) logic board. The micro-computer was powered by a portable generator. A filtering device was constructed to filter the ambiant noises and the electrical "kickback" to the instrument imposed by the computer. The testing progress was supervised by observing the data recorded on the TV monitor. Simultaneously the

Fig. 1 -- Photograph of the self-boring pressuremeter field testing operation.

data was stored on a standard floppy disk, and subsequent to the testing, the data from the diskette was transferred via a standard modem to a mainframe computer for processing and plotting.

A typical plot of the average pressure-strain relationship recorded by the micro-computer data acquisition system is shown in Figure 2. The large amount of data available generates a well defined pressuremeter curve from which the "lift-off" pressure (pressure at which the lateral stress in the ground is reached) is determined. This process is enhanced if the data is plotted on a large scale. Figure 2 illustrates how the expanded scale affects the lift-off pressure interpretation.

The micro-computer monitoring system also offers the advantage of helping assess the issue of disturbance such as that which might occur inadvertently with problems in probe insertion. There are two tell-tale characteristics in such cases. First, if the probe has clogged momentarily, and the soil around the probe is remolded as it advances, this will show up in the results in the form of an "S-shaped" phase. Having a large number of data points allows this to be exposed clearly as is shown in Figure 3. Secondly, if there is a localized disturbance, such as that caused by dragging a shell or gravel fragment down during probe advance, this can be defined by comparing the data from each of the three tracking arms. Figures 4 and 5 illustrate the importance of this. Figure 4 gives the conventional curve derived from the average of the three arms and no obvious problems are observed. However, when the three arms are plotted individually (Figure 5), it can be seen that arm 1 yielded data at odds to the other two. This allows highly inconsistent arm readings to be eliminated from the average. In only two cases was this found necessary since generally the results from the three arms were consistent.

PRINCIPAL HORIZONTAL STRESSES

Laboratory and field calibration tests were performed frequently during the course of the work, and in all cases the results showed the strain arms to have essentially identical response characteristics. However, careful examination of the field test data showed that there were often small differences in the lift-off pressures from the three arms, and depending upon which hole was examined, one arm tended to lift-off first. Figure 6 illustrates this behavior. In borehole JAB-3, strain arm number three consistently gave larger lateral pressures than arm number two. Similar consistent differences were observed in borehole JAB-4 where the probe orientation was changed. In this case, strain arm number two gave larger lateral pressures than arm number three. Because the behavior was repeatable, it is believed that this is a true indication of anisotropy in the lateral stress field in the ground.

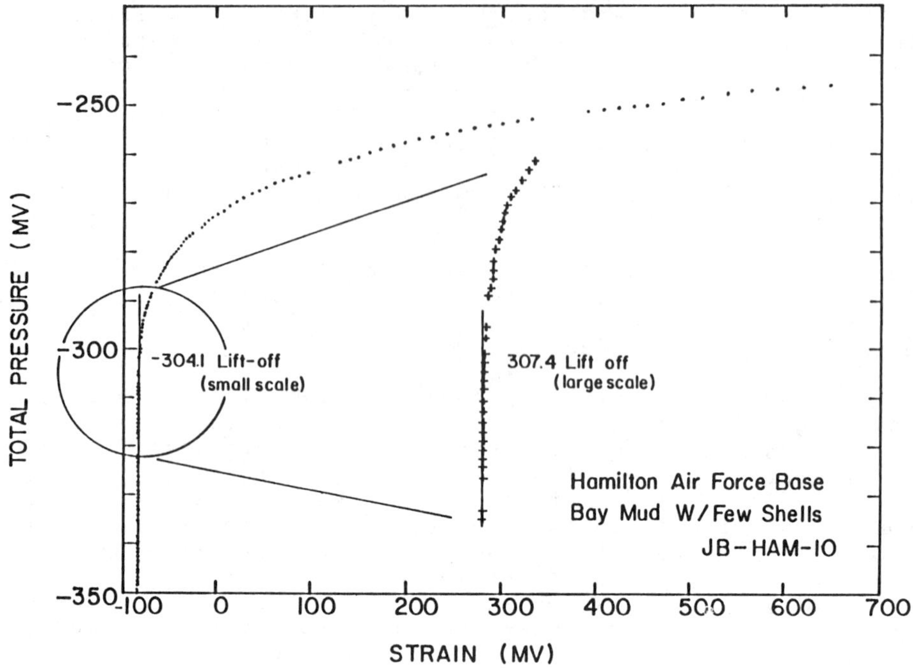

Fig. 2 -- Effect of scaling on the determination of the in-situ lateral stress.

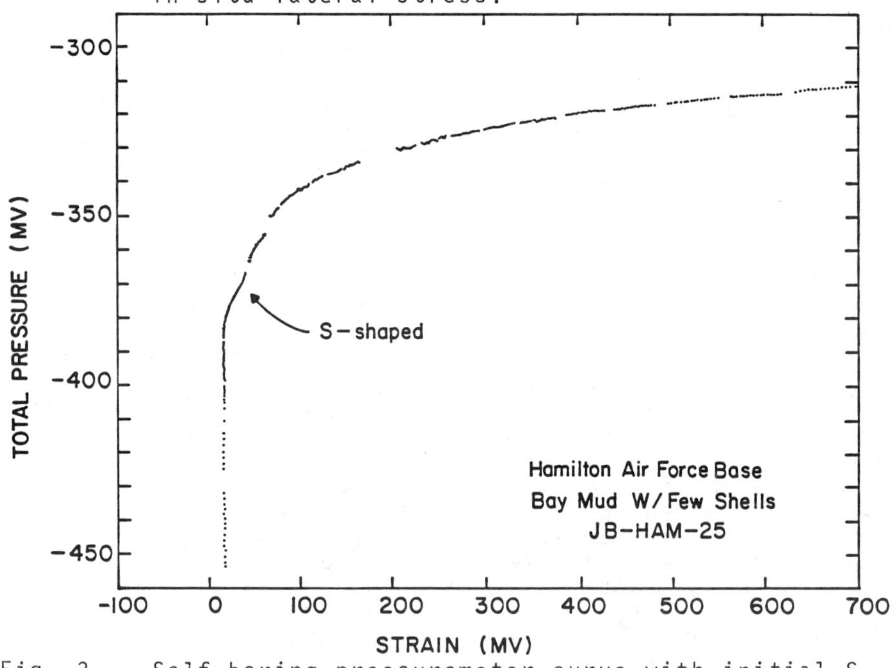

Fig. 3 -- Self-boring pressuremeter curve with initial S-shaped portion suggesting disturbance.

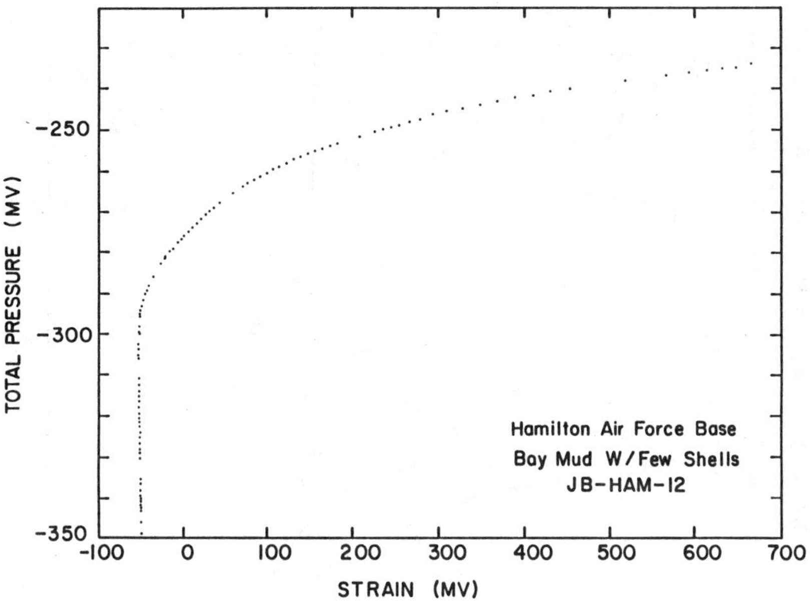

Fig. 4 -- Typical self-boring pressuremeter test curve.

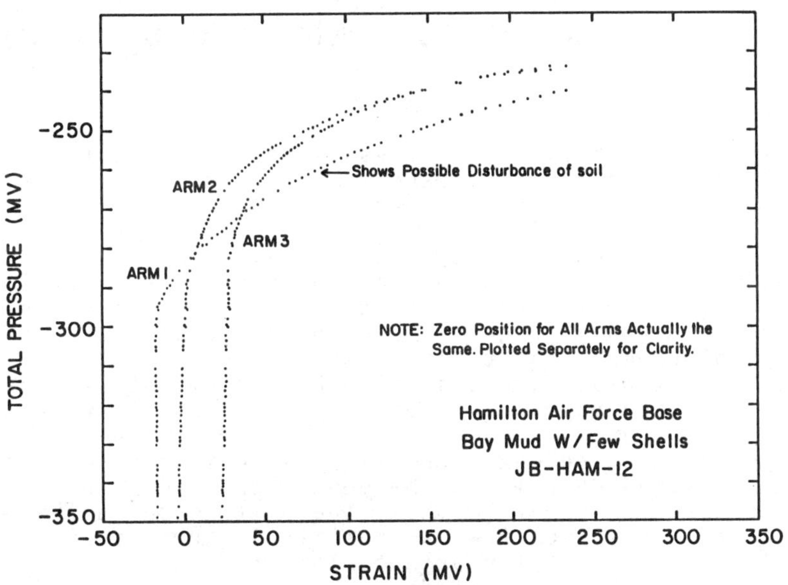

Fig. 5 -- Self-boring pressuremeter curves from all three individual strain sensors showing possible disturbance on one side of the probe.

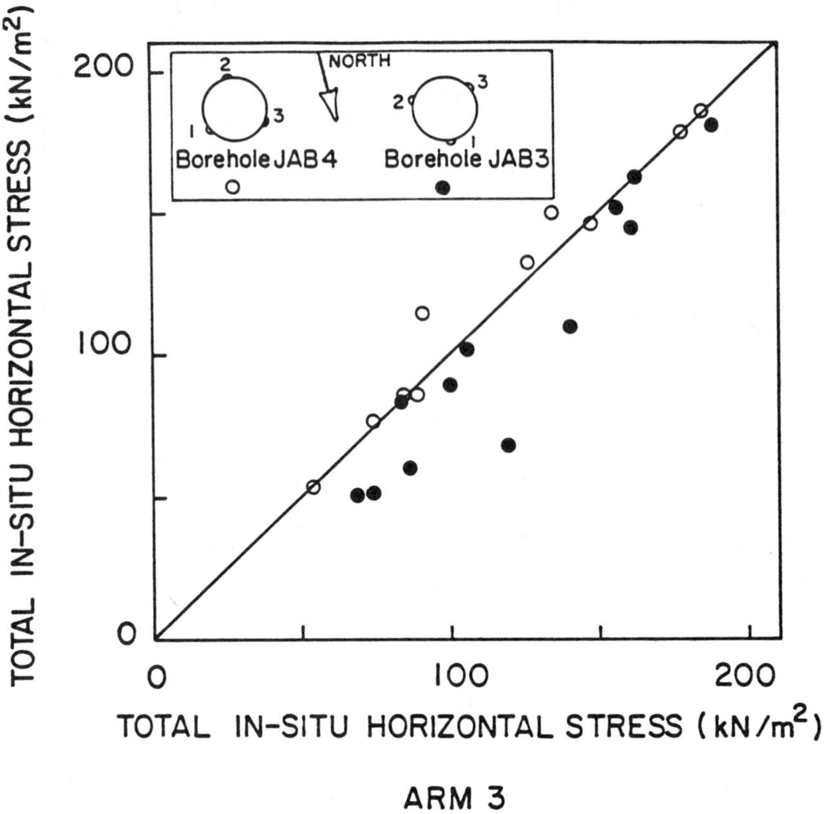

Fig. 6 -- In-situ horizontal stresses measured by strain arms 2 and 3 in boreholes JAB-3 and JAB-4 from self-boring pressuremeter tests.

Using the technique suggested by Dalton and Hawkins (2), the major and minor principal horizontal stresses from the three stress readings obtained in the SBPM tests were calculated. Given in Figure 7 are the principal horizontal stresses for boreholes JAB-3 and JAB-4. The major and minor principal horizontal stresses calculated from the results in the Young Bay Mud vary by as much as 20 percent.

As expected, the major and minor values tend to bracket the average value determined where only three combined results are used (Figure 7). While the difference in the principal stresses from the various strain arms is not great, it appears that the orientation of a stress measuring probe will affect the magnitude of the lateral stress which is indicated. Similar findings have been made in stiff clay (2,5,6), but not previously in soft clay (7). Given the vagaries involved in deposition of soil, it is perhaps not surprizing to find that differences exist in the lateral

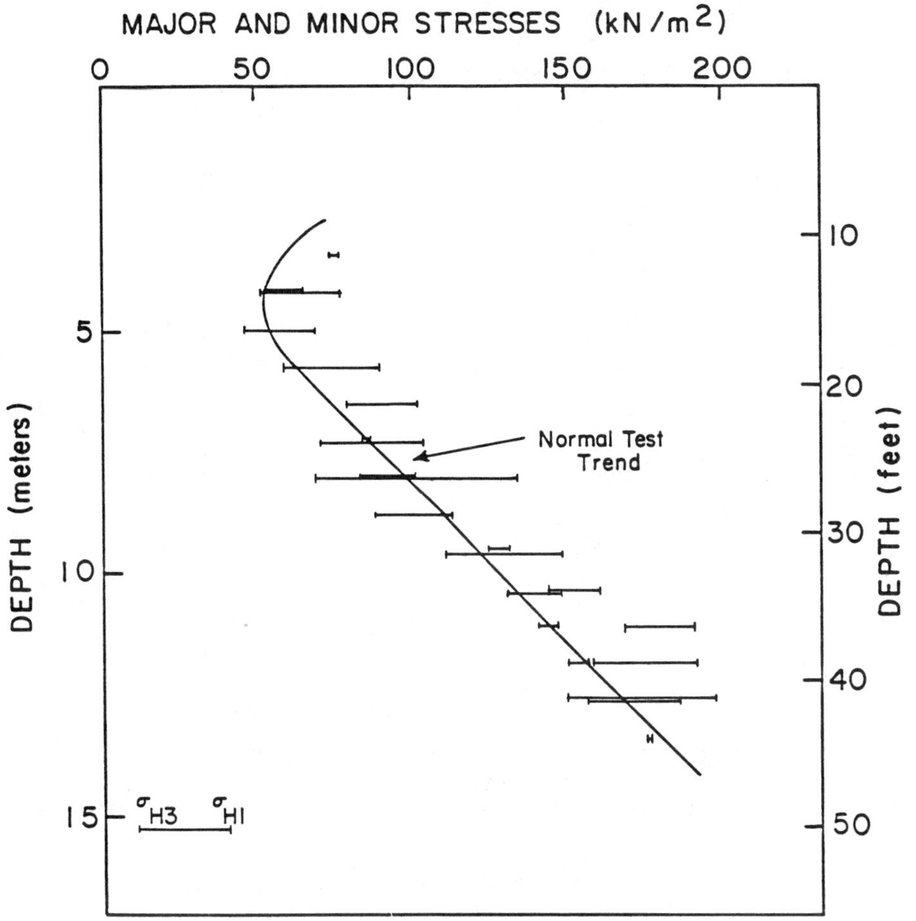

Fig. 7 -- Principal stresses measured in boreholes JAB-3 and JAB-4 from self-boring pressuremeter tests.

stress with direction, and this subject is an interesting one for further research.

COEFFICIENT OF EARTH PRESSURE AT REST

The coefficient of earth pressure at rest, K_o, which is defined as the ratio of the effective horizontal stress to the effective vertical stress was calculated for each of the three strain feeler arms and for the average value of strain from the SBPM tests. The results for borehole JAB-3 are shown in Figure 8. Since the arms all indicated small differences in lift-off pressure, the K_o values are different as well.

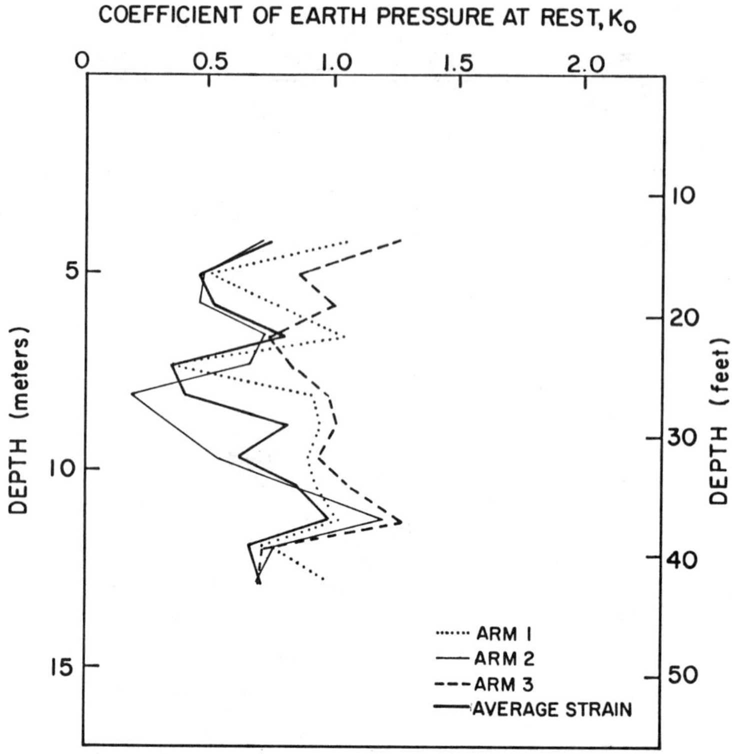

Fig. 8 -- Coefficient of earth pressure at rest from self-boring pressuremeter tests as calculated from each individual strain arm.

The K_o values calculated from the average of the arm in this investigation are compared to calculated K_o values from the previous SBPM tests of Denby (8) in Figure 9. The K_o value from this investigation yielded an average value of 1 for the strongly overconsolidated Bay Mud and of about 0.7 for the slightly overconsolidated clay below 5.5m (18 ft). The K_o values from this testing program are generally higher than those obtained by Denby. The exact reason for these differences is not known but possibly reflect the improved SBPM probe used in this investigation and the quantity of data collected by the micro-computer data acquisition system.

The K_o results are also compared to predicted K_o values obtained from two different accepted empirical relationships (9,10,11)(see Fig. 9). The first empirical approach relates K_o to the drained friction angle and the OCR for primary loading. The second approach relates K_o to the plasticity index and the OCR. The K_o values calculated for this testing program are within the range of values predicted by the empirical relationships. The agreement is fairly close with empirical methods based on the drained

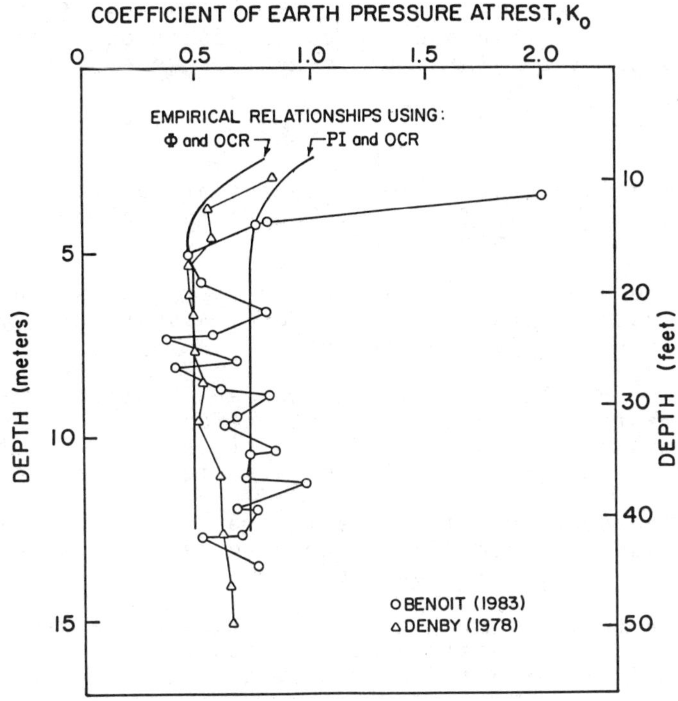

Fig. 9 -- Comparisons of coefficients of earth pressure at rest calculated from self-boring pressuremeter tests and empirical relationships.

friction angle for depths of 5.5 to 8.5m (18 to 28 ft) and based on the plasticity index for depths below 8.5m (28 ft).

EFFECTS OF DRILLING AND TESTING PROCEDURES

The effects of key insertion and testing parameters on the measured lateral stresses were evaluated in the present investigation. The results were compared against SBPM tests conducted during this investigation and that of Denby (4,8, 12,13) which followed a normalized set of procedures developed to minimize disturbance in Young Bay Mud. Factors considered were: (1) rate of rotation of the cutter; (2) oversizing of the hole by equipment imperfections; (3) membrane expansion rate; and (4) position of the cutter inside the driving shoe. In terms of test procedures and techniques, disturbance of the soft clay was found to be the most important factor that affected the results. Intentional disturbance induced by probe insertion techniques or by improper sizing of the probe parts caused deviations of up to 60 percent of the measured in-situ stresses relative to those expected (4,13). The influence of the membrane expansion rate and the cutter position inside the driving shoe on the lateral stress values were negligible.

CONCLUSIONS

The results obtained from a series of self-boring pressuremeter tests in soft clay led to the following conclusions:

1. The use of a micro-computer data acquisition system allowed a more careful assessment of the in-situ lateral stresses.

2. Lateral stresses in the three directions measurements were taken were not uniform; major and minor principal horizontal stresses calculated from the results varied by as much as 20 percent.

3. The coefficients of earth pressure at rest were found to be in fairly good agreement with accepted empirical relationships.

4. In terms of drilling and testing procedures in soft clay, disturbance was found to be the single most important factor that affected the results.

ACKNOWLEDGEMENTS

The authors would like to express their appreciation of Drs. R.C. Bachus and G.M. Denby for their help and advice during the course of the research. The research was sponsored by the National Science Foundation under the auspices of Grant No. CEE-8300464.

REFERENCES

[1] Lacasse, S. and Lunne, T., "In-Situ Horizontal Stress from Pressuremeter Tests," Norges Geotekniske Institutt, 1982.

[2] Dalton, J.C.P. and Hawkins, P.G., "Fields of Stress - Some Measurements of the In-Situ Stress in a Meadow in the Cambridgeshire Countryside," Ground Engineering, Vol. 15, No. 4, May 1982.

[3] Bonaparte, R. and Mitchell, J.K., "The Properties of San Francisco Bay Mud at Hamilton Air Force Base, California," Department of Civil Engineering, University of California, Berkeley, April 1979.

[4] Benoit J. and Clough, G.W., "Self-Boring Pressuremeter Tests in Soft Clay," Article accepted for publication in the Journal of the Geotechnical Engineering Division, ASCE.

[5] Ghionna, V.N., Jamiolkowski, M. and Lancellotta, R., "Characteristics of Saturated Clays as Obtained from SBP Tests," Symposium on the Pressuremeter and Its Marine Applications, Editions Technip, Paris, 1982.
[6] Ghionna, V.N., Jamiolkowski, M., Lacasse, S., Ladd, C.C., Lancellotta, R. and Lunne, T., "Evaluation of Self-Boring Pressuremeter," Symposium International In-Situ Testing, Paris, 1983.
[7] Clough, G.W. and Benoit, J., "Use of the Self-Boring Pressuremeter in Soft Clays," in Developments in Soil Mechanics and Foundation Engineering-2, P.K. Banerjee and R. Butterfield, Eds., Elsevier Applied Science Publishers Ltd., London, 1985, pp. 253-276.
[8] Denby, G.M., "Self-Boring Pressuremeter Study of the San Francisco Bay Mud," A Thesis Presented in Partial Fulfillment of the Ph.D. Degree, Stanford University, Stanford, California, 1978.
[9] Jaky, J., "The Coefficient of Earth Pressure at Rest," Journal for Society of Hungarian Architects and Engineers, Budapest, Hungary, October 1944, pp. 355-358.
[10] Ladd, C.C. and Foott, R., "New Design Procedures for the Stability of Soft Clays," Journal of the Geotechnical Engineering Division, ASCE, Vol. 100, No. GT7, July 1974, pp. 763-786.
[11] Mayne, P.W. and Kulhawy, F.H., "K_o-OCR Relationships in Soil," Journal of the Geotechnical Engineering Division, ASCE, Vol. 108, No. GT6, June 1982, pp. 851-872.
[12] Clough, G.W. and Denby, G.M., "Self-Boring Pressuremeter Study on San Francisco Bay Mud," Journal of the Geotechnical Engineering Division, ASCE, Vol. 106, No. GT1, January 1980, pp. 45-63.
[13] Benoit, J., "Analysis of Self-Boring Pressuremeter Tests in Soft Clay," A Thesis Presented in Partial Fulfillment of the Ph.D. Degree, Stanford University, Stanford, California, 1983.

Ilan Juran, J. F. Beech

EFFECTIVE STRESS ANALYSIS OF SOIL RESPONSE IN A PRESSUREMETER TEST

REFERENCE: Juran, I. and Beech, J.F., "Effective Stress Analysis of Soil Response in a Pressuremeter Test", The Pressuremeter and Its Marine Applications: Second International Symposium, ASTM STP 950, J.-L. Briaud and J.M.E. Audibert, Eds., American Society for Testing and Materials, 1986.

ABSTRACT: A simple rheological model for analyzing pressuremeter test results, which includes pore water pressure measurements, is presented. The soil is assumed to be an elasto-plastic material with a non-associated flow rule. The governing equation of the yield surface and the plastic flow potential are based on parameters which can be obtained from standard tests, such as triaxial compression tests. A parameter is introduced which accounts for the contracting and/or dilating behavior of the soils during expansion. The model is applied to pressuremeter test results obtained in a contracting, lightly overconsolidated clayey silt. The shear modulus, effective strength characteristics, and a parameter characterizing the contracting behavior of the soil are determined. The results of a numerical analysis, incorporating the proposed rheological model, simulating the pressuremeter tests are compared with the field test results.

KEYWORDS: elasto-plastic; contracting soil; pore water pressure; effective stress, in-situ tests; numerical analysis

INTRODUCTION

In the past, total stress analysis, assuming elastic perfectly plastic behavior, has been the principal means for modelling the cavity expansion which occurs during a pressuremeter test. The recent development of pressuremeter probes which measure the excess pore water pressure generated during cavity expansion suggests new

Dr. Juran is Associate Director of CERMES-ENPC, 2, rue Alfred Fouillee, 75013, Paris, France. J. Beech is a former Research Associate of CERMES-ENPC who is currently with GeoServices Inc., Consulting Engineers, Boca Raton, Florida.

methods, based on effective stress analysis, for interpreting pressuremeter test results. To analyze these new tests, a relatively simple rheological model has been developed which takes into account the contracting and/or dilating behavior of the soil during cavity expansion. The soil is assumed to be an elasto-plastic material with a non-associated flow rule. The governing equations of the yield surface and the plastic flow potential are based on parameters which can be obtained from standard triaxial compression tests results.

This paper demonstrates the application of the method to results of pressuremeter tests performed in a contracting, lightly overconsolidated clayey silt. First, the model is briefly presented along with the analysis procedure. Second, the site is briefly described and the results of typical pressuremeter tests with measurement of the pore water pressure are presented. Third, the proposed methodology is used to analyze the pressuremeter test results and to obtain the shear modulus, effective strength characteristics and a parameter which characterizes the contracting behavior of the soil. Finally, these soil characteristics are used in a numerical analysis incorporating the proposed rheological model simulating the pressuremeter tests, and the predicted results are compared with the actual field test data.

MODELING ASSUMPTIONS

The soil is considered to be a homogeneous, isotropic, strain-hardening elasto-plastic material with a non-associated flow rule. The strain-hardening is assumed to be isotropic with the hardening parameter defined by the shear strain γ:

$$\gamma = \frac{1}{\sqrt{2}} [(\varepsilon_1 - \varepsilon_2)^2 + (\varepsilon_2 - \varepsilon_3)^2 - (\varepsilon_3 - \varepsilon_1)^2]^{1/2} \tag{1}$$

in which: ε_1, ε_2, ε_3 are respectively the principal strains with $\varepsilon_1 > \varepsilon_2 > \varepsilon_3$.

Yield Function

A Mohr-Coulomb type yield criteria is considered to define the yield surface with an open elastic domain, and the assumed yield function can be written as:

$$f(\underline{\sigma};\gamma) = \frac{q - q_o(\gamma)}{p'} - h(\gamma) = 0 \tag{2}$$

where: $\underline{\sigma}$ is the stress tensor;

q is the deviatoric stress; $q = \frac{1}{2} [\sigma_1 - \sigma_3]$ with σ_1 and σ_3 being respectively the major and the minor principal stresses;

p' is the effective average stress; $p' = \frac{1}{2} [\sigma'_1 + \sigma'_3]$;

$q_o(\gamma)$ is the deviatoric stress at $p' = 0$.

For a contracting soil, $h(\gamma)$ is assumed to be a hyperbolic function:

$$h(\gamma) = \frac{\gamma}{a + b\gamma} \quad \text{with: } \frac{1}{a} = \frac{G}{\sigma_o'}, \text{ and } \frac{1}{b} = \sin \phi_{cv} \qquad (2a)$$

where: σ_o' is the initial isotropic consolidation stress;

G is the shear modulus;

ϕ_{cv} is the effective internal friction angle of the soil at the critical state.

This hyperbolic $h(\gamma)$ function is illustrated in Fig. 1.

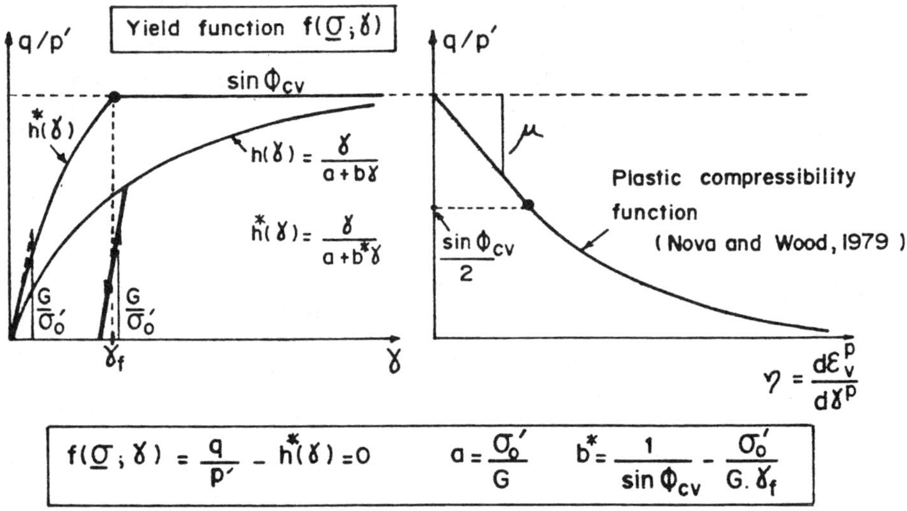

Fig. 1 Definition of the yield and plastic compressibility stress function.

Fig. 2 shows typical results of consolidated, undrained triaxial shear tests on a lightly overconsolidated clayey silt. In the (q; p') plane, the effective stress paths followed by the soil in different tests are similar and the "iso-γ" curves can be approximated as straight lines (Eq. 2). For normally consolidated soil, $q_o(\gamma) = 0$. Also, it is clearly illustrated in Fig. 2 that the stress paths actually reach the failure envelope at a failure shear strain $\gamma_f = 8\%$, which is less than $\gamma_f = \infty$ predicted with the hyperbolic model. At shear strains greater than γ_f, the soil is assumed to behave as a perfectly plastic material.

Considering these experimental results, the yield function was modified to take into account the perfectly plastic behavior of the soil for $\gamma > \gamma_f$. It is assumed that the failure shear strain γ_f is independent of the stress path and a "modified $h(\gamma)$ function" is considered which is defined as:

$$h^*(\gamma) = \frac{\gamma}{a + b^*\gamma} \quad \text{where: } \frac{1}{a} = \frac{G}{\sigma_o'}; \quad \frac{1}{b^*} = \frac{1}{\sin\phi_{cv}} - \frac{\sigma_o'}{G\gamma_f} \tag{2b}$$

This "modified $h(\gamma)$ function" is illustrated in Fig. 1.

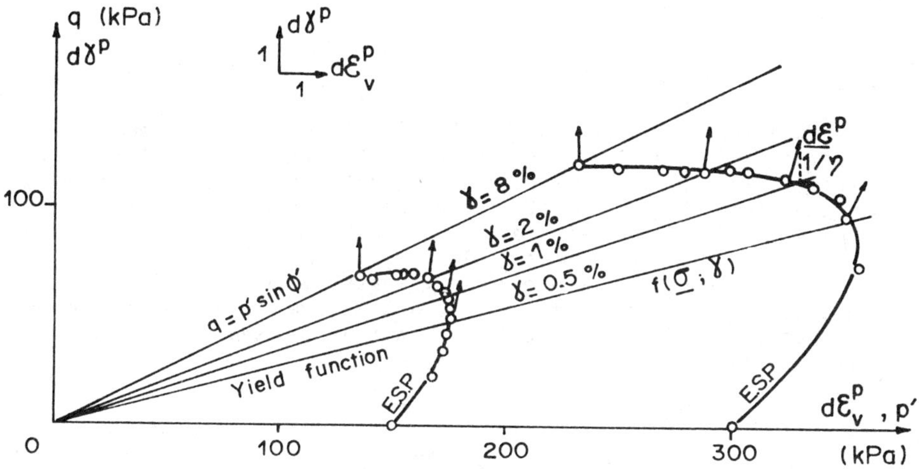

Fig. 2 Effective stress paths (ESP), yield functions, and plastic compressibility parameter from consolidated undrained triaxial tests on a normally consolidated silt [2].

Plastic Flow Function

It is assumed that a unique relationship that is independent of the actual stress path exists between the stress ratio q/p' and the value of the plastic compressibility parameter , characterizing the contracting (or dilating) soil behavior, which is defined by:

$$\eta = \frac{d\varepsilon_v^p}{d\gamma^p} = \eta\,(q/p') \tag{3}$$

where: $d\varepsilon_v^p$ and $d\gamma^p$ are the increments of the plastic volumetric strain and plastic shear strain, respectively.

This plastic compressibility parameter defines for each state of stress the corresponding direction of the plastic strain increment ($d\varepsilon^p$). This is illustrated in Fig. 2 assuming that principal axes of stresses and plastic strain increments coincide. Hence, a non-associated plastic potential function can be defined.

The projection of this plastic potential in the (q; p') plane is the curve g (q; p') which verified at each point:

$$\frac{\partial q}{\partial p'} = -\eta = -\frac{d\varepsilon_v^{\ p}}{d\gamma^p} \tag{4}$$

The plastic strain increments can then be calculated as:

$$d\varepsilon_{ij}^{\ p} = d\lambda \frac{g(\sigma)}{\partial\sigma_{ij}} \tag{5}$$

where λ is a proportionality constant.

Experimental results by Namy [1] illustrated that the variation of the plastic compressibility parameter η with the stress ratio q/p' is hyperbolic; the value of η goes to infinity as the stress ratio q/p' approaches zero. Similar results were obtained by Juran and Bensaid [2] who investigated soil response to a cavity expansion using a "hollow cylinder cell" which enabled them to simulate undrained expansion tests in annular soil samples. Consequently, a $\eta(q/p')$ function which is similar to that proposed by Nova and Wood [3] is considered in the proposed model, and is expressed as:

$$\eta = \begin{cases} \dfrac{\sin^2\phi_{cv}}{\mu} \cdot \dfrac{1}{q/p'} & ; \quad 0 \leq \dfrac{q}{p'} \leq \dfrac{\sin\phi_{cv}}{2} \\[2ex] \dfrac{1}{\mu}\left[\sin\phi_{cv} - \dfrac{q}{p'}\right] & ; \quad \dfrac{\sin\phi_{cv}}{2} \leq \dfrac{q}{p'} \leq \sin\phi_{cv} \end{cases} \tag{6}$$

This function is illustrated schematically in Fig. 1b. It can be observed that as the soil reaches its critical state of a perfect plastic flow ($\gamma = \gamma_f$), the value of the plastic compressibility parameter η is equal to zero and no volume change occurs. The plastic potential function g (q; p') can be calculated by using Eqs. 4, 5 and 6.

The proposed model defined by Eqs. 2b and 6 and shown in Fig. 1, requires an experimental determination of five soil characteristics: G/σ_o', ν (Poisson's coefficient); ϕ_{cv}; γ_f and μ. In the event the soil strength parameters include a cohesion intercept then the function $q_o(\gamma)$ is also required. These characteristics can be obtained by either triaxial shear tests (consolidated drained with measurement of the volumetric strain or consolidated undrained with measurement of the pore water pressure) or in-situ pressuremeter tests with measurement of the generated pore water pressures. The latter case is illustrated below.

EFFECTIVE STRESS ANALYSIS USING THE PROPOSED MODEL OF PRESSUREMETER TESTS WITH MEASUREMENT OF PORE WATER PRESSURES

The results of tests carried out by the French laboratory "LPC-Saint Brieuc" on the site of Cran are used to illustrate the proposed methodology for obtaining soil characteristics using an effective stress analysis.

Fig. 3 shows the geotechnical profile of the site and the soil characteristics as obtained by both in-situ vane shear tests and laboratory oedometer and triaxial tests on undisturbed soil samples [4]. The laboratory test results indicate a lightly overconsolidated plastic silt (MH : W_l = 35; I_p = 30) existing at a depth of 5 to 10m. The consolidated, undrained triaxial shear tests on undisturbed soil samples taken at the depths of 5.60m; 7.60m and 9.60m gave (Fig. 3) a small cohesion of c' ≃ 6.7 kPa and an internal friction angle of ϕ_{cv} = 33°.

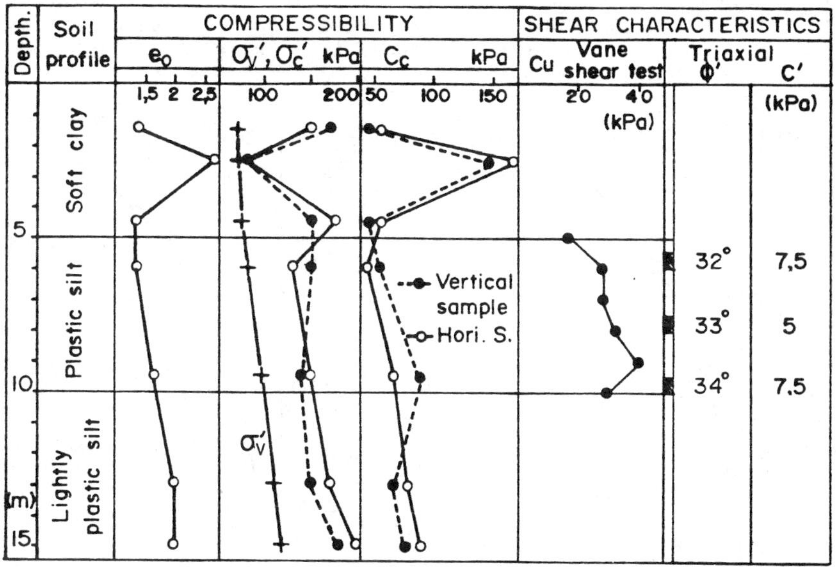

Fig. 3 Geotechnical characteristics of Cran.

Fig. 4 shows the results of two self-boring pressuremeter tests (PAF) at depths of 7 m and 10 m. These results include the expansion curves $\sigma_c - p_0' = f(2g_0 = \Delta V_c/V_0)$, (where σ_c is the applied pressure in the pressuremeter probe; p_0' is the initial effective horizontal stress; g_0 is the Green strain and ΔV_c and V_0 are respectively the volume change and the initial volume of the cavity) as well as the curves representing the generation of the pore water pressures measured at the cavity facing $\Delta u(r_0)$.

In order to analyze these test results, the following procedure is proposed:

- The shear curve of the soil $q(g_0)$ can be directly obtained by deriving the expansion curve $\sigma_c(g_0)$ according to a procedure proposed by Baguelin et al [4] which leads to the equation:

$$q=q(g_0) = \frac{d(\sigma_c - p_0')}{dg_0} \cdot g_0(1+2g_0) \tag{7}$$

The other parameter, p', for the effective stress path in the (q; p') plane is given by:

$$p' = \sigma_c - q - \Delta u(r_o) \tag{8}$$

The effective stress paths calculated for different depths within the considered, relatively homogeneous, zone (5m to 10m) enable an experimental determination of the effective shear strength characteristics c' and ϕ_{cv}.

"Iso-g_o" curves, corresponding to the state of stress attained in the pressuremeter tests at different depths for the same value of g_o, can be drawn in the (q; p') plane in order to obtain the yield function as defined by Eq. 2. It can be noted that the shear strain γ at the cavity facing is equal to: $\gamma(r_o)=\sqrt{3}[(2g_o + 1)^{1/2}-1]$.

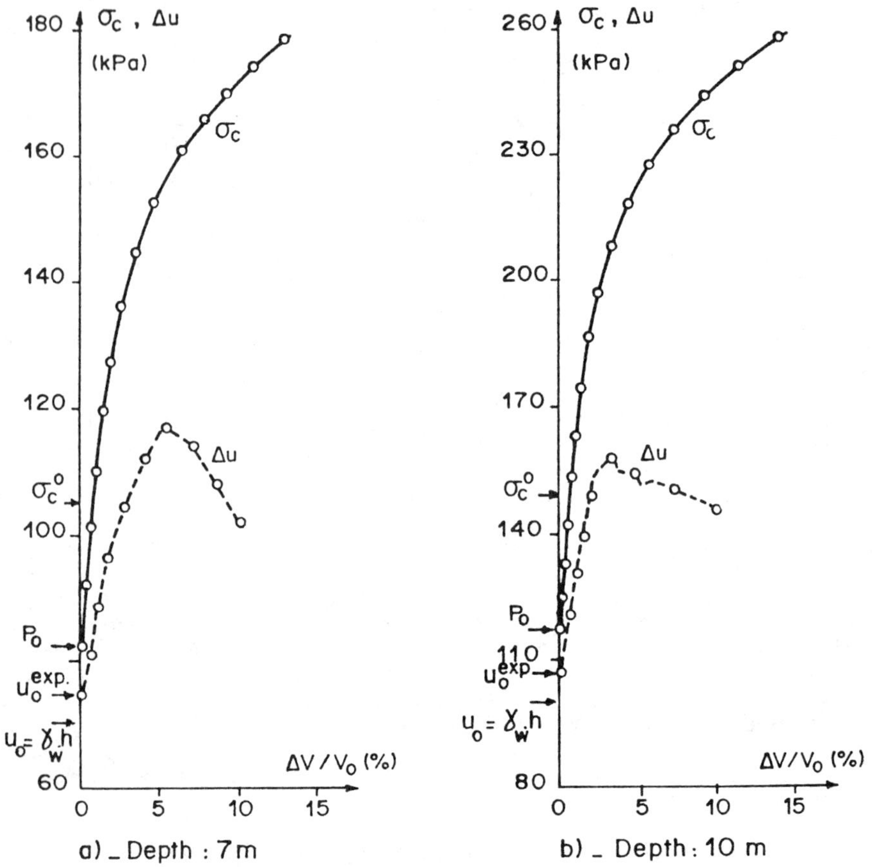

Fig. 4 Results of pressuremeter tests with measurement of pore water pressure.

The plastic compressibility-stress function $\eta(q/p')$ is obtained considering the specific test conditions of plane strain ($d\varepsilon_z = 0$) undrained expansion ($d\varepsilon_v = 0$) which leads to the following equation:

$$\eta = \frac{d\varepsilon_v^{\,p}}{d\gamma^p} = \frac{d\varepsilon_v - d\varepsilon_v^{\,e}}{d\gamma - d\gamma^e} = -\frac{\frac{1-2\nu}{G} \cdot \frac{dp'}{d\gamma}}{1 - \frac{1}{2G} \cdot \frac{d\tau oct}{d\gamma}} \tag{9}$$

where: $d\varepsilon_v$ and $d\varepsilon_v^{\,e}$ are respectively the total and elastic volumetric strains;

$d\gamma$ and $d\gamma^e$ are respectively the total and elastic strains;

τ_{oct} is the octahedral shear stress which is equal to:

$$\tau_{oct} = 1/\sqrt{2}[(\sigma_1 - \sigma_2)^2 + (\sigma_3 - \sigma_1)^2 + (\sigma_2 - \sigma_3)]^{1/2} \tag{10a}$$

In this application, the radial, tangential, and vertical stresses are principal stresses. The total vertical strain is equal to zero, and by assuming the plastic strain in the vertical direction to be equal to zero, the intermediate principal stress is: $\sigma_2 = \nu(\sigma_1 + \sigma_3)$. Using this expression for σ_2 and the definition of q and p', the following can be obtained:

$$\tau_{oct} = p' \, [3h^*(\gamma)^2 + (1 - 2\nu)^2]^{1/2} \tag{10b}$$

To derive the shear curve of the soil $q(g_o)$, it is necessary to estimate the initial effective radial stress σ_h'. As the soil is initially at a K_o state of stresses, σ_h' should be equal to:

$$\sigma'_{ho} = K_o(\gamma z - u_o) \tag{11}$$

where: γz is the overburden stress at a depth z;

u_o is the initial hydrostatic pore water pressure at the same depth.

As illustrated in Fig. 5, the measured initial radial stress $\sigma'_{ho} = p'_o(g_o = 0)$ corresponded to a K_o value which was significantly smaller than those typically observed in the laboratory or the field. Also, these low values of the measured p'_o stress led to unrealistic effective stress paths. It is uncertain why the self-boring pressuremeter would result in a lower K_o. In order to start the following analysis at a realistic stress state, the initial cavity pressure $\sigma_c^{\,o}$ is considered to be the effective radial stress which corresponds to the assumed K_o stress state, and the pore water pressures generated during the initial reloading to $\sigma_c^{\,o}$ have been neglected. The values of q_o corresponding to the stress at which an appropriate value of $\sigma_c^{\,o}$ was achieved are shown in Figure 5.

Fig. 6 shows the shear curves derived using Eq. 7 from the "corrected" expansion curves obtained at depths of 5m, 7m, 8m and

10m. Also shown are the shear curves derived from the "non-corrected" expansion curves obtained at 8m and 10m. This latter set of curves clearly illustrates that the low values of the measures p_o' stress predicts a higher shearing resistance of the soil. Fig. 7 shows that the values of the undrained cohesion c_u given by the shear curves at different depths correspond fairly well to those obtained by the vane shear tests. It also illustrates that the undrained cohesion derived from the "non-corrected" expansion curves is overestimated. The shear modulus G obtained from the shear curves at different depths is: $G - 67\sigma_o'$ (where σ_o' is the initial consolidation pressure: $\sigma_o' = [2 + 1/K_o]\sigma_c'$).

Fig. 8 shows the effective stress paths in the (q; p') plane calculated for different depths according to Eq. 7 and 8. The experimental results illustrate that the "iso-g_o" curves can be modelled by straight lines which correspond to the assumed yield function as defined by Eq. 2. As illustrated in Fig. 8, the effective shear strength characteristics predicted by this effective stress analysis method correspond fairly well to those obtained by consolidated undrained triaxial shear tests on undisturbed soil samples taken from the considered zone. Fig. 8 also illustrates that the effective stress paths actually reach the failure envelope at a cavity expansion (g_o) of about 0.8%; the failure shear strain γ_f is therefore taken to be about 2%.

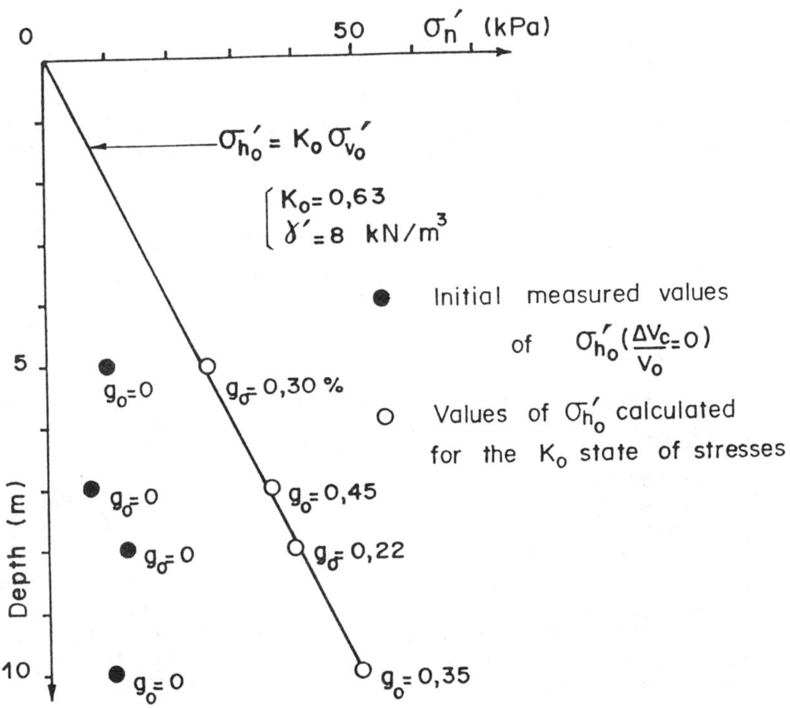

Fig. 5 Initial effective radial stresses at the cavity facing, measured and calculated for the K_o state.

Fig. 9 shows the evolution of the shear strength characteristics, $q_o(g_o)$ and $h(g_o) = [q - q_o(g_o)]/p'$, as a function of the cavity expansion g_o. The values of the $q_o(g_o)$ function shown in Fig. 9a correspond to the coordinates of the "iso-g_o" lines at the origin of the (q; p') plane. The values of the $h^*(g_o)$ function represent the slopes of the "iso-g_o" lines and therefore correspond to a mobilized effective internal friction angle. As shown in Fig. 9b, the $h^*(g_o)$ values can also be calculated from the analysis of each pressuremeter test considering the experimental values of $q_o(g_o)$ shown in Fig. 9a. The good agreement between the values of the $h^*(g_o)$ function obtained for the different tests further demonstrates that the "iso-g_o" curves can be modelled by straight lines which correspond to the assumed yield function. Fig. 9b also illustrates that the assumption of a hyperbolic $h(\gamma)$ function, that corresponds to a particular case for which γ_f is equal to infinity, leads to a significant underestimation of the the mobilized shear resistance of the soil.

Fig. 10 shows the plastic compressibility-stress function $\eta(q/p')$ determined experimentally from the analysis of the pressuremeter tests at different depths (7m, 8m and 10m) according to Eq. 9. The scatter between the results obtained at different

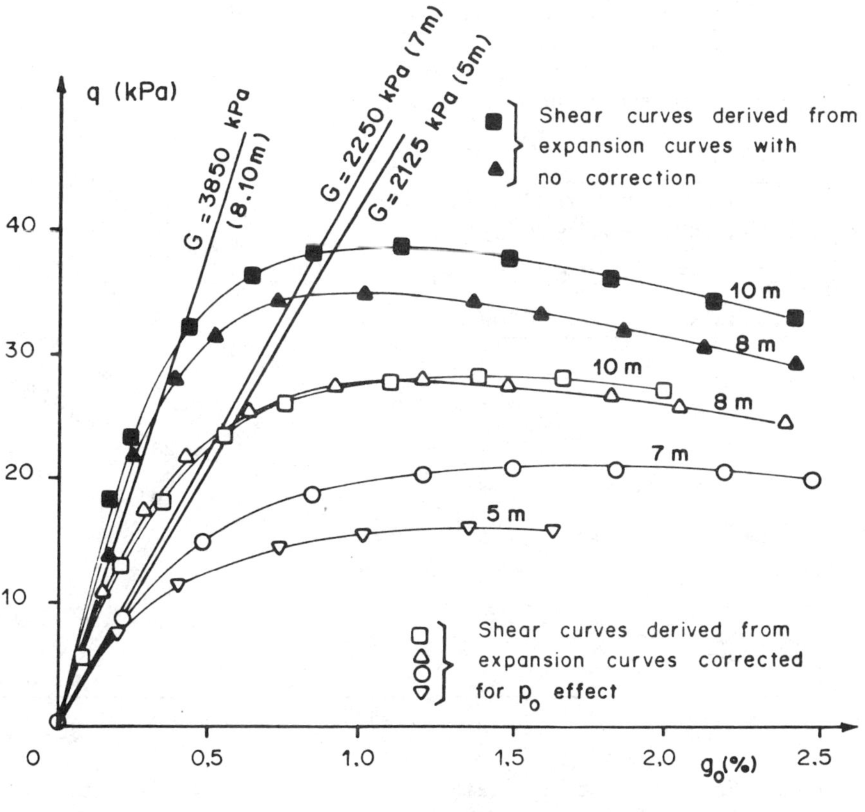

Fig. 6 Shear curves derived from expansion curves.

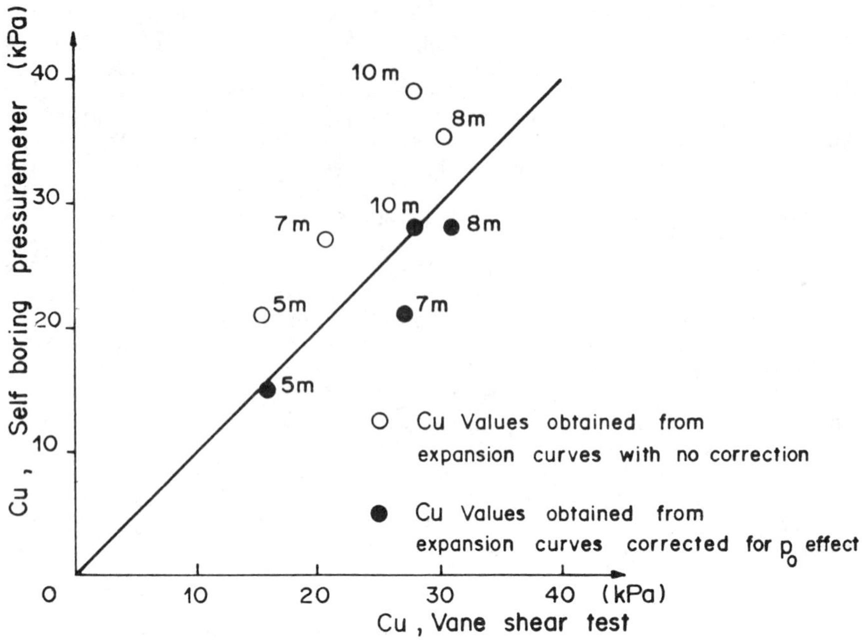

Fig. 7 C_u from shear curves versus C_u measured in vane shear test.

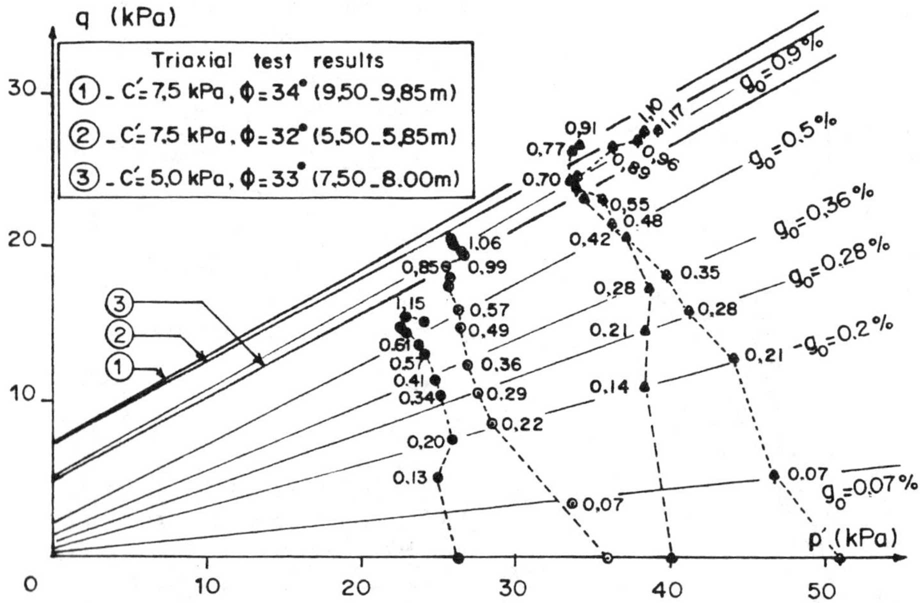

Fig. 8 Effective stress paths and "iso-g_0" lines calculated from pressuremeter test results.

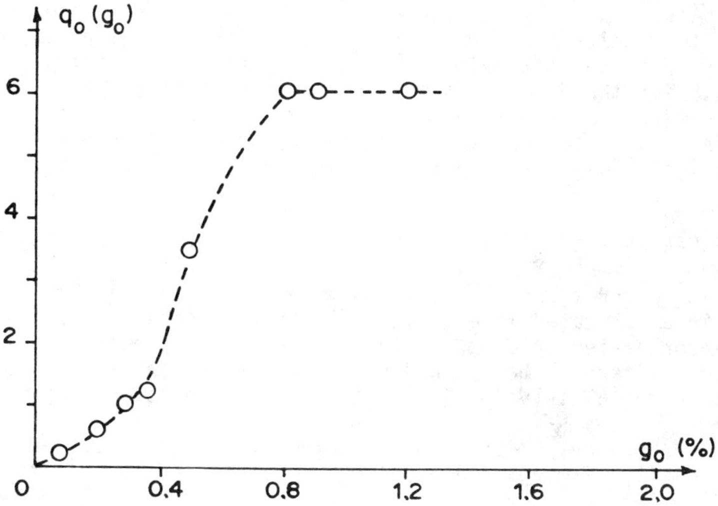

a. Experimental determination of the $q_o(g_o)$ function.

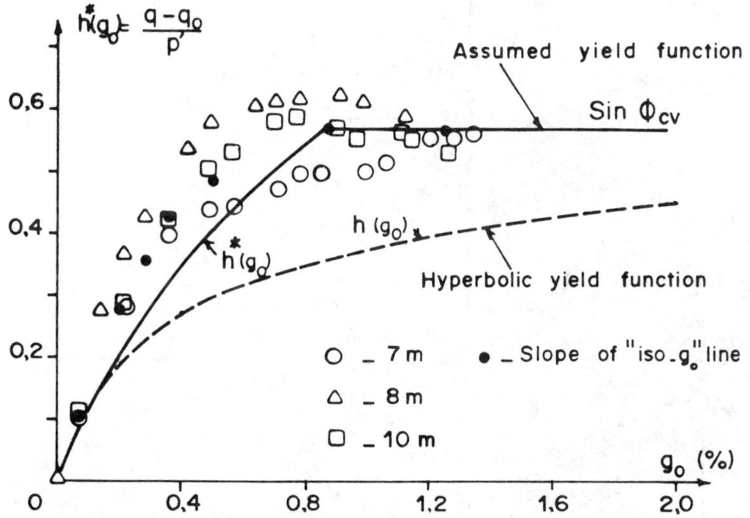

b. Experimental determination of the $h(g_o)$ function.

Fig. 9 Experimental determination of the yield function.

depths is relatively small but is highly dependent on the procedure used to derive the $\eta(q/p')$ function. This function corresponds fairly well to that defined by Eq. 6 with an initial μ modulus which, for the data analyzed here, varies from 2.6 to 7.5. It is interesting to note that a μ value of approximately 2.6 was obtained for the clayey silt used by Juran and Bensaid [2] to investigate soil response to an undrained cavity expansion.

NUMERICAL ANALYSIS OF THE SOIL RESPONSE TO A PRESSUREMETER TEST

In order to numerically simulate an undrained pressuremeter test in a saturated fine grain soil, the proposed rheological model was incorporated into a computer program. The program utilized the finite difference method and could be run on a personal computer. The incremental solution includes the following steps:

- a boundary displacement increment ΔX_o is applied at the cavity facing. The induced total strain increments are

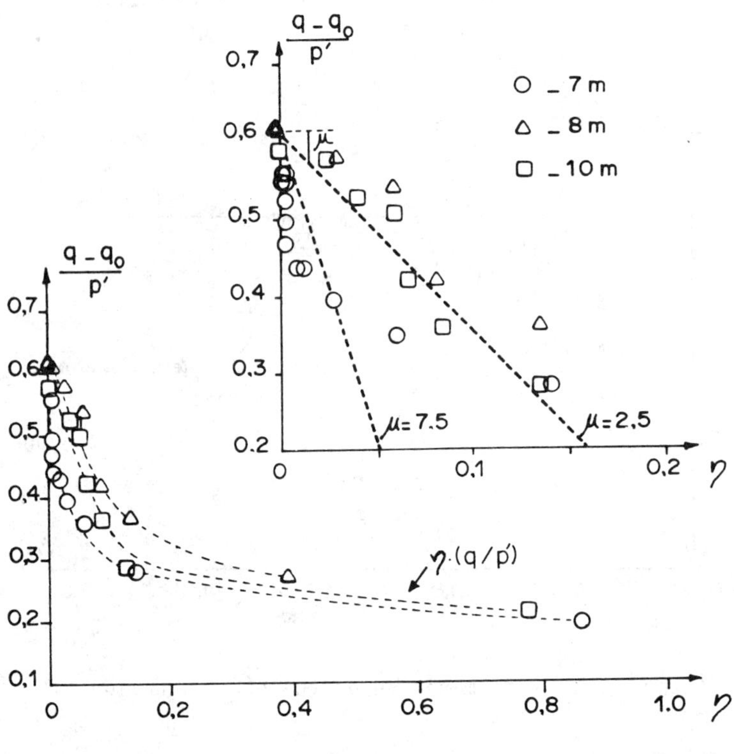

Fig. 10 Experimental determination of the plastic compressibility-stress function.

calculated from the compatibility equation considering the specific test conditions of plane vertical strain ($d\varepsilon_z = 0$) and undrained expansion ($d\varepsilon_v = 0$). The shear strain increment, $d\gamma$, can now be determined.

- the considered constitutive equations (2, 2b and 3) used in conjunction with Eqs 9 and 10b provide a solution for the effective stress increments $d\sigma'_r$ and $d\sigma'_0$ as a function of the shear strain increment $d\gamma$.

- the total radial stress increment $d\sigma_r$ is then calculated from the equilibrium conditions at the actual coordinates of the finite difference mesh as:

$$\frac{d\sigma_r}{d\rho} + \frac{2q}{\rho} = 0 \quad \text{(where: } \rho \text{ is the actual radius)} \tag{12}$$

considering the boundary conditions: $d\sigma_r(R) = 0$ with $R = 50\rho$.

The excess pore water pressure is given by:

$$u(\rho) = \sigma_r(\rho) - \sigma'_r(\rho) \text{ where: } \sigma'_r(\rho) = q(\rho) + p'(\rho) \tag{13}$$

σ_r and σ_r' are respectively the total and effective radial stresses.

- When the shear strain at the cavity facing attains and exceeds γ_f, a zone of confined plasticity propagates and its limit ρ_f at which $\gamma = \gamma_f$ separates a zone of perfect plasticity ($h^*(\gamma) = \sin\phi_{cv}$; $\eta = 0$) around the cavity ($\rho_0 < \rho < \rho_f$) from the surrounding elasto-plastic zone. In the latter zone, the total and effective stresses are calculated according to the procedure outlined above.

In the confined plastic zone, the shear resistance of the soil is entirely mobilized and the state of effective stresses is therefore constant, that is:

$$\begin{aligned} q(\rho) &= q(\rho_f) = q_f \\ p'(\rho) &= q_f/h^*(\gamma_f) = p'(\rho_f) \end{aligned} \tag{14}$$

The equilibrium equation is then used to calculate the total radial stress and the cavity pressure is given by:

$$\sigma_c = \sigma_r(\rho_0) = \sigma_r(\rho_f) + q_f \ln\left(\frac{\rho_f}{\rho_0}\right)^2 \tag{15}$$

The excess pore water pressure in this zone is given by: $\Delta u(\rho_0) = \sigma_c - \sigma_r(\rho_f)$.

In order to compare this numerical solution with the experimental results, the pressuremeter tests at depths of 7 m and 10 m are analyzed. Fig. 11 shows the effective stress paths at the cavity facing calculated according to Eq. 8. The numerical model does not take into account $q_0(\gamma)$ and, in order to model the failure envelope measured in triaxial tests, the following values were used: $c^* = 0$; $\phi^*_{cv} = 48°$. Consequently, the following soil characteristics are taken into account:

$$\frac{G}{\sigma'_0} = 67; \quad \phi^*_{cv} = 48°; \quad \gamma_f = 2\%; \quad \nu = 0.33$$

the (q/p') function is defined by Eq. 6: $\mu = 2.6$.

The comparison between the theoretical solution and the experimental results is shown in Fig. 12 for the test at 7 m and Fig. 13 for the test at 10 m. Good agreement between theoretical solution and experimental results is observed. It can also be observed that the assumption of a hyperbolic $h(\gamma)$ function leads to significant underestimation of the shear resistance of the soil, overestimation of the generated pore water pressures and, consequently, as illustrated in Fig. 11, to an unrealistic effective stress path. The predicted effective stress paths, calculated with $\gamma_f = 2\%$, correspond fairly well to the experimental results.

The experimental results described above are also compared in Figs. 12 and 13 with a "classical" solution assuming that the soil is an elastic/perfectly plastic material with Mohr-Coulomb's failure criterion [4]. It is clearly illustrated that this approach leads to an underestimation of the generated pore water pressure and consequently to overestimation of the soil resistance

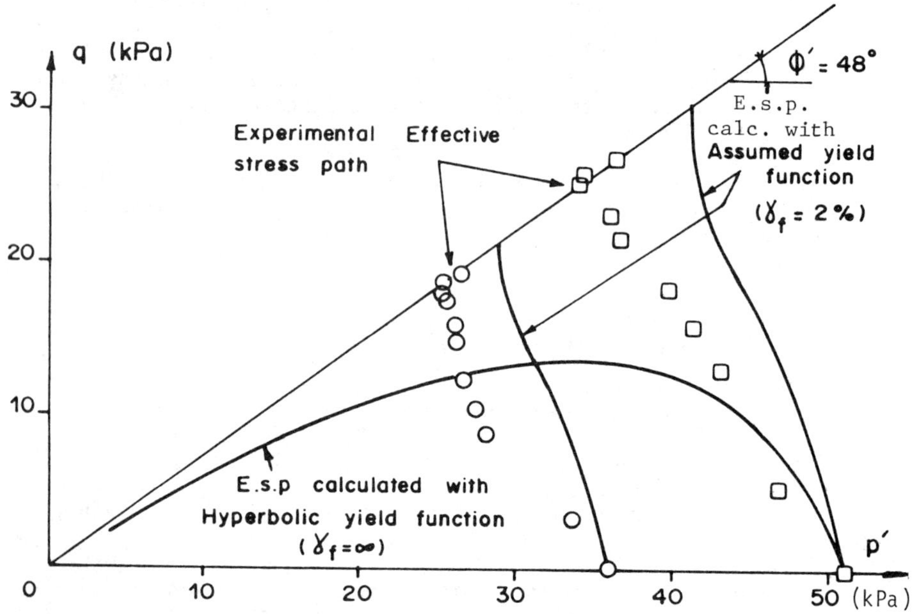

Fig. 11 Calculated and experimental effective stress paths for two typical pressuremeter tests.

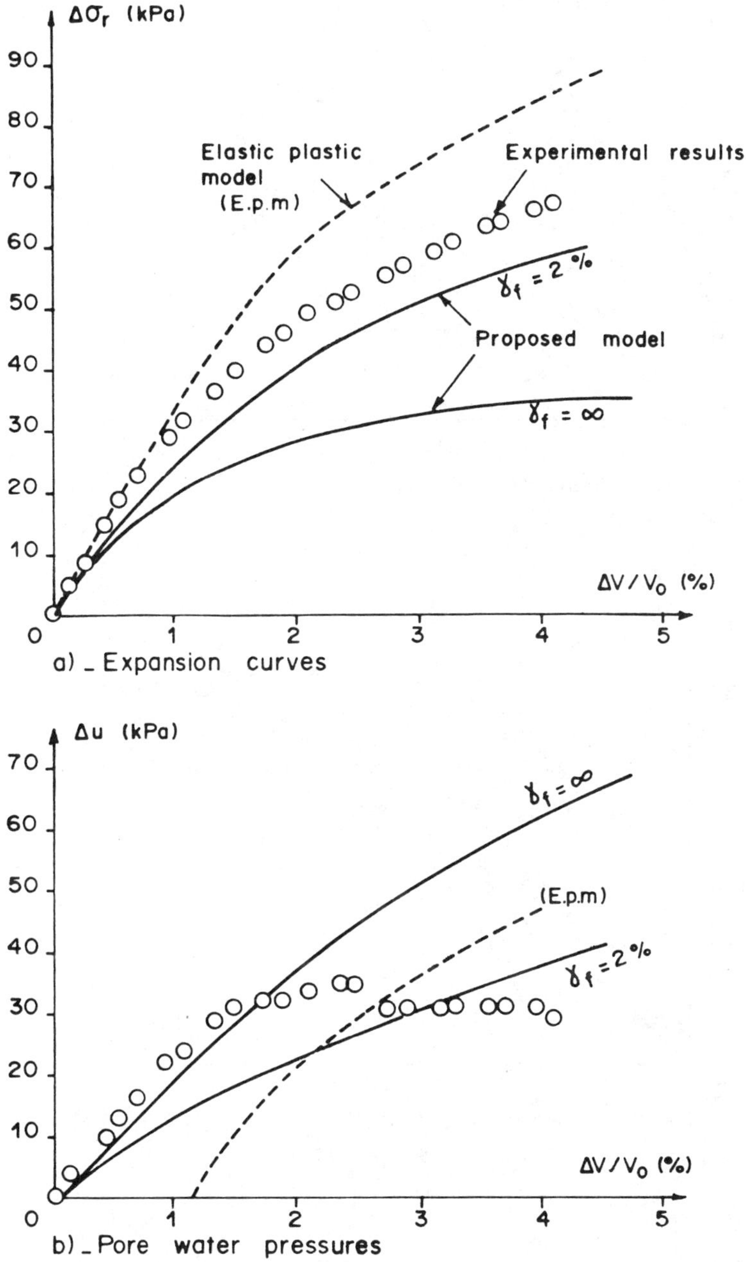

Fig. 12 Comparison of calculated and experimental results from pressuremeter test at 7 m.

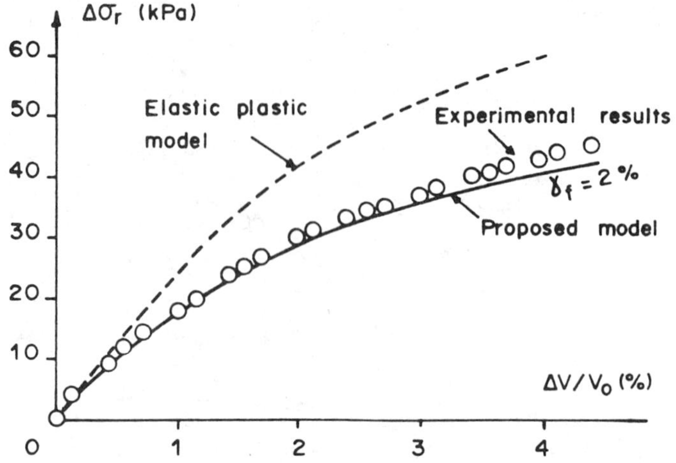

a) _ Expansion curves

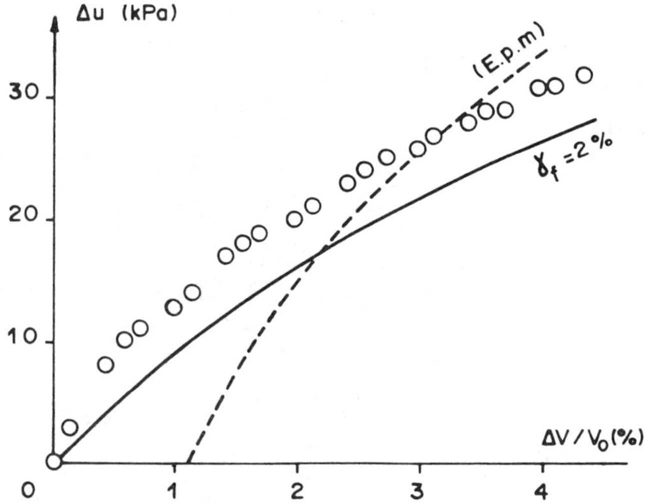

b) _ Pore water pressures

Fig. 13 Comparison of calculated and experimental results from pressuremeter test at 10 m.

to shearing. The plastic compressibility is therefore an essential parameter which significantly affects the soil response to the pressuremeter test. It is interesting to note, that, as the value of the μ modulus increases and the plastic compressibility parameter η decreases, the solution obtained with the proposed rheological model approaches that derived considering the elastic/perfectly plastic model.

CONCLUSIONS

A relatively simple method has been developed to perform an effective stress analysis of quasi-undrained pressuremeter tests in fine-grained, normally consolidated to lightly overconsolidated, contracting soils with measurement of the generated pore water pressures. The proposed approach provides an experimental determination of the properties of the soil, which is assumed to be an isotropic, elasto-plastic, strain-hardening material with a non-associated flow rule. These properties include: the ratio of the shear modulus to the consolidation pressure G/σ_o; the failure shear strain, γ_f; the effective shear strength characteristics c' and ϕ_{cv}; and the modulus μ characterizing the assumed plastic compressibility-stress hyperbolic function $\eta(q/p')$.

The proposed method was used to analyze the results of pressuremeter tests and to obtain the above mentioned soil properties. A numerical analysis to simulate the soil response to the pressuremeter test was performed. The comparison of the predicted soil behavior to the experimental results illustrates that there is relatively good agreement between the predicted and measured values of the generated pore water pressures and expansion curves.

The plastic compressibility and the failure shear strain $_f$ of the soil which have been introduced in this analysis are important parameters which can significantly affect the soil behavior during the test. It has also been shown that the "classical" solution, considering an elastic/perfectly plastic soil model with Mohr-Coulomb's failure criterion, does not take into account the effect of plastic compressibility and therefore leads to underestimation of the generated pore water pressure and to overestimation of the soil resistance to the cavity expansion and to shearing.

This model can be extended to encompass dilating soils such as overconsolidated clays. However, further in-situ verification of the proposed model in different types of soils is still required.

ACKNOWLEDGEMENTS

This research has been jointly sponsored by the French Administration (DAEI, DGRST), the Ecole Nationale des Ponts et Chaussées, FUGRO, and the University of Lille (EUDIL). The authors wish to thank Professor Robinet for the fruitful discussions during this research, and they also thank J.F. Jézéquel for providing the pressuremeter and laboratory test results of the site at Cran. The assistance of D. Barrier, E. Delaure and C. Pearce in the preparation of this paper is greatly appreciated.

REFERENCES

[1] Namy, D. "An Investigation of Certain Aspects of Stress-Strain Relationships for Clay Soils", Phd. Thesis, 1970, Cornell University.

[2] Juran, I., and Bensaid, A., "Etude experimentale en laboratoire du comportement du sol lors d'une expansion cylindrique", Rapport de Recherche presente a la DAEI, May, 1985.

[3] Nova, L., and Wood, D.M., "A Constitutive Model for Sand in Triaxial Compression", International Journal for Numerical and Analytical Methods in Geomechanics, Vol. 3, 1979.

[4] Baguelin, F., Jézéquel, J.F., and Le Méhauté, A.L., "Etude des pressions interstitielles développées lors de l'essai pressiometrique", Proceedings of the VIII ICSMFE, Moscow, 1973.

[5] Baguelin, F., Jézéquel, J.f., and Shields, D.H., "The Pressuremeter and Foundation Engineering", Trans. Tech. Publication, 1978.

[6] Juran, I., Bernardet, A., Schlosser, F., Gambin, M., Fine Soils consolidation by Radial Cyclic Loading", Proceedings of the VIII ECSMFE, Helsinki, 1983.

François J. Baguelin, Roger A. Frank, and Rafic Nahra

A THEORETICAL STUDY OF PORE PRESSURE GENERATION AND DISSIPATION AROUND THE PRESSUREMETER

REFERENCE: Baguelin, F.J., Frank, R.A., and Nahra, R., "A Theoretical Study of Pore Pressure Generation and Dissipation Around the Pressuremeter," The Pressuremeter and Its Marine Applications: Second International Symposium, ASTM STP 950, J.-L. Briaud and J.M.E. Audibert, Eds., American Society for Testing and Materials, 1986.

ABSTRACT: Results of finite element analyses of self-boring pressuremeter tests in theoretical two-phase soils are presented. In particular, pore pressures generated during an initial small strain expansion stage and their dissipation during a subsequent consolidation stage are analysed for different pressuremeter lengths. It is shown that the basic results of the reference expansion test are not affected by partial drainage. Pore pressure measurements during volume holding tests are proven to yield satisfactory values of radial coefficient of consolidation using the elastic plane strain solution of Randolph and Wroth (1979), even for the shortest probes. The self-boring pressuremeter thus provides, for offshore practice, interesting basic soil parameters measured in-situ, besides the limit expansion pressure for direct design of deep or shallow foundations.

KEYWORDS: pressuremeter, finite elements, elasto-plasticity, pore pressure generation, dissipation, consolidation, permeability.

In fine soils, a better understanding of the pressuremeter test as well as a better knowledge of the soil properties can be achieved through the interpretation of the generation and dissipation of pore pressures measured on the probe. The aim is to answer questions such as: do actual pore pressures alter the simple interpretation of the test obtained assuming that the soil is in an undrained condition?

Mr. F.J. Baguelin is Technical Director and Dr. R.A. Frank is Head of the Foundations Section at Laboratoire Central des Ponts et Chaussées (L.C.P.C.), 58, boulevard Lefebvre, 75732 Paris Cédex 15, France; Dr. R. Nahra is a former research student at L.C.P.C., now at Geodia, 16, rue Médéric, 75017 Paris, France.

Or, can information about consolidation parameters of the soil be obtained from pressuremeter surveys?

Results of finite element (FEM) analyses in theoretical saturated two-phase media [1] are presented here. In particular, pore pressures generated during an initial small strain expansion stage and their dissipation during a subsequent holding stage are examined.

This study is relevant only to self-boring pressuremeters, i.e. pressuremeters able to bore their own hole and to keep the soil disturbance to a minimum. As for standard (Ménard) pressuremeters the results of the initial part of the test are somewhat uncertain because of the soil disturbance due to the installation procedure (insertion into a prebore hole; driving or jacking). The application of refined test interpretations such as those involved here (in terms of elementary soil parameters), would lead to very hazardous results.

BACKGROUND

Self-boring pressuremeter tests

Self-boring pressuremeters started to be developed some fifteen years ago in France [2] and in England [3], in order to overcome the drawbacks due to the installation procedure with the Ménard pressuremeter. Their principle is now fairly well known. For offshore practice, the various existing devices as well as information on the operation methods can be found in the proceedings of the first symposium on the Pressuremeter and Its Marine Applications held in Paris, France, in 1982 [4]. Pore pressure measurements during expansion tests or during subsequent consolidation tests are reported, for instance, in [5] [6] and [7].

The numerical analyses described in the present paper primarily concern the reference test with the French self-boring pressuremeter, model 76 (PAF 76) [8]. This test is performed with a rate of volumetric expansion $\Delta\dot{V}/V_o$ = 2% per minute ($\Delta V/V_o = 2\varepsilon_o$ in small strains, where ε_o is the radial displacement of the cell divided by its radius r_o). Probes with measuring cells of length-to-diameter ratio $\ell/2r_o$ = 2 or 4 exist at present.

Numerical method

The finite element code 'Rosalie' of L.C.P.C. including Biot's coupled theory of consolidation, together with an elasto-plastic soil skeleton, is used. The algorithm is based on the classical Galerkin integration scheme for time discretization and on the modified Newton-Raphson or initial stress method for the treatment of the non linear skeleton behaviour [9] [10].

Calculations for pressuremeter probes of length-to-diameter ratios $\ell/2r_o$ of 2, 4, 6 and for an infinite cavity (plane strain) are performed. All the meshes are composed of axisymmetrical isoparametric 8-

noded elements with quadratic interpolation for the three unknowns (radial and vertical displacements and hydraulic head). External radius of the model is 50 r_o. A typical mesh is shown in Fig. 1. The plane strain calculations, needing only one element band, are run in order to vary some soil parameters at low computer costs (permeability k, coefficient K_o, OCR, etc.) and to estimate errors due to the numerical technique by comparing the results with known plane strain analytical solutions.

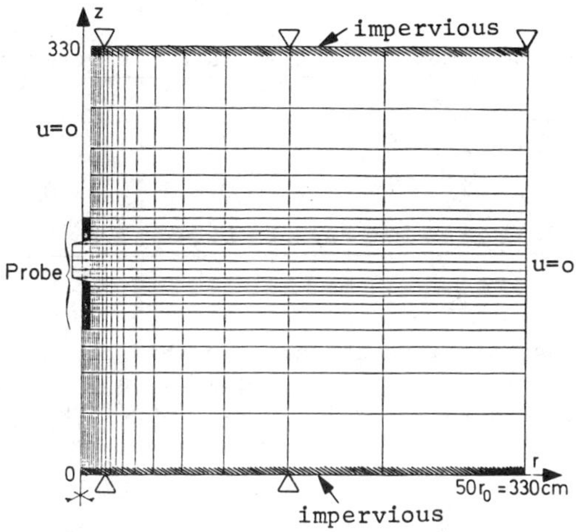

FIG. 1--Finite element mesh for $\ell/2r_o = 2$ and boundary conditions.

The expansions are simulated by imposing a uniform radial displacement over the length ℓ, and a linearly decreasing one over transition zones of one radius above and one radius below this length ℓ. Note that the definition of length-to-diameter ratio does not take these transition zones into account. In reality, the importance of the transition zones depends on the exact technology of the probe and on the soil characteristics, so that the relationship between the relative volume increase $\Delta V/V_o$ and the relative radial displacement ε_o can only be an approximation. The results reported here concern expansions until $2\varepsilon_o = 5\%$ at a rate $2\dot{\varepsilon}_o = 2\%/mn$ and dissipations with this deformation being kept constant (radius holding stage).

Soil model and other data

Two models are selected for the soil skeleton: the elastic perfectly-plastic model with Tresca's failure criterion (c_u is the undrained strength) and the modified Cam-Clay model [11] assuming linear isotropic elasticity inside the work-hardening yield surface. The pore water flow obeys Darcy's law (k is the coefficient of permeability).

Table 1 shows for the 'Tresca' soil, one basic set of parameter values, as for Cam-Clay (CC) two basic sets are used. The first Cam-Clay soil S1 is imaginary and has the same elastic and permeability characteristics as the 'Tresca' soil. The second set corresponds to the soft clay of Cubzac-les-Ponts at about 8 m depth [12]. For all three soils an isotropic state of initial stress is assumed ($K_o = 1$) in the basic calculations, with $\sigma'_{ro} = \sigma'_{\theta o} = \sigma'_{zo} = 100$ kPa for Cam-Clay S1 and 48 kPa for Cam-Clay Cubzac soil. The elastic shear modulus $G = E/2(1+\nu)$ is 5.77 MPa for 'Tresca' and S1 soils and 1.11 MPa for Cubzac.

TABLE 1--Basic values of parameters of selected soils.

Soil	E(MPa)	ν	k(m/s)	λ	κ	M	e_o	OCR	c_u(kPa)
'Tresca'	15	0.30	$2\ 10^{-10}$						50
CC-S1	15	0.30	$2\ 10^{-10}$	0.15	0.012	1.2	0.50	1	36.6
CC-Cubzac	3	0.35	$2.3\ 10^{-9}$	0.52	0.042	1.2	2.28	1	17.6

The basic values of the parameters are then changed or varied according to the problem under consideration. In the presentation of the results below, the changes will be mentioned, when necessary.

EXPANSION STAGE

Influence of permeability on expansion and derived shear curves

A self-boring expansion test yields a whole stress-strain curve, i.e. the expansion curve: pressure in the cell ψ vs relative radial displacement ε_o. Figure 2 gives the results obtained for two coefficients of permeability: $k = 5.10^{-11}$ m/s and $k = 5.10^{-9}$ m/s with the 'Tresca' soil. The expansion curves are practically identical, though the influence of permeability on the pore pressures generated at the probe $u_o(r_o)$ is quite evident. Note that for $k = 5.10^{-11}$ m/s the numerical solution is found to be within 5% of the undrained analytical solution (Eq. 1) obtained by assuming no variation of the mean effective stress ($\Delta p' = 0$) [13]:

$$u_o(r_o) = u_{max} = c_u \ell n(\frac{G}{c_u} \cdot 2\varepsilon_o) \tag{1}$$

One important consequence of the fact that ψ is fairly insensitive to permeability is that the subtangent method [2] which yields the shear curve $\tau(\varepsilon_o)$ and the undrained shear strengh c_u of the soil (Eq. 2) can still be applied, even if the expansion is not perfectly undrained:

$$\tau = \varepsilon_o \frac{d\psi}{d\varepsilon_o} \begin{cases} \nearrow 2G\varepsilon_o & \text{for elasticity} \\ \searrow c_u & \text{for plasticity} \end{cases} \tag{2}$$

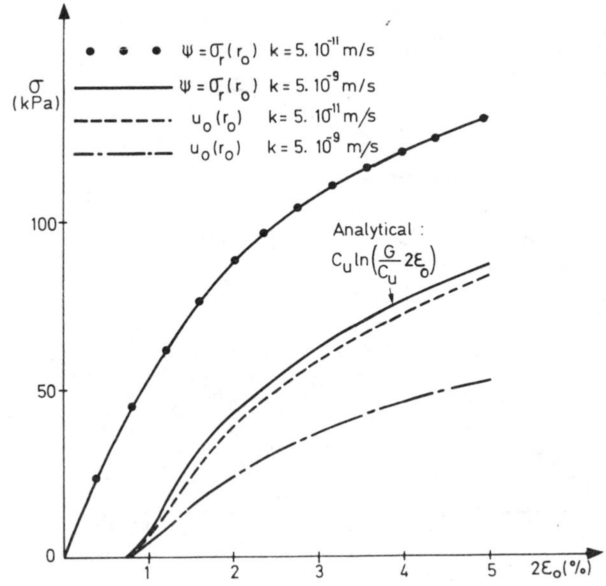

FIG. 2--'Tresca' soil. Expansion and excess pore pressure curves.

Figure 3 is for the Cam-Clay Cubzac soil with three different permeabilities. Again it can be seen that nearly no influence is detected on the total pressure curve $\psi(\varepsilon_0)$. Figure 3 also shows that the numerical results obtained are not accurate enough, - in the case of this work hardening soil - for deriving the shear curve (example shown for $k = 2.3\ 10^{-9}$ m/s).

Influence of length-to-diameter ratio

The expansion curves obtained for the different $\ell/2r_0$ ratios are shown in Fig. 4 for the 'Tresca' soil (together with the derived shear curves) and for the Cubzac soil (for which the derived shear curves, not presented, are again unreliable because of numerical reasons).

In the case of the 'Tresca' soil the overestimation of shear modulus G and undrained strength c_u due to the finite length of the pressuremeter agrees well with previously published analyses (see [14]). Table 2 compares the results obtained, to the plane strain computation ones (G_∞, $c_{u\infty}$).

TABLE 2--Numerical results for the 'Tresca' soil.

$\ell/2r_0$	2	4	6
G/G_∞	1.23	1.075	1.03
$c_u/c_{u\infty}$	1.22	1.11	1.065

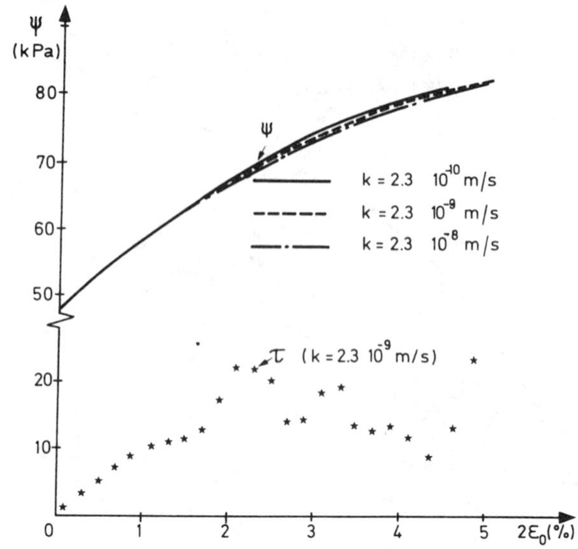

FIG. 3--Cubzac soil. Expansion curves.

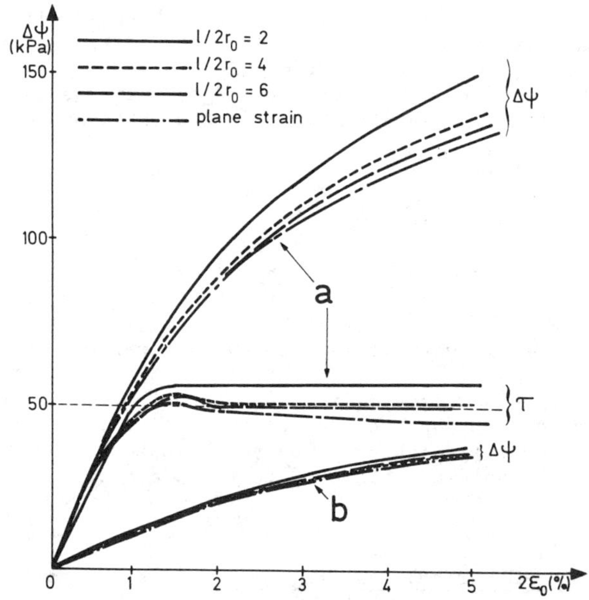

FIG. 4 --Expansion curves for different $\ell/2r_o$ ratios for the 'Tresca' soil (a) and the Cubzac soil (b).

For the Cam-Clay soil (Cubzac) Figure 4 shows that the initial shear modulus G obtained from the expansion curve is nearly the same for all length-to-diameter ratios, as already noticed elsewhere, for a similar work-hardening soil [15].

GENERATION AND DISTRIBUTION OF INITIAL EXCESS PORE PRESSURES

Theoretical solutions

For the elastic perfectly-plastic 'Tresca' soil under plane strain undrained conditions, the assumption of no variation of effective mean stress $\Delta p' = 0$ can be used [13] (or alternatively, the assumption of an associated flow rule, i.e. with no dilation), for determining the initial excess pore pressures generated by the expansion. This yields Eq. 1 and 3 for the plastic zone, while in the elastic zone the pore pressure is zero.

$$\Delta p(r) = u_o(r) = c_u \, \ell n \left(\frac{R_p}{r}\right)^2 \qquad (3)$$

where

R_p = radius of plastic zone (Eq. 4):

$$\left(\frac{R_p}{r_o}\right)^2 = \frac{G}{c_u} \cdot 2\varepsilon_o \qquad (4)$$

Note that this solution also assumes that the deviator stress in the horizontal plane $\sigma'_r - \sigma'_\theta$ is the maximum (σ'_z is always intermediary). In fact, for low values of K_o ($K_o < 1 - c_u/\sigma'_{zo}$) Tresca criterion is first reached, during the expansion, for σ'_z and σ'_θ. During this first stage, the distribution of pore pressures in the plastic zone is given by Eq. 5 [1]:

$$u_o(r) = \left(\frac{K_o-1}{4}\sigma'_{zo} + \frac{c_u}{2}\right)\left[1+\ell n\left(\frac{r_f}{r}\right)^2\right] - \frac{1}{12}\left(\frac{r_o}{r}\right)^2\left[(1-K_o)\sigma'_{zo} + 2c_u\right] \qquad (5)$$

where

r_f = radius of the first plastic zone (Eq. 6):

$$\left(\frac{r_f}{r_o}\right)^2 = \frac{G}{(K_o-1)\sigma'_{zo} + 2c_u} \cdot 2\varepsilon_o \qquad (6)$$

As soon as $\sigma'_r - \sigma'_\theta = 2c_u$ is reached, a second yielding stage develops near the probe. The distribution of the excess pore pressure generated is then given by Eq. 7:

$$u_o(r) = u_o(r_p) + c_u \ \ell n \left(\frac{r_p}{r}\right)^2 \tag{7}$$

where

r_p = radius of the second plastic zone created (Eq. 8):

$$\left(\frac{r_p}{r_o}\right)^2 = \frac{3G}{(1-K_o)\sigma'_{zo}+2c_u} .2\varepsilon_o \tag{8}$$

The corresponding stress distributions are detailed in reference [1]. For the numerical examples selected, it is mainly the stress σ'_z which varies in a manner very different from the solution assuming that it is always intermediary.

For the Cam-Clay model, the value of the mean effective stress varies during the undrained expansion and eventually reaches, near the probe, the value p'_f which corresponds to failure at critical state. Randolph et al. [16], for the problem of driven piles in clay, (i.e. infinite strain at the cavity wall) give the following effective stress and excess pore pressure distributions in the failed soil, near the pile (Eq. 9):

$$\left.\begin{aligned} \sigma'_r(r) &= \left(\frac{\sqrt{3}}{M} + 1\right) c_u \\ \sigma'_z(r) &= \frac{\sqrt{3}}{M} c_u = p'_f \\ \sigma'_\theta(r) &= \left(\frac{\sqrt{3}}{M} - 1\right) c_u \\ \Delta p - \Delta p' = u_o(r) &= c_u \ \ell n \left(\frac{R_p}{r}\right)^2 + (p'_i - p'_f) \end{aligned}\right\} \tag{9}$$

where

$$\left(\frac{R_p}{r_o}\right)^2 = \frac{G}{c_u} \quad \text{with } G \text{, appropriate secant modulus,}$$

and p'_i = initial mean effective stress.

Note that the distribution of the increase of mean total stress, and hence of excess pore pressure for the soil in critical state, is also logarithmic. This distribution has been obtained by finite element computations, where the driving of a pile of radius r_o, was modelled by doubling a cavity of initial radius $a_o = r_o/\sqrt{3}$ (see also Carter et al. [17]).

Numerical results

Figure 5 shows typical distributions obtained for $2\varepsilon_o$ = 5% expansion in the basic 'Tresca' soil. The difference between the theoretical fully undrained u_{max} (see Eq. 1) and the computed excess pore pressure in a pervious soil $u_o(r_o)$ increases with the permeability of the soil.

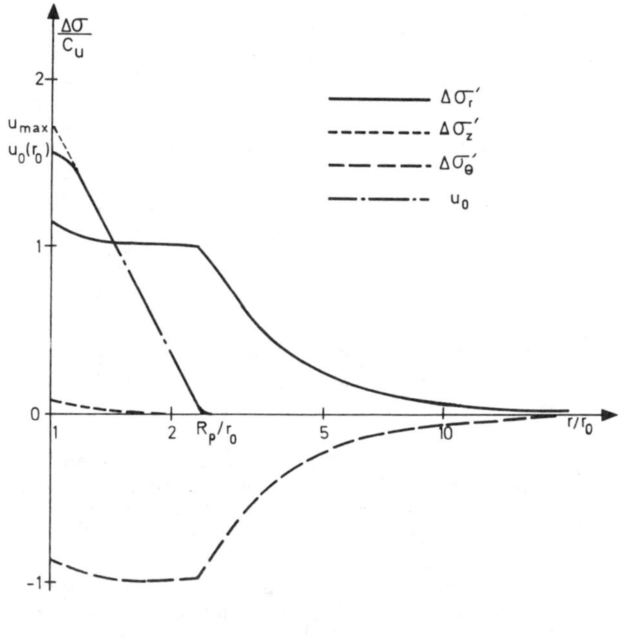

FIG. 5--'Tresca' soil. Effective stresses and pore pressure variations for $2\varepsilon_o$ = 5% expansion.

The influence of the coefficient of permeability is shown on Fig. 6 (the undrained value, given by Eq. 1, is $u_{max}/c_u = 1.75$).

The calculations performed for the Cam-Clay basic soil data yield, for the stresses, results in good agreement with the distributions of Eq. 9. The results for the Cam-Clay S1 soil, shown in Fig. 7, are quite representative, in this respect. σ'_r is usually fairly constant in the soil at critical state with a slight increase very near the cavity. σ'_θ is, in general, slightly decreasing with r. The linear decrease of Δp (and $\Delta\sigma_r$) with the logarithm of r (the slope being $2c_u$), is also well verified with a radius R_p given by Eq. 4. However, the logarithmic shape of the excess pore pressures generated by the expansion is altered in a great area of the failed soil. For pressuremeter

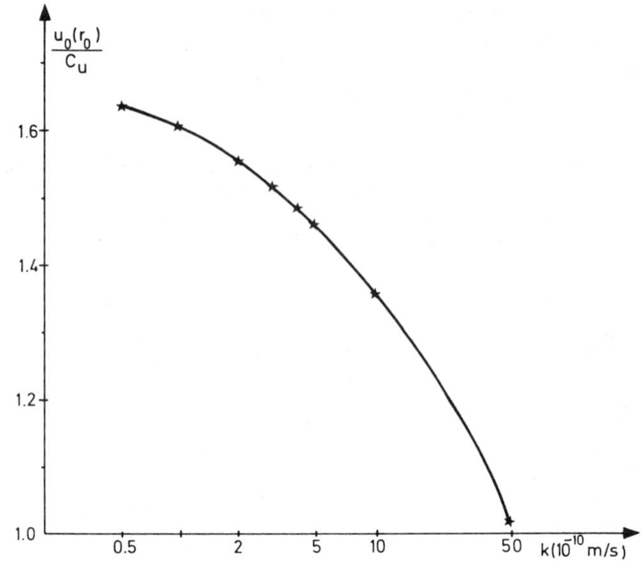

FIG. 6--'Tresca' soil. Influence of the coefficient of permeability on the excess pore pressure at the probe ($G.2\varepsilon_o/c_u = 5.77$).

tests with low volumetric strain like here (5%), the zone of soil at critical state is narrow and the final effective mean stress p'_{fin} starts deviating from p'_f around R_p in order to quickly join p'_i. Furthermore, the pore pressure does not vanish at R_p like for the 'Tresca' soil, but at a certain radius R_u. R_u is larger or smaller than R_p depending on the sign of $p'_i - p'_f$ (see Eq. 9). Computations with varying values of OCR show that $p'_i - p'_f$ becomes negative when OCR is greater than 2, in agreement with Randolph et al. [16]. (It can be shown that for κ very small with regard to λ, the modified Cam-Clay model yields $p'_f = p'_i.\mathrm{OCR}/2$).

Various calculations performed with $K_o < 1$ show a certain discrepancy with the logarithmic distribution of Δp given by Eq. 9. A better agreement is obtained with the corresponding logarithmic distribution for the variation of total radial stress $\Delta\sigma_r$, given by Eq. 10:

$$\Delta\sigma_r(r) = c_u \left(1 + 2c_u \ \ell n \ \frac{R_p}{r}\right) \tag{10}$$

By assuming that σ'_z is the mean of σ'_r and σ'_θ (classical postulate of the plasticity theory; it can be shown to be equivalent to $d\varepsilon_z^p = 0$ for the modified Cam-Clay model when the critical state is reached) the distribution given by Eq. 11 is obtained for Δp in

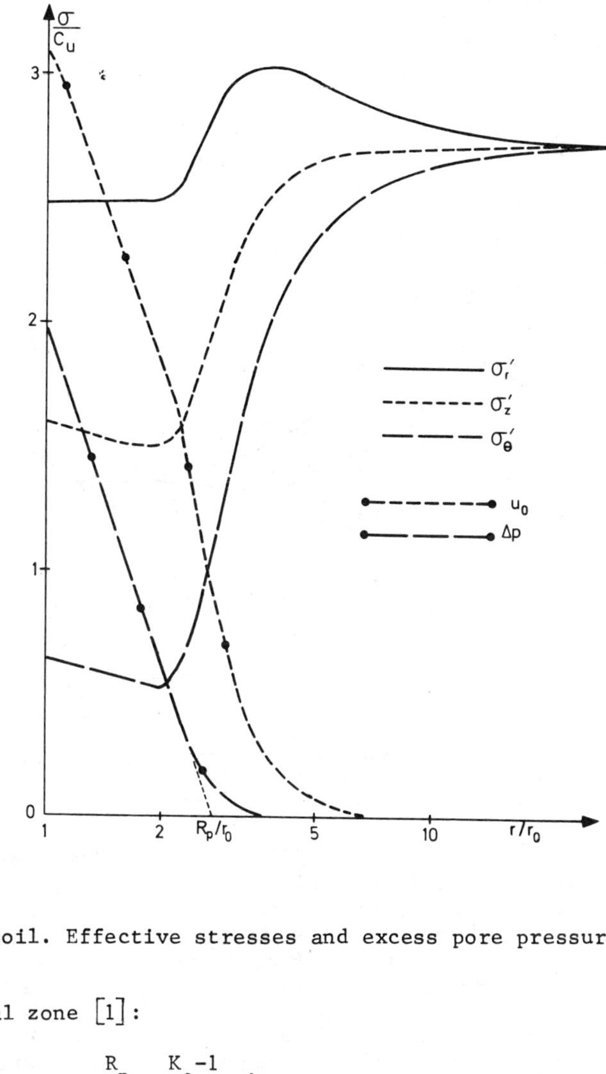

FIG. 7--S1 soil. Effective stresses and excess pore pressure for $2\varepsilon_o = 5\%$.

the critical zone [1]:

$$\Delta p = 2c_u \ \ell n \frac{R_p}{r} + \frac{K_o - 1}{3} \sigma'_{zo} \tag{11}$$

The excess pore pressures generated, for instance, at the wall of the cavity, follow Eq. 12:

$$u_{max} = c_u \ \ell n(\frac{G}{c_u} \cdot 2\varepsilon_o) + \frac{K_o - 1}{3} \sigma'_{zo} + p'_i - p'_f \tag{12}$$

<u>The influence of the finite length</u> of the pressuremeter on the distributions after the initial expansion stage, can be summarized by Fig. 8 and Table 3 for the elastic perfectly-plastic 'Tresca' soil.

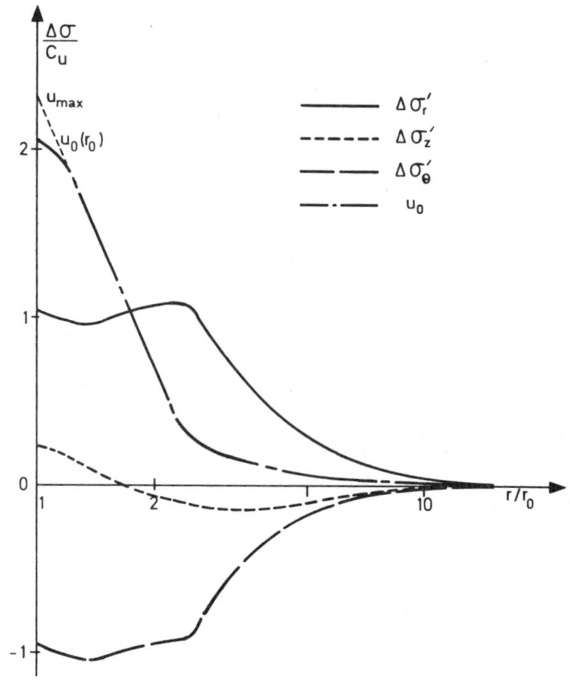

FIG. 8--'Tresca' soil. Increase in effective stresses and pore pressure at mid-probe of $\ell/2r_o = 2$ for $2\varepsilon_o = 5\%$.

Figure 8, for a probe of ratio $\ell/2r_o = 2$, shows by comparison with Fig. 5 (for an infinite cavity) a slight increase of $\Delta\sigma_z'$ near the wall, together with negative values in the soil mass. $\Delta\sigma_r'$ and $\Delta\sigma_\theta'$ show very small changes. Pore pressures increase when $\ell/2r_o$ decreases, though still exhibiting a logarithmic distribution near the probe. Excess pore pressures (positive) appear in the elastic zone where $\Delta p'$ increases. For the quarter depth of the probe the results are very near those for the center of the probe. For the extremity of the probe they tend to be fairly independent of the length-to-diameter ratio, showing great differences when compared to the infinite cavity case and resembling more those of a spherical cavity expansion. Table 3 compares, at the three locations, the excess pore pressures on the probe $u_o(r_o)/c_u$ with the infinite length (plane strain) result.

For the modified Cam-Clay Cubzac soil, the influence of the length-to diameter ratio on the stress distribution is less important than in the case of the elastic perfectly-plastic soil. Also, when moving from the center to the extremity of the cell, variations are smaller. The excess pore pressures generated are higher than for the plane strain situation. As in the case of the elastic perfectly-plastic soil, two reasons can be found: excess pore pressures already exist inside the

TABLE 3--'Tresca' soil. Excess pore pressures generated on the probe/c_u.

$\ell/2r_o$	2	4	6	Infinite (Plane Strain)
Mid-probe	2.04	1.75	1.64	1.55
Quarter-probe	2.11	1.84	1.72	|
Probe extremity	2.88	2.80	2.76	|

elastic zone and plastic yield is intermediary between cylindrical and spherical cavity expansion. Because of the narrowness of the critical state zone in the case of this soil ($\sqrt{R_p/r_o} = 1.78$ for $2\varepsilon_o = 5\%$), the distribution of excess pore pressures is not logarithmic in r, even in the plane strain case.

DISSIPATION OF EXCESS PORE PRESSURES

General shape of dissipation

Curves of Fig. 9 show, for the wall of the cavity, typical excess pore pressure dissipations with time. The modified Cam-Clay soil is soil S1 but with OCR = 2 and the same undrained strength as the 'Tresca' soil, so that initial excess pore pressure distributions match for both soils. It can be seen that pore pressure dissipation during consolidation is relatively unaffected by the soil model, in agreement with the results of Carter et al. [17] for pore pressures obtained by doubling the volume of a cavity.

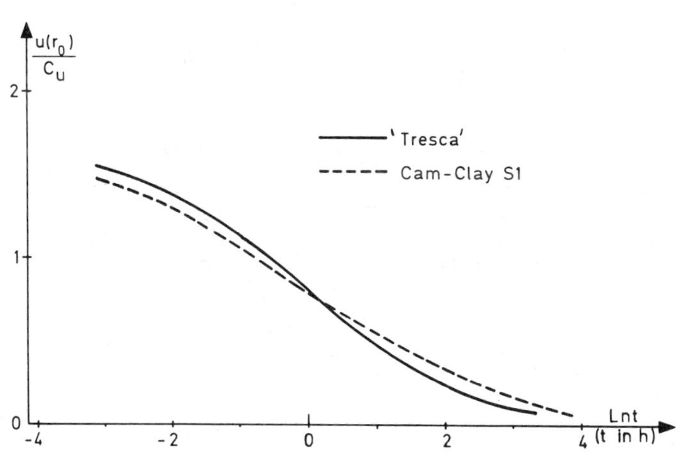

FIG. 9--Dissipations of excess pore pressure with time. (Initial expansion $2\varepsilon_o = 5\%$).

The dissipation curves of excess pore pressures for the various length-to-diameter ratios and for the basic 'Tresca' soil and the basic modified Cam-Clay Cubzac soil, when non-dimensionalized by the initial excess pore pressure, show that the shape of the dissipation does not depend significantly on the length of the pressuremeter. Numerical results obtained at the quarter probe location show a very good agreement with the centre of the probe, even for $\ell/2r_o = 2$.

Determination of the radial coefficient of consolidation

One of the main aims of this study is to know if a radial coefficient of consolidation can be derived from current self-boring pressuremeter tests, with a certain degree of confidence, and what would be the best procedure.

The only closed form solution which exists for this problem is the analytical solution of Randolph and Wroth [13] for the consolidation around a driven pile. The main assumptions are: plane strain; the soil skeleton is elastic, and the initial excess pore pressure distribution is logarithmic with a slope of 2 c_u (Eq. 3 for the Tresca soil in undrained condition). It is also suggested that this solution is suitable for pressuremeters inflated to a fixed radius. Carter et al. [17] find that the dissipation in elasto-plastic soils is very similar to the elastic one.

Figure 10 is the plot of the analytical results for 50% of consolidation ($T_{50} = c_{th} \cdot t_{50} / r_o^2$) versus u_{max}/c_u, u_{max} being the theoretical maximum excess pore pressure at the wall of the cavity (Eq. 1) and c_{th} the elastic coefficient of consolidation (Eq. 13) [16]:

$$c_{th} = \frac{Ek}{\gamma_w} \frac{1+\nu}{(1+\nu)(1-2\nu)} \tag{13}$$

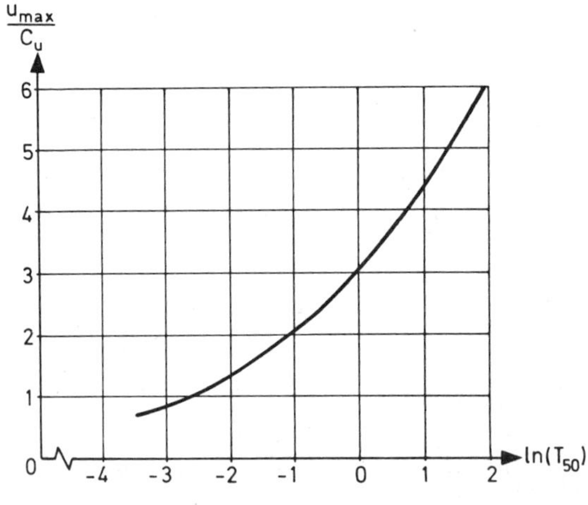

FIG. 10--Time factor for 50% of dissipation [16].

By measuring the pore pressures on the pressuremeter probe, the maximum excess u_{max} can be assessed as well as the time t_{50} required for 50% decay ($u_{max}/2$). c_u being deduced from the expansion curve, the chart of Fig. 10 can thus yield the horizontal coefficient of consolidation around the pressuremeter [6] [7]. The authors of the solution note that the coefficient of consolidation obtained is relevant to the elastic compressibility of the soil, as during consolidation most of the soil is unloading.

From the dissipation curves of the present calculations, depending on the assumption about the origin of time scale and about the value of pore pressure u_{max}, three different coefficients of consolidation are derived making use of this chart:

-u_{max} is a theoretical value (see below) and the origin of times for t_{50} determination is the end of the expansion; the coefficient is noted c_1.

-u_{max} is the same as for c_1 but the time origin is the beginning of the expansion; this coefficient is c_2.

- the actual computed pore pressure at the cavity wall $u_o(r_o)$ at the end of expansion is considered for u_{max} and the origin is at the end of expansion; this coefficient is c_3.

The theoretical value for u_{max} is taken, for plane strain, from Eq. 1 (Tresca) or Eq. 9 (Cam-Clay). For finite lengths, it is obtained by extrapolation at the cavity wall of the linear ℓnr law for the 'Tresca' soil (see Fig. 8). For the Cam-Clay Cubzac soil, an appropriate value for theoretical u_{max} is expected to be yielded by using 1.14 $u_o(r_o)$ for all lengths and the three locations on the probe [1]. Note that coefficient c_3 is the coefficient which is determined in practice.

The three different coefficients of consolidation obtained are checked, on Fig. 11 ('Tresca' soil) and 12 (Cam-Clay Cubzac soil), against the theoretical coefficient c_{th} of Eq. 13. It can be seen that the finite length of the pressuremeter increases the coefficients of consolidation as derived from Randolph's and Wroth's plane strain elastic solution (chart of Fig. 10). It is interesting to note that this counterbalances, to some extent, the intrinsic underestimation involved by the definition of c_3 which uses $u_o(r_o)$. (The slight discrepancy on Fig. 12 for Cubzac soil and $\ell/2r_o = 2$ is due to numerical inaccuracies for achieving satisfactory convergence). On the whole, the results are excellent considering the partial drainage during expansion and, in some cases, the very approximate logarithmic distributions for small strain expansions. (These results are for the center of the probe. At the quarter of the probe the results are, in substance, the same).

The main conclusion which can be drawn from the numerical simulations, as expressed by Fig. 11 and 12, is that a satisfactory estimate of the elastic coefficient of consolidation can be obtained for elasto-plastic soils by pressuremeters of length-to-diameter ratios, say, 2 or 4 and by using the analytical solution of Randolph and Wroth [13]. The result obtained (at the center or at the quarter of the probe) is the in-situ radial coefficient of consolidation for a drainage path of a few decimetres.

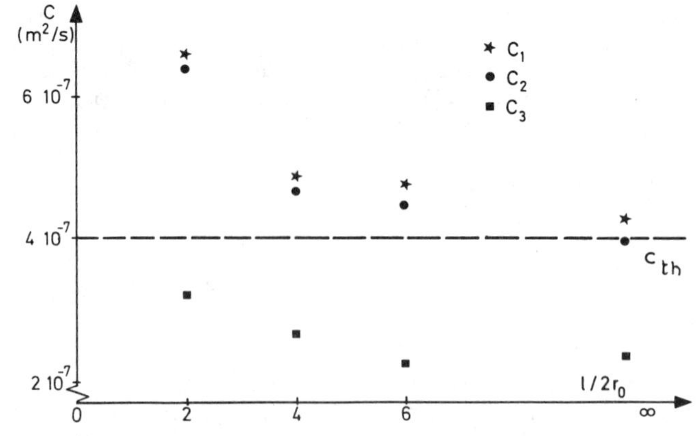

FIG. 11--'Tresca' soil. Coefficients c_1, c_2, c_3 and c_{th} for various length-to-diameter ratios and $2\varepsilon_o$ = 5%.

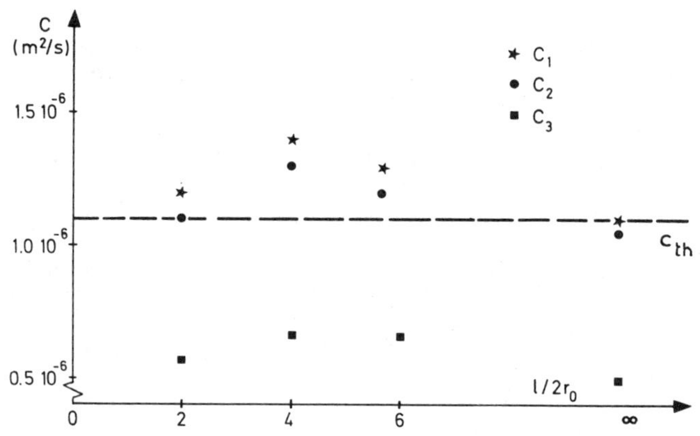

FIG. 12--Cam-Clay Cubzac. Coefficients c_1, c_2, c_3 and c_{th} for various length-to-diameter ratios and $2\varepsilon_o$ = 5%.

CONCLUSION

The measurement of pore pressures grants the self-boring pressuremeter with a new dimension. Not only can information on drainage conditions during the test be obtained, but also an assessment of consolidation parameters.

The numerical results obtained here show that the basic results (shear modulus G, undrained strength c_u) of the reference test are not affected by the partial drainage occuring during expansion in fine soils, for the range of permeabilities considered ($k \approx 10^{-11}$-10^{-8} m/s).

As for pore pressure measurements, after small strain expansion, the volume of the probe being fixed, it is found that they can yield reliable values of the in-situ radial coefficient of consolidation. The plane strain elastic solution of Randolph and Wroth can be used, even with the shortest probes.

The following recommended procedure is thus suggested by these results:

1. Expand the pressuremeter cell according to the reference test (volumetric rate of 2% per minute) until 5% of volumetric expansion.
2. Hold the volume constant and measure the pore pressures on the probe. In particular, the initial pore pressure $u_o(r_o)$ is noted.
3. Note the time t_{50} for which $u_o(r_o)/2$ is obtained.
4. Use chart of Fig. 10 to get c_3 which is an estimate for c_{th}.

For offshore practice, the advantage of determining in-situ such basic soil parameters is evident. The main fields of application concern slope stability on compressible layers or bearing capacity and settlement of driven piles. It can be noted that obtaining a coefficient of consolidation from pressuremeter tests is of particular interest when applying effective stress methods which, precisely, model pile driving as a cylindrical cavity expansion. The self-boring pressuremeter also yields the limit pressure of the soil, for which simple correlations can be used to obtain, directly, the bearing capacity of foundations whether shallow or deep, as shown in Ref. [14], for instance.

REFERENCES

[1] Nahra, R. and Frank, R., "Contributions Numériques et Analytiques à l'Etude de la Consolidation Autour du Pressiomètre," Report F.A.E.R. 1.05.10.4, Laboratoire Central des Ponts et Chaussées, Paris, 1985.

[2] Baguelin, F., Jézéquel, J.F., Lemée, E. and Le Méhauté, A., "Expansion of Cylindrical Probes in Cohesive Soils," Journal of the Soil Mechanics and Foundations Division, ASCE, Vol. 98, No. SM 11, November 1972, pp. 1129-1142.

[3] Wroth, C.P. and Hughes, J.M.O., "An Instrument for the In-Situ Measurement of the Properties of Soft Clays," Proceedings of the Eighth International Conference on Soil Mechanics and Foundation Engineering, Moscow, 1973, Vol. 1.2, 1/75, pp. 487-494.

[4] I.F.P. and L.C.P.C., Eds., Symposium on the Pressuremeter and Its Marine Applications, Editions Technip, Paris, 1982.

[5] Baguelin, F., Jézéquel, J.F. and Le Méhauté, A., "Etude des Pressions Interstitielles Développées Lors de l'Essai Pressiométrique," Proceedings of the Eighth International Conference on Soil Mechanics and Foundation Engineering, Moscow, 1973, Vol. 1.1, 1/3, pp. 19-24.

[6] Clarke, B.G., Carter, J.P. and Wroth, C.P., "In-Situ Determination of the Consolidation Characteristics of Saturated Clays," in Design Parameters in Geotechnical Engineering (VII ECSMFE 1979 Brighton), British Geotechnical Society, London, 1979, Vol. 2, d5, pp. 207-211.

[7] McKinlay, D.G. and Shwaik, R., "Pressuremeter Measurement of Consolidation Rate in a Glacial Till," Proceedings of the International Symposium on Soil and Rock Investigations by In-Situ Testing, Paris, 1983, Vol. II, pp. 341-346.

[8] Jézéquel, J.F., "The Self-Boring Pressuremeter," Symposium on the Pressuremeter and Its Marine Applications, Editions Technip, Paris, 1982, pp. 111-126.

[9] Magnan, J.P., Belkeziz, A., Humbert, P. and Mouratidis, A., "Finite Element Analysis of Soil Consolidation With Special Reference to the Case of Strain Hardening Elastoplastic Stress-Strain Models," Proceedings of the Fourth International Conference on Numerical Methods in Geomechanics, Edmonton, 1982, Vol. 1, pp. 327-336.

[10] Zawisza, W., "Consolidation Elastoplastique des Sols. Application de la Méthode des Eléments Finis," Report F.A.E.R. 1.06.54.2, Laboratoire Central des Ponts et Chaussées, Paris, 1982.

[11] Roscoe, K.H. and Burland, J.B., "On the Generalized Stress-Strain Behaviour of 'wet' clay," in Engineering Plasticity, Cambridge University Press, 1968, pp. 535-609.

[12] Belkeziz, A. and Magnan, J.P., "Analyse Numérique de la Consolidation Bidimensionnelle des Sols Elastoplastiques. Traitement par la Méthode des Eléments Finis et Application au Remblai Expérimental B de Cubzac-les-Ponts," Rapport de Recherche LPC No. 115, Laboratoire Central des Ponts et Chaussées, Paris, Juillet 1982.

[13] Randolph, M.F. and Wroth, C.P., "An Analytical Solution for the Consolidation Around a Driven Pile," International Journal for Numerical and Analytical Methods in Geomechanics, Vol. 3, No. 3, July-September 1979, pp. 217-229.

[14] Baguelin, F., Jézéquel, J.F. and Shields, D.H., The Pressuremeter and Foundation Engineering, Trans Tech Publications, Clausthal, FRG, 1978.

[15] Borsetto, M., Imperato, L., Nova, R. and Peano, A., "Effects of Pressuremeters of Finite Length in Soft Clay," Proceedings of the International Symposium on Soil and Rock Investigations by In-Situ Testing, Paris, 1983, Vol. II, pp. 211-215.

[16] Randolph, M.F., Carter, J.P. and Wroth, C.P., "Driven Piles in Clay - The Effects of Installation and Subsequent Consolidation," Géotechnique, Vol. 29, No. 4, December 1979, pp. 361-393.

[17] Carter, J.P., Randolph, M.F. and Wroth, C.P., "Stress and Pore Pressure Changes in Clay During and After the Expansion of a Cylindrical Cavity," International Journal for Numerical and Analytical Methods in Geomechanics, Vol. 3, No. 4, October-December 1979, pp. 305-322.

Joseph L. Kauschinger

RELATING UNDRAINED TRIAXIAL AND PRESSUREMETER STRESS-STRAIN CHARACTERISTICS BY USING MULTI-YIELD SURFACE MODEL PARAMETERS

REFERENCE: Kauschinger J.L., "Relating Undrained Triaxial and Pressuremeter Stress-Strain Characteristics By Using Multi-Yield Surface Model Parameters," The Pressuremeter and Its Marine Applications: Second International Symposium, ASTM STP 950, J.L. Briaud, and J.M.E. Audibert, Eds., American Society for Testing and Materials, 1986.

ABSTRACT: Multi-yield surface model parameters, which define the stiffness (H_i), size ($k^{(i)}$), and location ($\alpha_i^{(i)}$) of a nest of yield surfaces were fitted to triaxial compression and extension data for nine fine grained soils [1,2]. Each set of model parameters were used in a computer program to calculate the total stresses ($\sigma_{rr}, \sigma_{zz}, \sigma_{\theta\theta}$) and strains ($\varepsilon_{\theta\theta}$, ε_{rr}) for an idealized load/unload pressuremeter stress path. The pressuremeter stress and strain data for all nine soils were utilized as the basis for developing an empirical procedure which could be used to extract multi-yield surface model parameters from shear stress-strain data, ($\sigma_{rr} - \sigma_{\theta\theta}$) versus ε_{rr}. The empirical procedure was used to extract values of H_i, $k^{(i)}$, and $\alpha_i^{(i)}$ for each soil from the corresponding pressuremeter stress-strain data. The nine sets of model parameters fitted to ($\sigma_{rr} - \sigma_{\theta\theta}$) versus ε_{rr} data where specified as input for a computer program which were then used to calculate the corresponding load/unload triaxial compression and extension stress-strain data, ($\sigma_{yy} - \sigma_{xx}$) versus ε_{yy}.

KEYWORDS: constitutive theory, multi-yield surface model, model application

INTRODUCTION

The accuracy of finite element stress analyses are directly linked to how well the constitutive model used portrays soil's stress dependent behavior. In turn, stress-strain responses predicted using even an "appropriate" constitutive model can be greatly influenced by the test data used to fit the model's parameters. Many constitutive theories fit model parameters to triaxial test

Dr. Kauschinger is an Assistant Professor in Civil Engineering at Tufts University, Medford, Massachusetts 02155.

data which are often obtained using specimens with unknown degrees of disturbance.

Results from research conducted by Kauschinger [1], indicate that the multi-yield surface model proposed by Prevost [3,4] is quite capable of simulating the undrained, anisotropic, stress path dependent behavior of fine grained soils which do not have significant structural features such as varves or fissures (total of 11 soils in data base). Presently, the model parameters utilized in Prevost's Model are usually fitted to undrained triaxial compression and extension stress-strain data. Relative to conventional sampling and triaxial testing techniques, pressuremeter testing usually cause less disturbance to the soil structure. Therefore, in order to take advantage of the good predictive capabilites of the Prevost Model an empirical procedure was developed which can be used to fit multi-yield surface model parameters to values of stress-strain obtained from pressuremeter data.

FORMULATION OF PREVOST'S TOTAL STRESS MODEL

In the following section are presented only the essential equations needed to define the Prevost Model. A more complete description of the model has been previously presented by Prevost [3,4].

The formulation of the Prevost Model falls within the general framework of classical stress space plasticity. This approach is based upon defining functional relationships which define yielding, plastic flow, and hardening of a yield surface which is undergoing active plasticity. Instead of developing his constitutive theory for a single yield surface Prevost has selected a multi-yield surface model, which is often referred to as a "nested" yield surface theory or a sub-layer theory [5]. The constitutive relationship is based upon the assumption that the associative normality rule of plasticity is valid. It is further assumed that undrained yielding of saturated clays are only a function of the deviatoric stresses, $S_{ij} = \sigma_{ij} - 1/3\sigma_{kk}(\delta_{ij})$. Therefore, the yield function, $f^{(m)}$, which defines the stress state required to produce irrecoverable (plastic) strains is taken in the total stress version of the Prevost Model as follows:

$$f^{(m)} = 3/2 \left[(S_{ij} - \alpha_{ij}^{(m)})\ (S_{ij} - \alpha_{ij}^{(m)})\right] - [k^{(m)}]^2 = 0 \qquad (1)$$

When soils experience load/unload cycles of stress, elastic and plastic deformations usually occur prior to a complete stress reversal. Therefore, to model the reverse plastic flow in a more accurate manner, Prevost [3,4] selected a kinematic hardening rule for use in his model, in particular the Mroz Kinematic Hardening Rule

[6]. Although a combined isotropic/kinematic hardening rule can be used in the Prevost Model only kinematic hardening was considered during this research because only a simple loading and unloading of the soil was simulated. Therefore, each yield surface always maintained its initial size, $dk^{(m)} = 0$, while the hardening (incremental change, $d\alpha_{ij}^{(m)}$) of the model parameter which controls the location of the present active yield surface in the stress space, $\alpha_{ij}^{(m)}$, was calculated as follows:

$$d\alpha_{ij}^{(m)} = d\mu \ \ \mu_{ij} \tag{2}$$

In Eqn 2, $d\mu$ controls the scalar amount of hardening, while μ_{ij} dictates the movement of the present active yield surface through the stress space. The functional form of the equations which define μ_{ij} and $d\mu$ are given in several papers by Prevost [4,7].

The constitutive law used in the Prevost Total Stress Model is written in terms of deviatoric components and may be used to calculate the total stress response of saturated soils which deform in an incompressible manner ($d\varepsilon_{kk} = 0$). The constitutive law was presented as follows [3,4]:

$$dS_{ij} = 2G \ \ d\varepsilon_{ij} + \frac{3}{2} (H_m - 2G) \left[\frac{(S_{ij} - \alpha_{ij}^{(m)}) \ (S_{kl} - \alpha_{kl}^{(m)})}{[k^{(m)}]^2} \right] d\varepsilon_{kl} \tag{3}$$

where the deviatoric stresses are calculated as above, and the total strains ($d\varepsilon_{ij}$) in Eqn. 3 are equal to the deviatoric strains (de_{ij}) for incompressible soils. The elastic shear modulus, G, and elasto-plastic shear modulus, H_m, are related to the plastic shear modulus, H_m', as follows:

$$\frac{1}{H_m'} = \frac{1}{H_m} - \frac{1}{2G} \tag{4}$$

The yield function, Eqn. 1, can be rewritten in a form which will allow for a clearer examination of the stress path for both triaxial and pressuremeter tests [4]. The yield surfaces for the pressuremeter stress path plot as circles in the X-Y stress sub-space, given by Eqn. 5 and 6. Therefore, the model parameters $\alpha_i^{(i)}$ and $k^{(i)}$ can be respectively interpreted as the center location and radii of the ith yield surface circle. The triaxial compression and extension stress paths are confined totally along the X axis, where the X - Y coordinate system is defined by:

$$X = \sigma_{11} - 1/2\ (\sigma_{22} + \sigma_{33}) \tag{5}$$

$$Y = \frac{\sqrt{3}}{2}\ (\sigma_{33} - \sigma_{22}) \tag{6}$$

Since we are usually dealing with anisotropic soil, defining a fixed coordinate system becomes very important. Therefore, for the triaxial test the direction 11 is taken as the the vertical direction ,yy, and the two lateral directions, 22 and 33 as the xx and zz directions, respectively. For the pressuremeter stress path, the direction 11 is defined as the vertical direction, zz, while the 22 and 33 directions respectively correspond to the tangential and radial directions, $\theta\theta$ and rr, respectively.

Implementation of the Prevost Model

The total stress version of the Prevost Model presented in the above was implemented in a computer program called IMPMOD [1], which is an acronym for the Iwan - Mroz - Prevost Model. The six sets of equations represented by Eqn. 3 were rearranged so that the input for IMPMOD could be in terms of incremental changes in the total stress ($d\sigma_{ij}$) and/or strain ($d\varepsilon_{ij}$) components. Reforming the constitutive equations for this purpose requires the user to always specify one of the normal stress components as input. The output from IMPMOD was also in terms of total stress components, which were then used to calculate stress differences. The stress- strain calculations performed using program IMPMOD correspond to examining the equilibrium condition at a single point.

DATA BASE USED TO DEVELOP PRESSUREMETER FITTING PROCEDURE

Laboratory data were gathered from the literature for the eleven fine grained soils shown on Table 1 . Six soils were reconstituted in the laboratory from clay slurries, (Boston Blue Clay, Edgar Kaolinite, Gleason Clay, Grundite Clay, Kaolinite Hydrite UF, and Santa Barbara Silt). The remaining five soils were obtained using undisturbed sampling techniques (Connecticut Valley Varved Clay, Drammen Clay, East Atchafalaya Clay, Hackensack Valley Varved Clay, and Haney Clay).

With the exception of the isotropically consolidated Grundite Clay, all triaxial test specimens were consolidated under an anisotropic state of stress, where the ratio of the lateral stress, σ_{xx} , to vertical, σ_{yy}, is referred to hereafter as K_c. If the test specimens were anisotropically consolidated under the constraint of zero lateral strain ($\varepsilon_{xx} = \varepsilon_{zz} = 0$) then the consolidation process was classified as K_0 consolidation. Haney Clay was the only soil which was normally consolidated under K_0 conditions. Prevost [4] states that the Drammen Clay was anisotropically consolidated under K_0 conditions to an overconsolidation ratio of 4. This procedure re-

TABLE 1.0 Consolidation Histories and Undrained Shear Tests Performed on Eleven Soils For Evaluating the Prevost Total Stress Model

Soil Type	LL/PL	Shear Tests[a] Performed	Consolidation History Type[b]	OCR	Ref.
Boston Blue Clay	33/13	TC TE DSS PSC PSE	K_c= 0.54	1	15,16
Edgar Kaolinite	63/24	TC TE HCT	K_c= 1.0	?	18
Gleason Clay	54/25	TC TE DSS	K_c=0.57	1	19
Grundite Clay	56/30	TTT	Isotropic	1	20
Kaolinite Hydrite UF	74/34	TC TE DSS	K_c= 0.61	1	19
Santa Barbara Silt[c]	NP	TC TE DSS	K_c=0.50	1	8
Connecticut Valley Varved Clay	60/30	TC TE DSS PSC PSE	K_c=0.68	1	21
		TC TE DSS	K_c=0.81	2	21
		TC TE DSS	K_c= 1.0	4	21
Drammen Clay[c]	60/30	TC TE DSS	K_0= 1.0	4	4
East Atchafalaya Clay	100/70	TC TE DSS PSC	K_c=0.70	1	17
Hackensack Valley Varved Clay	60/41	TC TE DSS	K_c=0.65	1	22
Haney Clay	44/26	TC TE PSC PSE	K_0=0.55	1	23

Notes:

a. TC - Triaxial Compression
TE - Triaxial Extension
DSS - Direct Simple Shear
PSC - Plane Strain Compression
PSE - Plane Strain Extension
HCT - Hollow Cylinder Test
TTT - True Triaxial Test

b. $K_c = \sigma_{xx}/\sigma_{yy}$ (recorded from triaxial tests)

K_0 - anisotropic consolidation enforcing zero lateral strain

OCR - Overconsolidation Ratio

c. Model parameters published by Prevost

sulted in a final isotropic state of stress, i.e. final $K_0 = 1$. The procedures used to consolidate all the varved clays, Boston Blue Clay, East Atchafalaya Clay, Gleason Clay, and Kaolinite Hydrite UF resulted in "nominal" values of K_0, i.e. K_c consolidation which nearly produced one dimensional strain. Prevost [8] did not indicate if the K_c value of 0.50 for Santa Barbara Silt was from a one-dimensional consolidation test.

The multi-yield surface model parameters which define the stiffness (H_i), size ($k^{(i)}$), and location ($\alpha_1^{(i)}$) of each yield surface were fitted to triaxial compression and extension data for each soil. The parameters for Drammen Clay and Santa Barbara Silt were obtained from published literature [4,8]. The model parameters for the remaining nine soils shown on Table 1 were fitted by this writer. The failure model parameters, ($k^{(f)}$, $\alpha_1^{(f)}$, $H_f = 0$) for these nine soils were fitted to the maximum stress difference in triaxial compression and to the first pair of stress differences in triaxial extension which produced a slope which was nearly one percent of 2G. The model parameters fitted for all eleven soils are presented in the report by Kauschinger [9]. These parameters will be referred to in this paper as the triaxial model parameters.

The results from the evaluation studies [1,2,3,4] indicate that with the exception of the varved clays, the Prevost Model can be used to "accurately" calculate all stress-strain-strength responses for the shear tests shown on Table 1. Therefore, with the exception of the varved clays, all the soils listed on Table 1 were used to establish the data base of information employed for developing the pressuremeter fitting procedure. It is obvious from the above discussion that the pressuremeter fitting procedure is only valid for the particular class of soils used to establish the procedure.

FITTING MULTI-YIELD SURFACE MODEL PARAMETERS TO PRESSUREMETER DATA

Idealization of the Pressuremeter Test

The analysis conducted during this research are applicable to clays which are initially homogeneous and isotropic in the plane of expansion ($rr-\theta\theta$); also the in-plane stresses are initially equal and principal. Cavity expansion and deflation occur under plane strain, axisymmetric conditions, whereby the expansion/deflation occurs solely in the $rr - \theta\theta$ plane; and the strain in the vertical direction (zz) is always zero [10]. Additional restrictions were placed on the kinematics of the deformation whereby it was assumed that saturated clays respond in an incompressible manner during pressuremeter tests which are conducted rapidly.

The boundary conditions used to simulate the above idealized pressuremeter stress path during this research corresponded to increasing the radial strain during loading ($d\varepsilon_{rr} = 0.01\%$). When the yield condition for the failure yield surface was satisfied, the sign of $d\varepsilon_{rr}$ was changed to a negative and unloading calculations were performed. During the entire simulation the vertical strain increment was specified as zero, ($d\varepsilon_{zz} = 0$) and the incremental change in the tangential total stress was also zero ($d\sigma_{\theta\theta} = 0$). The tangential strain, $\varepsilon_{\theta\theta}$, was always calculated to be equal to $-\varepsilon_{rr}$. It should be noted that the shear stress-strain data, ($\sigma_{rr} - \sigma_{\theta\theta}$), ($\sigma_{zz} - \sigma_{rr}$), and ($\sigma_{zz} - \sigma_{\theta\theta}$) versus ε_{rr} generated using the above boundary conditions were exactly the same as the corresponding shear stress-strain data calculated when $d\varepsilon_{rr}$ varied as above, but $d\sigma_{zz} = 0$, and $d\varepsilon_{\theta\theta} = - d\varepsilon_{rr}$ were specified as input. The reasons for developing the empirical correlations based upon using the first set of boundary conditions (over the second) is discussed more fully in the section covering the calculation of $\alpha_i^{(f)}$, the location of the failure yield surface.

Derivation of the Stress-Strain Curve

Fitting multi-yield surface model parameters to pressuremeter data requires two steps:

1). A technique must be developed for deriving values of shear stress-strain, ($\sigma_{rr} - \sigma_{\theta\theta}$) versus $\varepsilon_{\theta\theta}$ (or ε_{rr} for incompressible soil), from the values of pressure (φ) and displacement (ρ) measured at the soil/probe interface.

2). A methodology must be formulated for fitting the multi-yield surface model parameters (H_i, $k^{(i)}$, $\alpha_i^{(i)}$) to the stress-strain data obtained in step one.

The shape of the shear stress-strain curve derived from values of φ versus ρ can be strongly influenced by the degree of disturbance of the soil around the probe. Prevost [11] performed finite element studies, using model parameters for undisturbed Drammen Clay, which indicate that shear stress-strain curves derived using the procedure advocated by Palmer [12] and others [13,14] results in ($\sigma_{rr} - \sigma_{\theta\theta}$) versus $\varepsilon_{\theta\theta}$ data which are the same as what Prevost describes as the "material" shear stress-strain curve. The "material" stress-strain curve corresponds to the pressuremeter shear stress-strain data calculated using program IMPMOD. Prevost further showed that when the probe causes disturbance of the Drammen Clay near the borehole, Palmer's procedure results in a derived shear stress-strain curve which exhibits strain softening, whereas the "material" stress-strain curve does not. Prevost presents a new procedure for deriving the "material" shear stress-strain curve which takes into account the extent of the radial disturbance around the borehole. Therefore, the shear stress-strain data needed to fit multi-yield surface model parameters should be obtained from pressuremeter values of φ versus ρ using Palmer's procedure [12] if the soil is undisturbed, and Prevost's procedure [11] if probe insertion causes disturbance of the soil in the borehole.

Procedure for Extracting Multi-Yield Surface Parameters

Extracting multi-yield surface model parameters from pressuremeter data involves the following steps:

1). the constitutive equations must be simplified so that the elasto-plastic shear moduli, H_i, can be calculated from values of $(\sigma_{rr} - \sigma_{\theta\theta})$ versus ε_{rr}.

2). the pressuremeter stress path for loading and unloading must be calculated and plotted in X - Y stress space (Eqn. 5 and 6), and a procedure found for calculating the initial size ,$k^{(i)}$, of each yield surface in the nest.

3). the initial yield surface configuration, as defined by the values of $\alpha_i^{(i)}$ obtained from triaxial data, must be reconstructed from pressuremeter data. The reconstruction involves first finding the location of the failure yield surface, $\alpha_i^{(f)}$, and then calculating the values of $\alpha_i^{(i)}$ which define the initial location for the inner nest of yield surfaces.

Calculation of the Scalar Stiffness, H_i

In the following derivation it was assumed that the scalar magnitude of the elasto-plastic shear modulus, H_i, for a particular yield surface remained constant during the loading and unloading portions of the pressuremeter stress path. The pressuremeter boundary conditions presented above were substituted into the constitutive relationship, which was then expanded in terms of the rr, zz, and θθ components as follows:

$$dS_{rr} = 2G\, d\varepsilon_{rr} - 3/2\,[2G-H_m] \left\{ \left[\frac{2/3\,\sigma_{rr} - 1/3\,\sigma_{zz} - 1/3\sigma_{\theta\theta} - \alpha_{rr}^{(m)}}{[\,k^{(m)}]^2}\right] \left[(\sigma_{rr} - \sigma_{\theta\theta}) - (\alpha_{rr}^{(m)} - \alpha_{\theta\theta}^{(m)})\right] d\varepsilon_{rr}\right\} \tag{7}$$

$$dS_{zz} = 2G\, d\varepsilon_{zz} - 3/2\,[2G-H_m] \left\{ \left[\frac{-1/3\,\sigma_{rr} + 2/3\,\sigma_{zz} - 1/3\,\sigma_{\theta\theta} - \alpha_{zz}^{(m)}}{[\,k^{(m)}]^2}\right] \left[(\sigma_{rr} - \sigma_{\theta\theta}) - (\alpha_{rr}^{(m)} - \alpha_{\theta\theta}^{(m)})\right] d\varepsilon_{rr}\right\} \tag{8}$$

$$dS_{\theta\theta} = 2G\, d\varepsilon_{\theta\theta}$$

$$- 3/2\ [2G-H_m] \left\{ \left[\frac{-1/3\ \sigma_{rr} - 1/3\ \sigma_{zz} + 2/3\ \sigma_{\theta\theta} - \alpha_{\theta\theta}^{(m)}}{[\ k^{(m)}]^2} \right] \right.$$

$$\left. \left[(\sigma_{rr} - \sigma_{\theta\theta}) - (\alpha_{rr}^{(m)} - \alpha_{\theta\theta}^{(m)}) \right] d\varepsilon_{rr} \right\} \tag{9}$$

By subtracting Eqn. 9 from 7, and noting that $d\varepsilon_{\theta\theta} = -d\varepsilon_{rr}$, an expression which relates the incremental changes in the stress difference, $d(\sigma_{rr} - \sigma_{\theta\theta})$, to the incremental change in the radial strain, $d\varepsilon_{rr}$, is obtained as:

$$d(\sigma_{rr} - \sigma_{\theta\theta}) = 4G\, d\varepsilon_{rr} \quad \text{(purely elastic response)} \tag{10}$$

and for elasto-plastic loading/unloading:

$$d(\sigma_{rr} - \sigma_{\theta\theta}) = 4G\, d\varepsilon_{rr} - 3/2\ [2G-H_m] \left[\text{Factor}_{rr} \right] d\varepsilon_{rr} \tag{11}$$

where:

$$\text{Factor}_{rr} = \frac{\left[(\sigma_{rr} - \sigma_{\theta\theta}) - (\alpha_{rr}^{(m)} - \alpha_{\theta\theta}^{(m)}) \right]^2}{[\ k^{(m)}]^2}$$

$\left[\text{Factor}_{rr} \right]$ was obtained for all nine soils by calculating the magnitudes of the variables required in Eqn. 11, using program IMPMOD. The data generated by IMPMOD, which was executed on a Digital VAX 11/780, was downloaded into an IBM PC where the values of $\left[\text{Factor}_{rr} \right]$ were calculated using a spreadsheet. For all nine soils the magnitude of $\left[\text{Factor}_{rr} \right]$ was calculated as being 4/3 always during <u>unloading</u>. Only the results from the Grundite Clay and Santa Barbara Silt studies produced values of $\left[\text{Factor}_{rr} \right]$ always close to 4/3 during the entire <u>loading</u> stage. The analyses conducted for the remaining seven soils typically resulted in values of $\left[\text{Factor}_{rr} \right]$ which ranged between 1.20 at low strains and increased to nearly 4/3 near failure. By substituting the value of $\left[\text{Factor}_{rr} \right]$ of 4/3 into Eqn. 11, the equation used to calculate the elasto-plastic shear moduli, H_i, was obtained as follows:

$$H_i = \frac{1}{2} \frac{d(\sigma_{rr} - \sigma_{\theta\theta})_i}{d\varepsilon_{rr_i}} \tag{12}$$

When calculating values of H_i, Eqn. 12 should be applied to the unloading portion of the pressuremeter shear stress-strain curve. It was assumed that the tangential total stress was constant during the pressuremeter test, Eqn. 12 can be re-written such that it can be used to calculate the incremental changes in the radial stress, as:

$$d\sigma_{rr_i} = 2\ H_i\ \ d\varepsilon_{rr_i} \tag{13}$$

Calculation of the $\alpha_i^{(i)}$ values requires equations which can be used to obtain not only $d\sigma_{rr_i}$ but also $d\sigma_{zz_i}$. An equation for calculating $d\sigma_{zz_i}$ was obtained by subtracting Eqn. 9 from 8 and noting that $d\varepsilon_{zz} = 0$, the following expressions were obtained:

$$d(\sigma_{zz}-\sigma_{\theta\theta}) = 2G\ d\varepsilon_{rr} \qquad \text{(elastic response)} \tag{14}$$

and for elasto-plastic loading/unloading,

$$d(\sigma_{zz}-\sigma_{\theta\theta}) = 2G\ d\varepsilon_{rr} - 3/2\ [2G-H_m]\ \left[\text{Factor}_{zz}\right]\ d\varepsilon_{rr} \tag{15}$$

where:

$$\text{Factor}_{zz} = \frac{\left[(\sigma_{zz}-\sigma_{\theta\theta}) - (\alpha_{zz}^{(m)}-\alpha_{\theta\theta}^{(m)})\right]}{[\ k^{(m)}]^2}$$

$$\left[\ \sigma_{rr}-\sigma_{\theta\theta}) - (\alpha_{rr}^{(m)}-\alpha_{\theta\theta}^{(m)})\ \right]$$

The magnitudes of $\left[\text{Factor}_{zz}\right]$ were calculated using the same procedure as outlined above for $\left[\text{Factor}_{rr}\right]$. Data generated using IMPMOD were substituted into Eqn. 15, and values of $\left[\text{Factor}_{zz}\right]$ for each of the nine soils in the data base were calculated as being equal to 2/3 always during unloading, and typically ranged between 0.90 to 0.66 during loading (especially near failure). Upon substituting a value for $\left[\text{Factor}_{zz} \right]$ of 2/3 into Eqn. 15 the following equation is obtained:

$$H_i = \frac{d(\sigma_{zz} - \sigma_{\theta\theta})_i}{d\varepsilon_{rr_i}} \tag{16}$$

Eqn. 16 can be simplified for the assumption of $d\sigma_{\theta\theta} = 0$, and therefore obtain the equation used to calculate the incremental change in the vertical total stress as follows:

$$d\sigma_{zz_i} = H_i \; d\varepsilon_{rr_i} \tag{17}$$

Calculation of the Scalar Size, $k^{(i)}$

The scalar "size" , $k^{(i)}$, of the ith yield surface in the nest may be calculated if we know what stresses satisfy the yield condition for the ith yield surface, and also the stress path followed. Program IMPMOD, which employed the Mroz Hardening Rule given by Eqn. 2, was used to calculate the pressuremeter stress path for all soils in the data base. The results from the Boston Blue Clay studies were typical of the results for all soils and will be used for illustrative purposes.

The initial yield surface configuration, projected in X-Y stress sub-space, for Boston Blue Clay is shown on Figure 1a. The yield surfaces in this figure project as circles, which have all their centers initially located along the Y = 0 axes. Superimposed upon the Boston Blue Clay's initial yield surface configuration is the pressuremeter stress path followed during the loading stage. Initially, the stress path is directed straight upward through the elastic region and then nearly intersects the apexes for those yield surfaces closest to the elastic region. Thereafter, the stress path cuts across the circles apexes for those yield surfaces closest to the failure (outermost) yield surface, i.e. the 13th yield surface on Figure 1a. Since the values of $k^{(i)}$ are equal to the radii of yield surface circles, projected in X - Y stress space, values of $k^{(i)}$ calculated from pressuremeter loading data would only be approximations to the corresponding values fitted with triaxial data because the loading stress path does not always cut across the apexes of all yield surfaces.

The pressuremeter unloading stress path for Boston Blue Clay, shown on the accompanying Figure 1b, is representative of all the unloading stress paths for the nine soils used during this study. The path starts at the apex of the failure yield surface, and continues in a downward (negative) direction intersecting the apexes of each and every yield surface in the nest. All values of $k^{(i)}$ were calculated using the unloading portion of the pressuremeter stress-strain curve and were identical to the corresponding values

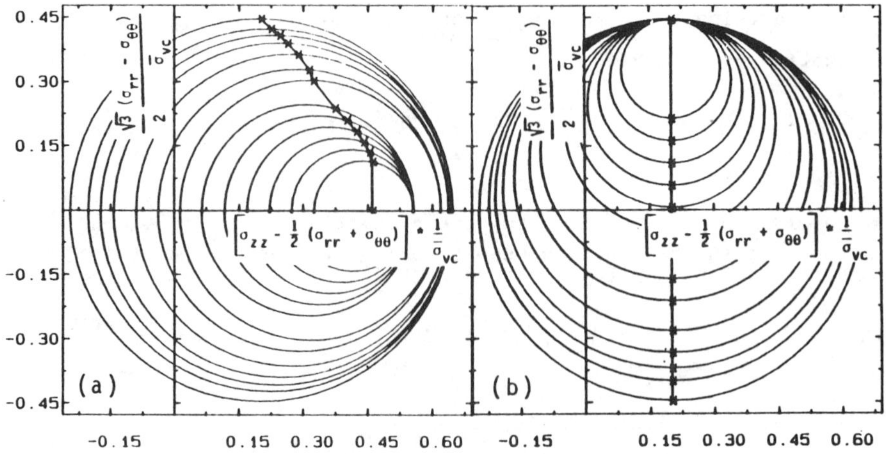

Figure 1 Stress Paths Followed During Pressuremeter Test for Boston Blue Clay, a). loading stage and initial yield surface configuration, b). unloading stage.

of $k^{(i)}$ obtained from triaxial data. The following equation was used to calculate all values of $k^{(i)}$, as:

$$k^{(i)} = \frac{\sqrt{3}}{4} \left\{ (\sigma_{rr} - \sigma_{\theta\theta})_f^L - (\sigma_{rr} - \sigma_{\theta\theta})_i^{UL} \right\} \tag{18}$$

where $(\sigma_{rr} - \sigma_{\theta\theta})_f^L$ is the maximum stress difference which satisfies the yield condition for the failure surface (f) during the loading (L) stage, while $(\sigma_{rr} - \sigma_{\theta\theta})_i^{UL}$ is the unloading (UL) stress difference at which the ith value of $k^{(i)}$ is fitted.

The results from all analysis indicated that the value of $(\sigma_{rr} - \sigma_{\theta\theta})_f^{UL} = -(\sigma_{rr} - \sigma_{\theta\theta})_f^L$. Therefore, the size of the failure yield surface, $k^{(f)}$, should be calculated as follows:

$$k^{(f)} = \frac{\sqrt{3}}{2} (\sigma_{rr} - \sigma_{\theta\theta})_f^L \tag{19}$$

Location of the Failure Yield Surfaces, $\alpha_1^{(f)}$

During this research it was assumed that the failure surface represented a fixed boundary. The location of the failure surface, $\alpha_1^{(f)}$, was obtained by observing that the stress point at failure during loading always resided at the apex of the failure yield surface, as illustrated in Figure 1a for Boston Blue Clay. Therefore, all values of $\alpha_1^{(f)}$ were calculated using the following equation:

$$\alpha_1^{(f)} = \sigma_{zz_f}^{L} - 1/2 \quad (\sigma_{rr_f}^{L} + \sigma_{\theta\theta_f}^{L}) \tag{20}$$

Prevost [11] presented equations, which were based upon the model formulation presented previously, which could be used to calculate the maximum stress differences predicted during the loading stage of the pressuremeter test as follows:

$$(\sigma_{rr} - \sigma_{\theta\theta})_f^L = \frac{2\,k^{(f)}}{\sqrt{3}} \tag{21}$$

$$(\sigma_{zz} - \sigma_{\theta\theta})_f^L = \alpha_1^{(f)} + \frac{k^{(f)}}{\sqrt{3}} \tag{22}$$

$$(\sigma_{zz} - \sigma_{rr})_f^L = \alpha_1^{(f)} - \frac{k^{(f)}}{\sqrt{3}} \tag{23}$$

If the simulations were conducted using the boundary condition of $d\sigma_{zz} = 0$, Eqns. 22 and 23 could not be simplified and used to calculate the tangential and radial stresses at failure because the value of $\alpha_1^{(f)}$ is not known at this point in the analysis. An empirical procedure would have to be developed for calculating $\sigma_{rr_f}^{L}$, whose value could then be substituted into Eqn. 21 for calculating $\sigma_{\theta\theta_f}^{L}$ (since $k^{(f)}$ is known at this point in the analyses). This option is presently being examined for calculating $\sigma_{rr_f}^{L}$.

For the assumption that $d\sigma_{\theta\theta} = 0$, the tangential stress is always constant during shear, Eqn. 21 can be simplified directly and used to calculate the radial stress at failure as follows:

$$\frac{\sigma^L_{\theta\theta_f}}{\bar{\sigma}_{vc}} = \frac{\sigma_{\theta\theta_i}}{\bar{\sigma}_{vc}} = K_0 \qquad (24)$$

$$\frac{\sigma^L_{rr_f}}{\bar{\sigma}_{vc}} = \frac{2}{\sqrt{3}} \frac{k^{(f)}}{\bar{\sigma}_{vc}} + K_0 \qquad (25)$$

However for $d\sigma_{\theta\theta} = 0$, the value of the vertical stress at failure cannot be calculated because it is a function of $\alpha_i^{(f)}$, as follows:

$$\frac{\sigma^L_{zz_f}}{\bar{\sigma}_{vc}} = \frac{\alpha_i^{(f)}}{\bar{\sigma}_{vc}} + \frac{1}{\sqrt{3}} \frac{k^{(f)}}{\bar{\sigma}_{vc}} + K_0 \qquad (26)$$

Therefore, employing the assumption of $d\sigma_{\theta\theta} = 0$, requires developing an empirical procedure for calculating the vertical stress at failure. This process was greatly simplified because during the analysis of all the pressuremeter data generated using $d\sigma_{\theta\theta} = 0$, it was observed that the following stress relationship was valid:

$$\sigma^{UL}_{zz_f} - \sigma^L_{zz_f} = -(\sigma_{rr} - \sigma_{\theta\theta})^L_f = -\frac{2\,[\,k^{(f)}]}{\sqrt{3}} \qquad (27)$$

By dividing Eqn. 27 by $1/\sigma^L_{zz_f}$, the following equation is obtained:

$$\frac{\sigma^{UL}_{zz_f}}{\sigma^L_{zz_f}} - 1 = -\frac{2\,[\,k^{(f)}]}{\sqrt{3}} \frac{1}{\sigma^L_{zz_f}} \qquad (28)$$

For each of the nine soils used to develop the pressuremeter fitting procedure, values of $\dfrac{\sigma^{UL}_{zz_f}}{\sigma^L_{zz_f}}$ $\dfrac{1}{\left[\dfrac{k^{(f)}}{\bar{\sigma}_{vc}}\right]^2}$ were calculated and the results plotted versus $\dfrac{1}{\left[\dfrac{k^{(f)}}{\bar{\sigma}_{vc}}\right]^2}$, as shown in Figure 2. The nine data points drawn on this figure tend to vary in a linear fashion. However, the data for all three undisturbed soils (Haney Clay (HC), East Atchafalaya (EA), and Drammen Clay (DC)) tend to plot farthest from the straight line draw through all the reconstituted soil's data plotted on Fig 2.

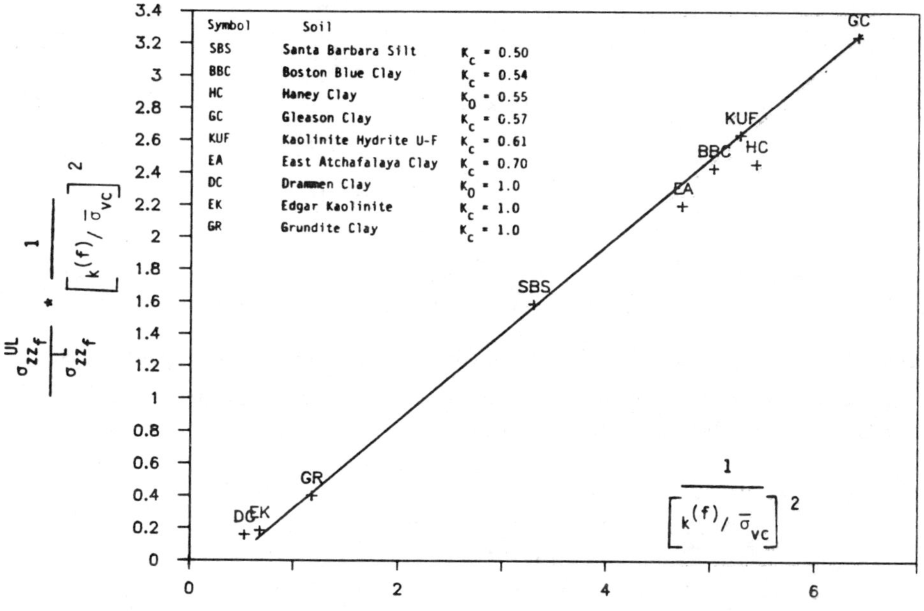

Figure 2 Ratio of Vertical Stresses at Failure Versus The Squared Inverse of the Failure Surface Sizes for Reconstituted and Undisturbed Soils

So that a clear distinction between reconstituted and undisturbed soils can be made, it was decided to use one equation for calculating values of $\sigma_{zz_f}^{L}$ for reconstituted soils, and another for undisturbed soils. The following equations were used to calculate the vertical stress at failure for each soil during this research:

$$\frac{\sigma_{zz_f}^{L}}{\bar{\sigma}_{vc}} = \frac{2}{\sqrt{3}} \frac{k^{(f)}}{\bar{\sigma}_{vc}} \frac{1}{\left[1 - F_1\right]} \qquad (29)$$

where for reconstituted soils, F_1 is equal to:

$$F_1 = 0.534 - 0.20 \left[\frac{k^{(f)}}{\bar{\sigma}_{vc}}\right]^2$$

and for undisturbed soils, F_1 is equal to:

$$F_1 = 0.475 - 0.10 \left[\frac{k^{(f)}}{\bar{\sigma}_{vc}}\right]^2$$

Upon substituting Eqns. 24, 25, and 29, into the expression for $\alpha_1^{(f)}$, Eqn. 20, which was normalized with respect to the vertical consolidation stress, $\bar{\sigma}_{vc}$, the following empirical relationship was obtained and used to calculate the location of the failure yield surface for all nine soils, as follows:

$$\frac{\alpha_1^{(f)}}{\bar{\sigma}_{vc}} = \frac{2}{\sqrt{3}} \frac{k^{(f)}}{\bar{\sigma}_{vc}} \frac{1}{\left[1 - F_1\right]} - \frac{1}{\sqrt{3}} \frac{k^{(f)}}{\bar{\sigma}_{vc}} - K_0 \qquad (30)$$

where $k^{(f)}$ is obtained using Eqn. 19, and F_1 is as defined before.

Once the location of the failure surface, $\alpha_1^{(f)}$ is obtained, the initial location of the inner nest of yield surfaces, $\alpha_1^{(i)}$ can be calculated.

<u>Location of the Inner Yield Surfaces, $\alpha_1^{(i)}$</u>

The location of the failure surface, $\alpha_1^{(f)}$, was obtained using

Eqn. 20. It was valid to use this equation because the failure stresses lie on the apex of the failure surface, when projected into X - Y stress space. During the analysis of the data for isotropically consolidated Grundite Clay, it was observed that the pressuremeter stress path always intersected the apexes of every yield surface during loading. This observation was partially true for all other eight soils examined. Therefore, an equation of the same form as Eqn. 20 was used to calculate values of $\alpha_i^{(i)}$ for the inner nest of yield surfaces, as follows:

$$\alpha_i^{(i)} = \sigma_{zz_i}^{\ L} - 1/2 \quad (\sigma_{rr_i}^{\ L} + \sigma_{\theta\theta_i}^{\ L}) \tag{31}$$

The values of the normal stresses in Eqn. 31 are obtained by calculating the incremental changes in the radial and vertical stresses, using Eqn. 13 and 17, and subtracting these incremental changes from the previous total stresses. When Eqn. 31 is used to calculate $\frac{\alpha_i^{(1)}}{\bar{\sigma}_{vc}}$, the location of the first yield surface normalized with respect to $\bar{\sigma}_{vc}$, the value of $\frac{\alpha_i^{(1)}}{\bar{\sigma}_{vc}}$ is equal to $(1 - K_c)$.

The development of the procedure for extracting multi-yield surface model parameters was quite cumbersome and lengthy. Therefore, so that a clearer understanding of the pressuremeter fitting procedure is made, a step by step procedure for fitting multi-yield surface model parameters to pressuremeter data is presented.

Step by Step Procedure For Fitting Multi-Yield Surface Parameters

For simplicity in presenting the following procedure, it is assumed that all stresses and multi-yield surface model parameters have been normalized with respect to the vertical consolidation stress, $\bar{\sigma}_{vc}$. The sequential steps to be followed are:

1). Derive values of $(\sigma_{rr} - \sigma_{\theta\theta})$ versus ε_{rr} from the pressure-radial displacement data obtained from the pressuremeter test, and evaluate the at-rest coefficient, K_0. Converting pressure-radial displacement data to shear stress-strain data was previously discussed

2). Calculate the size of the failure surface using the maximum stress difference as follows:

$$k^{(f)} = \frac{\sqrt{3}}{2} \quad (\sigma_{rr} - \sigma_{\theta\theta})_f^{\ L} \tag{32}$$

3). Select a convient strain increment size, $d\varepsilon_{rr_i}$, and calculate the elasto- plastic shear moduli, H_i, using the unloading portion of the pressuremeter data as follows:

$$H_i = \frac{1}{2} \quad \frac{d(\sigma_{rr} - \sigma_{\theta\theta})_i^{UL}}{d\varepsilon_{rr_i}} \tag{33}$$

4). Calculate the remaining values of $k^{(i)}$ using the unloading portion of the stress-strain curve at the data points where values of H_i were obtained. The $k^{(i)}$ values are calculated as:

$$k^{(i)} = \frac{\sqrt{3}}{4} \left\{ (\sigma_{rr} - \sigma_{\theta\theta})_f^{L} - (\sigma_{rr} - \sigma_{\theta\theta})_i^{UL} \right\} \tag{34}$$

5). Calculate the location of the failure surface as follows:

$$\alpha_1^{(f)} = \frac{2 \quad k^{(f)}}{\sqrt{3} \left[1 - F_1\right]} - \frac{k^{(f)}}{\sqrt{3}} - K_o \tag{35}$$

where for reconstituted soils:

$$F_1 = 0.534 - 0.2 \quad [k^{(f)}]^2$$

and for undisturbed soils:

$$F_1 = 0.475 - 0.1 \quad [k^{(f)}]^2$$

6). Calculate the normal stresses at failure for the loading stage:

$$\sigma_{\theta\theta_f}^{L} = \sigma_{\theta\theta_i} = K_o \tag{36}$$

$$\sigma_{rr_f}^{L} = \frac{2 \quad [k^{(f)}]}{\sqrt{3}} + K_o \tag{37}$$

$$\sigma_{zz_f}^{L} = \alpha_1^{(f)} + \frac{[k^{(f)}]}{\sqrt{3}} + K_o \tag{38}$$

7). Calculate the change in the radial and tangential stresses needed to go from the failure yield surface to the next inner surface (f-1), as follows:

$$d\sigma_{rr} = 2 \; H_{f-1} \; d\varepsilon_{rr_{(f-1)}} \tag{39}$$

$$d\sigma_{zz} = H_{f-1} \; d\varepsilon_{rr_{(f-1)}} \tag{40}$$

where H_{f-1} has been determined from the unloading data.

8). Calculate the total stresses which satisfy the yield condition for the (f-1) yield surface as follows:

$$\sigma_{rr_{(f-1)}} = \sigma_{rr_f}^{L} - d\sigma_{rr} \tag{41}$$

$$\sigma_{zz_{(f-1)}} = \sigma_{zz_f}^{L} \mp d\sigma_{zz} \tag{42}$$

where for isotropically consolidated soils, the - sign is selected in Eqn. 42, and for all other soils the + sign is selected.

9). Calculate the value of $\alpha_1^{(f-1)}$, which defines the location of the yield surface (f-1) which is just inside of the failure surface, as follows:

$$\alpha_1^{(f-1)} = \sigma_{zz_{f-1}}^{L} - 1/2 \; (\sigma_{rr_{f-1}}^{L} + \sigma_{\theta\theta_{f-1}}^{L}) \tag{43}$$

10). Repeat Steps 7 through 10 for the next inner yield surface changing the f parameters to (f-1), and the (f-1) parameters to (f-2). Repeat this process of sequentially decreasing the indices until all yield surface locations are obtained.

11). Adjust any values of $\alpha_1^{(i)}$ so that inner and outer yield surfaces do not overlap. This was typically required when fitting yield surfaces one and two, so that $\alpha_1^{(1)} = (1- K_c)$.

12). Calculate the elastic shear modulus, G, using the data which defines the steepest initial slope on the pressuremeter curve as follows:

$$G = 1/4 \; \frac{(\sigma_{rr} - \sigma_{\theta\theta})_i^{L}}{d\varepsilon_{rr}} \tag{44}$$

EVALUATION OF PRESSUREMETER FITTING PROCEDURE

In the following discussions model parameters referred to as triaxial model parameters were obtained from measured triaxial compression and extension data. Those model parameters referred to as pressuremeter model parameters were obtained using the aforementioned fitting procedure, which was applied to numerical data generated using program IMPMOD and triaxial model parameters.

Comparison Between Pressuremeter Stress-Strain Data

When each set of pressuremeter model parameters were used to calculate values of $\frac{(\sigma_{rr} - \sigma_{\theta\theta})}{\bar{\sigma}_{vc}}$ versus ε_{rr} nearly the exact same load/unload data were obtained as when the corresponding triaxial model parameters were used. The Boston Blue Clay stress-strain curve, $\frac{(\sigma_{rr} - \sigma_{\theta\theta})}{\bar{\sigma}_{vc}}$ versus ε_{rr}, shown in Figure 3, is representative of the results for all nine soils. It appears that the values of $(\sigma_{rr} - \sigma_{\theta\theta})$ versus ε_{rr} are strictly a function of the scalar parameters H_i and $k^{(i)}$, i.e. the triaxial and pressuremeter scalar values were exactly the same; any differences in model parameters were observed in the values of $\alpha_i^{(i)}$.

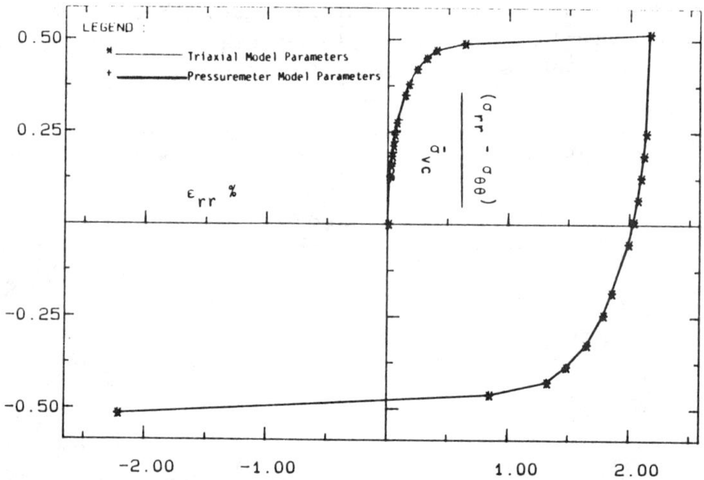

Figure 3 Comparison Between Pressuremeter Stress Differences Obtained Using Triaxial Versus Pressuremeter Model Parameters For Boston Blue Clay.

Comparison Between Triaxial Stress-Strain Data

The accuracy of the pressuremeter fitting procedure is evaluated by comparing the results of triaxial stress-strain-strength calculations, $\frac{(\sigma_{yy} - \sigma_{xx})^L_{max.}}{\bar{\sigma}_{vc}}$, and $\frac{(\sigma_{yy} - \sigma_{xx})}{\bar{\sigma}_{vc}}$ versus ε_{yy}, which were generated with pressuremeter model parameters to the corresponding stress-strain data obtained using model parameters fitted to triaxial compression and extension data.

The triaxial and pressuremeter model parameters, $(k^{(f)}, \alpha_1^{(f)})$ which define the failure surface were used to calculate the maximum stress difference for each of the nine soils by using the following equation:

$$(\sigma_{yy} - \sigma_{xx})^L_{max.} = \alpha_1^{(f)} \pm k^{(f)} \tag{45}$$

where in Eqn. 45, (+) is selected for triaxial compressive loading and (-) for triaxial extension.

The maximum stress differences calculated with model parameters obtained via the pressurmeter fitting procedure were in excellent agreement with the corresponding values calculated with triaxial parameters. The triaxial compressive values of $\frac{(\sigma_{yy} - \sigma_{xx})^L_{max}}{\bar{\sigma}_{vc}}$, calculated with pressuremeter model parameters, and shown in Figure 4a, are within $\pm$ 5% of the corresponding values obtained using triaxial model parameters. Similar excellent agreement in the failure stress differences is observed for the triaxial extension data plotted on the left hand side of the same figure.

The pressuremeter and triaxial model parameters for all nine soils in the data base were specified as input in IMPMOD, and load/unload triaxial compression and extension stress-strain data were calculated, $\frac{(\sigma_{yy} - \sigma_{xx})}{\bar{\sigma}_{vc}}$ versus ε_{yy}. The results from the Boston Blue Clay studies are typical of the good results obtained for the majority of the soils examined during this research, which also included Gleason Clay, Grundite Clay, Kaolinite Hydrite UF, East Atchafalaya Clay, and Haney Clay. The results from the Drammen Clay studies are typical of the poorer results obtained, which were also observed for Edgar Kaolinite and Santa Barbara Silt.

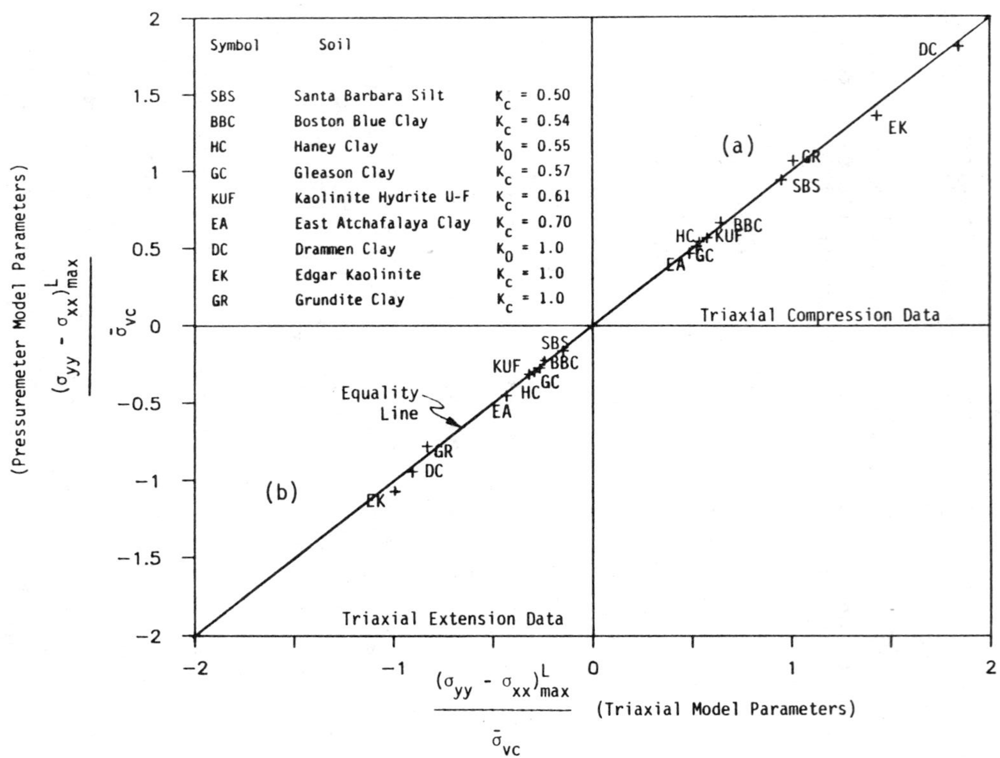

Figure 4. Comparison Between Normalized Maximum Stress Differences (Virgin Loading) Obtained Using Triaxial Versus Pressuremeter Model Parameters a). Triaxial Compression b). Triaxial Extension.

The triaxial compression stress-strain loading curves calculated using either triaxial or pressuremeter model parameters for Boston Blue Clay, and shown in Figure 5a, have failure stresses which peak at relatively low strains, $\varepsilon_{yy_f}^{L}$. The difference between the two values of $\varepsilon_{yy_f}^{L}$ calculated with triaxial versus pressuremeter parameters is approximately 0.25% strain, It appears that this strain difference propagates throughout the unloading stage and causes a rigid translation (left shift) of the unloading stress-strain curve calculated with pressuremeter parameters. In all other respects, the two unloading curves shown in Figure 5a can be considered very similar. When the same triaxial and pressuremeter model parameters were used to calculate Boston Blue Clay's triaxial extension stress-strain behavior, shown in Figure 5b, essentially the same triaxial extension stress-strain characteristics were obtained.

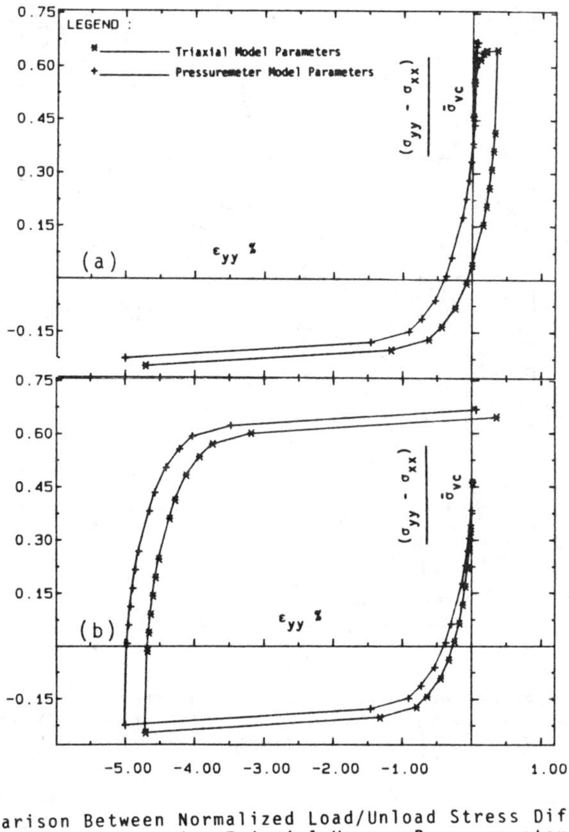

Figure 5. Comparison Between Normalized Load/Unload Stress Differences and Axial Strains Obtained Using Triaxial Versus Pressuremeter Model Parameters For Normally Consolidated Boston Blue Clay; a).Triaxial Compression b). Triaxial Extension.

The Drammen Clay's pressuremeter parameters were able to accurately simulate the triaxial compression loading behavior shown in Figure 6a for axial strains below 2.8%, which also corresponds to the axial (loading stage) strain at failure, $\varepsilon_{yy_f}^{L}$, obtained with triaxial parameters. With pressuremeter parameters $\varepsilon_{yy_f}^{L}$ (4.2%) is 1.4% greater than the corresponding triaxial value. The differences in the strain at failure always cause the unloading curve to plot to the right of (by 1.4 % strain) and below the stress-strain curve which was obtained with triaxial parameters. The pressuremeter

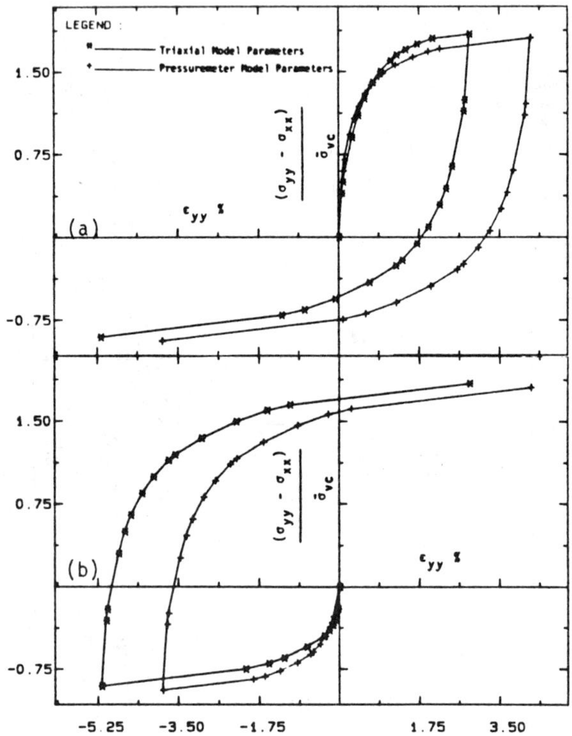

Figure 6. Comparison Between Normalized Load/Unload Stress Differences and Axial Strains Obtained Using Triaxial Versus Pressuremeter Model Parameters For Overconsolidated Drammen Clay; a). Triaxial Compression b). Triaxial Extension.

model parameters only reproduce the unloading triaxial compressive behavior in a fair manner. The two triaxial extension curves, $\frac{(\sigma_{yy} - \sigma_{xx})}{\bar{\sigma}_{vc}}$ versus ε_{yy}, plotted for Drammen Clay on Figure 6b are in good agreement during the unloading stage for axial strains below 3.8%. As previously observed for the compression test, the strain error associated with triaxial extension loading (pressuremeter parameters) is always 1.4%. This loading strain error causes the triaxial extension curve generated with pressuremeter parameters always to plot below and to the right of (+1.4%) of the corresponding curve obtained using triaxial parameters.

CONCLUSIONS

A unified empirical approach was developed which was used to extract multi-yield surface model parameters for nine fine grained soils from pressuremeter stress- strain data, $(\sigma_{rr} - \sigma_{\theta\theta})$ versus ε_{rr}. The resulting pressuremeter model parameters were able to accurately reproduce triaxial compressive and extensive load/unload behavior for six of the nine soils in the data base , while for the remaining three soils good to fair comparisons with triaxial data were observed.

The pressuremeter test "results" presented in this paper were generated using model parameters for soils proven to work well with the total stress version of the Prevost Model. It would have been much better to validate the described procedure using measured pressuremeter data. Work is presently being conducted whereby laboratory pressuremeter tests are being performed in a calibration chamber for purposes of gathering the experimental data needed to verify the pressuremeter fitting procedure. Results from these studies will be soon forthcoming.

ACKNOWLEDGEMENTS

This investigation was supported in part by the Engineering Foundation of the American Society of Civil Engineers under grant RI-A-84-2, and the College of Engineering, Tufts University. The support of these organizations is gratefully acknowledged.

REFERENCES

[1] Kauschinger, J.L. (1983a), "Evaluation and Implementation of Prevost's Total Stress Model", Dissertation presented to The University of Texas at Austin, in May 1983, in partial fulfillment of the requirement for the degree of Doctor of Philosophy.

[2] Kauschinger, J.L., and Roesset, J.M., (1983b), "Evaluation of a Multi-Yield Surface Elasto-Plastic Model", International Conference on Constitutive Laws for Engineering Materials: Theory and Application, January 10 - 14, Tucson, Arizona.

[3] Prevost, J.H., (1977), "Mathematical Modelling of Monotonic and Cyclic Undrained Clay Behavior", International Journal for Numerical and Analytical Methods in Geomechanics, Vol 1, No 2

[4] Prevost, J.H., (1978), "Anistropic Undrained Stress-Strain Behavior of Clays", Journal of the Geotechnical Engineering Division, American Society of Civil Engineers, Vol. 104, No. GT8, August.

[5] Iwan, W.D., (1967), "On a Class of Models for the Yielding Behavior of Continuous and Composite Systems", Journal of the Applied Mechanics Division, American Society of Civil Engineers, Vol. 34, Series E, No. 3, September.

[6] Mroz, Z., (1967), "On the Description of Anisotropic Work-hardening", Journal of the Mechanics and Physics of Solids, Vol. 15.

[7] Prevost, J.H., (1978), "Plasticity Theory For Soil Stress-Strain Behavior", Journal of The Engineering Mechanics Division, American Society of Civil Engineers, Vol. 104, No. EM 5, October.

[8] Prevost, J.H., Cuny, B., Hughes, T.J.R., and Scott, R.F.,(1981) "Offshore Gravity Structures: Analysis", Journal of the Geotechnical Engineering Division, American Society of Civil Engineers, Vol. 107, No. GT2, February.

[9] Kauschinger, J.L., (1985), "Interim Report to the Engineering Foundation of ASCE: Extracting Multi-Yield Surface Model Parameters From Pressuremeter Data", Contract No. RI-A-84-2, July.

[10] Wroth, C.P., and Hughes, J., (1973), "An Instrument for the In Situ Measurement of the Properties of Soft Clays", Proceedings of the Eighth International Conference on Soil Mechanics and Foundation Engineering, Moscow, Vol. 1-2.

[11] Prevost, J.H., (1979), "Undrained Shear Tests on Clays", Journal of the Geotechnical Engineering Division, American Society of Civil Engineers, Vol. 105, No. GT1, January.

[12] Palmer, A.C., (1972), "Undrained Plane Strain Expansion of a Cylindrical Cavity in Clay: a Simple Interpretation of the Pressuremeter Test", Geotechnique, Vol. 22, No. 3, September.

[13] Ladanyi, B., (1972), "In Situ Determination of Undrained Stress-Strain Behavior of Sensitive Clays with the Pressuremeter", Canadian Geotechnical Journal, Vol. 9, No. 3, August.

[14] Baguelin, F., Jezequel, J.F., LeMee, E., Le Mehaute, A. (1972) "Expansion of Cylindrical Probes in Cohesive Soils", Journal of the Soil Mechanics and Foundations Division, American Society of Civil Engineers, Vol. 98, No. SM11, November.

[15] Ladd, C.C., and Varallyay, J., (1965), "The Influence of Stress System on the Behavior of Saturated Clays During Undrained Shear", Research Report in Earth Science, Report No. R65-11, MIT, July.

[16] Ladd, C.C., Bovee, R.B., Edgers, L., and Rixner, J.J., (1971), "Consolidated Undrained Plane Strain Shear Tests on Boston Blue Clay", Research Report R71-13, Soils Publication 273, MIT, March.

[17] Ladd, C.C., Williams, C.E., Connell, D.H., and Edgers, L., (1972), "Engineering Properties of Soft Foundation Clays at Two South Louisiana Levee Sites", Research Report R72-26, Soils Publication 304, MIT, December.

[18] Yong, R.K., and Ko, H.Y., (1980), "Limit Equilibrium, Plasticity, and Generalized Stress-Strain in Geotechnical Engineering", Proceedings of the Workshop Sponsored by American Society of Civil Engineers, McGill University, May.

[19] Fennessey, T., (1983),"The Effect of Consolidation Procedure on the Undrained Shearing Behavior of Normally Consolidated Clays Thesis Submitted to the Department of Civil Engineering, The University of Texas at Austin.

[20] Lade, P.V., and Musante, H.M., (1976), "Three Dimensional Behavior of Normally Consolidated Cohesive Soils", Research Report UCLA- ENG-7626, Mechanics and Structures Department, UCLA, April.

[21] Sambhandharahsa, Surachat, (1978), "Stress-Strain- Strength Anisotropy of Varved Clay", Dissertation Submitted to the Department of Civil Engineering of the Massachusetss Institute of Technology in Partial Fulfillment of the Requirements for the Degree of Doctor of Philosophy.

[22] Saxena, S.K., Hedberg, J., and Ladd, C.C., (1974), "Results of Special Laboratory Testing Program on Hackensack Valley Varved Clay", Research Report R74-66, Soils Publication 342, MIT, June.

[23] Vaid, Y.P., and Campanella, R.G., (1972), "Influence of Stress Path on the Plane Strain Behavior of Sensitive Clay", Soil Mechanics Series No. 19, The University of British Columbia, Vancouver, April.

Chitta R. Gangopadhyay, and Athanacios N. Nasr

INTERPRETATION OF PRESSUREMETER TESTS USING LABORATORY SIMULATED TESTS.

REFERENCE: Gangopadhyay, C. R. and Nasr, A. N., Interpretation of Pressuremeter Tests Using Laboratory Simulated Tests, The Pressuremeter and its Marine Applications: Second International Symposium, ASTM STP 950, J.-L. Briaud and J. M. E. Audibert, Eds., American Society For Testing and Materials, 1986.

ABSTRACT: At least five different methods are available to interpret the results of field pressuremeter tests for saturated clays, each of which has been reported to yield different values of soil parameters when applied to the same pressuremeter test. In the present investigation, a laboratory pressuremeter model was developed, and simulated pressuremeter tests were run on a sedimented Kaolinite consolidated to different values of the maximum vertical pressure, $\bar{\sigma}_v$, and with overconsolidation ratios (OCR) ranging from 1 to 8. Analyses of the model pressuremeter tests indicate that the differences in normalized shear strength, $S_u/\bar{\sigma}_v$, using the five interpretation methods are large with the difference increasing with increasing overconsolidation ratio. The method by Prevost gave the best fit with the $S_u/\bar{\sigma}_v$ vs OCR relationship determined by the average of $\overline{CK_oU}$ Triaxial compression and extension strengths.

KEYWORDS: pressuremeter, clays, undrained shear strength, triaxial tests

INTRODUCTION

Because of the recognition of problems inherent in testing soils in the laboratory, in-situ testing procedures have gained favor in the last decade. The pressuremeter test is one such field test which is used to measure the lateral stress in the ground, the stress-strain behavior, and the strength of the soil. A considerable amount of study-both theoretical and experimental-has already been made by several investigators on the interpretation of the field pressuremeter test

Dr. Gangopadhyay is Professor and Chairman of the Civil Engineering Department at University of Detroit, MI, 48221. Mr. Nasr is an Engineer with Giffels Assoc., Inc. at Southfield, MI 48037.

[1 through 9 and others].

Although theoretical basis of the pressuremeter test is the old problem of expanding a cavity inside an infinite medium (Lame',1852), there are significant differences in derived soil parameters when different interpretation methods are applied to the same field pressuremeter data [2].

The known methods of analysis were established by Menard-1957[1], Gibson and Anderson-1961[1], Baguelin, Palmer, and Ladanyi-1972[1,3 and 4], Prevost and Hoeg-1976[5], and Denby and Clough-1980[2]. These approaches differ from each other in the assumptions made in soil condition and stress-strain behavior. However, these five methods made the same basic assumptions of plane-strain radial expansion and undrained condition for analyzing the problem (See Fig.1). In this investigation, a laboratory pressuremeter model was designed and built to simulate the field test. This model was used to study and verify the available methods of interpretation.

THE LABORATORY PRESSUREMETER MODEL

The pressuremeter model apparatus was designed in such a way as to be able to expand a 1.6 cm (5/8 in.) diameter cylindrical cavity within a 15 cm (6 in.) diameter and 10 cm (4 in.) long artificially sedimented cohesive sample in a state of plane-strain under undrained or drained condition. Prior to expanding the cavity, the soil is consolidated or consolidated-rebounded to the required stress level and stress history.

As shown in Fig. 2, the model apparatus consists of three cylinders: sedimentation cylinder (a), consolidation cylinder (b), and shearing cylinder (c). The vertical axes of these cylinders, each 15 cm (6 in.) in diameter, are coaxial with a 1.6 cm (5/8) O.D. perforated tube (d). These elements are all connected together finally to the base (e). A piston (f) with about 1.6 in. I.D. is used for consolidating the kaolin slurry until it has the required height.

After the sedimentation stage, the piston (f), the guide rod (g) and sedimentation cylinder (a) are disassembled and a consolidation cap (h) is placed on and fixed to the apparatus, as in Fig. 2 (C). This cap is connected to a compressed air system to control the stress inside the membrane (i). The maximum vertical applied stress on the sample and the overconsolidation ratio can be adjusted by controlling the pressure inside the consolidation cap. The drainage valve of the base is left open during both the sedimentation and consolidation stages. Finally, after a complete preparation of the sample, the consolidation cap (h) is fixed in position using the three clamps (j) around the three vertical legs (k) to ensure that there is no vertical strain during shearing of the sample. Two volume change measurement systems are connected to the system: one to the inner tube covered with membrane (ii) and the other to the outer chamber around membrane (iii), as shown in Fig. 3.

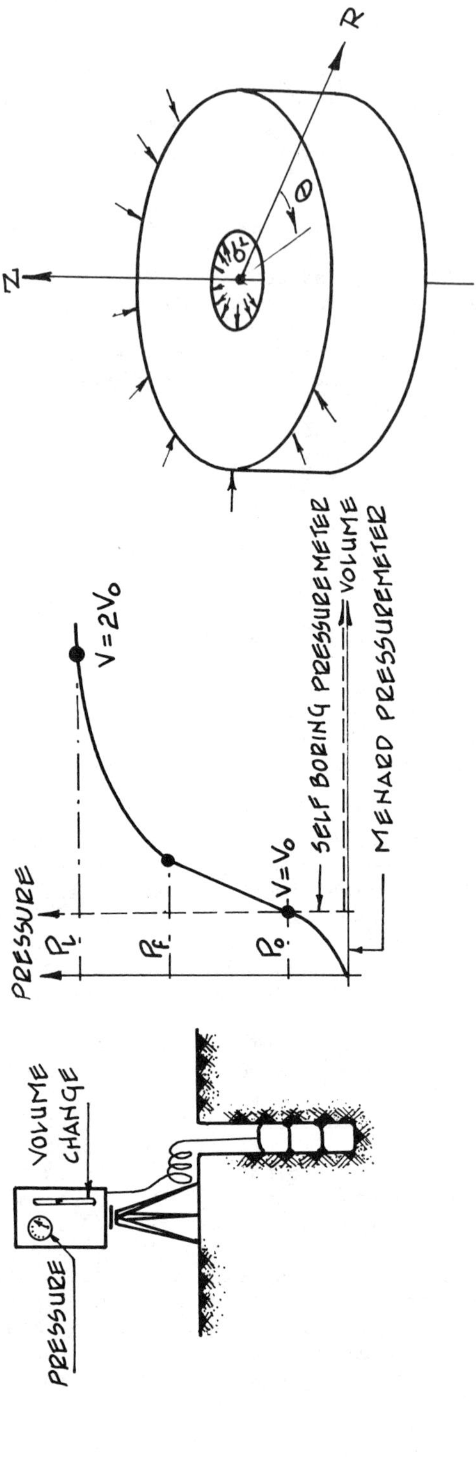

(A) (B) (C)

Fig. 1. (A) PRESSUREMETER APPARATUS, (B) PRESSUREMETER CURVE AND, (C) FIELD ELEMENT

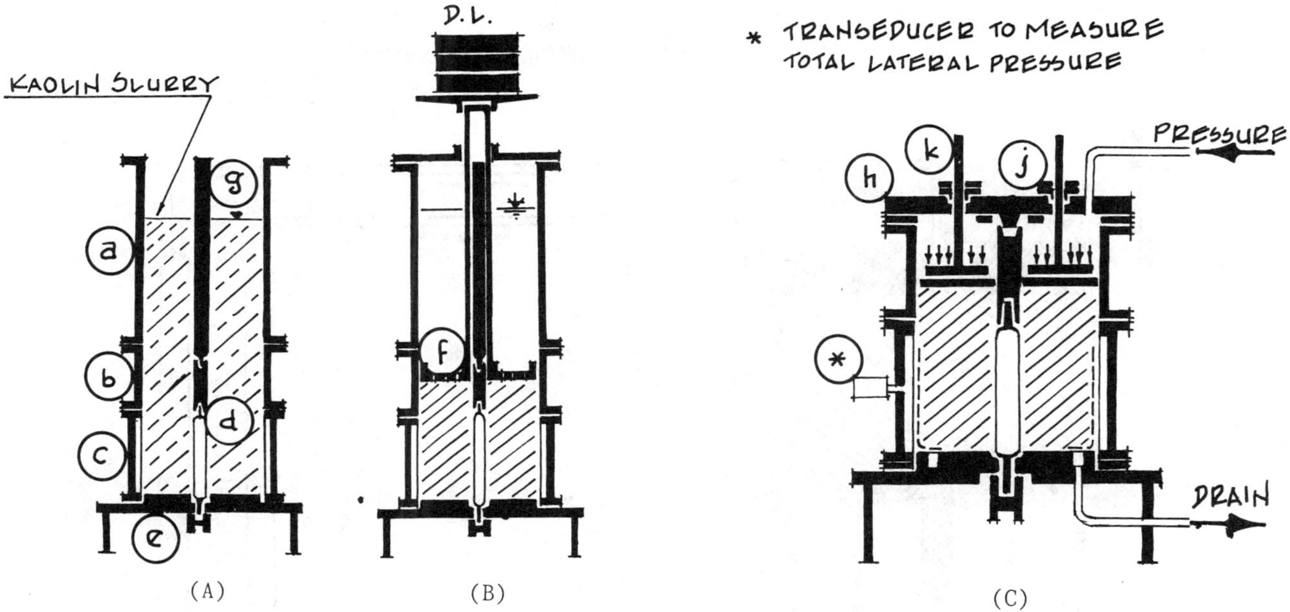

Fig. 2. SAMPLE PREPARATION

(A) & (B) SEDIMENTATION STAGE (C) CONSOLIDATION STAGE

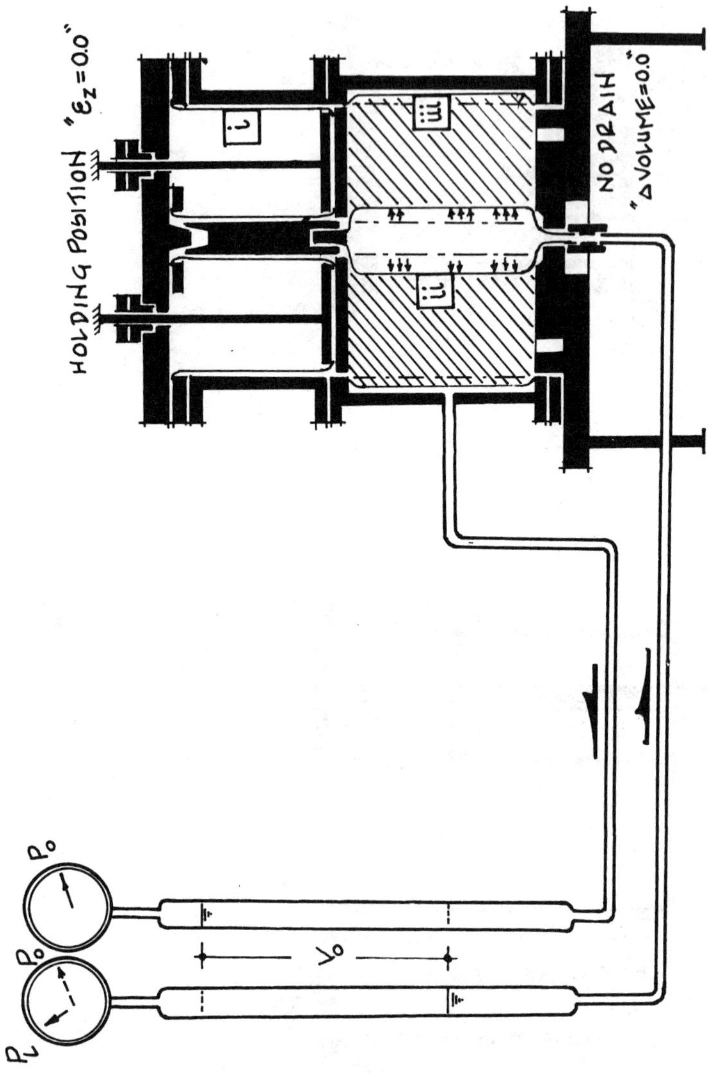

Fig. 3. SHEARING STAGE (FIELD TEST SIMULATION)

The Model Test

Before shearing the sample, the pressure inside both inner tube and outer chamber is checked. Both pressures are equal to: $Po = K_o \overline{\sigma_v}$, the lateral earth pressure under no lateral strain. Also, according to the drainage condition of the test, the drainage valve can be kept open or shut off during the shear test process.

The procedure of field test is applied to the model test: the pressure inside the inner tube membrane is increased in increments of 0.07 or 0.035 kg/cm^2, volume change of the tube is checked after 15, 30 and 60 seconds of each pressure increment and the test may be stopped when the volume of inner tube membrane is doubled.

It should be noted that the pressure inside outer chamber remains constant all the time with a value equal to Po. Also, the volume of water expelled from the outer chamber should be equal to the volume pushed in to the inner tube. Membrane resistance and back pressure are taken into account when pressuremeter curve is constructed.

A typical laboratory model test data is presented and reduced in Table 1 for pressure-volume change relationship. In Fig. 4, the pressuremeter curve and the creep curve of that test are shown and the pressures Po, the at rest pressure, P_f, the creep pressure, and P_1, limit pressure, are predicted.

The Tested Soil

The soil material used in this investigation was an artificially sedimented kaolinite by Georgia Kaolin Company under a brand name of Hydrite-PX. The prepared samples are reproducible and have the following properties:

Specific gravity	= 2.63
Liquid limit	= 60.7%
Plastic limit	= 35.7%
Compression index C_c	= 0.5
Overconsolidation index Cr	= 0.07
Void ratio e (at $\overline{P}$ = 1.0 kg/cm^2)	= 1.7

THE EXPERIMENTAL WORK

A series of 17 pressuremeter model tests were performed using different values of the maximum vertical pressure, $\overline{\sigma_v}$ and overconsolidation ratios (OCR = 1,2,4 & 8), as shown in Table 2.

For comparison purposes, two sets of Ko-consolidated undrained

TABLE 1--Pressuremeter test results (1psi = 0.07 kg/cm2)

Increment	Applied Pressure (psi)	60 Second Vol. RDG. V_{60} (cm^3)	Creep $V_{60}-V_{30}$ (cm^3)	Hydrostatic Pressure (psi)	Volume Correction (cm^3)	Corrected Volume (cm^3)	Membrane Resistance (psi)	Corrected Pressure (psi)
0	15.8	5.0	0	10	5.0	0	0	5.8
1	17	5.4	.1	"	"	.4	0	7.0
2	18	5.8	.2	"	"	.8	0	8.0
3	17	6.2	.1	"	"	.2	0	9.0
4	20	6.6	.1	"	"	1.6	.1	9.9
5	21	7.0	.2	"	"	2.0	.1	10.9
6	22	7.6	.2	"	"	2.6	.2	11.8
7	23	8.7	.4			3.7	.3	12.7
8	24	10.0	.7	"	"	5.0	.5	13.5
9	25	13.6	1.4	"	"	8.6	1.1	13.9
10	26	24.8	2.3	"	"	19.8	2.0	14.0
11	27	37.0	4.0	"	"	32.0	2.8	14.2

$\overline{\sigma_V}$ = 7.5 psi

$\overline{P_o}$ = 5.8 psi

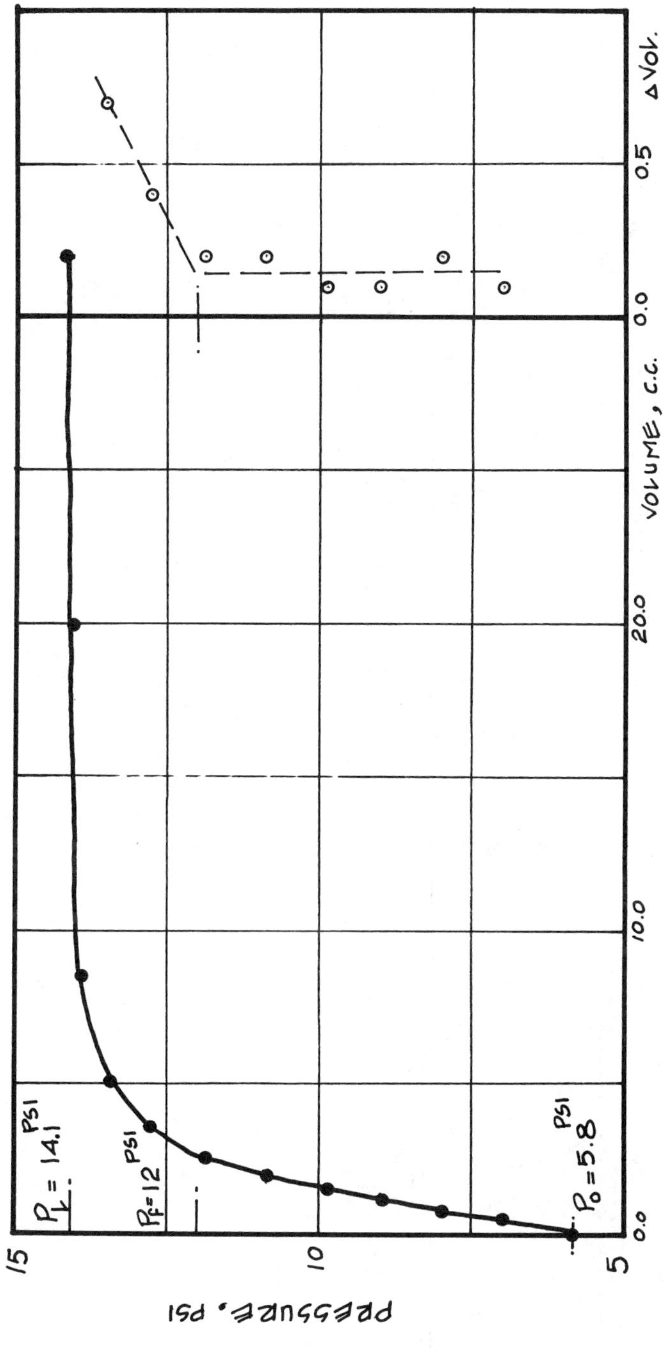

Fig. 4. TYPICAL PRESSUREMETER LABORATORY MODEL CURVE

(1 psi = 0.07 kg/cm^2)

TABLE 2—Summary of the normalized shear strength ($\frac{S_u}{\sigma_v}$) computed by different methods

OCR	Test No.	$\overline{\sigma_v}$	Normalized Undrained shear strength					
			Menard	Gibson	Palmer	Beguelin	Prevost	Denby
1	1	33	.09	.295	.27	.15	.24	.15
	2	19.4	.08	.265	.29	.14	.23	.15
	3	22	.06	.203	.21	.10	.16	.10
	4	40	.09	.28	.38	.226	.36	.215
	5	40	.085	.275	.325	.18	.30	.18
2	1	20	.09	.29	.31	.21	.28	.19
	2	15	.12	.39	.48	.30	.44	.26
	3	10	.102	.33	.41	.39	.57	.26
	4	16	.10	.33	.43	.23	.37	.22
4	1	8.0	.21	.68	.78	.34	.56	.42
	2	5.0	.19	.63	1.12	.53	.87	.44
	3	7.5	.20	.65	1.07	.59	.96	.53
	4	10.0	.20	.66	1.05	.59	.96	.54
8	1	7.8	.32	1.03	1.66	.78	1.59	.79
	2	3.75	.27	.87	1.33	.60	.97	.58
	3	3.0	.29	.95	1.47	.67	1.13	.60
	4	5.0	.29	.934	1.37	.68	1.14	.69

triaxial tests: compression ($\overline{CK_o\text{-}UC}$) and extension ($\overline{CK_o\text{-}UE}$) were carried out in a wide range of $\bar{\sigma}_v$ and OCR. The average of the compression and extension strength was used in the comparison with the undrained shear strength in the stress system of the pressuremeter test.

Results and analysis

The undrained shear strength of the laboratory model samples were predicted for each test by the different methods of interpretation. The normalized values of ($S_u/\bar{\sigma}_v$) for each test and method are presented in Table 2. The average values of ($S_u/\bar{\sigma}_v$) of each method as well as the normalized average values of the triaxial tests (average of compression and extension strengths) are plotted against the overconsolidated ratio (OCR) in Fig. 5. It can be seen from Fig. 5 that there are significant differences in the results by different methods of interpretation. Also there is a general disagreement between the results of the interpretation methods and the triaxial tests results. The method by Prevost fitted best with the triaxial test results.

The different assumptions made by each method in the stress-strain behavior can be considered to be the main reason for the significant differences in the results using different methods. The assumptions may be summarized as: elastic behavior (Menard), elastic-perfectly plastic (Gibson & Anderson), restrained to a certain elasto-plastic relationship (Prevost & Hoeg and Denby & Clough) and unrestrained to any particular stress-strain behavior (Baguelin, Ladanyi and Palmer).

On the other hand, the disagreement between the results of pressuremeter interpretation methods and the results of triaxial test could be due to the following reasons:

1. The different stress system of each test and the failure plane orientation. All methods of pressuremeter interpretation assume that the vertical stress does not influence the soil behavior. This assumption is good only for overconsolidated soil when the vertical stress, σ_v, at failure lies in between σ_r and σ_θ. For normally consolidated and highly over-consolidated soils, these methods would then be inappropriate since the vertical stress is major or minor principal stress and the plane of failure is not vertical any more [10].

2. The failure to include the effect of the excess pore water pressure developed due to the borehole expansion. None of the interpretation methods included the effect of the excess pressure in reducing the test data.

Conclusions

The early conclusions of this investigation can be summarized as follows:

1. The soil properties predicted by the pressuremeter field test not only depend on the soil condition and test procedure but also on the method used for the analysis.

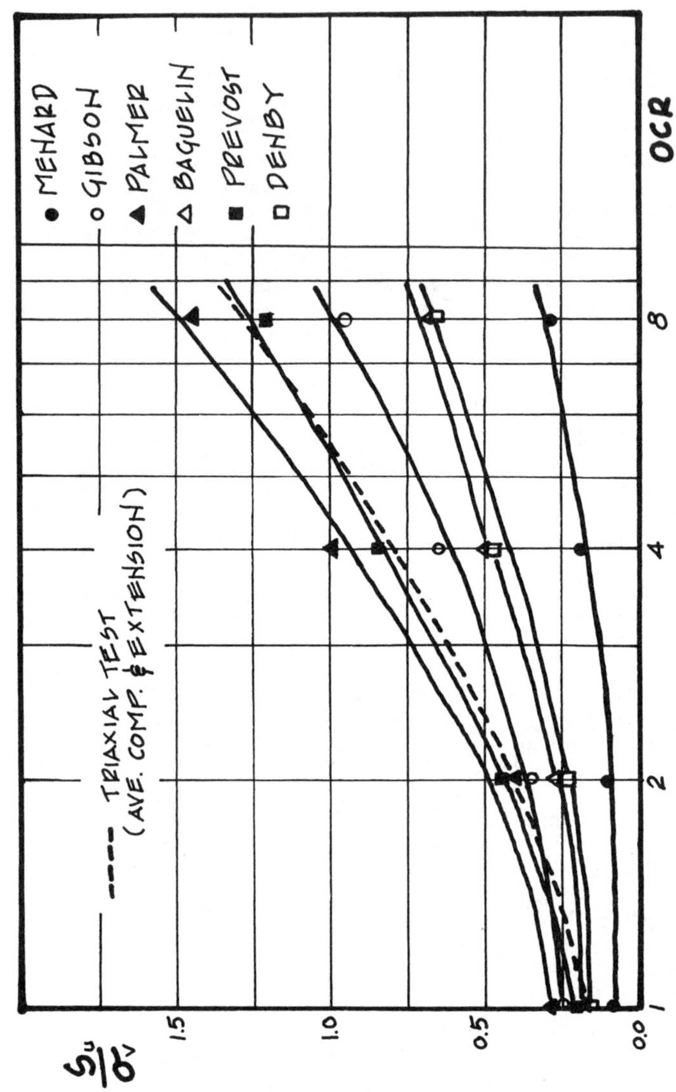

Fig. 5. NORMALIZED UNDRAINED SHEAR STRENGTH BY DIFFERENT METHODS VERSUS THE OVERCONSOLIDATION RATIO

2. Of the five methods reviewed, the method by Prevost gives the best agreement with the results of the triaxial test.

3. A new method taking into account the effect of the vertical stress should be developed considering vertical or non-vertical failure planes.

4. The use of pressuremeter with pore water pressure measurement capability should be considered.

APPENDIX

Typical Interpretation of Pressuremeter Test Results

Test No. 4-3 (presented in Table 1 & Fig. 4): OCR=4, $\sigma_{V max}$= 40 psi, σ_{vc} = 17.5 psi, Back Pressure = 10 psi, water content = 62.5%, P_o (at rest) = 5.8 psi, P_f (creep pressure) = 12.0 psi and P_l (limit pressure) = 14.1 psi

Menard (Method 1):

$$S_u = \frac{P_l - P_o}{5.5} = 1.51 \text{ psi}$$

Gibson and Anderson (Method 2):

$$S_u = \frac{P_l - P_o}{1 + \ln(1/2\alpha_f)}$$

in which α_f = Almansi strain at failure ($V_f = 2V_o$) = 0.25

$$S_u = 4.9 \text{ psi}$$

Palmer (Method 3):

$$\emptyset(Y_i) = \frac{2\, d\psi}{d[\log(\frac{\Delta V}{V})]}$$

in which $\emptyset(Y_i)$ = difference $(\sigma_r - \sigma_\theta)$ between principal stresses, ψ= the pressure applied to cavity wall, V=volume of the cell in the deformed state and ΔV = increase in the volume of cell.

$$S_u = \tfrac{1}{2}\, \emptyset(y_i)$$

The slopes have been computed from Fig. 6 and presented in Table 3, from which S_u = 8.0 psi

Baguelin (Method 3):

$$S_u = \frac{\sigma_r - \sigma_\theta}{2} = f(\epsilon); \quad f(\epsilon_o) = \epsilon_o \frac{dF}{d\epsilon_o}(1+\epsilon_o)(1+\epsilon_o/2)$$

in which $\epsilon_o \frac{dF}{d\epsilon_o}$ is the subtangent NT of the pressuremeter curve (Fig.7)

From Table 4, S_u = 4.44 psi

TABLE 3 - Data for Palmer method

Increment No.	Pressure P, PSI	ΔV cc	V_i cm	$\frac{\Delta V}{V_i} = \epsilon_v$ (V_o=27.6CC)	$\frac{\Delta P}{\Delta \log \frac{\Delta V}{V}}$
0	5.8	0	.0373	0	
1	7.0	.4	.9491	.014	4.5
2	8.0	.8	.951	.028	6.8
3	9.0	1.2	.957	.042	8.0
4	9.9	1.6	.964	.055	8.0
5	10.9	2.0	.971	.068	8.0
6	11.8	2.6	.981	.086	7.5
7	12.7	3.7	.998	0.118	6.5
8	13.5	5.0	1.019	.153	4.5
9	13.9	8.6	1.073	.238	1.5
10	14.1	19.8	1.228	.418	0.5
11	14.2	32.0	1.377	.537	0.1

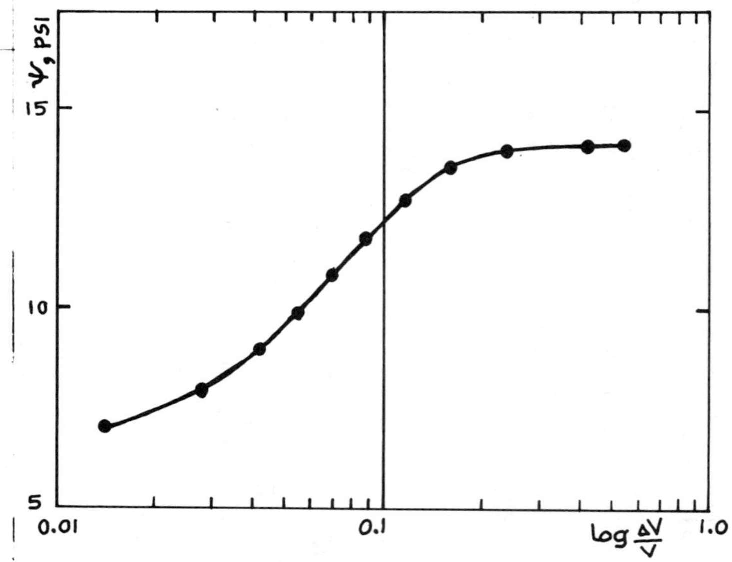

Fig. 6. ψ VERSUS LOG $\frac{\Delta V}{V}$ CURVE FOR PALMER METHOD (TEST 4-3)

TABLE 4 - Data for Baguelin Method

Increment No.	Pressure PSI	ΔV cm^3	V_i cm	ϵ_i	$1+\epsilon_i$	$1+\frac{\epsilon_i}{2}$	$\frac{\Delta p}{\Delta \epsilon}$	$[\epsilon_i(1+\epsilon_i)(1+\frac{\epsilon_i}{2}) \times \Delta p/\Delta\epsilon]$	$\sqrt{3}\,\epsilon_i \frac{\Delta p}{\Delta \epsilon}$	$\frac{\Delta \epsilon}{\Delta p}$
0	5.8	0	.9373	0	1	1	150	0	0	.0066
1	7.0	.4	.044	.0072	1.0072	1.0036	150	1.09	1.87	.0066
2	8.0	.8	.951	.015	1.015	1.0075	150	2.30	3.9	.0066
3	9.0	1.2	.957	.021	1.021	1.0105	150	3.25	5.46	.0066
4	9.9	1.6	.964	.0286	1.0286	1.0143	130	3.88	6.44	.0077
5	10.9	2.0	.971	.0356	1.0356	1.0178	110	4.13	6.78	.0091
6	11.8	2.6	.981	.0461	1.0461	1.023	90	4.44	7.19	.0111
7	12.7	3.7	.998	.0649	1.0649	1.0325	50	3.57	5.62	.020
8	13.5	5.0	1.019	.0868	1.0868	1.0434	30	2.95	4.51	.033
9	13.9	8.6	1.073	.145	1.145	1.0726	15	2.68	3.77	.066
10	14.1	19.8	1.228	.311	1.311	1.1552	5	2.35	2.68	.20
11	14.2	32.0	1.377	.568	1.568	1.284				

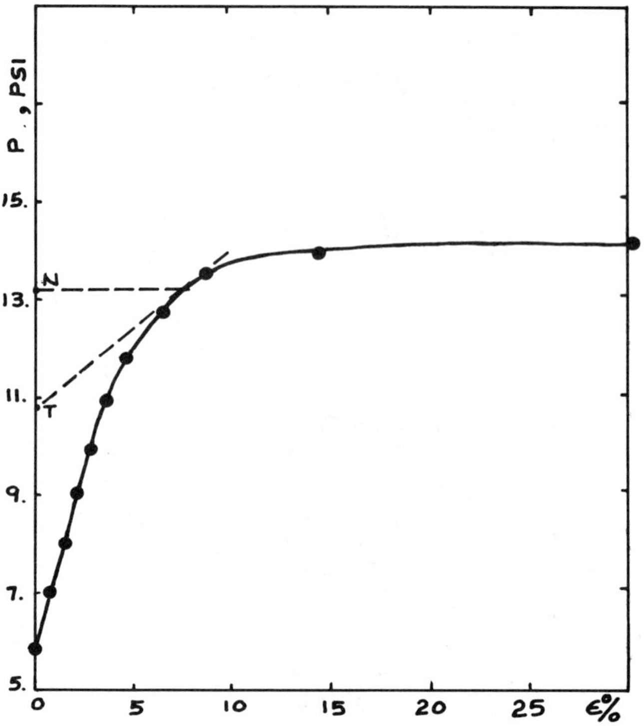

Fig. 7. P VS Є%

Prevost and Hoeg (Method 4): As presented in Reference (5), the stress-strain response was fitted for strain-hardening and strain-softening to the following equations

$$q = q_{ult} \frac{\epsilon}{\epsilon + a} \qquad \text{(strain-hardening)}$$

$$q = A \frac{B\epsilon^2 + \epsilon}{1 + \epsilon^2} \qquad \text{(strain-softening)}$$

where $\epsilon_{peak} = B + \sqrt{1 + B^2}$
and $q_{RES.} = AB$

The shear stress at any moment is computed as

$$S_u = \sqrt{3} \left(\frac{u_o}{r_o} \frac{d(\sigma_1)_o}{d(u_o/r_o)} \right)$$

in which $\frac{u_o}{r_o} = \epsilon$

= strain at the cavity walls and $\frac{d(\sigma_1)_o}{d(\frac{u_o}{r_o})} = \frac{\Delta p}{\Delta \epsilon}$ = slope of the pressure-meter curve, Fig. 7.

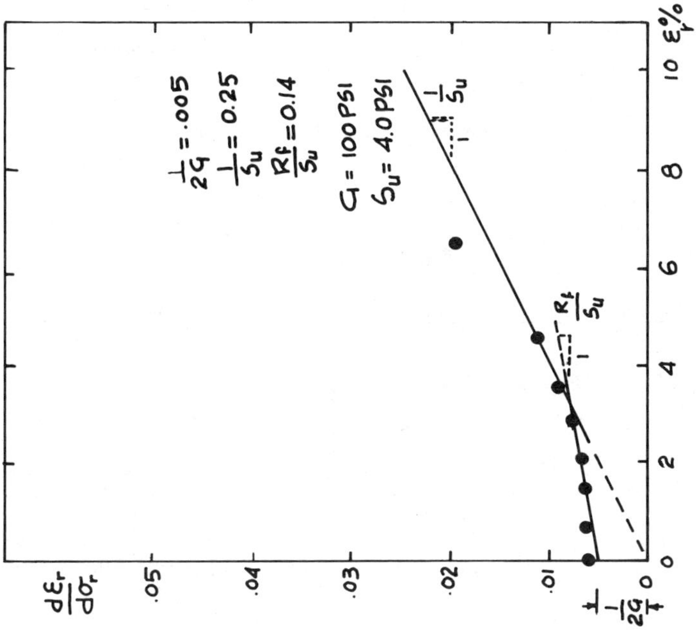

Fig. 9. $\frac{d\epsilon_r}{d\sigma_r}$ vs. ϵ_r FOR DENBY & CLOUGH

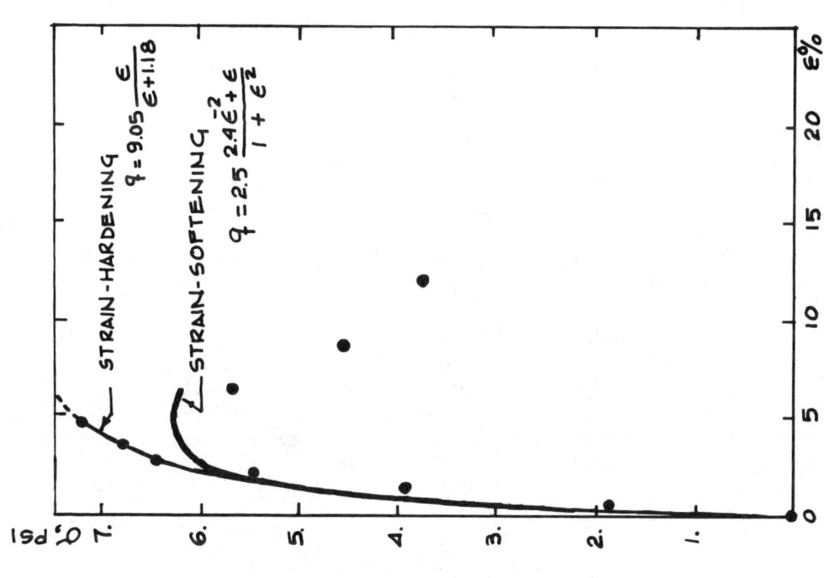

Fig. 8. σ VS ϵ% FOR PREVOST METHOD

Assuming strain-hardening condition ($0.0 < \epsilon < 5\%$), the stress-strain response can be presented by the following relationship

$$q = 9.05 \frac{\epsilon}{\epsilon + 1.18} \quad (0.0 < \epsilon < 5\%)$$

For strain-softening; the stress-strain relationship is

$$q = 2.5 \frac{2.4\epsilon^2 + \epsilon}{1 + \epsilon^2} \quad (0.0 < \epsilon < 6\%)$$

As shown in Fig. 8, the computed values are better fitted with the strain hardening curve. Therefore, S_u = 7.2 psi. ($\epsilon = 5\%$)

<u>Denby and Clough (Method 5)</u>: as presented in Reference (2) equation (2):

The shear shtress is computed as

$$S_u = \frac{\epsilon_r}{\frac{1}{2G_i} + \frac{R_f}{S_u} \epsilon_r}$$

From $\frac{d\epsilon_r}{dq_r}$ vs ϵ_r curve (Figure 9), the following parameters can be obtained

$\frac{1}{2G}$ is the intercept of the resulting straight line,

$\frac{R_f}{Su}$ is the initial slope of the curve,

and $\frac{1}{Su}$ is the slope of the second straight line,

where $\frac{d\epsilon_r}{d\sigma_r}$ vs ϵ_r curve is the inverse of the slope of the pressuremeter curve versus the radial strain.
from Figure 9, the shear strength S_u = 4.0 psi

REFERENCES

[1] Baguelin, F. & Jezequel, J.F. and Shields, D.H., <u>The Pressuremeter and Foundation Engineering</u>, Trans Tech Publications, Germany, 1978
[2] Denby, G. and Clough, G.W., "Self-Boring Pressuremeter Test in Clay," <u>Journal of Geotechnical Division</u>, ASCE, 1980, Vol. 106, No. GT 12, pp. 1369-1387.
[3] Palmer, A.C., "Undrained Plane-Strain Expansion of Cylindrical Cavity in Clay: A Simple Interpretation of the Pressuremeter Test," <u>Geotechnique</u>, 1972, Vol. 22, No. 3, pp. 451-457.
[4] Ladanyi, B., "In-Situ Determination of Undrained Stress Strain Behavior of Sensitive Clays with the Pressuremeter," <u>Canadian Geotechnical Journal</u>, 1972, Vol. 9, No. 3, pp. 313-319.
[5] Prevost, J.H. and Hoeg, K., "Analysis of Pressuremeter in Strain Softening Soil," <u>Journal of the Geotechnical Engineering Division</u>, ASCE, 1975, Vol. 101, No. GT8, pp. 717-731.
[6] Clough, G.W. and Denby, G., "Self-Boring Pressuremeter Study of San Francisco Bay Mud," <u>Journal of Geotechnical Division</u>, ASCE, 1980, Vol. 106. No. GT1, pp. 45-63.

[7] Baguelin, F., et al, "Expansion of Cylindrical Probes in Cohesive Soils," Journal of the Soil Mechanics and Foundations Division, ASCE, 1972, Vol. 98, No. SM11, pp. 1129-1142.

[8] Lukas, G.L. and Debussy, L., "Pressuremeter and Laboratory Test Correlations for Clays," Journal of the Geotechnical Engineering Division, ASCE, 1976, Vol. 102, GT9, pp. 945-963.

[9] Schmertmann, J.H., "The Measurement of In-Situ Shear Strength," Proc. of the Conference on In-Situ Measurement of Soil Properties, ASCE, Raleigh, NC, 1975.

[10] Jewell, R.J., Fahey, M., Wroth, C.P., "Laboratory Studies of the Pressuremeter Test in Sand," Geotechnique, 1980, Vol. 30, No. 4, pp. 507-531.

Michel Soulié, Branko Ladanyi, and Philippe Degenne

EXPANSION OF A CYLINDRICAL CAVITY IN A VERY DEFORMABLE MEDIUM: A THEORETICAL STUDY

REFERENCE: Soulié, M., Ladanyi, B., and Degenne, P., "Expansion of a Cylindrical Cavity in a Very Deformable Medium", The Pressuremeter and its Marine Applications: Second International Symposium, ASTM STP 950, J.L. Briaud and J.M.E. Audibert, Eds., American Society for Testing and Materials, 1986.

ABSTRACT: Most classical solutions for stress and strain distribution around an expanding cavity that have up to now been used for the interpretation of pressuremeter tests in soils, have been based essentially on a small strain assumption. This paper presents a finite element method solution of the problem of an expanding cylindrical cavity, which is based on a new incremental formulation, considered to lead to a more correct representation of large strain problems than some earlier ones. In addition, a new large-strain analytical solution is developed in the paper, which is shown to be able to determine the limit cylindrical expansion pressure, even for very deformable materials, which are outside the scope of classical solutions.

KEYWORDS: cylindrical cavity, expansion, large strain, low rigidity, finite element method, pressuremeter test interpretation.

INTRODUCTION

The classical solutions for stress and strain distribution around an expanding cavity that have up to now been used for the interpretation of pressuremeter tests in soils (as reviewed in [1]), have all been based essentially on small strain assumption. In the case of an "undrained" pressuremeter test in an imcompressible material, where the state of strain is usually characterized by the shear strain $\gamma = \Delta V/V$ (where ΔV denotes the volume increase of the cavity and V its current volume),

Drs. Soulié and Ladanyi are professors at the Civil Engineering Department of Ecole Polytechnique, C.P.6079, Station A, Montreal, H3C 3A7, Canada; Mr. Degenne, a former graduate student at the same department, is presently with Elf Aquitaine, Paris, France.

this strain attains usually a maximum value of about 0.5, occurring when the final cavity volume becomes twice the initial one. Although this strain is quite high, corresponding to an increase in the cavity radius of over 40% , this does not affect the results of conventional interpretation methods based on the pressuremeter curve [2,3,4]. However, when one wants to calculate the limit pressure, which corresponds to $\gamma = \Delta V/V \rightarrow 1$, the large strain problem comes into play. In that, as will be shown in the following, what really counts is the stress level with respect to the modulus of rigidity of the material.

THEORETICAL CONSIDERATIONS

A problem in continuum mechanics can be solved by applying the principle of virtual work, which in eulerian formulation can be expressed as

$$\int_S F_i\ \delta(v_i)dS + \int_V X_i\ \delta(v_i)dV = \int_V \sigma_{ij}\ \delta(\dot{\varepsilon}_{ij})dV \qquad (1)$$

where: V is the volume of the body, S is its surface, F_i are the components of a force acting on that surface, X_i are the components of a mass force per unit volume, $\delta(v_i)$ are the components of the virtual velocity vector, $\delta(\dot{\varepsilon}_{ij})$ is the virtual eulerian strain rate, and σ_{ij} is the eulerian stress tensor. In this formulation, all above defined magnitudes refer to the current deformed state of the body.

As the current state of the body is unknown, one can express the virtual work principle using a lagrangian formulation [5,6,7,8].

$$\int_{S_o} F_i^o\ \delta(v_i)dS_o + \int_{V_o} X_i^o\ \delta(v_i)dV_o = \int_{V_o} S_{ji}\ \delta(v_{i,j})dV_o \qquad (2)$$

where suffix "o" refers to the initial state of the body, v_i are the components of the velocity vector, $v_{i,j} = \delta v_i/\delta x^o_j$, while S_{ji} is the non-symmetric Boussinesq stress tensor (or Polia-Kirchhoff I).

Note that x_i are the coordinates of the solid in its initial state, and [S] is related to the eulerian stress tensor by

$$[\sigma] = \frac{\rho}{\rho_o}[\gamma][S] \qquad (3)$$

where ρ_o and ρ are the densities before and after deformation, and $[\gamma]$ is the coordinate transformation tensor,

$$\gamma_{ij} = \frac{\delta x_i}{\delta x^o_j} \qquad (4)$$

The last expression has the advantage of being formulated with respect to the initial state which is known. This formulation is necessary for large strain problems, as well as for incremental formulations used in FEM, when one wants to solve large strain problems by making small strain increments and by modifying the mesh at each stage. In fact, the need for using Eq. 2 in an incremental form applies also to the general case of a material which, in its initial state, is already submitted to a stress field.

In such an incremental form, starting from the state at the end of previous stage in an incremental process, Eq. 2 becomes

$$\int_{S_o} \dot{F}_i^o \, \delta(v_i) dS + \int_{V_o} \dot{X}_i^o \, \delta(v_i) dV_o = \int_{V_o} \dot{S}_{ji} \, \delta(v_{i,j}) dV_o \tag{5}$$

where V_o and S_o represent the state of the solid at the start of the increment (or the state at the end of the previous increment), which means that the solid is being continuously reshaped; $\dot{F}_i^o, \dot{X}_i^o$ are the rates of variation of surface- and mass-forces per unit area and volume, respectively, at the end of the preceding increment. Here, $\dot{F}_i^o$ is related to the rate of variation $\dot{F}_i$ per unit of current (deformed) surface by

$$\dot{F}_i^{\,o} \, dS_o = \dot{F}_i \, dS_o + F_i dS \tag{6}$$

Thus, for a constant force per unit area ($\dot{F}_i = 0$), the action of these forces is manifested by: $\dot{F}_i^{\,o} \, dS_o = F_i d\dot{S}$.

Further, in Eq. 5, $\dot{S}_{ij}$ represent the rate of variation of Boussinesq stresses (Polia-Kirchhoff I), which in an updated eulerian formulation has the form [10]:

$$[\dot{S}] = [\dot{\sigma}] + (v_{i,i})[\sigma] - [v][\sigma] \tag{7}$$

where $(v_{i,i})$ is the divergence of the velocity vector, and $[v]$ is the matrix of components

$$v_{i,j} = \delta v_i / \delta x_j^o$$

Moreover, in the case of a cylindrical expansion, the non-rotation of principal stress axes makes it possible to assimilate $[\dot{\sigma}]$ to the variation of the eulerian stress tensor at a fixed point.

The incremental form of the principle of virtual work can then be written as

$$\int_{V_o} \{(v_{i,i})[\sigma] + [\dot{\sigma}] - [\sigma][v]\} \, \delta(v_{i,j}) dV_o = \tag{8}$$

$$= \int_{V_o} \dot{X}_i^{\,o} \, \delta(v_i) dV_o + \int_{S_o} \dot{F}_i^{\,o} \, \delta(v_i) dS_o$$

Taking into account that the stress-strain relationship can be expressed by

$$[\dot{\sigma}] = D[\dot{\varepsilon}] \tag{9}$$

and that, because of absence of rotation during cylindrical cavity expansion,

$$v_{i,j} = \dot{\varepsilon}_{i,j} + \dot{w}_{i,j} = \dot{\varepsilon}_{i,j} \tag{10}$$

or

$$[v] = [\dot{\varepsilon}]$$

and considering also the absence of mass forces, the incremental formulation reduces to

$$\int_{V_o} \{[D-\sigma][\dot{\varepsilon}] + (v_{i,i})\sigma\}\ \delta(\dot{\varepsilon}_{ij})dV_o = \int_{S_o} \dot{F}_i\ \delta(v_i)dS_o \tag{11}$$

Only if the D-terms are much larger than those of $[\sigma]$, the first integral can be written as

$$\int_{V_o} D\,[\dot{\varepsilon}]\ \ \delta(\dot{\varepsilon}_{ij})dV_o \tag{12}$$

which is the classical way of expressing the incremental form of virtual work of deformations, corresponding to the principle of superposition of small strain solutions.

Thus, quite generally, especially if one wants to study the behavior of very deformable materials (which, by definition have the D-terms of the same order of magnitude as those of the state of stress), the formulation given by Eq. 11 should be used, as in this paper rather than that of Eq. 12, as it was done in some other papers, to be discussed later.

PROCEDURE

The following procedure was used for the "updated" eulerian formulation:

(1) In the incremental formulation, the stiffness matrix of each element is calculated following the geometry of the state of stress corresponding to the deformed state of previous stage.

(2) The solution of Eq. 11 by the finite element method furnishes the displacement increments in each node, as well as the variations of stresses given by

$$[\dot{\sigma}] = D[\dot{\varepsilon}]$$

(3) Since the equivalent matrix is a function of the state of stress, the system is again resolved by taking as initial stresses the average values between those just found and the former initial stresses $[\sigma]_i$.

(4) The displacements so determined are added to the coordinates, and the stress increments to the former initial stresses $[\sigma]_i$, in order to get a new initial geometry and a new initial state of stress.

(5) The process is repeated in the next increment.

The only existing analytical solutions, valid for any deformation level, furnish the value of internal pressure, p_{TY}, at the moment of the total yield. For that case, Mandel [7] gives the formula

$$(p_{TY})_{sph} = 4c \ln\frac{b}{a} \tag{13}$$

where b and a are respectively external and internal radius of the sphere at that moment. Table 1 shows, for three different values of G/c, the differences between theoretical and calculated values of $(p_{TY})_{sph}$.

TABLE 1

Total yield pressures for an expanding thick sphere

G/c	a (m)	b (m)	p_{TY}(FEM) (kPa)	p_{TY}(Theory) (kPa)
0.385	0.46946	0.47938	4.97	4.82
3.85	0.09895	0.13792	77.09	76.62
38.50	0.06553	0.12168	142.67	142.83

Moreover, as mentioned previously, several methods have been proposed for modifying the values of the stress tensor at each increment. Using two of these methods and our own, we have calculated the value of $(p_{TY})_{sph}$ for G/c = 0.962. The results are shown in Table 2.

TABLE 2

Total yield pressures for an expanding thick sphere.
Comparison of three different methods for G/c = 0.962

Method	a (m)	b (m)	p_{TY}(FEM) (kPa)	p_{TY}(Theory) (kPa)
Authors	0.18775	0.20925	25.65	25.02
Mandel	0.18733	0.20882	25.49	26.06
Yamada	0.18727	0.20870	22.94	25.00

In the same manner, the case of a thick cylinder was studied, with the same radii and characteristics as for the thick sphere ($b_o/a_o = 2$, $\nu = 0.3$, $c = \frac{1}{2}(\sigma_{max}-\sigma_{min}) = 57.7$ kPa. The finite element mesh comprised 20 quadratic 1-D elements.

For small strains, Hill [6] gives a complete analytical solution for stresses σ_r, σ_θ and σ_z both in the elastic ($a < r < d$) and in the plastic ($d < r < b_o$) region. In his equations, which are not repeated here, a denotes the deformed internal radius, d is the radius of the elastic-plastic boundary and b_o is the initial external radius.

A trial calculation by our method for a thick cylinder with $a = 0.060075$, $d = 0.08726$ and $b_o \approx 0.12$ m furnished the values of principal stresses that were in a perfect agreement with the Hill's solution [5], (not shown here because of lack of space).

Note that, in stages (3) and (4), a simple superposition was used for stresses. For example, for step i+1, this implies that $[\sigma]_{i+1} = [\sigma_i] + [\dot{\sigma}]$.

Some other authors have prefered to use other methods. For example, Mandel [7] calculates the variation of Boussinesq stresses and adds them to the initial stresses:

$$[S]_{i+1} = [\sigma]_i + [\dot{S}] \tag{14}$$

In order to pass to Euler stresses, he then transforms these latter stresses by means of the relationship

$$[\sigma]_{i+1} = \frac{\rho}{\rho_o} [S]^T_{i+1} [\gamma]^T \tag{15}$$

Yamada [8,9,10] and Banerjee [11] calculate the variations of Kirchhoff stresses and add them to the initial stresses

$$[t]_{i+1} = [\sigma]_i + [\dot{t}]$$

where

$$[\dot{t}] = [\dot{S}] - [\dot{e}]^T [S] \tag{16}$$

which they then transform into Euler stresses by the relationship

$$[\sigma]_{i+1} = \frac{\rho}{\rho_o} [\gamma][t]_{i+1} [\gamma]^T \tag{17}$$

A comparison of these three methods is shown later in this paper.

In the incremental calculation, one can use the increments of either loads or displacements. Since, for a spherical expansion, Mandel [7] has found the existence of a peak on the internal pressure vs. displacement curve, we have decided to use the displacement increments as the basis.

VALIDATION OF THE METHOD

In order to validate the proposed procedure, and compare it with an earlier solution [12] , an analysis was made of the case of spherical expansion in an elastic-plastic material, obeying the Tresca criterion ($\sigma_{max} - \sigma_{min} = 2c$), by considering a thick-walled sphere with an internal radius of $a_o = 0.06$ m and an external radius of $b_o = 0.12$ m. The Poisson coefficient ν was taken equal to 0.3, and the cohesion c = 57.7 kPa. In the calculation, the ratio G/c between the modulus of rigidity

$$G = E/2(1+\nu) \tag{18}$$

and the cohesion was varied to simulate the materials of various rigidities.

For a thick cylinder, Hill [6] gives also a formula for internal pressure at the moment of a total yield

$$(p_{TY})_{cyl} = 2c \ln\frac{b}{a} \tag{19}$$

with the notation as in Eq. 13. Table 3 shows a comparison between the values of $(p_{TY})_{cyl}$ obtained numerically and analytically.

TABLE 3

Total-yield pressures for an expanding thick cylinder

G/c	a (m)	b (m)	p_{TY}(FEM) (kPa)	p_{TY}(Theory) (kPa)
0.385	0.69893	0.71876	3.35	3.23
3.85	0.09565	0.14382	47.65	47.07
38.50	0.06385	0.12219	74.65	74.90

Further, in order to compare our method with that published earlier by Carter [12], we have varied G/c between 0.385 and 1000 and have plotted in Fig. 1 the results as p/c vs $G(b-b_o)/b_oc$. As will be seen in Fig. 1, all the curves show a distinct peak before complete plasticity, or for the most rigid case (G/c = 1000), at the moment of complete plasticity. For the most deformable case (G/c = 0.385), the maximum value is attained before any yield occurs.

The numerical solution shown in Fig. 1 is very different from that obtained by Carter [12], who solved the same problem using a different approach. As will be seen in Fig. 2, his internal pressure ratio p/c remains systematically higher than that obtained here (Fig. 1) and the curves do not present any peak.

In his approach, Carter [12] and Carter et al. [13,14] use the virtual work principle in an eulerian incremental form, giving for a time increment of t_o to t

$$\int_V \{[\sigma]_o + \int_{t_o}^{t} [\dot{\sigma}]dt\}\ \delta[\dot{\varepsilon}] = \int_V X_i\ \delta v_i\ dV + \int_S F_i\ \delta v_i\ dS \tag{20}$$

where: V and S are unknown values of the volume and the area of the solid at time t, $[\sigma]_o$ are the Euler stresses at time t, $[\dot{\sigma}]$ (with no rotation) is the differential of $[\sigma]$ with respect to time, corresponding to the constitutive Eq. 9.

Besides the difficulty of such an approach, because the integration domains corresponding to the deformed state of the solid are unknown, it seems difficult to justify, as done by Carter [13,14], the integration of Eq. 9 with respect to time for expressing that at time t

$$[\sigma] = \int_{t_o}^{t} D[\dot{\varepsilon}]dt + [\sigma]_o \tag{21}$$

The latter implies in fact that from t_o to t, the point where $[\sigma]$ has to be determined did not move.

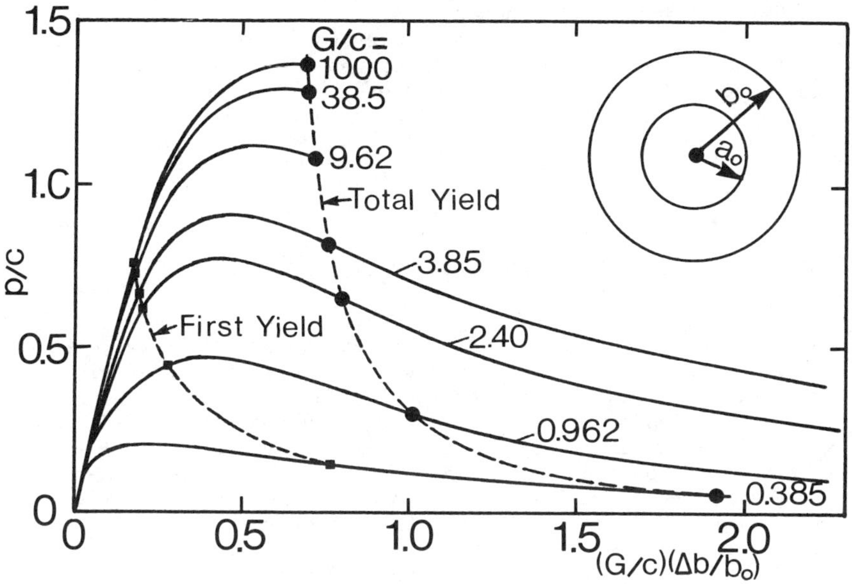

FIG. 1--Internal pressure, p, vs. external expansion, $\Delta b/b_o$, relationship for a thick cylinder with a_o = 6 cm, b_o = 12 cm, and for various G/c ratios (FEM calculation, this paper).

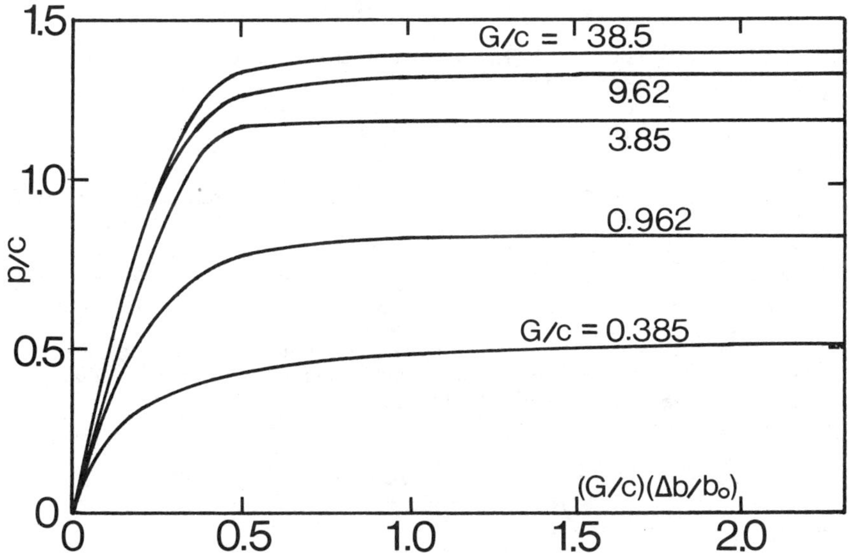

FIG. 2--Elasto-plastic cylinder behavior for the same case as in Fig. 1, according to Carter [12].

On the other hand, it is noted that for a cylinder of finite thickness, Eq. 19 remains still valid for calculating the total-yield pressure. If the expansion process is continued, it leads to an increase of both a and b, but the ratio b/a decreases and tends towards 1, resulting in a decrease of p . This behavior agrees well with that found by the proposed solution (Fig. 1), but differs from that shown by Carter (Fig. 2).

EXPANSION OF A CYLINDRICAL CAVITY IN AN INFINITE MEDIUM

In order to be able to interpret the cavity expansion corresponding either to a pressuremeter test or to the soil displacement during pile penetration, the expansion of a cylindrical cavity was approximated by that of a series of thick cylinders with increasingly thicker walls ($b_o/a_o \rightarrow \infty$). From this information, the limit pressure p_L for a cylindrical cavity was obtained by graphical extrapolation to $a_o/b_o \rightarrow 0$ of a line (Fig. 3), representing the variation of the peak pressure, $p_{max} = p_L$ with a_o/b_o, for cylinders with stress-free outer surface. From a number of such results for different rigidities, the relationship between p_L/c and G/c was determined and plotted in Fig. 4, for comparison with some other solutions. For example, the solution for p_L by Hill [6], which takes into account all elastic volume strains in the material, given by

$$p_L = c[1 + \ln \frac{2(1+\nu)}{(5 - 4\nu)} \frac{G}{c}] \tag{22}$$

is shown in Fig. 4 for $\nu = 0.3$, as well a more conventional one, to which Eq. (22) reduces when $\nu = 0.5$

$$p_L = c[1 + \ln \frac{G}{c}] \tag{23}$$

It will be seen in Fig. 4 that the large strain FEM solution deviates considerably from the analytical small-strain solutions, the difference increasing up to about 50% when G/c falls below unity. On the other hand, the analytical solutions are quite accurate when G/c exceeds about 50 to 100.

As for the interpretation of a pressuremeter test in very deformable materials under undrained conditions, it can be demonstrated that the method proposed in 1972 [2,3,4], relating the shear stress $\tau_m = \frac{1}{2}(\sigma_r - \sigma_\theta)$ at the cavity wall with the pressuremeter curve ($\sigma_r, \Delta V/V$)

$$\tau_m = \frac{d\sigma_r}{d \ln(\Delta V/V)} \tag{24}$$

where: V is the current cavity volume, ΔV is its variation, and σ_r is the pressure applied inside the cavity, remains still valid for large strains, as long as $\Delta V/V$ does not exceed about 0.5. However, extrapolation to the limit pressure by a straight-line extrapolation to $\Delta V/V = 1$ in a (σ_r, ln $\Delta V/V$)-plot, as proposed in [15] is not allowable for very deformable materials, as shown previously.

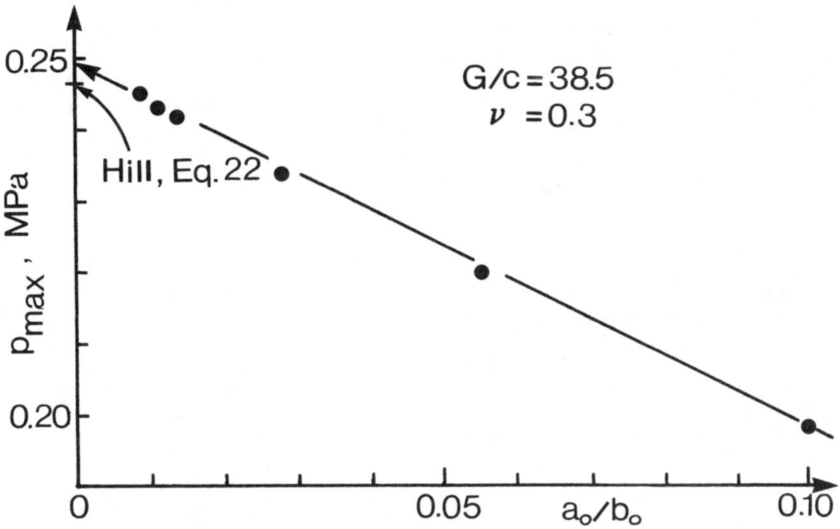

FIG. 3--Determination of the limit pressure, $p_{max} = p_L$ for an expanding cylindrical cavity, by extrapolating the FEM thick cylinder data to a_o/b_o = 0. Case: G/c = 38.5.

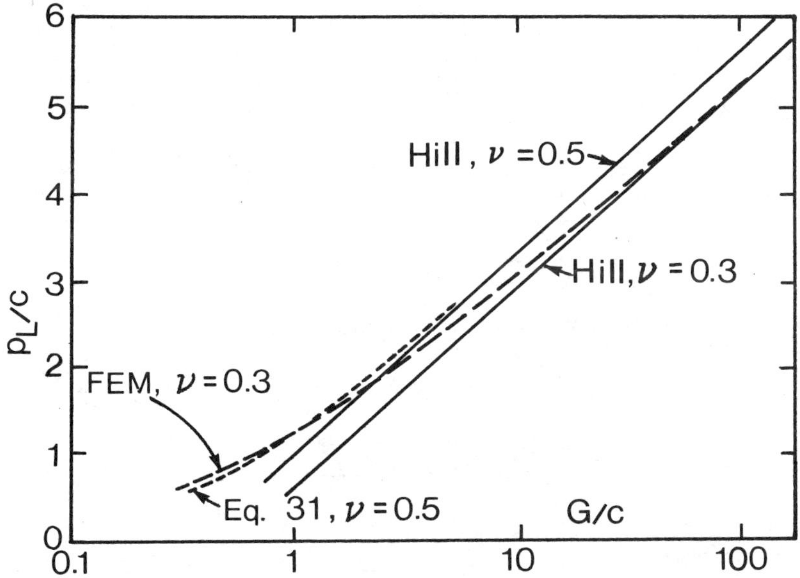

FIG. 4--Comparison of limit pressures calculated by the finite element method with those obtained from theoretical solutions.

To improve on this situation, an attempt was made in [16] to find an analytical solution for the limit pressure that would hold also for such very deformable materials. This was done by combining Eqs. 23 and 24. The resulting differential equation made it possible to prove that Eq. 23 remains valid for any soil rigidity, as long as $G/c > 1$ and the shear stress τ_m is linearly related to the shear strain γ, i.e.

$$2\tau_m = (\sigma_r - \sigma_\theta) = 2G\gamma \tag{25}$$

for $$\sigma_r - \sigma_\theta \leq 2c$$

On the other hand, for very deformable materials, Eq. 25 should be replaced by its incremental form

$$\delta(\sigma_r - \sigma_\theta) = 2G\ \delta(\Delta V)/V \tag{26}$$

for $$\sigma_r - \sigma_\theta \leq 2c$$

which leads to

$$\sigma_r - \sigma_\theta = 2G \ln(V/V_o) \equiv -2G \ln(1-\gamma). \tag{27}$$

Considering Eq. 27, integration of Eq. 24 gives

$$\sigma_r = G\ L(\gamma) \tag{28}$$

$$\sigma_\theta = G\ L(\gamma) + 2G \ln(1-\gamma) \tag{29}$$

where

$$L(\gamma) = -\int_o^\gamma \frac{\ln(1-z)}{z}\, dz = \sum_{n=1}^{\infty} \frac{\gamma^n}{n^2} \tag{30}$$

If, for $0 \leq \gamma \leq 1$, the validity of Tresca failure criterion is assumed, the limit pressure is given by

$$p_L = G\ L(\gamma_F) + c \ln \frac{1}{\gamma_F} \tag{31}$$

where γ_F is the failure strain, given by

$$\gamma_F = (\Delta V/V)_F = 1 - \exp(-\frac{G}{c}) \tag{32}$$

and $$L(\gamma_F) = \sum_{n=1}^{\infty} \frac{\gamma_F^{\,n}}{n^2} \tag{33}$$

It will be seen that, when

$$\frac{G}{c} \to \infty, \quad \frac{p_L}{c} \to (1 + \ln \frac{G}{c}) \tag{34}$$

and when $$\frac{G}{c} \to 0, \quad \frac{p_L}{c} \to \frac{G}{c}\frac{\pi^2}{6} \to 0. \tag{35}$$

Thus, it is found not only that the limit pressure depends on the value of γ_F (function of G/c) at the moment when τ_m attains its limit value at the cavity wall, but also that it depends on two terms, one related to the deformability and another to the maximum shear strength of the material.

Some results obtained by Eq. 31 are also plotted in Fig. 4. They are seen to come quite close to those calculated by the FEM solution shown previously.

CONCLUSIONS

A new "updated" eulerian formulation was used for solving by the finite element method the problem of an expanding cylindrical cavity in a very deformable material. The results show that classical closed-form solutions remain valid for relatively high G/c ratios, but become increasingly inaccurate when G/c falls below about 10. However, for $G/c < 1$, they become inapplicable, and should be replaced by numerical methods. The new FEM formulation shown in the paper is considered to be more correct for such very deformable materials than some earlier ones.

As for the interpretation of a pressuremeter test in weak soils, it is found that the current method based on pressuremeter curve discretization is valid for such soils, as long as $\Delta V/V < 0.5$, which is usually the case. However, for finding the ultimate limit pressure, the usual method of extrapolation of a pressuremeter curve to $\Delta V/V = 1$ in a semi-log plot is not recommended in such soils, but that pressure should be calculated by using the large-strain solution developed in the paper.

ACKNOWLEDGEMENTS

Financial assistance received for this study from the National Sciences and Engineering Research Council of Canada is gratefully acknowledged.

REFERENCES

[1] Baguelin, F., Jézéquel, J.F. and Shields, D.H., "The Pressuremeter and Foundation Engineering", Trans Tech Publications, Clausthal, 1978.

[2] Palmer, A.C., "Undrained Plane-Strain Expansion of a cylindrical Cavity in Clay: a Simple Interpretation of a Pressuremeter Test", Géotechnique, Vol. 22, Sept., 1972, pp. 451-457.

[3] Ladanyi, B., "In Situ Determination of Undrained Stress-Strain Behavior of Sensitive Clays with the Pressuremeter", Canadian Geotechnical Journal, Vol. 9, No. 3, 1972, pp. 313-319.

[4] Baguelin, F., Jézéquel, J.F., Lemée, E. and Le Méhauté, A., "Expansion of Cylindrical Probes in Cohesive Soils", Journal of the Soil Mechanics and Foundations Division, American Society of Civil Engineers, Vol. 98, No. SM11, Nov. 1972, pp. 1129-1142.

[5] Degenne, P., "Expansion d'une cavité cylindrique dans des sols très compressibles et très déformables", M.Sc. Thesis, Dept. of Civil Engineering, Ecole Polytechnique, Montreal, 1981.

[6] Hill, R., "The Mathematical Theory of Plasticity", Oxford University Press, Oxford, 1950.

[7] Mandel, J., "Cours de mécanique des milieux continus", Gauthier-Villars Editeur, Vol. I and II, Paris, 1966.

[8] Yamada, Y. and Wifi, A.S., "Large Strain Analysis of Some Geomechanics Problems by the Finite Element Method", International Journal for Numerical and Analytical Methods in Geomechanics, Vol. 1, 1977, pp. 299-318.

[9] Yamada, Y., Takatsuka, K. and Iwata, K., "Non-linear Analysis by the Finite Element Method and some Expository Examples", in "Theory and Practice in Finite Element Structural Analysis", University of Tokyo Press, Tokyo, 1973, pp. 125-138.

[10] Yamada, Y., and Sakurai, T., "Basic Formulation and a Computer Program for an Elastic Large Strain Analysis", Proceedings, 3rd International Conference on Pressure Vessel Technology, Part I: Analysis, Design and Inspection, Tokyo, 1977, pp. 341-352.

[11] Banerjee, P.K. and Fathallat, R.C., An Eulerian Formulation of the Finite Element Method for Predicting the Stresses and Pore-Water Pressures around a Driven Pile", Proceedings, 3rd International Conference on Numerical Methods in Geomechanics, Aachen, 1979, pp. 1053-1060.

[12] Carter, J.P., "The Expansion of a Cylinder under Conditions of Finite Plane Strain", Nuclear Engineering and Design, No. 47, 1978, pp. 101-106.

[13] Carter, J.P., Booker, J.R. and Davis, E.H., "Finite Deformation of an Elasto-plastic soil", International Journal of Numerical Analysis Methods in Geomechanics, Vol. 1, No. 1, 1977, pp. 1-25.

[14] Carter, J.P., Booker, J.R. and Small, J.C. "The Analysis of Finite Elasto-Plastic Consolidation", International Journal of Numerical Analysis Methods in Geomechanics, Vol. 3, No. 2, 1979, pp. 107-129.

[15] Gibson, R.E. and Anderson, W.F., "In Situ Measurements of Soil Properties with the Pressuremeter", Civil Engineering and Public Works Review, May 1961, pp. 615-618.

[16] Ladanyi, B., Soulié, M. and Degenne, P., "Interpretation de l'essai pressiométrique dans les sols très déformables", Preprints, 34th Canadian Geotechnical Conference, Fredericton, Sept. 1981.

Guy T. Houlsby, Barry G. Clarke and C. Peter Wroth

ANALYSIS OF THE UNLOADING OF A PRESSUREMETER IN SAND

REFERENCE: Houlsby, G. T., Clarke, B. G. and Wroth, C. P., "Analysis of the Unloading of a Pressuremeter in Sand", The Pressuremeter and Its Marine Applications: Second International Symposium, ASTM STP 950, J.-L. Briaud and J. M. E. Audibert, Eds., American Society for Testing and Materials, 1986.

ABSTRACT: The self-boring pressuremeter offers a particular advantage in that it can be installed in the ground with minimal disturbance. The loading stage of a pressuremeter test in sand has been analysed previously in terms of the shear modulus, angle of friction, angle of dilation and the initial horizontal stress and pore pressure. A new analysis is presented of the unloading stage of the pressuremeter test; this can provide a separate check on the parameters measured during the loading stage. An optimisation routine is used to fit the properties to the results of a series of tests at a site in Wales. Various important effects are discussed, including the effect of stress level on the shear modulus and the relevant value of the angle of friction to be used for the unloading stage.

KEYWORDS: analysis, angle of dilation, angle of friction, case history, in situ stress, pressuremeter test, sand, shear modulus, unloading

INTRODUCTION

The self-boring pressuremeter is a site investigation device which is gaining wide acceptance as a means of determining accurately strength and deformation properties of soils. The test involves the installation of a cylindrical pressuremeter into the ground with minimal disturbance, and subsequent inflation of the pressuremeter membrane with measurement of the associated strain. Hughes, Wroth and Windle [1] presented an analysis of the loading of a pressuremeter in sand using a simplified model of material behaviour. In the model the sand is assumed to be isotropic linear elastic until yield occurs when the Mohr-

Prof. Wroth is Professor of Engineering Science at the Department of Engineering Science, Oxford University, Parks Road, Oxford OX1 3PJ, U.K. Dr. Houlsby is a University Lecturer in the same department. Dr. Clarke is a University Lecturer at the Department of Geotechnical Engineering, University of Newcastle upon Tyne, Newcastle NE1 7RU, U.K.

Coulomb condition is satisfied; the material then flows plastically at a fixed angle of dilation. The analysis represents a refinement of the analysis presented by Gibson and Anderson [2]. From this analysis the shear modulus, angle of friction and angle of dilation can be determined.

During a typical pressuremeter test one or two unload-reload cycles are carried out, and at the end of the test, data may also be obtained from the final unloading. These data contain valuable information about the soil properties, which may be used for comparison with the results obtained from the loading section. There are certain advantages in making use of these data in that they are less sensitive to any initial disturbance caused by the installation of the pressuremeter. This paper is concerned with an analysis of the unloading section of the test, using the same model as that used by Hughes, Wroth and Windle [1]. The key feature of the new analysis is that it accounts for the fact that the unloading takes place from an initially non-homogeneous stress state which has been set up by the preceding expansion phase. The results of the analysis are compared with a set of high quality pressuremeter test results in sand.

The importance of the new analysis lies in the fact that the loading analysis usually requires some knowledge of the value of ϕ_{cv} for the sand. If independent information can be obtained from unloading then this necessity may be bypassed. A problem arises in the choice of the appropriate ϕ value to describe the unloading, and this is discussed further in a later section.

THEORY

Using a cylindrical coordinate system let the axial, radial and circumferential (hoop) stresses be denoted by σ_z, σ_r and σ_θ (compressive positive) and the corresponding strains by ϵ_z, ϵ_r and ϵ_θ. In the initial part of the analysis pore pressures will be assumed zero, so that total and effective stresses are equal; in the final discussion pore pressure will be denoted by u and effective stresses by the usual dash notation. The radial coordinate is r and radial displacement (assumed small compared to r) is ξ.

Plane strain conditions are assumed in the axial direction, so that $\epsilon_z = 0$, and the definitions of the other strains are:

$$\epsilon_r = -\frac{d\xi}{dr}$$

$$\epsilon_\theta = -\frac{\xi}{r}$$

The radial equilibrium equation for the stresses is written:

$$\frac{d\sigma_r}{dr} + \frac{\sigma_r - \sigma_\theta}{r} = 0$$

At the outer boundary ($r = \infty$) the boundary conditions $\sigma_r = \sigma_\theta = \sigma_{ho}$ (where σ_{ho} is the in situ horizontal stress) and $\xi = 0$ will be used. At the inner boundary the radius of the pressuremeter is denoted by a, the value of σ_r at a by ψ and the value of $\frac{\xi}{r}$ $(= -\epsilon_\theta)$ at a by ϵ. Prior to cavity expansion $\xi = 0$ and $\sigma_r = \sigma_\theta = \sigma_{ho}$ at all r.

Loading

The soil is assumed to behave initially in an isotropic elastic manner. Under the conditions of plane strain in the axial direction and axial symmetry the result for the case of the outer boundary at infinity is well known at any radius as:

$$d\sigma_r = -2Gd\epsilon_\theta$$

i.e.

$$\sigma_r = \sigma_{ho} - 2G\epsilon_\theta$$

(Note that ϵ_θ is negative on loading). In particular at the pressuremeter:

$$d\psi = 2Gd\epsilon$$

$$\psi = \sigma_{ho} + 2G\epsilon$$

Note that the only elastic property which enters the analysis is the shear modulus G.

Additionally it can easily be shown that during the elastic phase $d\epsilon_r = -d\epsilon_\theta$ and $d\sigma_r = -d\sigma_\theta$. It follows that in the elastic region there is no volumetric strain and also no change in mean stress $\sigma = \frac{\sigma_r + \sigma_\theta}{2}$. It is useful before proceeding further to eliminate ξ between the definitions for the strain rates to obtain the strain compatability relation:

$$\frac{d\epsilon_\theta}{dr} + \frac{\epsilon_\theta - \epsilon_r}{r} = 0$$

The plastic behaviour of the soil is assumed to conform to the following conditions. The soil is assumed to behave as an elastic - plastic material in which the yield surface is given by the (cohesionless) Mohr-Coulomb condition:

$$\frac{\sigma_1}{\sigma_3} = \frac{(1 + \sin\phi)}{(1 - \sin\phi)} = \frac{1}{N} > 1$$

where the parameter N is introduced to simplify some of the subsequent mathematics. Thus on loading, assuming that the axial stress is the intermediate stress:

$$\frac{\sigma_\theta}{\sigma_r} = N$$

and the equilibrium equation may be re-written as:

$$\frac{d\sigma_r}{dr} + \frac{\sigma_r}{r}(1 - N) = 0$$

which has the solution

$$\sigma_r = Ar^{(N - 1)}$$

It follows that the stresses at two radii r_o and r_1 are related by:

$$\frac{\sigma_{ro}}{\sigma_{r1}} = \left[\frac{r_o}{r_1}\right]^{(N - 1)} \qquad (1)$$

and in particular:

$$\frac{\psi}{\sigma_{r1}} = \left[\frac{a}{r_1}\right]^{(N - 1)}$$

During the plastic behaviour it is assumed that the soil dilates with an angle of dilation ν:

$$\frac{d\epsilon_1}{d\epsilon_3} = - \frac{(1 - \sin \nu)}{(1 + \sin \nu)} = - n \geqslant -1$$

[Note that, for simplicity of the mathematics, the angle of dilation refers to the total strains not the conventionally defined plastic strains. In other words once plasticity occurs the elastic strains are ignored. It is possible to carry out all the following analysis accounting for elastic and plastic strains in the plastic region but the mathematics is more complex.]

Thus on loading, the strains in the plastic region are given by:

$$\frac{d\epsilon_r}{d\epsilon_\theta} = - n$$

so that:

$$\frac{(\epsilon_r - \epsilon_r^{ip})}{(\epsilon_\theta - \epsilon_\theta^{ip})} = - n \qquad (2)$$

where the superscript 'ip' refers to the strain causing initiation of plasticity for that particular soil element. The compatability equation becomes:

$$\frac{d\epsilon_\theta}{dr} + \frac{\epsilon_\theta}{r}(1 + n) = \epsilon_r^{ip} + n\,\epsilon_\theta^{ip} = f(r) \qquad (3)$$

This may be integrated to give:

$$\epsilon_\theta = Br^{-(1+n)} + g(r)$$

It follows that the hoop strains at two radii are related by:

$$\frac{\epsilon_{\theta o} - g(r_o)}{\epsilon_{\theta 1} - g(r_1)} = \left[\frac{r_o}{r_1}\right]^{-(1 + n)} \qquad (4)$$

and in particular:

$$\frac{-\epsilon - g(a)}{\epsilon_{\theta 1} - g(r_1)} = \left[\frac{a}{r_1}\right]^{-(1 + n)}$$

where $g(r)$ is the particular integral for $f(r)$. In particular when $f(r)$ takes the form of the power expression ar^b then $g(r)$ is equal to $\frac{ar^b}{(b + (1 + n))}$. For the special case $b = 0$ (i.e. $f(r) = a$) this expression is still true and $g(r) = \frac{a}{(1 + n)}$.

The expressions describing the plastic behaviour may now be combined to determine the relationship between ψ and ϵ during plasticity.

Since during elastic behaviour $d\sigma_r = -d\sigma_\theta$, yield of the soil at any radius occurs when:

$$\sigma_r = \sigma_{ho} + \sigma_{ho}\sin\phi$$

$$\sigma_\theta = \sigma_{ho} - \sigma_{ho}\sin\phi$$

Thus at the edge of the plastic zone $\sigma_r - \sigma_{ho} = \sigma_{ho}\sin\phi$ and $\epsilon_\theta = -\frac{\sigma_{ho}\sin\phi}{2G}$. The elastic portion of the loading curve in (ϵ, ψ) space is therefore from $(0, \sigma_{ho})$ to $(\frac{\sigma_{ho}\sin\phi}{2G}, \sigma_{ho} + \sigma_{ho}\sin\phi)$. The pressuremeter strain at the initiation of plasticity will be referred to as ϵ^{ip} which is given by:

$$\epsilon^{ip} = \frac{\sigma_{ho}\sin\phi}{2G}$$

Note that the strain at any radius as plasticity is initiated at that radius is given by:

$$\epsilon_r^{ip} = \epsilon^{ip}$$

$$\epsilon_\theta^{ip} = -\epsilon^{ip}$$

Consider now the case when plastic deformation has extended to a radius r_1. The function f(r) in equation (3) is given by:

$$f(r) = \epsilon^{ip}(1 - n)$$

so that:

$$g(r) = \epsilon^{ip}\frac{(1 - n)}{(1 + n)} = \epsilon^{ip}\sin\nu$$

The relationship between strains, equation (4) now becomes:

$$\frac{-\epsilon - \epsilon^{ip}\sin\nu}{-\epsilon^{ip} - \epsilon^{ip}\sin\nu} = \left[\frac{a}{r_1}\right]^{-(1 + n)} \qquad (5)$$

and the relationship between stresses gives:

$$\frac{\psi}{\sigma_{ho} + \sigma_{ho}\sin\phi} = \left[\frac{a}{r_1}\right]^{(N - 1)} \qquad (6)$$

Eliminating r_1 between equations (5) and (6) and re-arranging gives:

$$\psi = \sigma_{ho}(1 + \sin\phi)\left[\frac{\frac{\epsilon}{\epsilon_{ip}} + \sin\nu}{1 + \sin\nu}\right]^{\left[\frac{1 - N}{1 + n}\right]} \qquad (7)$$

This result is exactly as obtained by Hughes, Wroth and Windle [1] but has been derived using a **slightly** different method here. The new method is introduced because it proves to be easier to use in the analysis of the unloading.

Figure 1 shows schematically the state of the soil at the end of loading. The soil between A and E has been loaded elastically along the line OFE in Figure 1(b) and then plastically along ECA. Points with the same letters in Figures 1(a) and 1(b) refer to the same soil element at the end of loading. Soil outside the radius E is still elastic.

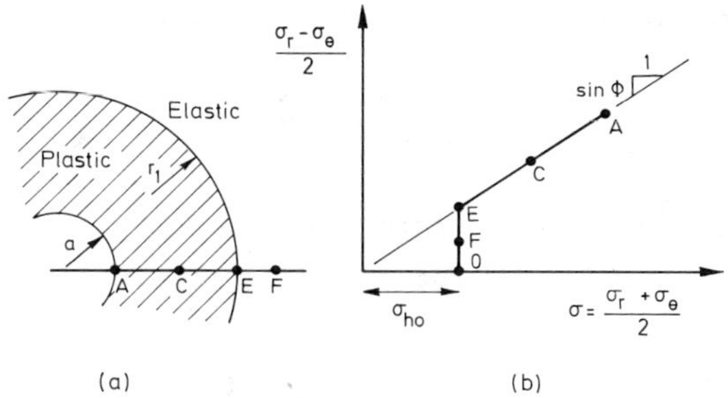

FIG. 1--Stress state at the end of loading

Unloading

The unloading stage of the test begins by a purely elastic stage. This continues until the radial stress is reduced sufficiently so that plastic deformation in the opposite direction from the loading occurs. Note that at the end of loading at the pressuremeter surface:

$$\psi = \psi_1 = \sigma_r = \sigma + \sigma \sin \phi$$
$$\sigma_\theta = \sigma - \sigma \sin \phi$$

during the elastic unloading there is no change in σ, so that when reverse plasticity is initiated:

$$\psi = \sigma_r = \sigma - \sigma \sin \phi$$
$$\sigma_\theta = \sigma + \sigma \sin \phi$$

so that $\psi = \frac{\psi_1}{N}$.

If ϵ_1 is the pressuremeter strain at ψ_1 the elastic unloading is therefore from (ϵ_1, ψ_1) to $(\epsilon_1 - \frac{\psi_1(1 - \frac{1}{N})}{2G}, \frac{\psi_1}{N})$.

Reverse plasticity then occurs in a region which gradually spreads out from the pressuremeter. In this region the stresses are related by:

$$\frac{\sigma_\theta}{\sigma_r} = \frac{1}{N}$$

so that the analysis of the stresses is exactly as before except with N replaced by $\frac{1}{N}$. In particular one may write:

$$\frac{\psi}{\sigma_{r2}} = \left[\frac{a}{r_2}\right]^{(\frac{1}{N} - 1)} \qquad (8)$$

During the reverse plasticity the strain rates are related by:

$$\frac{d\epsilon_r}{d\epsilon_\theta} = -\frac{1}{n}$$

so that the analysis of the strains is also as before, except that n is replaced by $\frac{1}{n}$. In particular the result is obtained:

$$\frac{-\epsilon - g^*(a)}{-\epsilon_{\theta 2} - g^*(r_2)} = \left[\frac{a}{r_2}\right]^{-(1 + \frac{1}{n})} \qquad (9)$$

Where $g^*(r)$ is the particular integral for $f^*(r)$ where:

$$f^*(r) = \epsilon_r^{irp} + \frac{1}{n}\epsilon_\theta^{irp}$$

and the superfix 'irp' indicates the strain at the initiation of reverse plasticity.

If $f^*(r)$ is of the form of the power expression ar^b then $g^*(r)$ is equal to $\dfrac{ar^b}{b + (1 + \frac{1}{n})}$.

The analysis may now proceed on exactly similar lines to the analysis of the loading, except that the determination of the strains on the initiation of unloading is slightly more complex.

The radial stress at radius r in the plastic zone at the end of loading is derived from equation (1).

$$\sigma_r = \sigma_{ho}(1 + \sin\phi)\left[\frac{r}{r_1}\right]^{(N-1)}$$

and the hoop strain is given by comparison with equation (4) as:

$$\epsilon_\theta = \epsilon^{ip}\left[\sin\nu - (1 + \sin\nu)\left[\frac{r}{r_1}\right]^{-(1+n)}\right]$$

Making use of equation (2) one may derive:

$$\epsilon_r = \epsilon^{ip}\left[\sin\nu + (1 - \sin\nu)\left[\frac{r}{r_1}\right]^{-(1+n)}\right]$$

During the elastic unloading the mean stress σ remains constant, so that the value of σ_r when reverse plasticity occurs at r is given by:

$$\sigma_r = \sigma_{ho}(1 - \sin\phi)\left[\frac{r}{r_1}\right]^{(N - 1)}$$

and that the associated increment of the hoop and radial strains result in expressions for the strains at the beginning of reverse plasticity:

$$\epsilon_r^{irp} = \epsilon^{ip}\left[\sin\nu + (1 - \sin\nu)\left[\frac{r}{r_1}\right]^{-(1 + n)} - 2\left[\frac{r}{r_1}\right]^{(N - 1)}\right]$$

$$\epsilon_\Theta^{irp} = \epsilon^{ip}\left[\sin\nu - (1 + \sin\nu)\left[\frac{r}{r_1}\right]^{-(1 + n)} + 2\left[\frac{r}{r_1}\right]^{(N - 1)}\right]$$

So that $f^*(r)$ is given by:

$$f^*(r) = \epsilon^{ip}(1 - n)\left[1 - 2\left[\frac{r}{r_1}\right]^{-(1 + n)} + 2\left[\frac{r}{r_1}\right]^{(N - 1)}\right]$$

and $f^*(r)$ has the particular integral:

$$g^*(r) = \epsilon^{ip}\left[\sin\nu - \frac{2}{(1 + n)}\left[\frac{r}{r_1}\right]^{-(1 + n)} + \frac{2(1 - N)}{(1 + nN)}\left[\frac{r}{r_1}\right]^{(N - 1)}\right]$$

Note that at $r = a$ the expression simplifies to:

$$g^*(a) = -\epsilon_1 + \frac{\Psi_1(1 - n)(1 - N)}{2G(1 + nN)}$$

If, after plasticity on loading to a radius of r_1, reverse plasticity occurs to a radius of r_2 $(a < r_2 < r_1)$ then the expression for the stress, equation (8) gives:

$$\frac{\Psi}{\sigma_{ho}(1 + \sin\phi)\left[\frac{r_2}{r_1}\right]^{(N - 1)}} = \left[\frac{a}{r_2}\right]^{(\frac{1}{N} - 1)}$$

and the expression for the strain, equation (9), gives:

$$\frac{\left[-\epsilon - \epsilon_1 - \frac{\Psi_1(1 - n)(1 - N)}{2G(1 + nN)}\right]}{\epsilon^{ip}\left[-(1+\sin\nu)\left[\frac{r}{r_1}\right]^{-(1+n)} + 2\left[\frac{r}{r_1}\right]^{(N-1)} + \frac{2}{(1+n)}\left[\frac{r}{r_1}\right]^{-(1+n)} - \frac{2(1-N)}{(1+nN)}\left[\frac{r}{r_1}\right]^{(N-1)}\right]}$$

$$= \left[\frac{a}{r_2}\right]^{-(1 + \frac{1}{n})}$$

This may be considerably simplified to:

$$2G(1 + nN)(\epsilon_1 - \epsilon) - \psi_1(1 - n)(1 - N) = \psi_1 n(1 - N^2)\left[\frac{r_2}{a}\right]^{(N + \frac{1}{n})}$$

Finally r_2 may be eliminated to give:

$$\psi = n\psi_1\left[\frac{\dfrac{2G(1 + nN)(\epsilon_1 - \epsilon)}{(1 - N)\psi_1} - (1 - n)}{n(1 + N)}\right]^{\left[\frac{N - \frac{1}{N}}{N + \frac{1}{n}}\right]} \qquad (10)$$

which is the relationship between ψ and ϵ on plastic unloading. Figure 2 shows diagrammatically the state of the soil after plastic unloading to a radius of r_2. Points A and B have initially been loaded elastically, then plastically to A_1, and B_1. They have then unloaded elastically at constant σ and finally unloaded plastically to A and B. Points C and D have been through a similar history, except that C has just reached the beginning of plastic unloading and D has not yet reached it. Points E and F have been loaded and unloaded elastically.

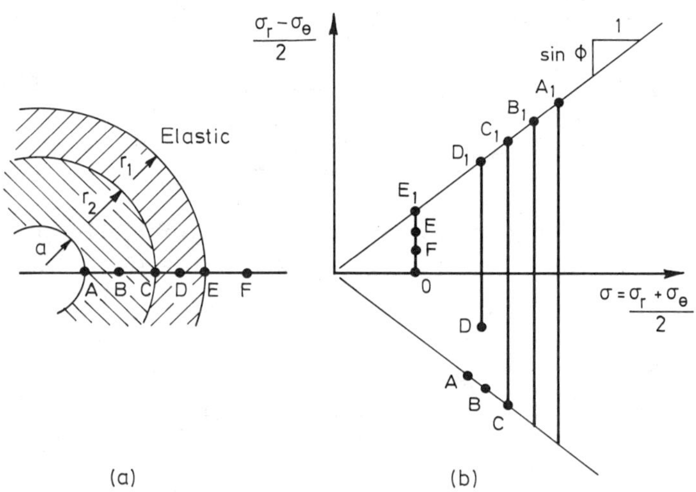

FIG. 2—Stress state after unloading

The above solution applies until r_2 becomes equal to r_1, at which stage ϵ is given by ϵ_2 where:

$$\epsilon_2 = \epsilon_1 - \frac{\psi_1(1 - n)(1 - N)}{2G(1 + nN)} - \frac{\epsilon^{ip}2n(1 + N)}{1 + nN}\left[\frac{\psi_1}{\sigma_{ho} + \sigma_{ho}\sin\phi}\right]^{\left[\frac{1 + \frac{1}{n}}{1 - N}\right]}$$

Although it has not been possible to show the following result in general, it has been shown for a wide variety of parameter values that ϵ_2 is negative. It follows that r_2 remains smaller than r_1 unless the pressuremeter diameter can be reduced to less than the original installation diameter. The solution for $r_2 > r_1$ is not therefore relevant to existing pressuremeter designs.

THE STRESS-DILATANCY RELATIONSHIP

The two parameters ϕ and ν are not independent, since for a given sand it is known that as the density increases both the angle of dilation and the angle of friction increase together. The relationship between ϕ and ν is conveniently expressed by using Rowe's stress-dilatancy relation [3], which is known to fit the behaviour of many sands well. The relation may be expressed as:

$$\left[\frac{1 + \sin\phi}{1 - \sin\phi}\right] = \left[\frac{1 + \sin\phi_{cv}}{1 - \sin\phi_{cv}}\right]\left[\frac{1 + \sin\nu}{1 - \sin\nu}\right]$$

where ϕ_{cv} is the angle of friction at constant volume. The value of ϕ_{cv} may be determined from tests on disturbed samples, so that by making an independent measurement of ϕ_{cv} the stress dilatancy relation can be introduced and only one of the parameters ϕ and ν in the pressuremeter analysis needs to be treated as an unknown.

PORE PRESSURES

All the above analysis assumes that the pore pressures are zero. In practice the pore pressure in a sand will not be zero, but if the sand is highly permeable it may be assumed that the pore pressure throughout the test remains at the initial pore pressure u_o (which will often be equal to the hydrostatic pore pressure). In this case the analysis is unaltered and all the above stresses should be interpreted as effective stresses, to obtain total stresses u_o must be added to the values of each of the stresses in the above analysis. In the remainder of the paper the pore pressure will be taken as u_o and the stresses will be denoted as effective stresses by a dash notation e.g. σ'_{ho}. Since on unloading the value of the effective pressuremeter pressure ψ' rapidly falls to a very small value, the total pressure ψ on unloading may be used as an estimate of u_o.

ANALYSIS OF FIELD DATA

The analysis described above has been used to interpret data obtained from self-boring pressuremeter (SBP) tests carried out as part of an investigation of the soil types and properties at the site of the Shotton Paper Mill, North Wales, U.K. The site was the former British Steel Shotton Works. It is level with uniformly graded fine to coarse sand extending from ground level to at least 20m depth. The ground water level is at a depth of 3.8m and the pore pressure profile is hydrostatic.

The instrument used, similar to that described by Windle and Wroth [4], was 83.4mm in diameter and 1050mm long, with the expanding section being 500mm long. The importance and uniqueness of the SBP is that, using correct drilling techniques, it is possible to insert it into the ground causing minimal disturbance to the surrounding soil. Although it is difficult to drill the SBP into sand, improved techniques mean it is now possible to insert the SBP into the ground causing minimal disturbance (Clarke and Wroth, [5]).

Figure 3 shows a typical test carried out at 5.8m below ground level. The total applied membrane pressure ψ corrected for the membrane stiffness is plotted against the cavity strain ϵ. The total lateral stress σ_{ho} is taken directly from the curve and is equal to 96kN/m^2 giving a value of 0.90 for the coefficient of earth pressure at rest K_o. The shear modulus of 52MN/m^2 is taken from the unload/reload loop. The loading portion of the curve was analysed using the method of Hughes et al. [1] to give a value of the angle of friction of 37°.

The field data were recorded with the aid of a microcomputer. An optimisation routine was written for a microcomputer to analyse the data using the above analysis. The analysis involves five soil parameters: pore pressure u_o, total lateral stress σ_{ho}, shear modulus G, angle of friction ϕ, and angle of dilation ν. The pore pressure profile at this site was known, and the total lateral stress taken directly from the test curve. Thus there are only three variables to be optimised, and two of those are related by Rowe's stress dilatency law. It may be expected at many sites that the ϕ_{cv} value will be known, but unfortunately no data were available from this site.

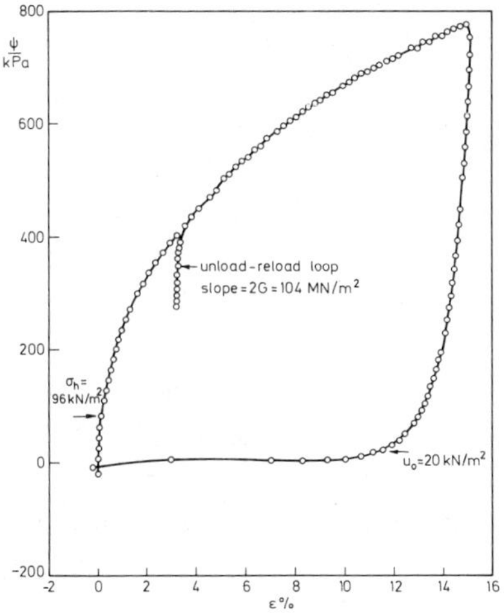

FIG. 3--Pressuremeter curve for test at 5.8m

In the analysis it is assumed that G is constant both for loading and unloading. There is evidence that G varies throughout the test (Clarke and Wroth [5]), and Figures 4 and 5 show the G values taken from the unload/reload loops for the tests at this site. Fig. 4 shows the values plotted against depth and Fig. 5 shows them plotted against the **mean** effective stress at the pressuremeter surface at which they were measured. There is considerably less scatter in Fig. 5. However, for the purposes of this analysis it will be assumed G is constant over each of four sections of the test, initial elastic loading (G_{EL}), plastic loading (G_{PL}), initial elastic unloading (G_{EU}) and plastic unloading (G_{PU}). Separate moduli values will be reported for each of these sections.

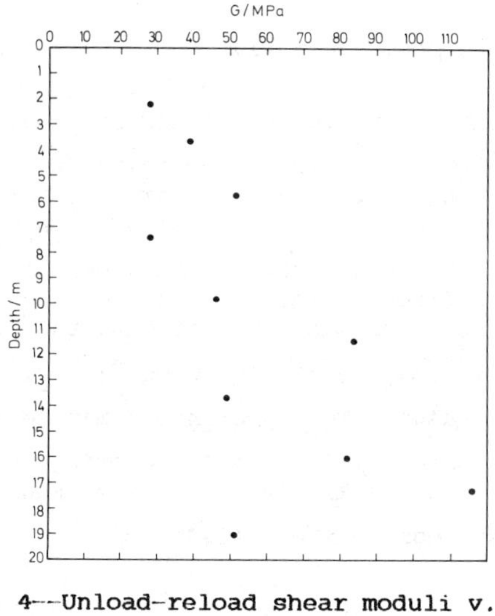

FIG. 4--Unload-reload shear moduli v. Depth

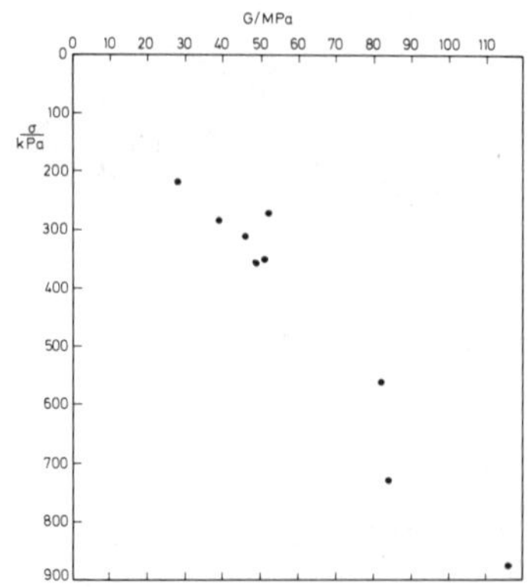

FIG. 5--Unload-reload shear moduli v. mean stress at pressuremeter

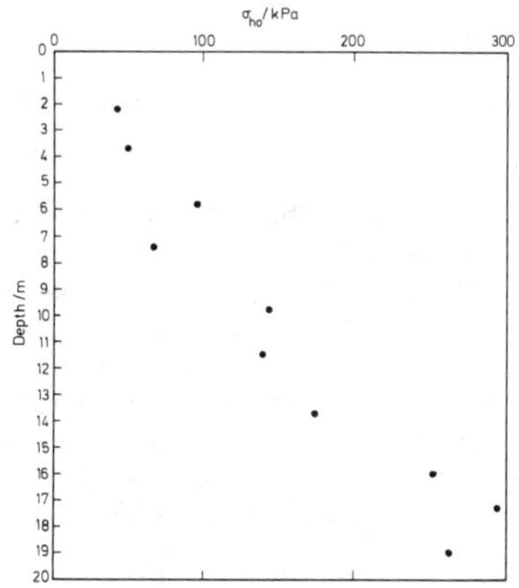

FIG. 6--Variation of total lateral stress σ_{ho} with depth

Figure 6 shows the observed variation in total lateral stress with depth. Using a unit weight of $18 kN/m^3$ and the known pore pressure profile the variation of K_o with depth can be calculated. The implication is that the sand is overconsolidated, with the degree of overconsolidation varying with depth. Therefore the sand is likely to exhibit a peak ϕ on loading. The value of ϕ on unloading (ϕ_U) will depend on the amount the membrane has expanded before unloading. If the membrane expansion is sufficient for the sand to reach critical state, so that the angle of friction is that at constant volume ϕ_{cv}, then on unloading the appropriate ϕ_U is likely to equal ϕ_{cv}. This represents a lower bound on the ϕ_U value. The upper bound may be appropriate if the membrane has just expanded sufficiently to bring the sand to the peak deviator stress, then ϕ_U is likely to equal ϕ_L (the loading ϕ). Thus three cases for analysis can be identified.

(a) $\phi_U = \phi_L$

(b) $\phi_U = \phi_{cv}$

(c) $\phi_L \neq \phi_U \neq \phi_{cv}$

For all cases it is assumed that ϕ_{cv} is constant and Rowe's stress dilatancy is assumed to apply. An outline of the program for the microcomputer used for analysing the test is now described:

(a) input u_o, σ_{ho}, ϕ_{cv}

(b) label the loading portion of the curve as that from σ_{ho} to the maximum applied stress

(c) label the unloading portion of the curve as that from the maximum applied stress to u_o

(d) using an initial estimate $\phi_L = \phi_{cv}$ determine a yield point in compression

(e) optimize the plastic loading expression to give values of G_{PL} and ϕ_L (Eq. 7)

(f) repeat step (e) with the new yield point until the optimum values of G_{PL} and ϕ_L are achieved

(g) calculate G_{EL} from the loading data from before the yield point

(h) using an initial estimate of $\phi_U = \phi_{cv}$ determine a yield point in extension

(i) optimize the plastic unloading expression to give values of G_{PU} and ϕ_U (Eq. 10) (this stage is simplified if $\phi_U = \phi_{cv}$ or $\phi_U = \phi_L$ is used)

(j) repeat step (i) with a new yield point until the optimum values of G_{PU} and ϕ_U are achieved

(h) calculate G_{EU} from the unloading section before yield

The optimisation routine was carried out by calculating ψ for the measured values of cavity strain and comparing them with the measured values of ψ. Table 1 gives the results for test 3 at 5.8m depth using each of the three cases above. Figure 7 shows a comparison between the optimum curve and the field curve for Case (b), $\phi_U = \phi_{cv}$.

TABLE 1--Test at 5.8m depth interpreted using different assumptions

Case	Loading			Unloading		
	G_{EL}	G_{PL}	ϕ_L	G_{EU}	G_{PU}	ϕ_U
(a) $\phi_U = \phi_L$	9050	4300	60°	20940	21250	60°
(b) $\phi_U = \phi_{cv}$	9060	10310	40°	37830	35050	29°
(c) $\phi_U \neq \phi_L \neq \phi_{cv}$	9050	10040	43°	37830	35050	29°

The following points can be noted:

(a) The Case (a) assumption $\phi_U = \phi_L$ results in very high ϕ values and implies that $G_{EL} > G_{PL}$, which is not consistent with the notion that G increases with effective stress. This case therefore seems unrealistic.

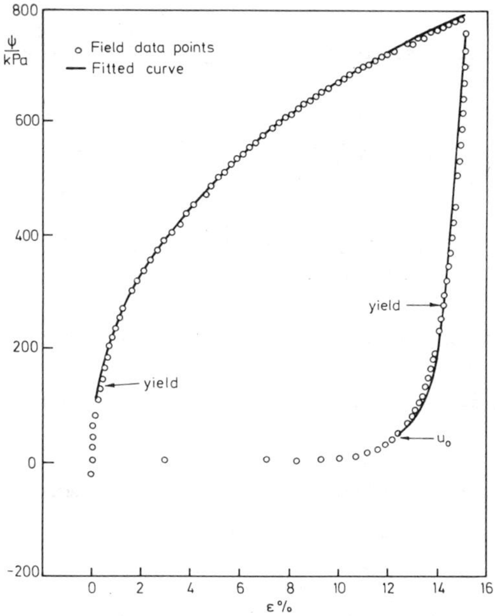

FIG. 7--Measured and fitted pressuremeter curve for test at 5.8m

(b) Case (c), $\phi_L \neq \phi_U \neq \phi_{cv}$ results in moduli which are consistent with the notion that G increases with effective stress. However, for this case $\phi_{cv} = 35°$, which is inconsistent with the analysis which requires that $\phi_U \geq \phi_{cv}$.

(c) Comparison between Cases (b) and (c) shows that ϕ_U is insensitive to the value of ϕ_{cv}. This was found to be the case in the majority of the tests. This comparison also shows that a decrease in ϕ_{cv} of

6° implies a reduction in ϕ_L of 3°. This is consistent with the analysis of Hughes et al. [1].

(d) Small differences in G are a result of the numerical calculations.

(f) There are many data points for the plastic loading of the soil, whereas for the plastic unloading there are few.

As ϕ_U is insensitive to the value of ϕ_{cv}; ϕ_U is unlikely to be equal to ϕ_L and using $\phi_U = \phi_{cv}$ gives a conservative value of ϕ_L, it would seem reasonable that the test should be analysed assuming $\phi_U = \phi_{cv}$. This has the advantage that further laboratory data are not required. These observations were found to be similar in all the tests analysed.

Table 2 gives the result of the interpretation of all the tests assuming $\phi_U = \phi_{cv}$. There is scatter in the results. This is not inconsistent with field and laboratory measurements on other soils and reflects the non-homogeneity of sands. Further inspection of the total stress measurements indicate that the geological history of the site may be complex with the sand being deposited in layers at different times and possibly from different sources.

TABLE 2--Results of tests assuming $\phi_U = \phi_{cv}$

Depth, m	σ_{ho}	u_o	G_{EL}	G_{PL}	ϕ_L	G_{EU}	G_{PU}	$\phi_U = \phi_{cv}$
2.2	43	0	3650	5040	48°	37480	40310	22°
5.8	96	20	9060	10310	40°	37830	35050	29°
9.8	130	59	5780	6690	49°	28820	29130	34°
13.7	174	97	4700	4950	47°	44010	42950	24°
17.3	222	132	–	27400	47°	96010	88080	26°
19.0	262	149	2930	3090	49°	52730	50530	23°

The effect of varying σ_{ho} was also investigated. A change of 25% induced a change of approx 1° in ϕ_L. If the correct drilling techniques are used is should be possible to determine σ_{ho} accurately. However, in many sand tests there is some disturbance. No investigation has been undertaken of the effects of analysing disturbed tests.

Inspection of Fig. 4 shows that the value of modulus derived from the unload/reload loop is greater than G_{EU}, but the effective stress is less. The value of G varies not only with the effective stress at which it is measured but also the stress range over which it is measured.The stress range for the unload/reload loop is less than the difference between the maximum applied stress and the yield point in extension therefore the results may not be inconsistent.

Unlike the case at this site, the pore pressure is often unknown. Wroth [4] describes how the pore pressure can be estimated from the point on unloading at which the membrane begins to contract rapidly due to the pressure of the water balancing the applied pressure. Inspection

of Fig. 8 shows the pore pressure estimated by this method is up to 25% less than that derived from the hydrostatic profile. The test at 5.8m was analysed for different values of pore pressure and it was found that a 25% change in pore pressure changed ϕ_L by approx 1°. This change is small compared to all the other unknowns and assumptions, and therefore the estimated value could be used.

ACKNOWLEDGEMENT

The authors are grateful to the Shotton Paper Company Ltd. and Laing-Loy Management Contracting Ltd. for permission to use the field data presented in this paper.

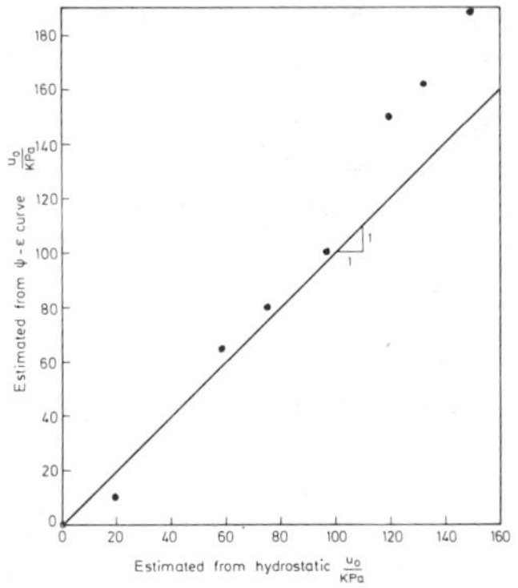

FIG. 8—Pore pressures estimated from hydrostatic profile and pressuremeter tests

REFERENCES

[1] Hughes, C.M.O., Wroth, C.P., and Windle, D.J., "Pressuremeter Tests in Sands", Geotechnique, Vol. 27, No. 4, pp. 455-478, Dec. 1977

[2] Gibson, R.E. and Anderson, W.F., "In Situ Measurement of Soil Properties with the Pressuremeter", Civ. Eng. Pub. Works Review, Vol. 56, No. 658, pp. 615-618, May 1961

[3] Rowe, P.W., "The Stress-Dilatancy Relation for Static Equilibrium of an Assembly of Particles in Contact", Proc. Roy. Soc., Vol. 269A, pp. 500-527, 1971

[4] Windle, D.J. and Wroth, C.P., "The Use of the Self-Boring Pressuremeter to Determine the Undrained Properties of Clays", Ground Engineering, Vol. 10, No. 1, pp.37-46, Jan. 1977

[5] Clarke, B.G. and Wroth, C.P., Discussion on "Effect of Disturbance on Parameters Derived from Self-Boring Pressuremeter Tests in Sand", Geotechnique, Vol. 35, No. 2, pp. 219-222, June 1985

[6] Wroth, C.P., "The Interpretation of In Situ Soil Tests", Geotechnique, Vol. 24, No. 4, pp. 447-492, Dec. 1984

Soil Properties from Pressuremeter Data

Guy Y. Felio, and Jean-Louis Briaud

CONVENTIONAL PARAMETERS FROM PRESSUREMETER TEST DATA: REVIEW OF EXISTING METHODS

REFERENCE: Felio, G.Y., and Briaud, J. -L., "Conventional Parameters from Pressuremeter Test Data: Review of Existing Methods," The Pressuremeter and Its Marine Applications, Second International Symposium, STP 950, J.-L. Briaud and J.M.E. Audibert, Eds, American Society for Testing and Materials, 1986.

ABSTRACT: The solution for the expansion of an infinitely long cylindrical cavity in a soil mass, based on given constitutive laws, allows to obtain "conventional" soil parameters such as the undrained shear strength or the angle of internal friction based on pressuremeter test results. The pressuremeter test also allows to obtain adequate values of the in-situ horizontal stress in the soil when proper insertion of the probe (little disturbance) is achieved. The existing methods to obtain these parameters are discussed for preboring and selfboring pressuremeter units.

KEYWORDS: pressuremeter, in-situ horizontal stress, undrained shear strength, cohesion, angle of internal friction

The design of foundations based on pressuremeter (PMT) test results can performed 1) by using pressuremeter design rules [1-12] that require soil parameters such as the limit pressure and a modulus of deformation or 2) by deriving "conventional" parameters such as the undrained shear strength, the cohesion , or the angle of internal friction from the PMT test data; these parameters are in turn used in the usual bearing capacity and settlement equations.

The "conventional" soil properties derived from the PMT tests are based on theoretical models that simulate the expansion of an

Dr. Felio is assistant professor in the Department of Civil Engineering, University of California, Los Angeles, CA 90024. Dr. Briaud is associate professor in the Civil Engineering Department, Texas A&M University, College Station, TX 77843.

infinitely long cylindrical cavity. The accuracy of the predictions depends on the rheological equations used to model the soil behavior. A number of researchers have produced constitutive laws applicable to the pressuremeter test [13-26]. The application of these constitutive laws to derive soil properties from PMT test data is discussed in this article.

The methods to derive soil properties from PMT results described here are based on corrected PMT curves. Procedures to correct the field PMT data for membrane resistance and volume losses (in the case of hydraulically filled PMT probes) can be found in the literature [1,27].

Some differences exist in the interpretation of preboring pressuremeter (PBPMT) data and selfboring pressuremeter (SBPMT) test results. When appropriate, the methods to interpret both the PBPMT and the SBPMT tests are given.

AT-REST STATE OF STRESS

Preboring Pressuremeter Tests

Fig. 1 shows a typical corrected PBPMT curve. The total horizontal pressure at rest, p_{oh}, is obtained from the beginning of the straight portion of the pressuremeter curve (point A in Fig. 1). The value of the at rest coefficient of earth pressure, K_o, is given by:

$$K_o = \frac{P_{oh} - u_o}{P_{ov} - u_o} \tag{1}$$

where p_{oh} = total horizontal pressure at rest, point A in Fig. 1
u_o = estimated hydrostatic pressure at the level of the probe, from ground water level observations.
p_{ov} = estimated total vertical pressure at the level of the probe, based on unit weight measured on samples.

The accuracy of this method is doubtful and the results should be used with caution. The following must be kept in mind:

1. In a stress-controlled test, very few points are obtained within the "K_o" pressure range.
2. It is assumed that $p_{oh} = p_{oh}' + u_o$ where u_o is the hydrostatic pressure at the testing depth. Excess pore pressures generated during the test are neglected.
3. The value of p_{oh} is affected by the quality of the borehole, including the amount of inward yielding and subsequent recompression of the borehole wall.

Briaud et al. [27] proposed to plot the initial portion of the PMT curve on a p versus $\log(\frac{\Delta R}{R})$ scale (Fig. 2) to accentuate the curvature. A procedure similar to that devised by Casagrande [28] to obtain the preconsolidation pressure from an oedometer test is then used to obtain p_{oh}. This procedure is purely hypothetical and has not yet been compared with reliable measurements of K_o.

In general, even in the case of carefully conducted tests in high quality boreholes, the accuracy of K_o from PBPMT tests is uncertain and the results should be used with caution.

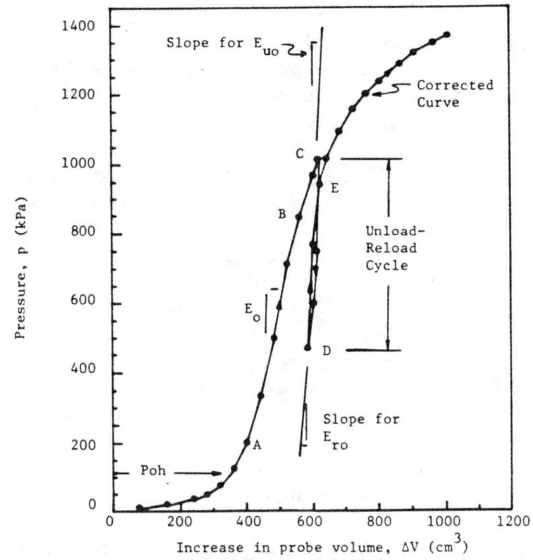

FIG. 1--Typical PBPMT test curve

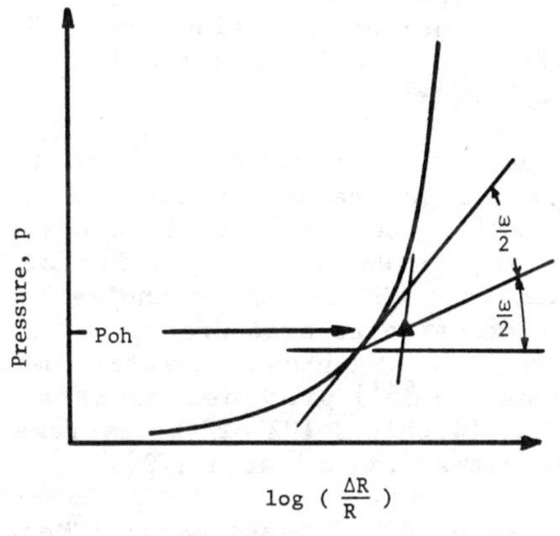

FIG. 2--Proposed method to obtain p_{oh}. (After Briaud et al. [27])

Selfboring Pressuremeter Tests

A study by Lacasse and Lunne [29] reported the interpretation of 21 SBPMT tests to obtain the in-situ horizontal stress in two deposits of soft Norwegian marine clay. The pressuremeter tests were conducted with the Cambridge SBPMT Mark IV unit (80mm diameter and 480mm long measuring cell). Three classes of interpretation methods were used to evaluate the in-situ horizontal stress:

Direct Measurement Methods: these procedures are based on the direct measurement of the "initial" or "lift-off" pressure.

1. Visual inspection method [1,29]
2. Initial development of excess pore pressure method [30]
3. Modified lift-off pressure method [31]
4. Logarithm of strain method [32]

Empirical Method: procedure based on an elasto-plastic model for soil behavior for which the in-situ horizontal stress, p_{oh}, is given by [29]:

$$p_{oh} = p_l - N_p S_u \tag{2}$$

where p_l = limit pressure
S_u = undrained shear strength
N_p = empirical factor given by:

$$N_p = 1 + \ln \frac{E_u}{3S_u}$$

E_u = undrained Young modulus

Methods Using the PMT Expansion Curve: these procedures use the complete pressuremeter expansion curve to determine p_{oh}.

1. Graphical iteration [33]
2. Hyperbolic-Plastic model of expansion curve [34]
3. Double zero shift of hyperbolic model [35]
4. Origin adjustment [36]

Lacasse and Lunne [29] concluded that, due to the uncertainties that exist with respect to the interpretation of the in-situ horizontal stress from PMT tests, it is best to use several methods to obtain p_{oh}, including results based on direct measurements and derived from the complete PMT curve. Of the methods evaluated, five procedures (initial excess pore pressure method [30], modified lift-off pressure method [31], graphical iterative method [33], and the hyperbolic approaches [34,35]) provided satisfactory estimates of the horizontal stress (within ± 10% of the average plausible in-situ conditions) for the sites investigated [29].

A comparison between PBPMT tests using a Menard type unit (tricellular) and SBPMT tests with the PAF-68 (selfboring pressuremeter developed by the Laboratory of Ponts et Chausees, France) showed that, in general, the standard Menard test leads to larger values of p_{oh} than those measured with the PAF-68 [1].

MODULI OF DEFORMATION

Preboring Pressuremeter Tests

First Loading Modulus: The first loading modulus, E_o (also referred in the literature as E_M [1]) is obtained from the pseudo-elastic or straight portion of the PMT curve (between points A and B in Fig. 1). The expansion of a cylindrical cavity in a linear elastic material yields:

- for hydraulically filled probes:

$$E_o = 2\ (1 + \nu)\ \frac{\Delta p}{\Delta V}\ V_m \tag{3}$$

where ν = the soil's Poisson ratio

$\frac{\Delta p}{\Delta V}$ = slope of pseudo-elastic portion of PMT curve (between points A & B in Fig. 1)

V_m = average volume of the cavity (measured at midpoint of straight line AB in Fig. 1)

- for units measuring the radius of the probe:

$$E_o = (1 + \nu)\ \frac{\Delta p}{\Delta R}\ R_m \tag{4}$$

where $\frac{\Delta p}{\Delta R}$ = slope of pseudo-elastic portion of PMT curve

R_m =average radius of borehole within pseudo-elastic range

Some investigators [37] have found that in heavily overconsolidated clays, two distinct slope were apparent (Fig. 3); the investigators logically proposed that p_c in the figure be considered as a preconsolidation pressure. In this case, two first loading moduli can be determined: one pre-p_c and one post-p_c.

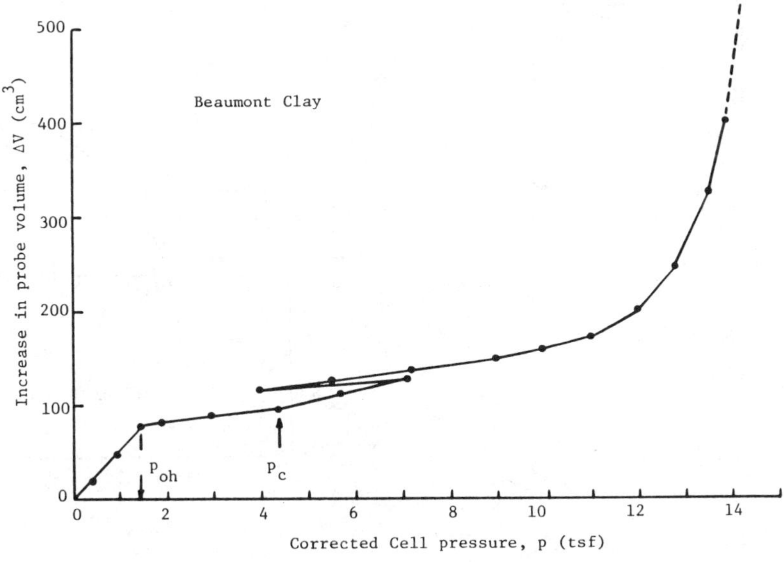

FIG. 3--PBPMT curve in Heavily overconsolidated clay. (After Merritt et al. [37])

The first loading modulus, E_o, is generally low ($E_o = 50\ S_u$); one reason for this phenomenom is that E_o is measured over a large strain range. In design, it is suggested [38] to devide E_o by μ, where μ varies from 0.25 for gravels to 1.0 for overconsolidated clays.

Reload Modulus: If an unload/reload cycle is conducted towards the end of the pseudo-elastic part of the PMT curve, an unload modulus, E_{uo}, and a reload modulus, E_{ro}, can be obtained (Fig. 1). The ratio E_{ro}/E_o was found to average 1.75 in two deposits of clay and 2.65 in a deposit of sand [39]. Others have reported similar ratios [1].

When performing the unload/reload cycle, the unloading phase must not extend beyond the unloading elastic behavior of the material [40]. For clays, the maximum stress release, p_u, is $2S_u$ (Fig. 4a); for sands, it is given by:

$$p_u = (p - u_o)\ \frac{2 \sin \phi'}{1 + \sin \phi'} \tag{5}$$

where ϕ' = effective angle of internal friction
p = unloading pressure in the probe at which the cycle begins (Fig. 4b)
u_o = hydrostatic pressure at probe level

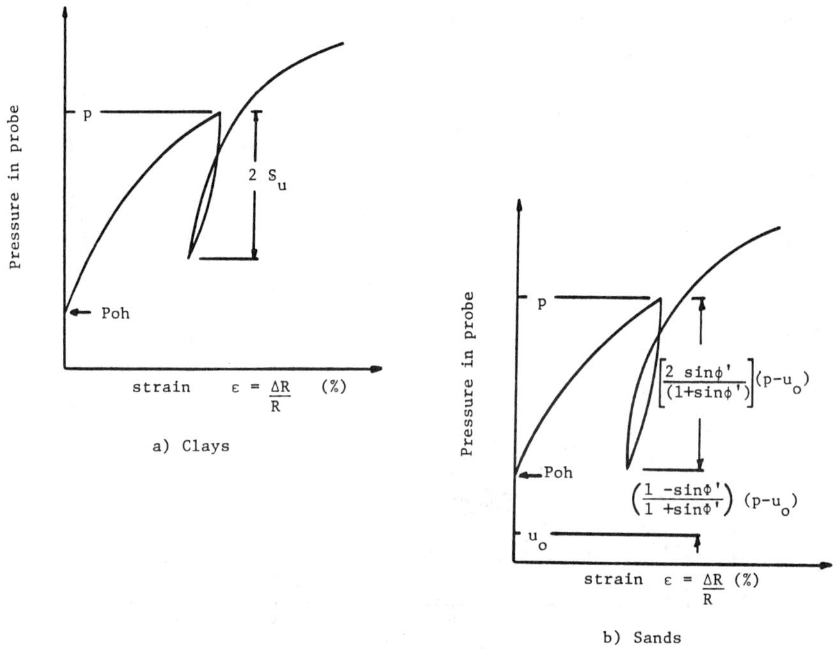

FIG. 4--Unloading Limits for PMT cyclic tests:
a) clays, b) sands (After Wroth [40])

Zero Strain Modulus: it was found [41] that the reload portion of the cycle can be modeled successfully by a hyperbolic model [42], which, applied to the pressuremeter, yields:

$$1/E_s = a + b\varepsilon_v \tag{6}$$

where E_s = secant modulus of deformation at volumetric strain
a,b = constants found by best fit regression
ε_v = volumetric strain

The quantity 1/a represents the modulus of deformation at a strain equal to zero (Fig. 5). The model also allows to obtain the secant modulus at any strain level.

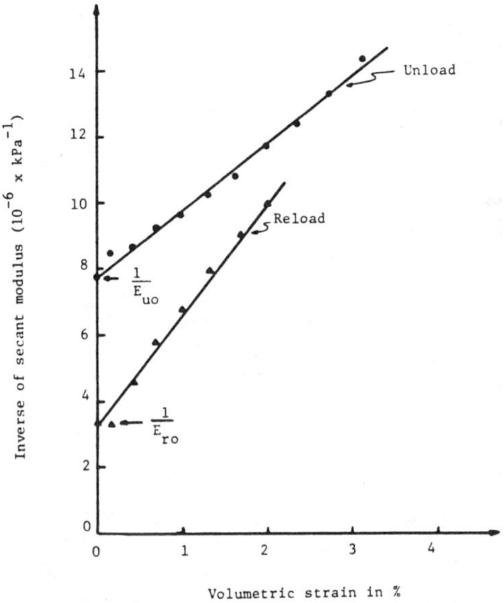

FIG. 5--Hyperbolic Model regression to obtain zero strain modulus. (after Briaud et al. [41])

Selfboring Pressuremeter Tests

A typical SBPMT test curve is shown on Fig. 6. As for PBPMT tests, several moduli can be defined.

Initial Tangent Modulus: in practice, the initial tangent shear modulus, G_o, is based on the straight line between p_{oh} and the first experimental point on the curve [1]. The modulus of deformation E_o follows from:

$$E_o = 2\ (1 + \nu)\ G_o \tag{7}$$

Secant Moduli: secant shear moduli can be measured at given

strain levels as shown on Fig. 6. For example, G_{p2}, the secant shear modulus corresponding to $\frac{\Delta V}{V_o} = 2\ \%$ is calculated as [1]:

$$G_{p2} = \frac{p_2 - p_{oh}}{2} \times 100 \qquad (8)$$

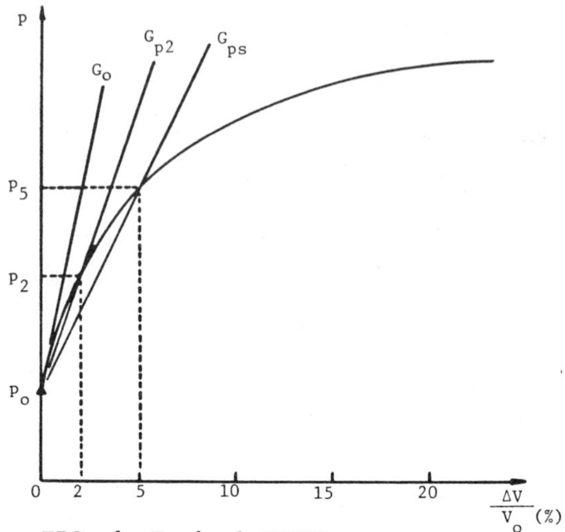

FIG. 6--Typical SBPMT curve.

Comparisons

In a comparative study [43] between PBPMT (Menard), SBPMT (Camkometer), and UU triaxial tests in Seattle stiff clay, it was found that the unload/reload cycle moduli from PBPMT tests were comparable to SBPMT results (a similar conclusion was reported by Baguelin et al. [1]). The UU triaxial tests in the Seattle clay gave initial tangent moduli lower than the Menard initial moduli values.

In soft cohesive soil along the northern coast of Tokyo Bay [44], initial moduli from CU triaxial tests and SBPMT tests were in good agreement; Menard type PMT tests yielded results consistently lower (2 to 3 times) than the SBPMT or the triaxial tests.

Small strain moduli from crosshole seismic tests compared with SBPMT initial moduli were in good agreement for OC clay [45]. Moduli measured with a resonant column were lower (approximately 2/3) than those measured by the SBPMT or the crosshole seismic tests.

Other comparisons have been reported [46-51].

UNDRAINED SHEAR STRENGTH

Several approaches exist to obtain the undarined shear strength, S_u, from a pressuremeter curve.

Undrained Shear Strength from the Limit Pressure

Based on ideal elastic-plastic assumptions but with different volume change properties, three theoretical solutions have been reported [14,17,18]. If the Poisson's ratio for the soil is assumed to be 0.5 (undrained loading), the equations in [14,17,18] all reduce to [1]:

$$p_1 - p_{oh} = S_u \left[1 + \ln \left(\frac{E}{3\,S_u}\right)\right] = \alpha S_u \tag{9}$$

The parameter α typically ranges between 6.5 and 12 for stiff to very stiff clays [1].

Empirical relationships have also been proposed [52,53]:

- Amar and Jezequel [52]

$$S_u = \frac{p_1 - p_{oh}}{10} + 25 \text{ kPa} \tag{10}$$

- Centre d'Etudes Menard [53]

$$S_u = \frac{p_1 - p_{oh}}{5.5} \tag{11}$$

Gibson and Anderson Method

The analysis of Gibson and Anderson [54] for the interpretation of an undrained PMT test in clay assumes that the soil behaves as a perfectly elastic/perfectly plastic material with a Poisson's ratio of 0.5. After failure has been initiated, the applied pressure is linearly related to the logarithm of the current volumetric strain, $\frac{\Delta V}{V}$. The gradient of the line (Fig. 7) is equal to S_u.

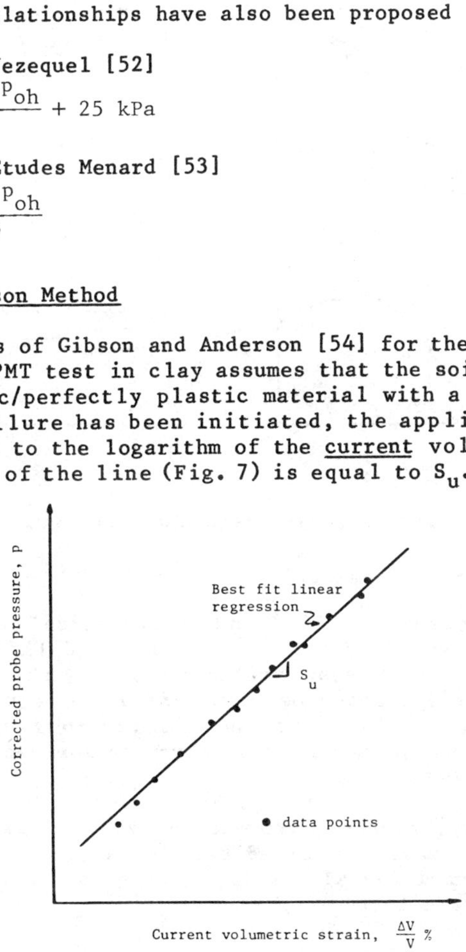

FIG. 7--Determination of S_u by Gibson and Anderson Method [54].

Derivation of the Shear Stress-Strain Curve

A more general interpretation of undrained PMT tests in cohesive soils was achieved in 1972 [20-22]. The assumption of incompressibility during undrained loading allows to obtain the entire shear stress-strain curve from the corrected PMT curve. The equations resulting from the analysis lead to a graphical construction of the shear stress-strain curve [1] as shown in Fig. 8. It must be emphasized that, if the soil is disturbed during the insertion of the probe, the undrained shear strength measured will be higher than if the soil was undisturbed [1]. On the other hand, the shear modulus will be lower in the case of disturbance of the soil.

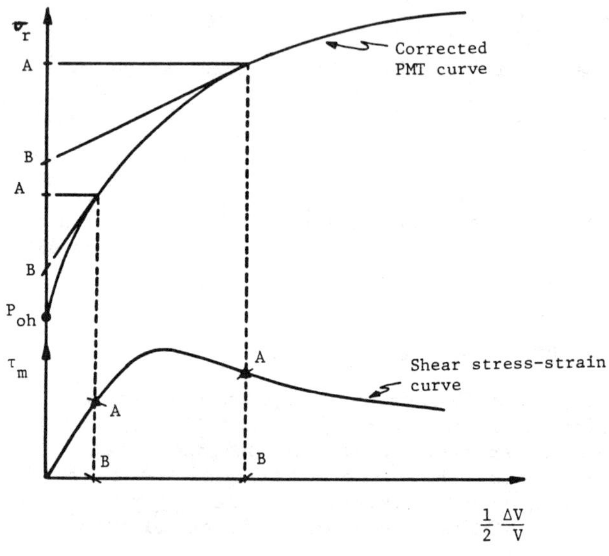

FIG. 8--Shear stress-strain curve construction [1].

Comparisons

A study comparing SBPMT and anisotropically consolidated undrained (CU) tests in San Francisco Bay mud [55] showed that S_u evaluated from the pressuremeter results by the Baguelin-Ladanyi-Palmer [20,21,22] methods were in good agreement with the CU test values. Mori [44] also reported similar conclusions between CU triaxial tests and undrained shear strength derived from PMT results using Palmer's method.

Roy et al. [56] found that the peak undrained shear strength from PBPMT test results (derived from the shear stress-strain curve) was, for the sensitive clay studied, up to 3 times higher than that measured with vane tests.

Other studies reporting comparisons between laboratory and PMT measurements of S_u are numerous [16,31,33,52,57-66].

DRAINED SHEAR STRENGTH - FRICTION ANGLE

An estimate of the effective angle of internal friction, ϕ', can be obtained from the PMT curve either theoretically or empirically.

No Volume Change Theory

When no volume change is assumed to exist during the expansion of the pressuremeter probe, the expressions for the creep pressure and the limit pressure are respectively [1]:

$$p_f = p_{oh} (1 + \sin \phi') \tag{12}$$

$$p_L = p_{oh} (1 + \sin \phi') \left(\frac{G}{p_{oh} \sin \phi'} \right)^{\frac{1 - K_a}{2}} \tag{13}$$

where ϕ' = effective angle of internal friction

$K_a = \tan^2 \left[\frac{\pi}{4} - \frac{\phi'}{2} \right]$

These expressions allow to calculate a value of ϕ'; however, the assumption of no volume change for sands is not appropriate. Because of dilatancy (which will be discussed later), obtaining ϕ' from the limit pressure and no volume change theory tends to overestimate ϕ' [1].

Dilatancy Theory

Several attempts have been made to include realistic volume change laws in the solution of the expansion of a cyclindrical cavity [1,67,68]. Most of these laws require additional testing, such as triaxial tests, to be performed in order to obtain the necessary parameters.

The method recommended by Wroth [40] is based on the assumption that before failure, the soil behaves in a linear elastic manner, and fails at a constant rate of volume change [68]. With these assumptions, the following equations are obtained [68]:

$$\sin \phi' = \frac{(K+1)s}{(K-1)s+2} \tag{14}$$

$$\sin \theta = \frac{2Ks - (K-1)}{(K+1)} \tag{15}$$

$$s = \frac{\sin\phi'(1 + \sin \theta)}{(1 + \sin \phi')} \tag{16}$$

where θ = angle of dilation

ϕ' = angle of internal friction

K = earth pressure coefficient

$$K = \tan^2 \left[\frac{\pi}{4} + \frac{\phi_{cv}}{2} \right]$$

ϕ_{cv} = angle of friction at the end of a test where dilation has ceased and the sand has reached a critical state, thus deforming at constant volume.

If ϕ_{cv} is not known, a value of $\phi_{cv} = 35°$ can be used [40].

<u>Empirical Correlations</u>

Empirical correlations between the limit pressure p_l and have been developed [1]. One of them is given in Fig. 9.

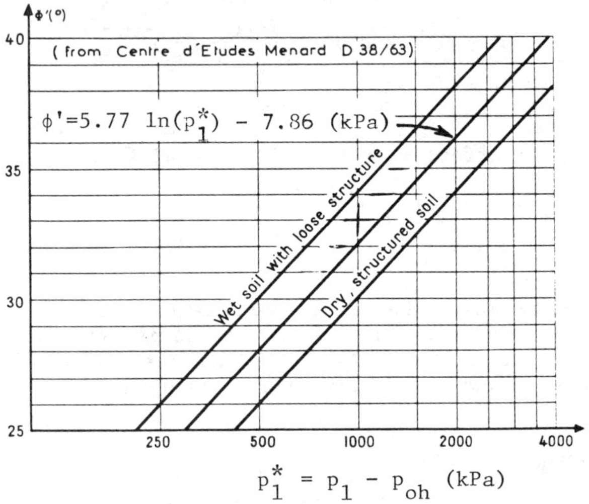

FIG. 9--Correlation between ϕ' and p_l^* (from Centre d'Etudes Menard - in [1])

<u>Comparisons</u>

Little data is available on drained strength from PMT test results. Hughes, Wroth, and Windle [68] showed good agreement between SBPMT friction angles using their analysis (re. "Dilatancy Theory" section) and triaxial tests in hydraulic fill at Kernik (UK) and wash sand.

CONCLUSIONS

Due to the disturbance of the soil during the preparation of the borehole, the values of K_o derived from PBPMT test results are uncertain and should be used with caution. Carefully performed SBPMT tests can yield to adequate measurements of p_{oh} by several methods of interpretation.

The Baguelin-Ladanyi-Palmer method for obtaining the undrained shear strength based on PMT tests provides comparable values to those measured in CU triaxial tests.

Although adequate comparisons of the angle of internal friction from PMT and triaxial tests have been found, the method of interpreting the PMT requires parameters that are measured in the laboratory. Additional work is necessary to examine problems such as the influence of pore pressure generation around the probe or the effects of partial drainage during the PMT test.

REFERENCES

[1] Baguelin, F., Jézéquel, J.-F., and Shields, D.H., The Pressuremeter and Foundation Engineering, Trans Tech Publications, Rockport, Mass., 1978.

[2] Bustamante, M., and Gianesselli, L., "Portance Réelle et Portance Calculée des Pieux Isolés, Sollicités Verticalement," Revue Française de Géotechnique, No. 16, August 1981.

[3] Ménard, L., and Rousseau, J., "L'Evaluation des Tassements, Tendances Nouvelles," Sols - Soils, No. 1, 1962.

[4] Ménard, L., "Calcul de la Force Portante des Fondations sur la Base des Résultats Préssiometriques," Sols - Soils, No. 5, 1963.

[5] Ménard, L., "Etude Experimentale du Tassement et de la Force Portante des Fondations Superficielles," Sols - Soils, No. 10, 1964.

[6] Ménard, L., "Le Tassement des Fondations et les Téchniques Préssiometriques," Annales ITBTB, Paris, Supplement No. 288, Decembre 1971.

[7] Baguelin, F., "Rules for the Structural Design of Foundations Based on the Selfboring Pressuremeter Test," Symposium on the Pressuremeter and Its Marine Applications, IFP, Paris, April 1982, pp. 347-360.

[8] Briaud, J.-L., Smith, T.D., and Meyer, B.J., "Using the Pressuremeter Curve to Design Laterally Loaded Piles," 15th Offshore Technology Conference, Houston, Texas, May 1983.

[9] GAI Consultants, Inc., "Laterally Loaded Drilled Pier Research. Volume 1: Design Methodology. Volume 2: Research Documentation," Research Reports EPRI EI-2197, January 1982.

[10] Gambin, M., "Application of the Menard Pressuremeter Test Results to Foundation Design," Symposium on the Pressuremeter and Its Marine Applications, IFP, Paris, April 1982, pp. 337-346.

[11] Suyama, K., Ohya, S., and Imai, T., "Development of the LLT Pressuremeter and its Application to the Prediction of Pile Behavior under Horizontal Loads," Symposium on the Pressuremeter and Its Marine Applications, IFP, Paris, April 1982, pp. 61-76.

[12] Ménard, L., and Bourdon, G., and Gambin, M., "Méthode Generale de Calcul d'un Rideau ou Pieu Sollicité Horizontalement en Fonction des Résultats Préssiometriques," Sols - Soils, No. 20/23, 1969.

[13] Lamé, G., Leçons sur la Theorie Mathématique de l"Elasticité des Corps Solides, Bachelier, Paris, France, 1852.

[14] Bishop, R.F., Hill, R., and Mott, N.-F., "The Theory of Indentation and Hardness Test," Proceedings of the Physical Society, London, No. 57, 1945, pp. 147-159.

[15] Ménard, L., "Mesure In Situ des Propriétés Physiques des Sols," Annales des Ponts et Chaussées, Paris, No. 14, Mai-Juin, 1957, pp. 357-377.

[16] Cassan, M., "Méthode Préssiometrique d'Etude de Sols," Revue de l'Ingenieur Constructeur, Paris, Mai 1960, pp.3-16.

[17] Hill, R., The Mathematical Theory of Plasticity, Oxford University Press, New York, 1950.

[18] Salençon, J., "Expansion Quasi-Statique d'une Cavité a Symmetrie Sphérique ou Cylindrique dans un Milieu Elasto-Plastique," Annales des Ponts et Chaussées, Paris, Vol. 3, 1966, pp. 175-187.

[19] Vesic, A.-S., "Expansion of Cavities in Infinite Soil Mass," Journal of the Soil Mechanics and Foundation Division, ASCE, Vol. 98, No. SM3, March 1972, pp. 265-290.

[20] Baguelin, F., Jézéquel, J.-F., Lemée, E., and Le Méhaute, A., "Expansion of Cylindrical Probes in Cohesive Soils," Journal of the Soil Mechanics and Foundation Division, ASCE, Vol. 98, No. SM11, November 1972, pp. 1129-1142.

[21] Ladanyi, B., "In Situ Determination of Undrained Stress-Strain Behavior of Sensitive Clays with the Pressuremeter,"Canadian Geotechnical Journal, Vol. 9, No. 3, August 1972, pp. 313-319.

[22] Palmer, A.-C., "Undrained Plane-Strain Expansion of a Cylindrical Cavity in Clay: a Simple Interpretation of the Pressuremeter Test," Geotechnique, 22, No. 3, September 1972, pp. 451-457.

[23] Frank, R., "Etude Théorique du Comportement des Pieux Sous Charge Verticale. Introduction de la Dilatance," Thèse de Docteur Ingénieur, Université de Paris VI, France, 1974.

[24] Ladanyi, B., "Etude Théorique et Expérimentale de l'Expansion dans un Sol Pulverulent d'une Cavité Présentant une Symétrie Sphérique ou Cylindrique," Annales des Travaux Publics de Belgique, Bruxelles, No. 2 et 4, 1961.

[25] Wroth, C.-P., and Windle, D., "Analysis of the Pressuremeter Test Allowing for Volume Change," Geotechnique, London, Vol. 25, No. 3, 1975, Technical notes, pp. 598-604.

[26] Baguelin, F., Frank, R., and Jézéquel, J.-F., "Quelques Résultats Théoriques sur l'Essai d'Expansion dans les Sols et sur le Frottement Latéral des Pieux," Bulletin de Liaison des Ponts et Chaussées, No. 78, Juillet-Aout 1975, pp. 131-136.

[27] Briaud, J.-L., Tucker, L., and Felio, G.Y., Pressuremeter, Cone Penetrometer, and Foundation Design, Short Course Notes, Vol.1, Texas A&M University, College Station, Texas, 1983.

[28] Casagrande, A., "The Determination of the Preconsolidation Load and its Practical Significance," Proceedings of the 1st International Conference on Soil Mechanics and Fondation Engineering, Cambridge, Mass., Vol.3, 1936, pp. 60-64.

[29] Lacasse, S., and Lunne, T., "In Situ Horizontal Stress from Pressuremeter Tests," Symposium on the Pressuremeter and Its Marine Applications, IFP, Paris, April 1982, pp. 187-208.

[30] Wroth, C.P., "Cambridge In-Situ Probe," Symposium on Site Exploration in Soft Ground Using In Situ Techniques, Alexandria, VA, 1978. Also in Federal Highway Administration Report FHWA-TS-80-202, pp. 97-135.

[31] Lacasse, S., Jamiolkowski, M., Lancellotta, R., and Lunne, T., "In Situ Characteristics of Two Norwegian Clays,"Proceedings of the 10th International Conference on Soil Mechanics and Foundation Engineering, Stockholm, Vol. 2, 1981, pp. 507-511.

[32] Law, K.T., and Eden, W.S., "Effects os Soil Disturbance in Presuremeter Tests," Proceedings of the Engineering Foundation Conference on Updating Subsurface Sampling of Soils and Rocks on their In Situ Testing,Santa Barbara, CA, 1982.

[33] Marsland, A., and Randolph, M.F., "Comparison of the Results from Pressuremeter Tests and Large In Situ Plate Tests in London Clay," Geotechnique, 27, No. 2, 1977, pp. 217-243.

[34] Denby, G.M., "Self-Boring Pressuremeter Study of the San Francisco Bay Mud," Ph.D. Thesis, Stanford University, Department of Civil Engineering, 1978.

[35] Arnold, M., "An Empirical Evaluation of Pressuremeter Test Data," Canadian Geotechnical Journal, Vol. 18, No. 3, 1981, pp. 455-459.

[36] Denby, G.M., and Hughes, S.M.O., "Horizontal Stress Interpretation of Pressuremeter Tests," Proceedings of the Engineering Foundation Conference on Updating Subsurface Sampling of Soils and Rocks on their In Situ Testing, Santa Barbara, CA, 1982.

[37] Merritt, B.K., Davidson, R.R., and Baker, G.L., "Menard Pressuremeter Experience in Beaumont Clay," Spring Texas Section Meeting, ASCE, Austin, Texas, 1979.

[38] Ménard, L., "Interpretation and Application of Pressuremeter Test Results to Foundation Design,"Sols - Soils, No. 26, 1975.

[39] Briaud, J.-L., "The Pressuremeter: Application to Pavement Design," Ph.D. Dissertation, University of Ottawa, Civil Engineering Department, 1979.

[40] Wroth, C.P., "British Experience with the Selfboring Pressuremeter," Symposium on the Pressuremeter and Its Marine Applications, IFP, Paris, April 1982.

[41] Briaud, J.-L., Lytton, R.L., and Hung, J.-T., "Obtaining Moduli from Cyclic Pressuremeter Tests," Journal of the Geotechnical Engineering Division, ASCE, Vol. 109, No. 5, May 1983, pp. 657-665.

[42] Duncan, J.M., and Chang, C.Y., "Nonlinear Analysis of Stress and Strain in Soils," Journal of the Soils Mechanics and Foundation Division, ASCE, Vol. 96, No. SM5, 1970, pp. 1629-1653.

[43] Davidson, R.R., and Perez, J.-Y., "Pressuremeter Research: I-90 Project," Woodward Clyde Consultants, Research Report, February 1980.

[44] Mori, H., and Tajima, S., "The Application of the Pressiometre Method to the Design of Deep Foundations," Soil and Foundations, Vol. 4, No. 2, Tokyo, 1964, pp. 34-44.

[45] Johnson, G.W., Stokoe, K.H., II, and Reese, L.C., "Comparison of Dynamic Moduli Determined by Self-Boring Pressuremeter and Conventional Dynamic Methods," Contribution, Symposium on the Pressuremeter and Its Marine Applications, IFP, Paris, April 1982, pp. 229-236.

[46] Shields, D.H., and Bauer, G.E., "Determination of the Modulus of Deformation of Sensitive Clay Using Laboratory and In Situ Tests," Proceedings of the Specialty Conference on In Situ Measurements of Soil Properties, ASCE, Raleigh, NC, Vol. 1, 1975, pp. 395-421.

[47] Tavenas, F., Chapeau, C., LaRochelle, P., and Roy, M., "Immediate Settlement of Three Test Embankments on Champlain Sea Clay," Canadian Geotechnical Journal, Vol. 11, No. 11, 1974, pp. 109-141.

[48] Greenland, S., "Economic Loading Tests with the Menard Pressuremeter," Proceedings of the Symposium on the Economic Use of Soil Testing, Birmingham, 1964.

[49] Burgess, N., and Eisenstein, Z., "Determination of Deformation Characteristics of Soils and Rocks in Western Canada by the Pressuremeter," Proceedings of the 28th Canadian Geotechnical Conference, Montreal, 1975, pp. 180-198.

[50] Briaud, J.-L., and Shields, D.H., "Use of a Pressuremeter Test to Predict the Modulus And Strength of Pavement Layers," Transportation Research Record No. 810, 1981.

[51] Briaud, J.-L., Lytton, R.L., and Hung, J.-T., "Using a Pressuremeter for Pavement Design and Evaluation," Proceedings of the International Symposium on Bearing Capacity of Roads and Airfields, Norwegian Institute of Technology, June 1982.

[52] Amar, S., and Jézéquel, J.-F., "Essais en Place et en Laboratoire sur Sols Cohérents: Comparaison des Résultats," Bulletin de Liaison des Ponts et Chaussées, No. 58, Mai-Avril, 1972, pp.97-108.

[53] Centre d'Etudes Ménard, "Interprétation d'un Essai Préssiometrique, " Publication D 31/67, 1967.

[54] Gibson, R.E., and Anderson, W.F., "Measurement of Soil Properties with the Pressuremeter," Civil Engineering and Public Works Review, Vol. 3, No. 6, May 1961.

[55] Clough, G.W., and Denby, G.M., "Self-Boring Pressuremeter Study of San Francisco Bay Mud," Journal of the Geotechnical Engineering Division, ASCE, Vol. 106, No. GT1, January 1980, pp. 45-64.

[56] Roy, M., Juneau, R., La Rochelle, P., and Tavenas, F., "In Situ Measurement of the Properties of Clays by Pressuremeter Tests," Proceedings of the Specialty Conference on In Situ Measurement of Soil Properties, ASCE, Raleigh, NC, Vol. 1, 1975, pp.350-372.

[57] Higgins, C., "Pressuremeter Correlation Study," Highway Research Record, Highway Research Board, Washington, No. 284, 1969, pp. 51-62.

[58] Thorley, A., Broise, Y., Calhoon, M.L., Zeman, Z.P., and Watt, W.G., "Borehole Instruments for Economical Strength and Deformation In Situ Testing," Proceedings of the Conference on In Situ Investigation of Soils and Rocks, British Geotechnical Society, London, 1969, pp.155-165.

[59] Livneh, M., and Uzon, S., Discussion, "A study of In Situ Testing with the Pressuremeter," Proceedings of the Conference on In Situ Investigation of Soils and Rocks, British Geotechnical Society, London, 1969, pp.212-213.

[60] Eisenstein, Z., and Morrison, N., "Prediction of Foundation Deformations in Edmonton Using an In Situ Pressure Probe," Canadian Geotechnical Journal, Vol. 10, 1973, pp. 193-210.

[61] Schmertmann, J.H., "Measurement of In Situ Shear Strength," Proceedings of the Specialty Conference on In Situ Measurement of Soil Properties, ASCE, Raleigh, NC, Vol. 2, 1975, pp.57-138.

[62] Amar, S., Baguelin, F., Jézéquel, J.-F., and Le Méhaute, A., "In Situ Shear Resistance of Clays," Proceedings of the Specialty Conference on In Situ Measurement of Soil Properties, ASCE, Raleigh, NC, Vol. 1, 1975, pp. 22-44.

[63] Lukas, G.-L., and Leclerc de Bussy, B., "Pressuremeter and Laboratory Test Correlations for Clays," Journal of the Geotechnical Engineering Division, ASCE, Vol. 102, No. GT9, September 1976, pp. 945-963.

[64] Ladd, C.C., Foot, R., Ishihara, K., Schlosser, F., and Poulos, H.G., "Stress-Deformation and Strength Characteristics," State-of-the-Art Report, Proceedings of the 9th International Conference on Soil Mechanics and Foundation Engineering, Tokyo, Vol. 2, 1977, pp. 421-494.

[65] Ghionna, V.N., Jamiolkowski, M., and Lancellotta, R., "Characteristics of Saturated Clays as Obtained from SBP Tests," Proceedings of the Symposium on the Pressuremeter and Its Marine Applications, IFP, PAris, 1982, pp. 165-186.

[66] Mori, H., "Pressuremeter Test for Soft Cohesive Soil," Proceedings of the Symposium on the Pressuremeter and Its Marine Applications, IFP, PAris, 1982, pp. 77-90.

[67] Ladanyi, B., "Evaluation of Pressuremeter Tests in Granular Soils," Proceedings of the 2nd Panamerican Conference on Soil Mechanics and Foundation Engineering, 1963, pp. 3-20.

[68] Hughes, J.M.O., Wroth, C.P., and Windle, D., "Pressuremeter Tests in Sand," Geotechnique, 27, No. 4, 1977.

Peter K. Robertson and John M.O. Hughes

DETERMINATION OF PROPERTIES OF SAND FROM SELF-BORING PRESSUREMETER TESTS

REFERENCE: Robertson, P.K. and Hughes, J.M.O., "Determination of Properties of Sand from Self-boring Pressuremeter Tests", The Pressuremeter and Its Marine Applications: Second International Symposium, ASTM STP 950, J.-L. Briaud and J.M.E. Audibert, Eds., American Society for Testing and Materials, 1986.

ABSTRACT: The self-boring pressuremeter test is ideally suitable for determining in-situ engineering properties of sand. A new method for the determination of the friction angle is presented that accounts for the non-linear volume change behaviour of loose sands. The determination of shear modulus from small unloading-reloading cycles is discussed. A method is presented that allows the shear modulus values to be normalized with respect to stress level. A rational method is also presented to allow the 'elastic' shear modulus measured from small unloading-reloading cycles to be applied to engineering design problems where the loading may induce strains larger than the 'elastic' level.

KEYWORDS: sand, friction angle, modulus, self-boring pressuremeter, disturbance.

INTRODUCTION

It is generally accepted that in-situ tests offer one of the best means of determining the engineering properties of soils. This is particularly true for sands since it is frequently very difficult to retrieve and test undisturbed samples, especially in very loose sands.

Dr. Robertson is a NSERC University Research Fellow in the Department of Civil Engineering, 2324 Main Mall, The University of British Columbia, Vancouver, B.C., Canada, V6T 1W5; Dr. Hughes is a consultant at 3009 St. Andrews St., North Vancouver, B.C., Canada, V7N 1Z5.

The self-boring pressuremeter appears to be ideally suitable for determining in-situ engineering properties of sand. The self-boring pressuremeter can be installed with minimized initial disturbance and results can be evaluated analytically using cavity expansion theories. This paper shows how self-boring pressuremeter test (SBPMT) results in sand can be analysed to obtain engineering properties, such as friction angle (ϕ) and shear modulus (G).

FRICTION ANGLE

Theoretical methods for determination of the friction angle of sands from pressuremeter test data have been proposed by several authors, i.e. Gibson and Anderson [1], Ladanyi [2], Vesic [3] and Hughes et al. [4]. Each method relies on a model for the sand behavior. All the above methods consider that sand has a constant friction angle at failure. However, not all methods allow for the fact that sand changes in volume during shearing. In Ladanyi's method the volume change is considered to be constant at the point the failure stress ratio is reached. This volume change is introduced into the assessment of the friction angle by a trial and error method.

Vesic's solution uses the results of laboratory tests directly. However, the problem of determining the appropriate laboratory density to perform tests at, is not easy to resolve. Also, the laboratory tests may not produce reliable volume change behaviour because the in-situ structure and fabric cannot be reproduced in the laboratory.

The solution developed by Hughes et al. [4] relies on the fact that the volume changes are occurring during the expansion of the cavity and the amount of volume change (dilation) is closely related to the current friction angle developed. This approach brings together the stress dilatancy concept of Rowe [5] and the observed behavior of sand in simple shear as observed by Vaid et al. [6] and Stroud [7].

Consider the results of simple shear tests conducted by Shroud [7] shown in Fig. 1(a). Several important features can be observed from such results. When sand reaches failure, at peak principal stress ratio, the volume change is approximately linear with shear strain and the rate of volume change is at a maximum. The volumetric strains are increasing even though the sand is failing at a constant stress ratio. For loose sand the peak strength is not reached until shear strain levels of between 20 to 30 percent. Before failure the volume change behaviour is non-linear, particularly for loose sands. Most self-boring pressuremeter probes expand to a maximum of about 10 to 15 percent circumferential strain (ε_θ) (i.e. approximately 20 to 30 percent shear strain, γ). Therefore, in loose sands, only the elements of soil immediately adjacent to the probe reach failure.

The model proposed by Hughes et al. [4] assumes that the ideal soil behaves as shown in Fig. 1(b). The rate of volume change is proportional to the friction angle, as suggested by Rowe [5].

In the method proposed by Hughes et al. [4], it was shown that:

$$\log\left(\frac{\Delta R}{R_o} + \frac{c}{2}\right) = \frac{n+1}{1-N} \cdot \log(P-u_o) + \text{constant} \qquad (1)$$

where R_o = initial radius of pressuremeter
ΔR = change in radius of pressuremeter
$\frac{\Delta R}{R_o}$ = circumferential strain, ε_θ.
c = intercept shown on Fig. 1(d)
P = total pressuremeter pressure
u_o = pore water pressure
$\frac{1-N}{n+1} = \frac{(1+\sin\nu)}{(1+\sin\phi')} \cdot \sin\phi'$ = slope s
$\sin\nu$ = maximum dilation rate.

In the above method a plot of the pressuremeter data in terms of $\log(P-u_o)$ (effective pressure) against $\log\left(\frac{\Delta R}{R_o} + \frac{c}{2}\right)$ will tend towards a straight line with a slope s. This slope is related to the in-situ friction angle (ϕ') and the maximum dilation rate (sine $\nu°$).

For very dense sands the correction term "c" is negligible and for all practical purposes can be ignored. The results of the laboratory studies conducted by Jewel et al. [8] in very dense sands (D_r = 90%) using the self-boring pressuremeter probe show that the above technique works very successfully. In loose materials the method is not so convenient as the pressuremeter does not expand sufficiently for the sand around the probe to reach the linear portion of its volumetric strain/shear strain curve, (see Fig. 1(a)).

If self-boring pressuremeter results, where the maximum circumferential strain is 10%, are analysed with the assumption that c = 0 the results obtained will be accurate for very dense sands, as discussed above, but for medium dense and loose sand artificially low dilation and friction angles will be obtained. To illustrate this point, consider the behavior of an elastic soil surrounding the probe when expanded to maximum capacity, i.e. in the region of circumferential strain, $\varepsilon_\theta = \frac{\Delta R}{R_o}$ = 10 percent (approximately 20 percent shear strain, γ). The shear strains in the soil surrounding the probe will follow approximately the form shown in Fig. 2, i.e. they will be large near the face of the probe and decrease rapidly with distance away from the probe. The shear strain one radius away from the probe will have dropped to about one quarter of the shear strain at the face. Beyond one radius the strains are probably not sufficient to fail medium dense to loose soil, however inside this zone the soil can be assumed to be failing at constant stress ratio, which is related to the friction angle (ϕ'). The average shear strain around the probe in the failed region is approximately 10 percent, as shown on Fig. 2.

The volume changes in the shear strain range of 5 to 20 percent are non-linear, as shown by the simple shear test data on Ottawa sand

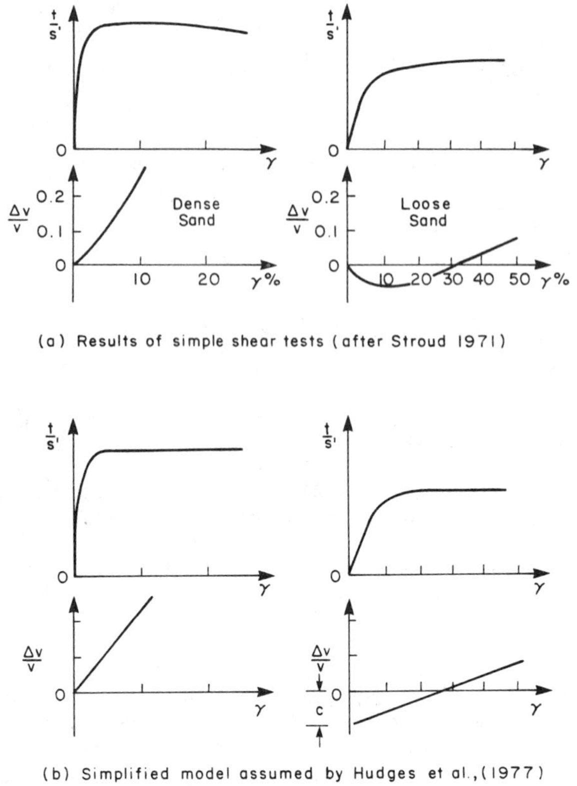

Fig. 1. Stress-Strain and Volumetric Strain-Shear Strain Curves for (a) Simple Shear Test Results (Stroud, 1971), (b) Idealized by Hughes et al. (1977).

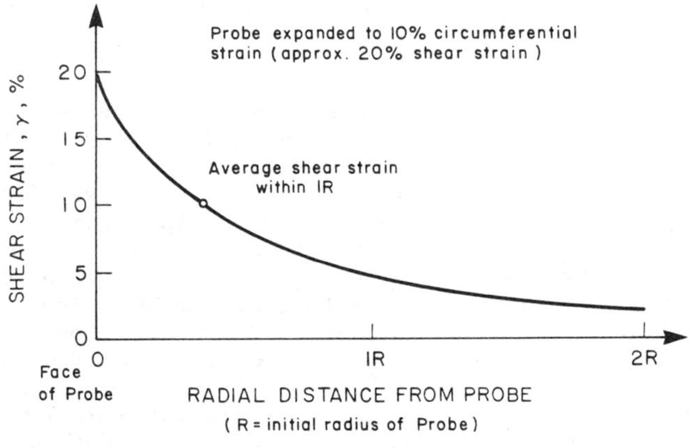

Fig. 2. Approximate Variation of Shear Strain with Radial Distance from Pressuremeter Probe.

conducted by Vaid et al. [6], shown on Fig. 3. Consider the simple shear results obtained by Vaid et al. [6] on a loose sand at a relative density of 26.9 percent, as shown on Fig. 4. The volume changes around the probe follows a path ABCDE. If the pressuremeter data is analysed assuming the "c" intercept is zero then the calculated dilation rate will be based on a straight line passing through the origin A and going through the average shear strain point at C. The dilation angle derived by this method will be considerably smaller than the large strain level dilation angle that is associated with the peak friction angle and that is normally anticipated for the sand at that density and stress level. For this particular sand the anticipated dilation rate would probably be based on the straight line section DEF, since it is this maximum dilation rate that relates to the peak friction angle (see Fig. 3).

If the data shown in Fig. 3 were available for any particular sand then it would be possible to correct the calculated dilation angle. Fig. 5(b) is a plot of relative density against dilation angle for Ottawa Sand obtained from the data by Vaid et al. [6] presented in Fig. 3. This plot shows two lines representing different definitions of dilation. Line AB is the maximum rate of dilation, i.e. related to the slope of the straight portion of the volume shear curve, GH in Fig. 5(a). The maximum dilation rate is associated with the peak friction angle. Line CD is the average dilation rate around the probe (i.e. the secant IJ, in Fig. 5(a)) when the probe is expanded to a maximum of about 10 percent circumferential strain (ε_θ).

The calculated dilation angles using the method of Hughes et al. [4] are based on a linear ratio between volume change and shear strain, i.e. Line CD. By combining pressuremeter results at a circumferential strain level of 10% with the data on Ottawa Sand (Fig. 5), the maximum rate of dilation and peak friction angle can be determined. For example, if the slope of the log log curve at $\varepsilon_\theta \approx$ 10% is measured to be 0.34 the calculated dilation angle will be 0° (based on ϕ_{cv}= 32°). The corrected maximum dilation will be + 6°.

Design charts have been developed by the writers for calculating the peak friction angle and maximum dilation rate that incorporate the proposed strain level correction discussed above and are shown in Figs. 6 and 7.

To carry out the above analyses a knowledge of the friction angle at constant volume, ϕ'_{cv}, is required. The friction angle at constant volume is the friction angle of the material when all dilation has ceased. The range of values of ϕ'_{cv} for most sands usually lies in the range of 30 to 40 degrees.

An estimate of ϕ'_{cv} can be made from Table 1.

If a better estimate of ϕ'_{cv} is required it should be possible to estimate ϕ_{cv} (for plane strain conditions) from drained triaxial compression tests or direct shear box tests [9] carried out on loose samples reconstituted from disturbed samples.

The strain level correction proposed here is applicable to self-boring pressuremeter data where the maximum circumferential strain is

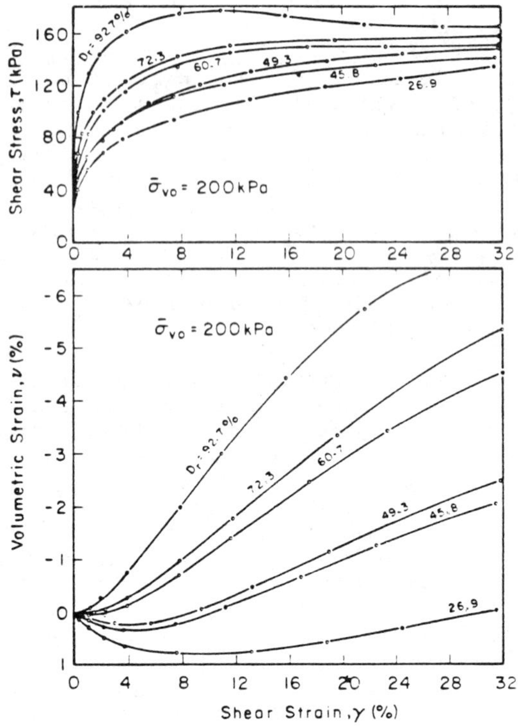

Fig. 3. Stress-Strain Behavior of Ottawa Sand in Drained Simple Shear. (After Vaid et al. 1981)

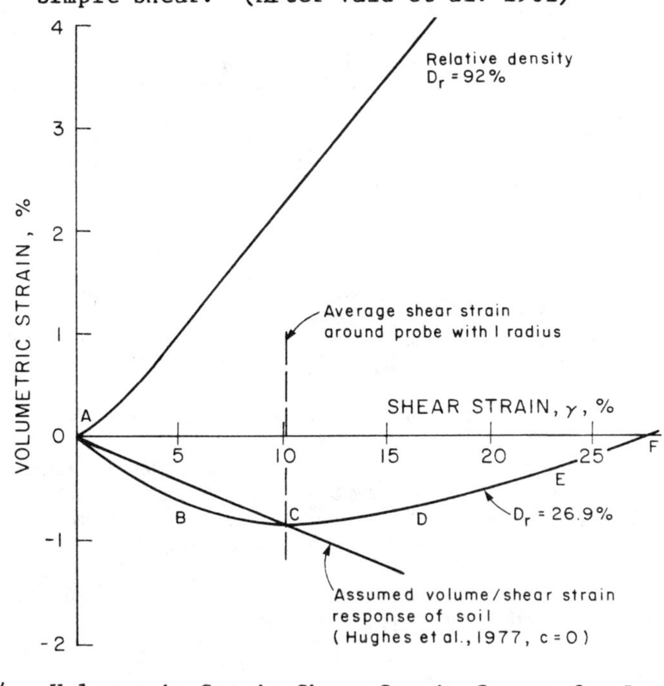

Fig. 4. Volumetric Strain-Shear Strain Curves for Loose and Dense Ottawa Sand in Drained Simple Shear. (Data from Vaid et al. 1981)

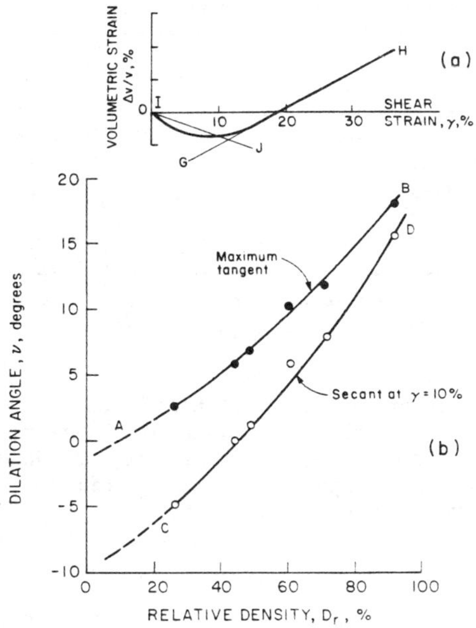

Fig. 5. Variation of Dilation Angle with Relative Density for Ottawa Sand in Drained Simple Shear.

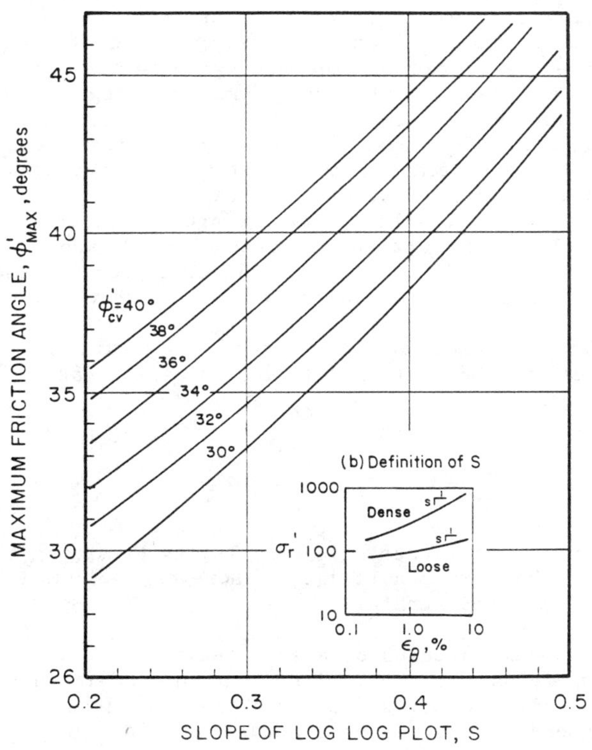

Fig. 6. Proposed Correlation Between Self-Boring Pressuremeter Data and Peak Friction Angle, Corrected for Strain Level.

TABLE 1--Preliminary estimates of ϕ_{cv}.

	ϕ'_{cv}
Well graded gravel-sand-silt	40°
Uniform coarse sand	37°
Well graded medium sand	37°
Uniform medium sand	34°
Well graded fine sand	34°
Uniform fine sand	30°

Assign lower values for well rounded particles.
Assign higher values for angular particles.

about 10%. The correction would be significantly less in loose sands if the pressuremeter were able to expand to a circumferential strain of about 40%. A comparison between the original Hughes et al. [4] method and the proposed strain level correction method is shown in Fig. 8. This confirms the work by Jewel et al. [8] that shows the Hughes et al. [4] method to work successfully for very dense sands (i.e. S > 0.55).

The dilation angle (ν) measured by the pressuremeter is related to the mean effective stress level during the test. Values of ν corrected for strain level are related to the average effective stress level around the probe during the test. When the soil reaches failure the mean effective stress increases in proportion to the effective radial stress, σ'_r, since failure is taking place at approximately constant principle stress ratio. The average effective stress around the probe during the test is therefore somewhat larger than the in-situ mean effective stress before the test. Since the angle of friction decreases with increasing stress level, the friction angle calculated from the pressuremeter data will be somewhat less than the maximum potential friction angle at the in-situ stress level. However, the friction angle values from the pressuremeter are obtained under approximately plain strain conditions which would result in values generally higher than under triaxial conditions. These effects may partially cancel one another with the result that the calculated friction angle, ϕ, and dilation angle, ν, obtained during a pressuremeter test in sand and corrected for strain level may be approximately equal to that obtained from a triaxial test.

SHEAR MODULUS

One of the most common uses of a self-boring pressuremeter is for the derivation of the soil moduli [10]. However, soil moduli varies with both stress level and strain.

The "elastic" shear modulus of a soil can be measured by performing an unloading-reloading cycle during a pressuremeter expansion test. If the soil is perfectly elastic in unloading then the unloading-reloading cycle will have a gradient of 2G. In carrying out

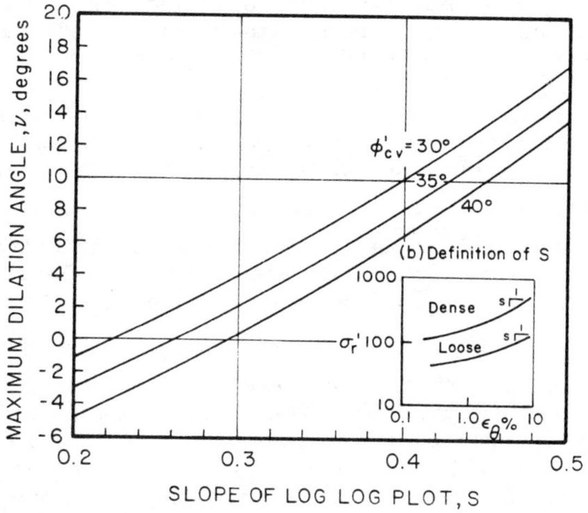

Fig. 7. Proposed Correlation between Self-Boring Pressuremeter Data and Maximum Dilation Angle Corrected for Strain Level.

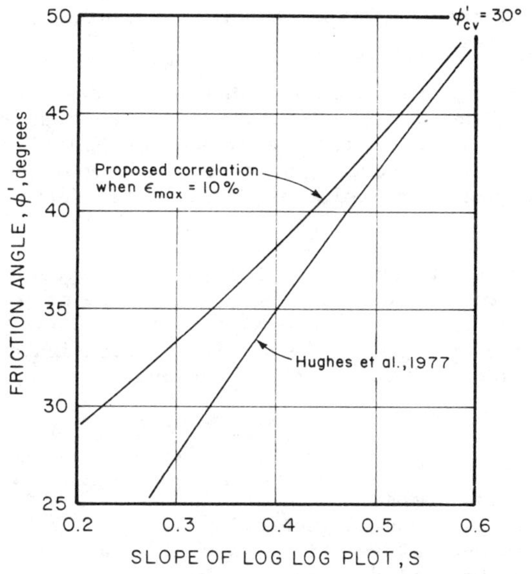

Fig. 8. Comparison between Proposed Correlation and Hughes et al. (1977) Correlation for Peak Friction Angle from Self-Boring Pressuremeter Test Data in Sands.

such a cycle care must be taken not to exceed the elastic limit of the soil during the unloading phase; this restricts the amplitude of the stress cycle that can be carried out.

The idealised behaviour of a free draining sand during a pressuremeter test is shown in Fig. 9. The stress path of the sand in terms of the shear stress $\tau = \frac{1}{2}(\sigma'_r - \sigma'_\theta)$ and a mean effective stress $\frac{1}{2}(\sigma'_r + \sigma'_\theta)$ is represented in Fig. 9(b). AX is the effective stress path during elastic unloading at constant mean effective stress. Full details of the theory behind determination of elastic shear moduli from unloading-reloading cycles are given in Hughes [11] and Wroth [10].

A typical self-boring pressuremeter test result in a medium to fine sand is shown in Fig. 10. The test was performed at a depth of 10.9 m in a sand deposit at the University of British Columbia (UBC) research site (McDonald's Farm). Full details of the site are given by Campanella et al. [12]. The self-boring probe was 75 mm in diameter. The measured value of the elastic shear modulus for cycle AB is 33 MPa (340 kg/cm^2). For the example shown in Fig. 10, another unloading-reloading cycle (EF) was performed on the decreasing pressure side of the pressure expansion curve. The stress levels for the two cycles are approximately equal and the remarkable similarity of the slopes of the cycles is clearly evident. It is also interesting to note that the slope of the unloading (CD) at the end of the increasing pressure side of the test is initially linear with a slope slightly greater than the other two cycles, which were carried out at a lower stress level.

Since sand is free draining and no excess pore pressures are recorded during the test, it follows that the mean effective stress (σ'_m) increases as the test progresses. Therefore, it would be expected that the measured shear modulus would increase as the position of the unloading-reloading cycle moves further from the start of the test.

It is well established that for granular materials the value of elastic moduli are dependent on the value of the mean effective stress (σ'_m), and is usually proportional to $(\sigma'_m)^n$, where n is typically 0.5. The moduli may be expressed in a dimensionless form that is independent of stress level by the equation:

$$G = k_G \cdot Pa\left(\frac{\sigma'_m}{Pa}\right)^n \qquad (2)$$

where k_G = modulus number
n = modulus exponent, typically 0.5
Pa = reference stress (i.e. Pa = 100 kPa)
σ'_m = mean effective stress
$= \frac{1}{3}(\sigma'_r + \sigma'_\theta + \sigma'_z)$.

Hence, results are more clearly expressed in dimensionless form using the modulus number, k_G, or by normalizing the shear modulus, G, to a known stress level.

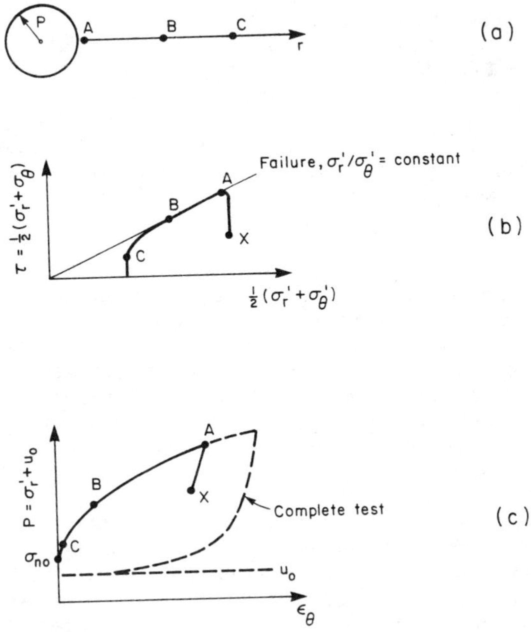

Fig. 9. Idealised Behavior of Free Draining Sand during Self-Boring Pressuremeter Test. (After Wroth, 1982)

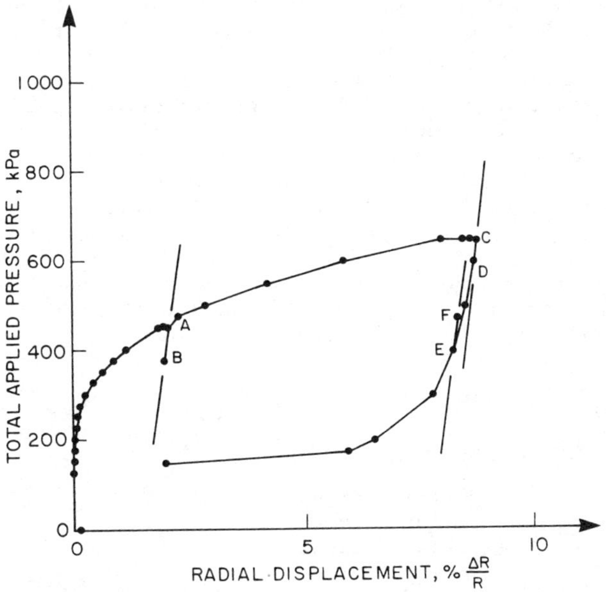

Fig. 10. Typical Self-Boring Pressuremeter Test Result in Sand at a Depth of 10.9 m at McDonald's Farm, B.C.

During the very early stages of a pressure expansion test when the soil is approximately elastic the mean effective stress is unchanged (see Fig. 9). However, when the soil around the pressuremeter reaches failure the mean effective stress increases in proportion to the effective radial stress (σ_r'), since failure is taking place at constant stress ratio (σ_r'/σ_θ'). Therefore, comparing shear modulus from different unload-reload cycles performed at different stresses can be misleading. The shear modulus values should, therefore, be normalized to a consistent known stress level for realistic comparison of data.

The value of the mean effective stress in the failed or plastic zone decreases rapidly with radius from the probe, becoming equal to the in-situ stress at some distance from the probe. However, it is reasonable to assume that the unload-reload response is dominated by the maximum stresses immediately around the probe. Therefore, the value of G obtained from unload-reload cycles can be correlated to the maximum mean effective stress, i.e. the mean effective stress at the face of the probe.

The writers have reviewed a large number of SBPMT results in sand and have found that, in general, the mean effective stress at the face of the pressuremeter, when the soil is at failure is approximately one half the effective radial stress. Numerical studies using finite element analysis [13] have also confirmed this observation. Therefore, the writers believe that, as a first step, shear modulus data should be normalized to a known stress level by assuming the mean effective stress equal to one half the effective radial stress (σ_r') at the start of the unloading cycle and the modulus exponent, n, equal to 0.5. Further research is needed to fully evaluate these numbers, but the writers believe they represent a reasonable first approximation.

As the behaviour of soil is non-linear the appropriate modulus for design will also depend on the strain range expected in the design problem. For computer analyses it may be more appropriate to fit a mathematical curve to the pressure expansion curve so the modulus can automatically be calculated at any strain level. Denby [14] used this approach for calculating clay moduli. However, for pressure expansion tests in clay the mean effective stress is essentially unchanged.

Many static non-linear computer analyses used in design employ a non-linear stress strain curve represented by the hyperbolic function [15] introduced by Kondner [16], and of the form,

$$\tau = \frac{\gamma}{\dfrac{1}{G_i} + \dfrac{\gamma}{\tau_{max}}} \qquad (3)$$

where G_i = initial tangent modulus
τ_{max} = shear strength
γ = shear strain.

The writers consider it reasonable to assume that the elastic shear modulus obtained from the unload-reload cycle is equivalent to the initial tangent modulus, G_i, since the strain levels are comparable at about $\gamma = 10^{-1}\%$, and since most natural sand deposits have experience some past loading.

Thus, the full stress strain curve for a sand deposit can be quickly estimated by taking the unload-reload modulus and applying it to the hyperbolic function (eq. 3). The variation of stress strain characteristics with stress level could be incorporated by using the modulus number, k_G, to calculate the initial shear modulus. The relevant secant modulus at the appropriate working stress or strain level can then be derived from the stress strain curve, as shown on Fig. 11. This provides a rational approach to allow the 'elastic' shear modulus measured from the SBPMT to be applied to any engineering problem where the loading may induce strains significantly larger than the 'elastic' strains.

A similar approach can be applied to obtain the shear modulus from pressuremeter tests in clay. However, since the clay is sheared in an undrained manner, the effective stress is constant at failure, so that the deduced shear modulus should remain constant and be independent of the stage of the test at which it is measured. In reality some drainage will take place during the test. The radial distribution of excess pore pressure is approximately linear with the logarithm of radial distance. This gives rise to substantial hydraulic gradients in the radial direction. Some consolidation will occur, leading to an increase in effective stress near the pressuremeter, with the consequence that the observed values of G may increase marginally with the value of strain at which the unloading cycle is commenced.

EFFECT OF DISTURBANCE

The major advantage of the self-boring pressuremeter test is that it enables testing of undisturbed soil, which is very important for cohesionless soils. However, research and experience has shown that some disturbance is always present. The amount of disturbance depends on the skill and experience of the operator during installation. With good equipment and careful installation, this disturbance can be minimizied. However, it is useful to be aware of the influence different amounts of disturbance can have on the interpreted geotechnical parameters. In the author's experience, self-boring pressuremeters have been installed into cohesionless soils which contain particle sizes up to 10% of the diameter of the probe. However, if significant numbers of particles are greater in size than 1% of the diameter of the probe some disturbance will occur.

The measurement most sensitive to initial disturbance is the in-situ horizontal stress, σ_{ho}. Several studies [14,17] have shown that reasonably good estimates of σ_{ho} can be obtained from good self-boring pressuremeter tests in soft, normally to lightly overconsolidated clay deposits. However, for tests in clay deposits a relaxation time is required to allow all excess pore pressures generated during installation to dissipate. This relaxation time may take several days for very low permeability deposits.

There appears to be very little published data concerning the success of the self-boring pressuremeter in estimating σ_{ho} in sands. This may, in part, be due to the extreme difficulty of obtaining other estimates of σ_{ho} for sands.

Hughes [18] showed that, under controlled laboratory conditions, the radial displacement of the soil immediately adjacent to the surface of the probe in sand was less than 0.5% of the radius of the probe in an outward direction. Thus, under controlled conditions, the error in the circumferential strain, ε_θ, axis is about 0.5%. This is approximately equivalent to 1% shear strain (γ). Therefore, for stiff soils, i.e. soils with a steep initial stress strain curve, the error in estimating σ_{ho} from self-boring pressuremeter tests is likely to be significant. Most dense sands are considerably stiffer than normally consolidated clays. Thus, the estimation of σ_{ho} in dense sands from a SBPMT is likely to be in error even with 'minimal' disturbance. However, reasonable estimates of σ_{ho} do appear to be possible from SBPMT in loose sands. The size of the error in σ_{ho} will depend on the density of the sand deposit and the degree of disturbance generated during installation.

However, despite the foregoing comments, it would appear that in the Beaufort Sea sands and silts, the σ_{ho} estimated from SBPMT does give values which correlate remarkably well with other in-situ and laboratory measurements.

The calculation of friction angle and dilation angle from SBPMT results appears to be less sensitive to initial disturbance than the measurement of in-situ stress. This is mainly because the calculation for shear strength uses the large strain portion of the pressuremeter curve (ε_θ= 10%), where initial minor disturbance is likely to have less influence. Significant initial disturbance, on the other hand, can be expected to have an influence on the calculated friction angle.

From the discussion given in the early part of the paper, it is clear that every element of soil that exists radially away from the pressuremeter probe is experiencing different levels of shear strain, stress and volumetric strains at any particular time during the pressure expansion test. It is, therefore, unlikely that the log-log plot proposed by Hughes et al. [4] should necessarily be linear, since each element of soil is experiencing different levels of dilation. Therefore, methods to correct for disturbance that rely on the existence of a linear log-log plot [9] will not be correct.

The influence of a large initial disturbance can be shown by performing two pressure expansion tests at the same location. Fig. 12 summarizes two pressure expansion tests performed in sand at a depth of 9.3 m at the UBC research site (McDonald's Farm). The first expansion test (Fig. 12) was performed after the instrument was self-bored to the required depth. The expansion curve is typical of a self-boring pressuremeter test in sand. After completion of the test the pressuremeter membrane was allowed to come back to its initial position with no internal pressure. Another pressure expansion test was performed without moving the instrument, Fig. 12. The re-expansion test shows the effects of considerable initial disturbance.

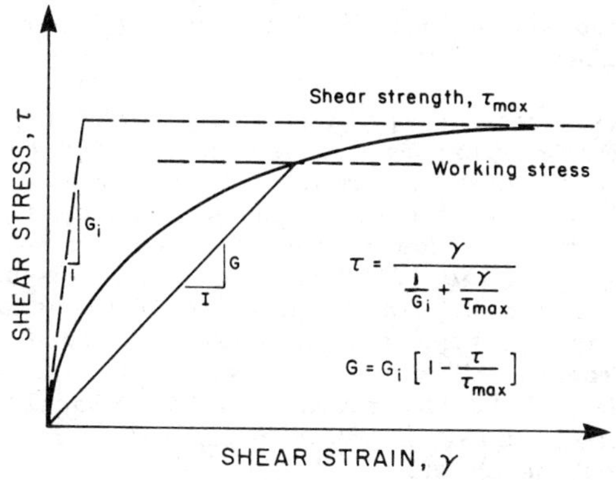

Fig. 11. Hyperbolic Stress-Strain Curve.

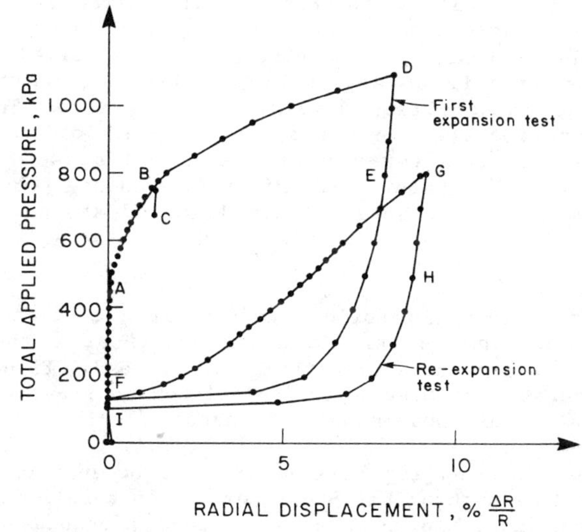

Fig. 12. Effects of Large Initial Disturbance on Pressuremeter Data in Sand at McDonald's Farm Site, B.C.

The calculated in-situ stress and shear strength are considerably different for the re-expansion test as compared to the first self-boring expansion test. It is also interesting to note that the slope of the unload-reload and unload section of the two curves in Fig. 12 are remarkably similar.

Hughes and Robertson [19] have shown that the shear modulus determined from unload-reload cycles are insensitive to the method of installation. Fig. 13 shows a comparison between the shear modulus, normalized to the in-situ effective mean stress, measured from small unloading-reloading cycles from both self-bored and full-displacement pressuremeter tests. The full-displacement pressuremeter tests (FDPMT) were carried out by placing a solid 60° cone on the base of the pressuremeter and pushing the pressuremeter into the soil in a full-displacement manner. The tests, shown in Fig. 13, were performed in the sand deposit at the UBC research site (McDonald's Farm). The modulus values from the FDPMT show remarkable agreement with the values obtained from the SBPMT.

A profile of the dynamic shear modulus (G_{max}) determined from CPT seismic shear wave velocities is also given in Fig. 13. Full details of the CPT seismic method are given by Robertson et al. [20]. The dynamic shear modulus (G_{max}) values were measured in-situ at the in-situ stresses, whereas the modulus derived from the pressuremeter tests are at a much higher average stress. Thus, normalization of the pressuremeter modulus values to the in-situ effective stress is important for a realistic comparison. The fact that the pressuremeter shear modulus, when normalized to the in-situ stress level, are about one-third of the dynamic shear moduli (G_{max}) is not surprising when one considers the shear strain levels for each test. The seismic method records the moduli at a shear strain levels of $1{\times}10^{-4}\%$ or less whereas the pressuremeter records the moduli at a shear strain level of about $1{\times}10^{-1}\%$. Laboratory and field evidence [21] has shown that the shear moduli at a shear strain level of about $1{\times}10^{-1}\%$ is about one-third of the shear moduli (G_{max}) at a shear strain level of about $1{\times}10^{-4}\%$.

The data in Figure 13 provides considerable confidence in the reliability of measuring the static shear moduli at a shear strain level of about $10^{-1}\%$ from pressuremeter tests in sand. Figure 13 also shows the considerable promise a full-displacement pressuremeter has for the determination of shear modulus in sands.

However, the authors [19] have shown that because of the large disturbance caused by pushing the pressuremeter in a full-displacement manner, it is not possible to use the method proposed here or by Hughes et al. [4] to estimate the friction angle in sands.

OFFSHORE INSTALLATION EXPERIENCE

The insertion of a self-boring pressuremeter in sands can be achieved very simply and tests can be conducted continuously without withdrawing the instrument between tests. It is common to complete 10 to 20 tests in a profile. The limitation on the number of tests is

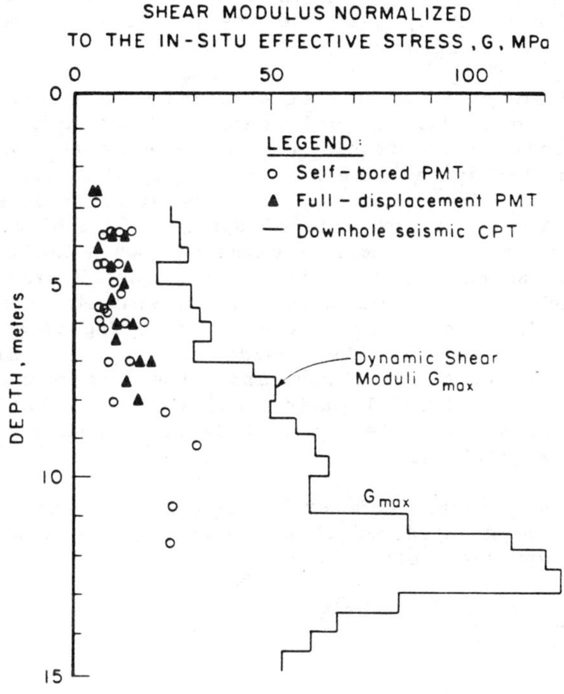

Fig. 13. Comparison of Shear Modulus, Normalized to the In-situ Effective Stress, from Self-Bored and Full-Displacement Pressuremeter Tests and Dynamic Shear Modulus at McDonald's Farm Site, B.C.

the thickness of the formation, the pushing capacity of the drill rig and the grain size of the soil.

The technique of insertion is different from that described by Wroth [22]. The double rod system or a down hole motor to drive a rotating cutter is not required to operate the instrument in sands or silty clays.

The technique described below was developed for operating self-drilled pressuremeters off ships in the shallow Beaufort Sea. Since its introduction in 1982, over 350 tests have been completed using this method in water depths of up to 45 m.

A 75 mm diameter instrument is deployed on a single string of BQ or AW rods. The umbilical cord containing the hydraulic pressure control tube and the electronic signals goes up the outside of the rods in the annular space between that left by the instrument and the rods. Drilling mud is pumped down the rods at a pressure of approximately 2,000 kPa with a flow rate of between 45 to 65 ℓ/min. At the back of the instrument, the mud is diverted down a small rod, concentric with the instrument, to cutting jets just inside the cutting shoe. The high pressure mud cuts the undisturbed sands inside the cutting shoe. Thus, the principle of self-boring is achieved. The flushings are returned up the inside of the instrument and are exhausted outside the AW or BQ rods above the instrument. It is very important that the jetting fluid is a drilling mud capable of holding the sand in suspension while the test is being conducted. This can take from 3 to 20 minutes.

For the above system to operate from a ship, the pushing system must be wave compensated so that the instrument can be advanced independent of any wave motion.

SUMMARY

For loose sands, the peak strength is not reached until shear strains reach between 20 to 30 percent. Before failure, the volume change behaviour is non-linear, particularly for loose sands. Most selfboring pressuremeter probes expand to a maximum of about 20% shear strain. Therefore, in loose sands, only the elements of soil immediately adjacent to the probe reach failure. A method has been presented that shows how the method by Hughes et al. [4] has been modified to account for this observed behaviour in loose sands.

A method has also been presented that allows shear modulus values determined from small unloading-reloading cycles during a pressuremeter test to be normalized with respect to stress level. This is considered by the authors to be extremely important to allow correct use and comparison of modulus data. The method proposed, however, is approximate in nature and further research is needed to evaluate the approach.

A rational method has been presented to allow the 'elastic' shear modulus measured from small unloading-reloading cycles to be applied to engineering design problems where the loading may induce strains significantly larger than the 'elastic' strains.

The effect of disturbance has been shown to have little influence on the shear modulus determined from unloading-reloading cycles. Thus, the pressuremeter can be pushed into sand in a full-displacement manner and the elastic shear modulus measured from small unload-reload cycles. The installation of the pressuremeter in this manner can considerably simplify the testing procedure, especially in the deep offshore environment. The combination of an electric cone and a pressuremeter has significant potential. Initial work on this approach in which the pressuremeter is vibrated into the soil has been initiated in France. Research and development is currently underway at UBC to combine a 15 cm^2 probe with a standard electronic cone on the tip.

The calculation of friction angle and dilation angle from SBPMT results is less sensitive to small initial disturbances than the measurement of in-situ stress. However, this is mainly because the calculation for friction angle uses the large strain portion of the pressuremeter data, where small initial disturbance is likely to have less influence. However, large initial disturbance has a significant influence on the calculated friction angle.

ACKNOWLEDGEMENTS

The assistance of the Natural Sciences and Engineering Research Council of Canada and the technical staff of the Civil Engineering Department, University of British Columbia, is very much appreciated.

REFERENCES

[1] Gibson, R.E. and Anderson, W.F., 1961, "In-situ Measurement of Soil Properties with the Pressuremeter", Civil Engineering and Public Works Review, Vol. 56, pp. 615-618.

[2] Ladanyi, B., 1963, "Evaluation of Pressuremeter Tests in Granular Soils", 2nd Panamerican Conference Soil Mechanics and Foundation Engineering, Vol. 1.

[3] Vesic, A.S., 1972, "Expansion of Cavities in Infinite Soil Masses", Journal of the Soil Mechanics and Foundation Division, ASCE, Vol. 98, SM3, pp. 265-290.

[4] Hughes, J.M.O., Wroth, C.P. and Windle, D., 1977, "Pressuremeter Tests in Sands", Geotechnique 27, No. 4, pp. 455-477.

[5] Rowe, P.W., 1962, "The Stress-Dilatancy Relation for Static Equilibrium of an Assembly of Particles in Contact", Proceedings of the Royal Society, Vol. 269, Series A, pp. 500-527.

[6] Vaid, Y.P., Byrne, P.M. and Hughes, J.M.P., 1981, "Dilation Angle and Liquefaction Potential", Geotechnical Engineering Division, ASCE, Vol. 107, GT7, July 1981.

[7] Stroud, M.A., 1971, "Sand at Low Stress Levels in the Simple Shear Apparatus", Ph.D. Thesis, University of Cambridge.

[8] Jewell, R.J., Fahey, M. and Wroth, C.P., 1980, "Laboratory Studies of the Pressuremeter Test in Sand", Geotechnique, 30, No. 4, pp. 507-531.

[9] Fahey, M. and Randolph, M.F., 1984, "Effect of Disturbance on Parameters Derived from Self-boring Pressuremeter Tests in Sand", Geotechnique 34, No. 1, pp. 81-97.

[10] Wroth, C.P., 1982, "British Experience with the Self-Boring Pressuremeter", Proceedings of the Symposium The Pressuremeter and its Marine Applications, Paris.

[11] Hughes, J.M.O., 1982, "Interpretation of Pressuremeter Tests for the Determination of the Elastic Shear Modulus", Engineering Foundation Conference on Updating Subsurface Sampling of Soils and Rocks and Their In-situ Testing, Santa Barbara, California.

[12] Campanella, R.G., Robertson, P.K. and Gillespie, D., 1983, "Cone Penetration Testing in Deltaic Soils", Canadian Geotechnical Journal, Vol. 20, No. 1, February, pg. 23-35.

[13] Eldridge, T.L., 1983, "Pressuremeter Tests in Sand: Effects of Dilation", A thesis submitted in partial fulfillment of the requirements for the M.A.Sc., Department of Civil Engineering, University of British Columbia.

[14] Denby, G.M., 1978, "Self-boring Pressuremeter Study of the San Francisco Bay Mud", Ph.D. Thesis, Department of Civil Engineering, Stanford University, California.

[15] Duncan, J.M. and Chang, C-Y, 1970, "Nonlinear Analysis of Stress and Strain in Soils", Journal of the Soil Mechanics and Foundation Division, ASCE, Vol. 96, No. SM5, pp. 1629-1651.

[16] Konder, R.L., 1963, "Hyperbolic Stress-Strain Response: Cohesive Soils", Journal of the Soil Mechanics and Foundations Division, ASCE, Vol. 89, No. SM1, pp. 115-143.

[17] Ghionna, V.N., Jamiolkowski, M. and Lancellotta, R., 1982, "Characteristics of Saturated Clays as Obtained from SBP Tests", Proceedings Symp. The Pressuremeter and its Marine Applications, Paris.

[18] Hughes, J.M.O., 1973, "An Intrument for the In-situ Measurement of the Properties of Soft Clays", Ph.D. Dissertation, University of Cambridge.

[19] Hughes, J.M.O. and Robertson, P.K., 1985, "Full Displacement Pressuremeter Testing in Sands", Canadian Geotechnical Journal, Vol. 22, No. 3.

[20] Robertson, P.K., Campanella, R.G., Gillespie, D. and Rice, R., 1985, "Seismic CPT to Measure In-situ Shear Wave Velocity", Specialty Session on Measurement and Use of Shear Wave Velocity for Evaluating Dynamic Soil Properties, ASCE Spring Convention, Denver, Colorado.

[21] Seed, H.B. and Idris, I.M., 1970, "Soil Moduli and Damping Factors for Dynamic Response Analyses", Report to EERC 70-10, Earthquake Engineering Research Center, University of California, Berkeley.

[22] Wroth, C.P., 1984, "The Interpretation of In-situ Soil Tests", Geotechnique 34, No. 4, pp. 449-489.

Jean-Louis Briaud, Larry M. Tucker, and Chaidir A. Makarim

PRESSUREMETER STANDARD AND PRESSUREMETER PARAMETERS

REFERENCE: Briaud, J.-L., Tucker, L.M., "Pressuremeter Standard and Pressuremeter Parameters," The Pressuremeter and Its Marine Applications: Second International Symposium, ASTM STP 950, J.-L. Briaud and J.M.E. Audibert, Eds., American Society for Testing and Materials, 1985.

ABSTRACT: Remarks are made in a first part on key issues in the performance of a pressuremeter test. These include the length to diameter ratio for the probes, monocell versus tricell probes, pressure versus volume increment procedure, volume and membrane calibrations, data reduction. In a second part a data base is presented of pressuremeter, cone penetrometer, standard penetration and undrained shear strength tests results. Correlation equations between the various parameters are given and the large scatter is shown. In a third part the magnitude of the pressuremeter modulus is discussed in light of experimental findings and theoretical developments. It is shown that considering the weakness of soils in extension changes the moduli values obtained by the theory of elasticity.

KEYWORDS: pressuremeter, clay, sand, standard, modulus, tension.

INTRODUCTION

ASTM is progressing in the establishment of a standard for the preboring pressuremeter test. The effort is divided into two parts: the preparation of the borehole and the performance of the test. A standard cannot exist for the preparation of the borehole simply because such preparation is as much an art as it is a science. Only a suggested practice can be written; such has been done (1) and the issue should be brought to rest. A standard can be written for the performance of the test; such has been done for the Menard

Dr. Briaud is an Associate Professor in the Civil Engineering at Department at Texas A&M University, College Station, TX 77843; Mr. Tucker is a Research Associate in the Civil Engineering Department Texas A&M University, College Station, TX 77843. Mr. Makarim is a Research Assistant in the Civil Engineering Department at Texas A&M University, College Station, TX 77843.

pressuremeter (2) but the issue has been complicated by the recent commercial development of new pressuremeters such as the LLT (3), the Elastmeter 100 (4), and the TEXAM pressuremeter (5,6). The issue is further complicated by the fact that the standard for the preboring pressuremeter should be written in such a way that it can be readily extended to selfboring pressuremeters such as the PAF (7) and the Camkometer (8), to offshore pressuremeters such as the PIP (9) and the PAM (10), and to even more recent developments such as the cone pressuremeter (11, 12).

In the first part of this article, remarks are made on key issues in the performance of the PMT test. In the second part a data base of PMT, CPT, SPT and laboratory data is presented and the results are summarized. In the third part it is shown how the PMT modulus is influenced by the consideration, in the theories, of the soil weakness in extension.

MONOCELL VERSUS TRICELL PROBES AND THE LENGTH TO DIAMETER RATIO

The Menard probe has three cells, one measuring cell between two guard cells. The idea is to isolate the true plane strain/cylindrical expansion in the central part of the probe and record it with the measuring cell. The LLT and the Texam pressuremeter are monocell pressuremeters which record the soil response over the entire probe length including end effect. The monocell pressuremeters are much simpler to use than the tricell pressuremeters; the monocell probe is also much easier to repair than the tricell probe. The concern is, however, whether or not monocell and tricell pressuremeters give the same results namely the modulus and the limit pressure.

Tranter (13) gave a close form solution to the problem of an infinitely long cylindrical hole with a diameter D pressurized over a length L in an isotropic homogeneous elastic space. Hartman (14) used Tranter's solution to calculate the ratio of the volume displaced V_f for the finite length L pressuremeter to the volume displaced V_i over the same length L for the infinitely long pressuremeter. Hartman shows that the error is small even for relatively short probes (Fig. 1). The Texam NX probe has an L/D of 6.6, the LLT NX probe has an L/D of 7.5.

In fact the maximum error would be obtained if one would inflate a probe with an L/D of 1, essentially a sphere, and used the equation for the cylindrical expansion in order to calculate the modulus. The equations are:

for the cylinder $$G = \frac{\Delta p}{\Delta V} V_m \quad (1)$$

for the sphere $$G = \frac{3}{4}\frac{\Delta p}{\Delta V} V_m \quad (2)$$

where G is the shear modulus, Δp is the increase in pressure at the cavity wall, ΔV is the increase in volume of the cavity, V_m is the volume of the cavity. Therefore if one inflates a sphere and uses the cylinder equation, the calculated modulus will be equal to 1.33 times the true soil modulus. This shows that the maximum error for the shortest monocell probe will be 33%. Monocell probes have L/D ratios

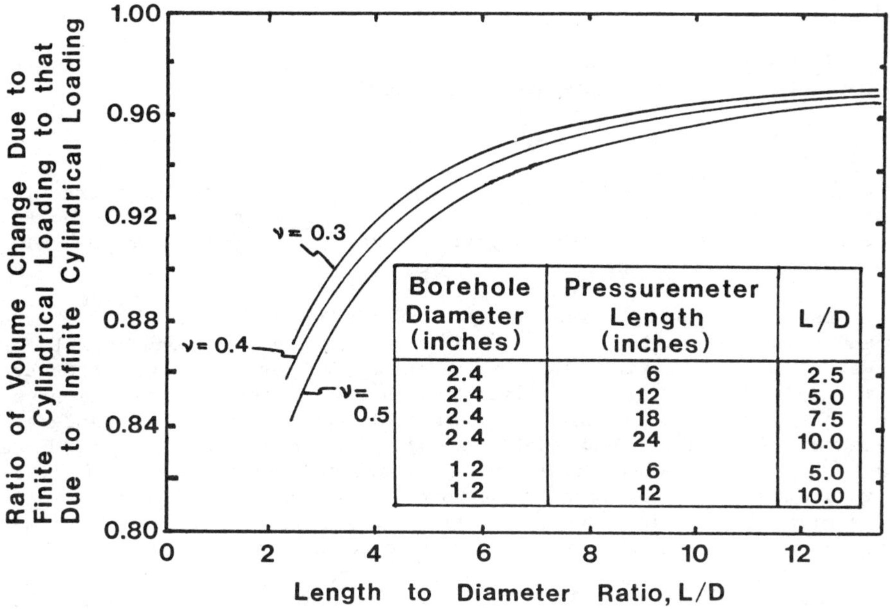

Figure 1. Ratio of Volume Displaced for Finite and Infinite PMT. (from Hartman, 1974)

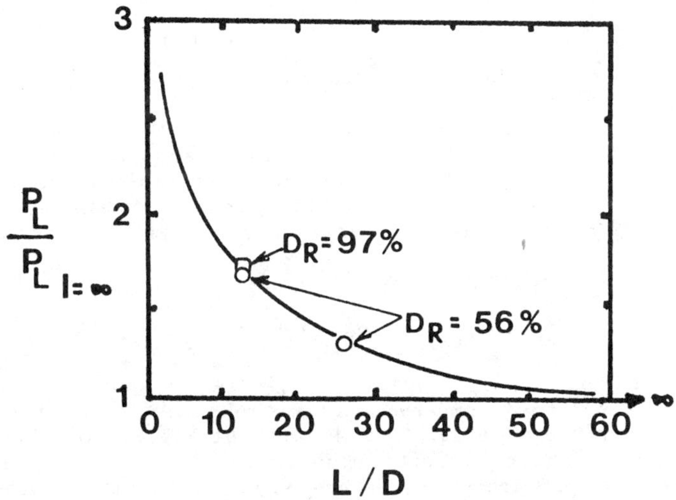

Figure 2. Influence of Probe Length on Limit Pressure Measurement in Dry Sand (from Laier, 1973)

much higher than 1 and therefore the error is much less than 33% (Fig. 1).

It is proposed that the modulus calculations for any pressuremeter include the Hartman-Tranter correction factor:

$$E = \frac{2(1+\nu)}{I_o} \frac{\Delta p}{\Delta V} V_m \quad \ldots\ldots\ldots\ldots \quad (3)$$

where E is Young's modulus, ν Poison's Ratio, I_o is the correction factor from Figure 1. If the factor I_o is ignored for monocell pressuremeters with L/D ratios larger than 6, the moduli obtained will be at most 5% too high.

Laier (15) using tricell probes of various L/D ratios in dry sand found almost no influence of L/D on the pressuremeter modulus. Laier however found a significant influence of L/D on the limit pressure (Fig. 2). He obtained the limit pressure by extrapolating the pressuremeter curve to infinite expansion using the Gibson and Anderson technique (16) and found that when the L/D ratio is decreased by a factor of 2 the limit pressure is increased by a factor of 1.28. Suyama et al. (17,18) compared a monocell and a tricell pressuremeter of similar L/D ratios and found very little difference in the limit pressure and the modulus obtained from both apparatus.

The influence of L/D on the limit pressure p_L can be evaluated by comparing the theoretical expression of p_{Lcyl} for the cylinder (L/D = ∞) and of p_{Lsph} for the sphere (L/D = 1). In sand, the expressions are:

$$p_{LcyL} = p_o\ (1+\sin\phi)\left[\frac{G}{p_o \sin\phi}\right]^{0.5(1-K_a)} \quad \ldots\ldots\ldots \quad (4)$$

$$p_{Lsph} = p_o\ \frac{3(1+\sin\phi)\cos\phi}{(3-\sin\phi)\sin\phi}\left[\frac{G(3-\sin\phi}{3\ p_o \sin\phi}\right]^{0.66(1-K_a)} \quad \ldots\ldots \quad (5)$$

where p_o is the horizontal stress at rest, ϕ is the friction angle, G is the shear modulus, K_a is the coefficient of active earth pressure. For reasonable values of the above variables, the ratio of p_{Lsph} over p_{Lcyl} varies from 4 for loose sands to 3 for dense sands. For the undrained behavior of clays, the expressions are:

$$p_{L\ cyl} = p_o + S_U\ (1 + \mathrm{Ln}\ \frac{G}{S_U}) \quad \ldots\ldots \quad (6)$$

$$p_{L\ sph} = p_o + \frac{4}{3} S_U\ (1 + \mathrm{Ln}\ \frac{G}{S_U}) \quad \ldots\ldots \quad (7)$$

Here the ratio of $p_{L\ sph}$ over $p_{L\ cyl}$ is at most 1.33. In sand the ratio of 3 to 4 compares well with Laier's curve and allows to extend it to L/D of 1 (Fig. 2). It must be kept in mind that the above p_L values are for infinite probe expansion while in practice p_L is defined as the pressure reached when the cavity reaches two times its initial volume.

Briaud et al. (19) ran a number of PMT tests at two sites, one in sand, one in clay. The two pressuremeters used were the Menard pressuremeter, and the Texam pressuremeter. The Menard PMT had a tricell probe with a length to diameter ratio of 5. The Texam pressuremeter had a monocell probe with a length to diameter ratio of 6.6. The sand deposit is made of a silty sand with an average SPT blow count of 15 blows per 30 cm (1 ft), water content of 14%, unit weight 19.3 kN/m^3 (123 pcf). The clay deposit is made of a high plasticity clay with an average undrained shear strength measured in unconfined compression tests of 118 kPa (1.2 tsf), water content 20%, unit weight 20.4 kN/m^3 (130 pcf). The results of the series of comparative tests are shown on Figures 3 and 4. The results exhibit scatter due to the natural heterogeneity of the soil. Careful interpretation shows that no definite trend exists which would allow to conclude to a difference between the two pressuremeter test results.

Considering all the above evidence it appears that the length to diameter ratio has little influence on the modulus and some influence on the limit pressure. It is therefore important for the ASTM standard that a unique length to diameter ratio be chosen. A value of 6.5 is recommended. For this length to diameter ratio and for all practical purposes, there does not seem to be any difference between monocell and tricell pressuremeters. For pressuremeters which measure radial expansion through electrical feeler arms located at midheight of the probe, the influence of the L/D ratio on the results is less; for the sake of uniformity however the same L/D ratio is recommended.

PRESSURE VERSUS VOLUME INCREMENT TEST PROCEDURE

The pressuremeter test can be performed by increasing the probe pressure in equal increments or by increasing the probe volume in equal increments or by increasing the probe volume at a constant rate of volume increase. It is the authors opinion that a standard pressuremeter test should be performed by doubling the initial size of the probe in 10 minutes. This is then used to calculate the rate of loading, pressure increments, volume increments or constant rate of inflation, at which the test is run.

There are drawbacks and advantages to both pressure and volume increments procedures. The drawback of the pressure increment procedure is that the limit pressure p_L must be estimated before running the test; this is the advantage of the volume increments procedure where no estimate of p_L is required. The drawback of the volume increment procedure is that, if the volume increments are not small enough, the modulus part of the curve may be defined by too few data points. This is the advantage of the pressure increment procedure where the modulus is usually well defined. Considering these comments, ten 1 minute increments equal to $p_L/10$ are recommended for the pressure increment procedure (Fig. 5.) and forty 15 seconds increments equal to $V_o/40$ are recommended for the volume increment procedure where V_o is the deflated volume of the probe (Fig. 6). These volume increments are sufficiently small to ensure a proper definition of the modulus, and allow a very precise description of the entire curve including the early part which is of interest for the at rest earth pressure.

There is a difference between the two procedures. With the volume increments the rate of loading during the modulus portion of the curve may be 3 times faster than with the pressure increments. The

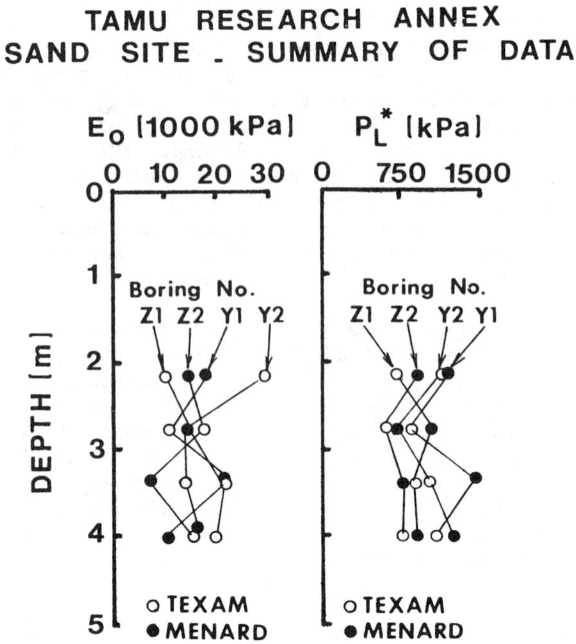

Figure 3. Comparison of the Texam and Menard Pressuremeter at a Sand Site.

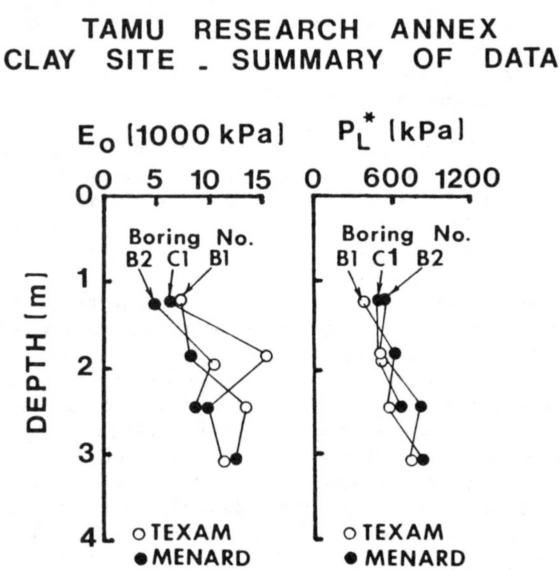

Figure 4. Comparison of the Texam and Menard Pressuremeter at a Clay Site.

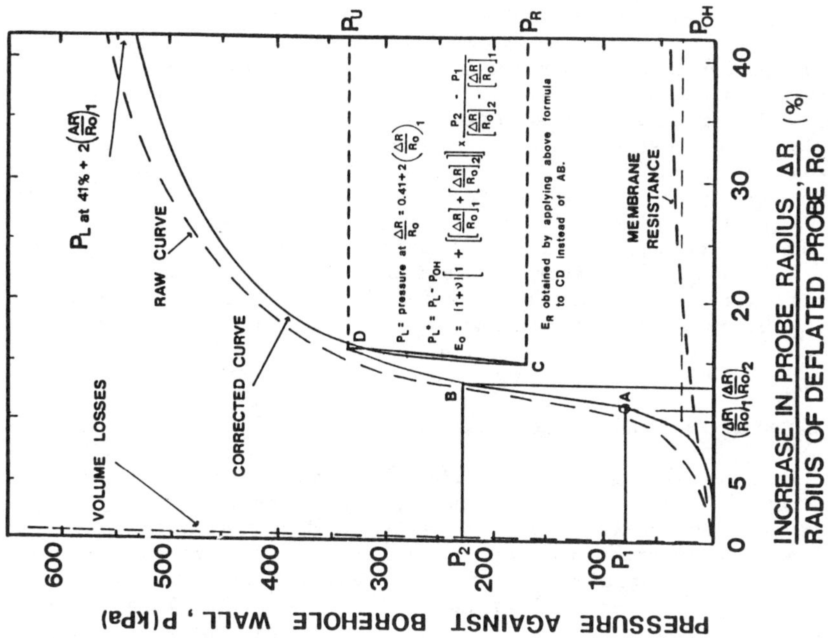

Figure 7. Standard Plot for Pressuremeter.

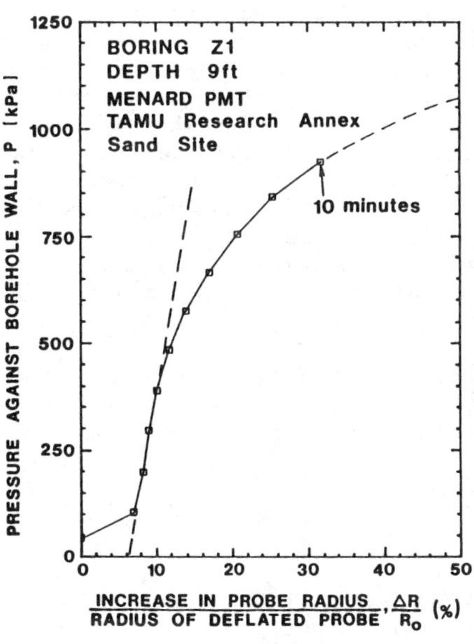

Figure 5. Pressure Increment Pressuremeter Test.

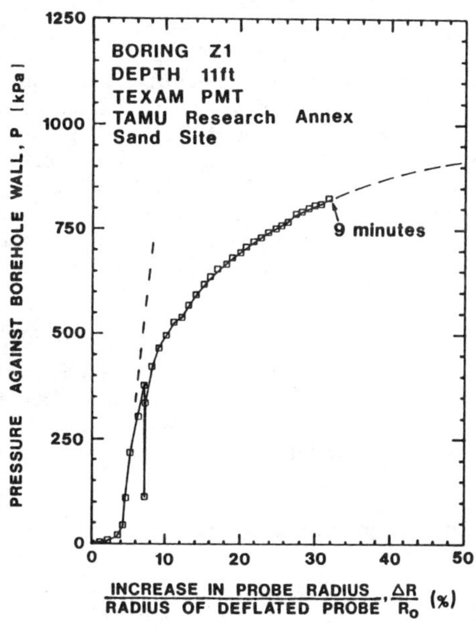

Figure 6. Volume Increment Pressuremeter Test.

modulus will therefore be slightly higher; it will be higher by approximately $3^{0.05} - 1 = 0.056$ or 5.6% (20). The contrary is true for the limit pressure p_L. Concerning p_L it may also be noted that with the volume increments the probe initial volume will be doubled in precisely 10 minutes whereas with the pressure increments it will typically be doubled in 7 to 14 minutes depending on how accurately p_L has been estimated. The tests shown on Figure 3 and 4 were performed using both procedures. The Menard tests were pressure increment tests (between 7 and 15 one minute increments) while the Texam tests were volume increment tests (40 fifteen second increments). Again little difference is found between the results of the two tests. Considering the above evidence and for all practical purposes, it appears that the recommended pressure and volume increment procedures can be used interchangeably.

VOLUME LOSSES AND MEMBRANE CALIBRATION

Calibration for membrane stiffness is generally performed for all pressuremeters. It is essential that the membrane be inflated at the same rate as for the test: 10 minutes to double the initial size of the probe. Calibration for volume losses in the case of hydraulic pressuremeters is not always performed (Menard PMT) but it should be done, and should be reported as part of the results. Then and only then, can it be argued that correction for volume losses is unnecessary.

The best way to perform a volume calibration test is to insert the deflated probe of diameter D_p in a tight fitting steel tube of diameter D_T (say $D_T = 1.005\ D_p$) and to inflate the probe in ten 1 minute or twenty 30 seconds equal pressure increments up to the anticipated limit pressure for the deposit tested. Again the volume calibrations must satisfy the 10 minute criterion. The volume correction curve is given by the slope of the expansion curve when that expansion curve becomes linear. With this tight fitting steel tube it becomes very simple to determine a consistent zero probe volume. On deflation from the volume correction test the zero probe volume is defined as the volume of the probe obtained when it first becomes possible to slide out by hand the steel tube from over the probe. The borehole is then calibrated to the size of the steel tube and not to the size of the probe.

It is strongly recommended that such a tight fitting steel tube become an integral part of any PMT equipment sold or used commercially. Steel tubes somewhat larger in diameter than the probe can be used but not as conveniently.

PRESSUREMETER DATA REDUCTION

First the raw curve is corrected for volume and membrane correction. This is best done on a personnal computer (21). In that respect the corrections can be done by finding equations for the curves that best fit the volume and membrane correction curves or by describing the volume and membrane correction curves point by point. The curve fitting process does not work equally well for all pressuremeters. The point by point description is to be preferred as it is much more flexible and can, with an interpolation routine handle the data correction process efficiently. The correction for the unload-reload cycles are obtained by performing membrane and volume calibrations with an unload-reload cycle at the same level as in the test.

The corrected curve can be plotted in a number of ways. It is recommended that it be plotted as the radial stress at the cavity wall versus the relative increase in probe radius (Fig. 7). Indeed this type of curve allows to compare any pressuremeter data and calculate the pressuremeter parameters without any additional information on the probe dimensions. The first load modulus E_o is obtained from the straight part of the curve (AB on Figure 7):

$$E_o = (1+\nu)\left[1+\left[\left[\frac{\Delta R}{R_o}\right]_1+\left[\frac{\Delta R}{R_o}\right]_2\right]\right] \times \frac{p_2 - p_1}{\left[\frac{\Delta R}{R_o}\right]_2 - \left[\frac{\Delta R}{R_o}\right]_1} \quad \ldots\ldots \quad (8)$$

where ν is Poison's ratio and all other parameters are defined on Figure 7. The limit pressure p_L is defined as the pressure reached when the initial volume of the cavity has been doubled. This corresponds to a value of $\Delta R/R_o$ equal to $0.41+2\ [\Delta R/R_o]_1$. The net limit pressure p^*_L is obtained by subtracting from p_L the estimated horizontal pressure at rest p_{OH}:

$$p^*_L = p_L - p_{OH} \quad \ldots\ldots\ldots\ldots \quad (9)$$

Note that the horizontal pressure at rest p_{OH} can be obtained from the beginning of the preboring PMT curve provided a sufficient number of data points are collected to define the recompression curve (OA on Figure 7). The pressure p_{OH}, by analogy with the determination of the preconsolidation pressure from a consolidometer test curve, is found at the point of maximum curvature along OA. This point is more clearly defined after plotting OA as pressure versus log $(\Delta R/R_o)$. A reload modulus E_R can also be obtained by applying equation 8 to the unload-reload loop on the curve; points C and D on Figure 7 are used instead of A and B when using equation 8.

DATA BASE OF PMT PARAMETERS

A data base of preboring pressuremeter test data and other test data was formed. The pressuremeter data was collected over the last 10 years on various research and consulting projects. The pressuremeters used were the Menard pressuremeter, the Texam pressuremeter and the pavement pressuremeter all inserted in a prebored hole. The 82 pressuremeter borings were located in the south, southwest, west and central United States with 36 sand sites, 44 clay sites and 2 silt sites. Next to the PMT borings other borings were performed leading to data on undrained shear strength S_U, effective stress friction angle ϕ, SPT blow count N, cone point and friction resistance q_c and f_s. At each depth in a boring a record was created which consisted of E_o, E_R, p_L, S_U, ϕ, q_c and f_s. A total of 426 records were accumulated. A list of contributors is shown in Table 1 and the data is described in details in Reference 22.

Table 1. Contributors to the Data Base

Amoco Corporation
Black & Veatch Consulting Engineers
Briaud Engineers
Earth Technology Corporation
Electric Power Research Institute
Federal Highway Administration
Fugro International, Inc.
GAI Consultants, Inc.
Gulf States Utilities
Houston Lighting & Power Company
Los Angeles Department of Water & Power
Lower Colorado River Authority
McClelland Engineers, Inc.
Mississippi State Highway Department
Raba-Kistner Consultants, Inc.
Salt River Project
Sargent & Lundy Engineers
Shell Development Company, Shell Oil Company
Southwestern Laboratories
Kenneth E. Tand & Associates, Inc.
Texas A&M University
Texas State Department of Highways and Public Transportation
University of Houston
U.S. Army Corps of Engineers, Ft. Worth District
U.S. Army Corps of Engineers, St. Louis District
U.S. Army Corps of Engineers, Vicksburg District
U.S. Army Engineer, Waterways Experiment Station

Table 2. Correlation results for clay (Column A = Number in table x Row B).

A \ B	E_O tsf	E_R tsf	p^*_L tsf	q_c tsf	f_s tsf	S_u tsf
E_O tsf	1	0.278	14	2.5	56	100
E_R tsf	3.6	1	50	13	260	300
p^*_L tsf	0.071	0.02	1	0.2	4	7.5
q_c tsf	0.40	0.077	5	1	20	27
f_s tsf	0.079	0.0038	0.25	0.05	1	1.6
S_u tsf	0.010	0.0033	0.133	0.037	0.625	1

Table 3. Correlation results for sand (Column A = Number in table x Row B).

A \ B	E_O tsf	E_R tsf	p^*_L tsf	q_c tsf	f_s tsf	N bl/ft
E_O tsf	1	0.125	8	1.15	57.5	4
E_R tsf	8	1	64	6.25	312.5	22.7
p^*_L tsf	0.125	0.0156	1	0.11	5.5	0.5
q_c tsf	0.87	0.16	9	1	50	5
f_s tsf	0.0174	0.0032	0.182	0.02	1	0.1
N bl/ft	0.25	0.044	2	0.2	10	1

The clay deposits had undrained shear strengths as low as 9.6 kPa (0.1 tsf), as high as 2490 kPa (26 tsf), and averaged 1.5 tsf. The sand deposits had blow counts as low as 1, as high as 100 and averaged 43. Best fit regressions were performed for the entire data base and for combinations of any two parameters. The results are shown in Tables 2 and 3. The scatter in the correlations was very large as exemplified by Figures 8 and 9. This drastic scatter makes these correlations essentially useless in design.

The value of the data base is to give a relative idea of the magnitude of the pressuremeter parameters. Among the relationships are:

In clay

$$p_L = 7.5\ S_U \quad (10)$$
$$p_L = 0.2\ q_c \quad (11)$$
$$p_L = 0.071\ E_o \quad (12)$$
$$E_o = 100\ S_U \quad (13)$$
$$E_o = 2.5\ q_c \quad (14)$$
$$E_o = 0.278\ E_R \quad (15)$$

In sand

$$p_L\ (\text{tsf}) = 0.5\ N\ (\text{blows/ft}) \quad (16)$$
$$p_L = 0.11\ q_c \quad (17)$$
$$p_L = 0.125\ E_o \quad (18)$$
$$E_o\ (\text{tsf}) = 4\ N\ (\text{blows/ft}) \quad (19)$$
$$E_o \neq 1.15\ q_c \quad (20)$$
$$E_o = 0.125\ E_R \quad (21)$$

Equation 10 indicates that the pressuremeter limit pressure is somewhat larger than the ultimate bearing pressure of a shallow footing and somewhat smaller than the ultimate bearing pressure under the point of a pile. Equation 13 gives an idea of the magnitude of the preboring pressuremeter modulus; the scatter of the data on Figure 8 tends to indicate that obtaining elastic moduli from undrained shear strength is unreliable. Comparison of equations 12 and 18 shows that the stiffness to strength ratio is much higher for clays than it is for sand and that the E_o/p_L ratio can serve as a soil classification index. Comparison of equation 15 and 21 shows that the reload to first load modulus ratio is much higher for sands than it is for clay and that the E_R/E_o ratio can also serve as a soil classification index. Equation 20 can be compared to the relationship proposed by Schmertmann (23) for calculating the settlement of spread footings on sands using elasticity:

$$E = 2.5 \text{ to } 3.5\ q_c \quad (22)$$

This would indicate that the pressuremeter modulus E_o is 2 to 3 times too low for use in elasticity settlement formulas. There may be several reasons for this discrepancy, one of which is the strain level corresponding to E_o. The straight part of the curve from which E_o is obtained exists over a range of 2% to 5% radial strain at the cavity wall, with clays tending towards the lower values and sands towards the higher ones. Another reason for the low values of E_o is explained in the following.

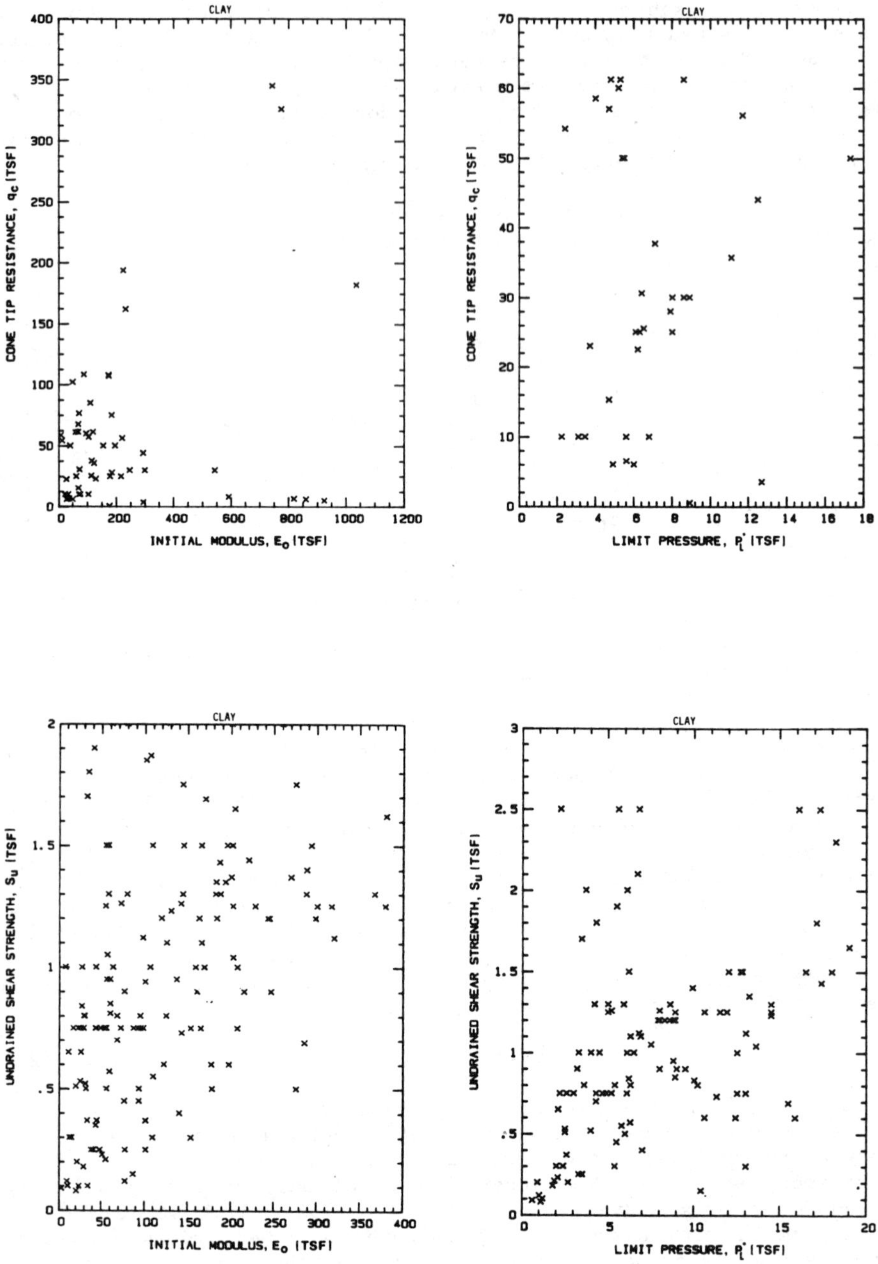

Figure 8. Examples of Correlations in Clay from PMT Data Base

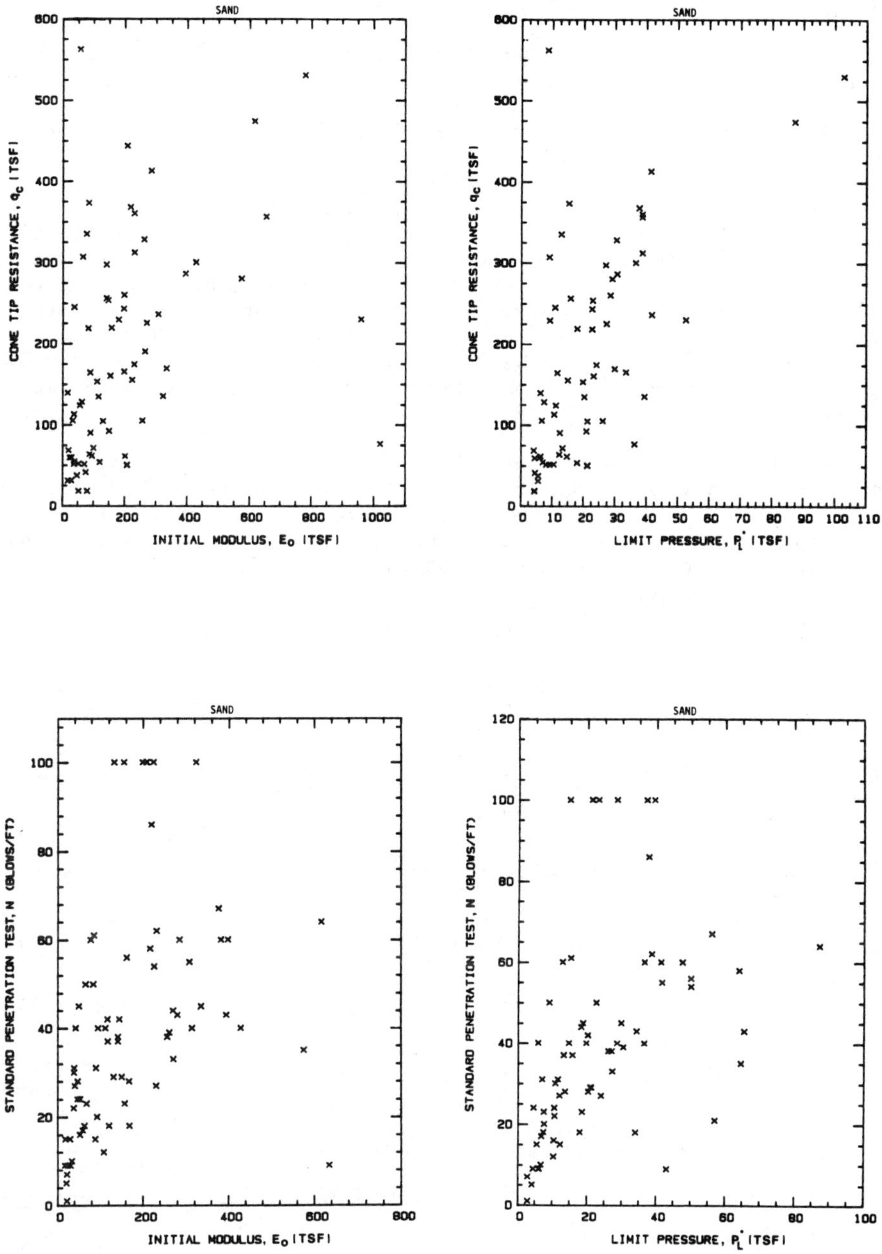

Figure 9. Examples of Correlations in Sand from PMT Data Base.

WHY IS THE PRESSUREMETER MODULUS TOO LOW?

Consider the pressuremeter test result of Figure 5. The elastic response of the soil goes up to a radial yield pressure p_y of 400 kPa (4.18 tsf). At the test depth (2.75 m or 9 ft), the total horizontal stress at rest p_{OH} is 27.5 kPa (574 psf) if an at rest earth pressure coefficient K_o value of 0.5 is used. In elasticity the increase in radial stress above p_{OH} is equal to the decrease in circumferential stress below p_{OH}. At p_y in the case of Figure 5, the radial stress is 400 kPa (4.18 tsf) and the elastic circumferential stress is -245 kPa (-2.56 tsf). It is very unlikely that this sand can resist such a large tensile stress; yet the elasticity theory, which predicts this unreasonable circumferential tensile stress, is used over the radial stress range of 100 kPa to 400 kPa (1.04 to 4.18 tsf) on Figure 5 in order to obtain the modulus E_o. A modified elasticity theory is necessary in order to model more properly the soil up to p_y.

The following elastic approach is presented in order to study the influence of having a modulus E^- in the circumferential direction (θ) different from the modulus E^+ in the radial (r) and vertical (z) directions instead of having the same modulus E_o in all directions. The general orthotropic elastic constitutive equations are (24):

$$\varepsilon_r = \frac{\sigma_r}{E_r} - \frac{\nu_{\theta r}\sigma_\theta}{E_\theta} - \frac{\nu_{zr}\sigma_z}{E_z} \quad (23)$$

$$\varepsilon_\theta = -\frac{\nu_{r\theta}\sigma_r}{E_r} + \frac{\sigma_\theta}{E_\theta} - \frac{\nu_{z\theta}\sigma_z}{E_z} \quad (24)$$

$$\varepsilon_z = -\frac{\nu_{rz}\sigma_r}{E_r} - \frac{\nu_{\theta z}\sigma_\theta}{E_\theta} + \frac{\sigma_z}{E_z} \quad (25)$$

Because of the symmetry rules the following equations must also be satisfied:

$$E_r\nu_{\theta r} = E_\theta\nu_{r\theta} \quad (26)$$

$$E_\theta\nu_{z\theta} = E_z\nu_{\theta z} \quad (27)$$

$$E_z\nu_{rz} = E_r\nu_{zr} \quad (28)$$

For the problem considered the parameters are:

$$E_z = E_r = E^+ \quad (29)$$

$$E_\theta = E^- \quad (30)$$

$$\nu_{rz} = \nu_{zr} = \nu_1 \quad (31)$$

$$\nu_{\theta z} = \nu_{\theta r} = \nu_2 \quad (32)$$

$$\nu_{z\theta} = \nu_{r\theta} = \nu_3 \quad (33)$$

Equations 26 to 28 reduce to:

$$\nu_3 = \frac{E^+}{E^-}\nu_2 = a\nu_2 \quad (34)$$

and equations 23 to 25 become:

$$\varepsilon_r = \frac{\sigma_r}{E^+} - \nu_2 \frac{\sigma_\theta}{E^-} - \nu_1 \frac{\sigma_z}{E^+} \quad (35)$$

$$\varepsilon_\theta = -\frac{\nu_2\sigma_r}{E^+} + \frac{\sigma_\theta}{E^-} - \nu_2 \frac{\sigma_z}{E^+} \quad (36)$$

$$\varepsilon_z = -\frac{\nu_1\sigma_r}{E^+} - \frac{\nu_2\sigma_\theta}{E^-} + \frac{\sigma_z}{E^+} \quad (37)$$

Using the plane strain condition ($\varepsilon_z = 0$):

$$\sigma_r = \frac{E^+}{(1-2\nu_2^{\,2} - \nu_1)}\left(\frac{(1-\nu_2^{\,2})}{1+\nu_1}\varepsilon_r + \nu_2\varepsilon_\theta\right) \quad (38)$$

$$\sigma_\theta = \frac{E^-}{(1-2\nu_2^{\,2} - \nu_1)}\left(\nu_2\varepsilon_r + (1-\nu_1)\ \varepsilon_\theta\right) \quad (39)$$

or $$\sigma_r = S_{11}\ \varepsilon_r + S_{12}\ \varepsilon_\theta \quad (40)$$

$$\sigma_\theta = S_{21}\ \varepsilon_r + S_{22}\ \varepsilon_\theta \quad (41)$$

The strain to displacement relationships for cylindrical expansion are::

$$\varepsilon_r = \frac{du}{dr} \quad (42)$$

$$\varepsilon_\theta = \frac{u}{r} \quad (43)$$

where u is the radial displacement of a point at a radius r from the axis of the cylinder; the equilibrium equations reduce to:

$$\frac{d\sigma_r}{dr} + \frac{\sigma_r - \sigma_\theta}{r} = 0 \quad (44)$$

Equations 40 to 44 lead to the governing differential equation:

$$r^2\frac{d^2u}{dr^2} + r\left(1 + \frac{S_{12} - S_{21}}{S_{11}}\right)\frac{du}{dr} - \frac{S_{22}}{S_{11}}u = 0 \quad (45)$$

which, considering the boundary conditions, has for solution:

$$u = u_o\left[\frac{r_o}{r}\right]^{\frac{1}{2S_{11}}\left\{(S_{21}-S_{12}) - \sqrt{(S_{12}-S_{21})^2 + 4S_{11}S_{22}}\right\}} \quad (46)$$

where u_o is the displacement at the cavity wall, and r_o is the radius of the cavity. The exponent of the term (r_o/r) in equation 46 is called n_2. The expression for σ_r is:

$$\sigma_r = \frac{u_o}{r} \left(\frac{r_o}{r}\right)^{n_2} (S_{12} - n_2 S_{11}) \qquad (47)$$

and at the cavity wall

$$\sigma_{ro} = (S_{12} - n_2 S_{11}) \frac{u_o}{r_o} \qquad (48)$$

where S_{12} and S_{11} are given in equations 38 to 41. Equation 48 is to be compared with the isotropic elasticity solution ($E^+ = E^- = E_0$):

$$\sigma_{ro} = -\frac{E_o}{1+\nu} \frac{u_o}{r_o} \qquad (49)$$

Consider the case where $E^+/E^- = 10$ and $\nu_1 = 0.33$; then $\nu_2 = 0.033$ from equation 34 and $\sigma_{ro} = -0.309\, E^+ u_o/r_o$ from equation 48. This is to be compared with $\sigma_{ro} = -0.7519\, E_o u_o/r_o$, therefore since r_{ro}, u_o, r_o are unique measured values, $E^+ = 2.43\, E_o$. Using this approach figure 10 was prepared which indicates that for the values shown of E^+/E^-, E_o is 2 to 3 times too low as found by comparing equation 20 to Schmertmann's recommendations.

Another approach to the problem is to consider the soil as a no-tension material ($E^- = 0$) and the elastic medium as a ring of radius R (Fig. 11). In this case, S_{21}, S_{12} and S_{22} are zero and S_{11} reduces to

$$S_{11} = \frac{E^+}{1-\nu_1^2} \qquad (50)$$

Equilibrium constitutive and strain definition equations are:

$$\frac{d\sigma_r}{dr} + \frac{\sigma_r}{r} = 0 \qquad (51)$$

$$\sigma_r = S_{11} \varepsilon_r \qquad (52)$$

$$\varepsilon_r = \frac{du}{dr} \qquad (53)$$

Combining the above equations leads to:

$$\sigma_{ro} = \frac{E^+}{(1-\nu_1^2) \operatorname{Ln} \frac{r_o}{R}} \frac{u_o}{r_o} \qquad (54)$$

Again this is to be compared with equation 49. For a Poisson's ratio of 0.33 and $R = 20\, r_o$, the relation between E^+ and E_o is $E^+ = 2.53\, E_o$. Using this approach Figure 11 was prepared which indicates that for the values shown of R/r_o, E_o is 2 to 3 times too low. This theoretical finding explains why Menard needs to multiply E_o by

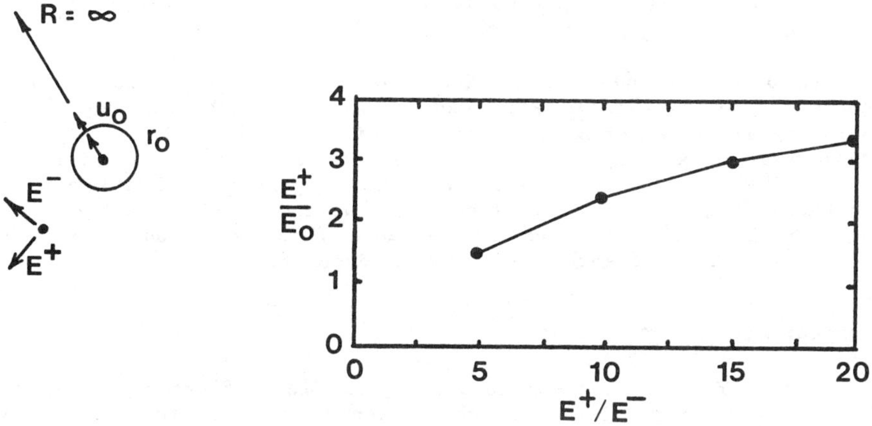

Figure 10. Influence of Tension Consideration on the Moduli Values for an Orthotropic Soil

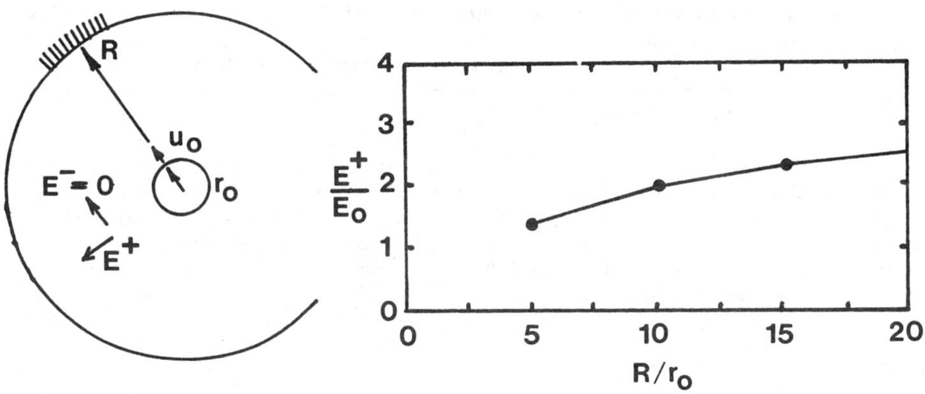

Figure 11. Influence of Tension Consideration on the Moduli Values for a No-Tension Soil

a correction factor $1/\alpha$ in order to match the settlement results of shallow footing tests (25). Indeed the factor $1/\alpha$ varies from 1 for overconsolidated clays for which it is reasonable to assume that $E^+ = E^-$ to 4 for very dense sands where E^+ is much larger than E^-. It is therefore logical to state that E^+/E_o is $1/\alpha$ (26).

In order to investigate experimentally if tension was truly a problem in pressuremeter expansion at least at shallow depth, the following simple experiment was performed. A 74 mm (2.91 in) diameter pressuremeter was placed at the shallowest depth possible in a clay deposit.The probe was inflated and the propagation of a crack was monitored (Figs. 12 to 14). The fact that the crack turned at right angle at $R = 12\ r_o$ tends to validate this value as a zone of influence of the pressuremeter for that deposit.

CONCLUSIONS:

A number of recommendations are made in a first part for the standard way to perform a pressuremeter test. A probe length to diameter ratio of 6.5 is recommended. For this ratio, there does not seem to be any theoretical nor experimental difference between monocell and tricell pressuremeters. Pressure increments or volume increments test procedures can be used provided the probe volume is doubled in 10 minutes: ten 1 minute pressure increments and forty 15 seconds volume increments are recommended. For volume correction the probe of diameter D_p should be inserted in a steel tube of diameter 1.005 D_p and the calibration performed in ten 1 minute pressure increments. The results should always be plotted as pressure against the cavity wall versus relative increase in probe radius.

In a second part correlations are presented between pressuremeter parameters, undrained shear strength, standard penetration test blow count, cone point and friction resistances. The data base is a collection of 82 pressuremeter borings in the South, West and Central United States for a total of 426 records. The resulting correlations are shown on tables 2 and 3.

In a third part it is shown that the pressuremeter modulus calculated by the conventional elasticity equation is too low because that equation does not acknowledge the weakness of soils in extension. Treating the soil as an orthotropic material shows that the pressuremeter modulus calculated by the common elasticity equation is two to three times too low.

REFERENCES

[1] Briaud, J.L., Gambin, M., "Suggested Practice for Drilling Boreholes for Pressuremeter Testing," ASTM Geotechnical Testing Journal, Vol. 7, No. 1, March, 1984.

[2] Winter, E., "Suggested Practice for Pressuremeter Testing in Soils,"ASTM Geotechnical Testing Journal, Vol. 5, No. 3/4 Sept/ Dec., 1982.

[3] OYO Corporation, "Lateral Load Tester (LLT) Operation Manual," Internal Document, Houston, TX, 1983.

Figure 12. Pressuremeter Probe before Inflation.

Figure 13. Pressuremeter Probe at Full Inflation (limit pressure).

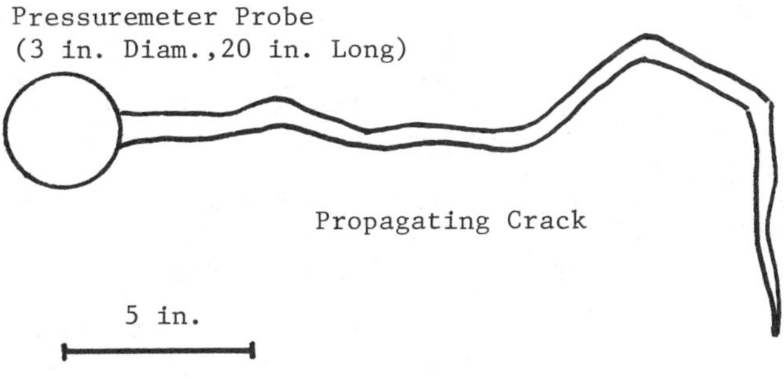

Figure 14. Tension Crack at Full Inflation

[4) OYO Corporation, "Elastmeter 100 Operation Manual," Internal Document, Houston, Texas, 1983.

[5] Briaud, J.L., Tucker, L.M., Olsen, R.S., "Pressuremeter, Cone Penetrometer and Foundation Design," Short Course Notes, Vol. No. 1, Civil Engineering, Texas A&M University, 1985.

[6] Roctest, Inc., "Texam Operation Manual," Internal Document, Plattsburg, N.Y., 1983.

[7] Jezequel, J.F., "The Selfboring Pressuremeter, "Symposium on the Pressuremeter and Its Marine Applications, Editions Technip, Paris, 1982.

[8] Wroth, C.P., Hughes, J., "An Instrument for the In Situ Measurement of the Properties of Soft Clays," Int. Conf. Soil Mechanics and Foundation Engineering, Moscow, 1973.

[9] Reid, W.M., Fyffe, S., St. John, J.D., Rigden, W.J., "The Push in Pressuremeter," Symposium on the Pressuremeter and Its Marine Applications, Editions Technip, Paris, 1982.

[10] Fay, J.B., LeTirant, P., "Offshore Selfboring Pressuremeter for Deep Water," Symposium on the Pressuremeter and its Marine Applications, Editions Technip, Paris, 1982.

[11] Roctest, Inc., "The Cone Pressuremeter Pencell: Operation Manual," Internal Document, Plattsburg, N.Y., 1984.

[12] Withers, N.J., Schaap, L.H.J., Kolk, K.J., Dalton, J.C.P., "The Development of the Full Displacement Pressuremeter," The Pressuremeter and Its Marine Applications; Second International Symposium, ASTM STP 950, Texas A&M University, 1986.

[13] Tranter, C.J., "On the Elastic Distortion of a Cylindrical Hole by a Localized Hydrostatic Pressure," Quarterly of Applied Mathematics, QAMAA, Vol. 4, No. 3, 1946.

[14] Hartman, J.P., "Finite Element Parametric Study of Vertical Strain Influence Factor and the Pressuremeter Test to Estimate the Settlement of Footings in Sand," Ph.D. dissertation, Civil Engineering, University of Florida, 1974.

[15] Laier, J.E., "Effects of Pressuremeter Probe Length/Diameter Ratio and Borehole Disturbance on Pressuremeter Test Results in Dry Sand, Ph.D. dissertation, Civil Engineering, University of Florida, 1974.

[16] Gibson, R.E., Anderson, W.F., "In Situ Measurement of Soil Properties with the Pressuremeter," Civil Engineering and Public Works Review, London, May 1961.

[17] Suyama, K., Ohya, S., Imai, T., "Development of LLT pressuremeter and Its Application in Prediction of Pile Behavior Under Horizontal Load," Symposium on the Pressuremeter and Its Marine Applications," Paris, 1982.

[18] Suyama, K., Ohya, S., Imai, T., Matsubara, M., Nakayama, E., "Ground Behavior During Pressuremeter Testing," Int. Symp. Soil and Rock Investigations by In Situ Testing, Paris, 1983.

[19] Briaud, J.L., Cosentino, P., Tucker, L.M., Terry, T., "Comparison Between the Texam and the Menard Pressuremeter," Research Report, Civil Engineering, Texas A&M University, 1985.

[20] Briaud, J.L., Garland, E., "Loading Rate Method for Pile Response in Clay," ASCE Geotechnical Engineering Journal, Vol. 111, No. 3, March, 1985.

[21] Briaud, J.L., Tucker, L.M., "PRESRED: an IBM-PC Program For Pressuremeter Data Reduction and Plotting Routine," User's Manual, Civil Engineering, Texas A&M University, 1985.

[22] Briaud, J.L., Noubani, A., Kilgore, J., Tucker, L.M.,

"Correlation between Pressuremeter Data and Other Parameters," Research Report, Civil Engineering, Texas A&M University, 1985.

[23] Schmertmann, J.H., "Improved Strain Influence Factor Diagram," Technical Note, ASCE Geotechnical Engineering Journal, Vol. 104, No. GT8, 1978.

[24] Jones, R.M., Mechanics of Composite Materials," McGraw-Hill, 1975.

[25] Menard, L., "The Menard Pressuremeter: Interpretation and Application of the Pressuremeter Test Results to Foundation Design," Sols-Soils No. 26, 1975.

[26] Leblanc, J., "Menard Pressuremeter Testing," Symposium on the Pressuremeter and Its Marine Applications, Editions Technip, Paris, 1982.

William F. Anderson and Ian C. Pyrah

UNDRAINED STRENGTH AND DEFORMATION PARAMETERS FROM PRESSUREMETER TEST RESULTS

REFERENCE: Anderson, W. F. and Pyrah, I. C., "Undrained Strength and Deformation Parameters from Pressuremeter Test Results", The Pressuremeter and Its Marine Applications: Second International Symposium, ASTM STP 950, J.-L. Briaud and J. M. E. Audibert, Eds., American Society for Testing and Materials, 1986.

ABSTRACT: Pressuremeter tests in cohesive soils are generally analysed to give undrained strength parameters. Differences in pressuremeter test techniques have been shown to affect test results and derived parameters, and it has been established that these variations are due to the combined time-dependent effects of local consolidation and deviatoric creep. Various sizes of pressuremeter are used offshore. The influence of consolidation and creep on the results of tests carried out using pressuremeters with different initial diameters has been examined using a numerical analysis. This has shown that the initial size of the expanding cavity can have a significant effect on the test results. The numerical simulations have also shown that the influence of test technique cannot simply be expressed as the 'time to failure'; the way in which the test is performed also affects the results.

KEYWORDS: pressuremeter, clay, consolidation, creep, numerical modelling, size effects

INTRODUCTION

Because of the problems of recovering and testing undistrubed samples, in situ tests have always been an attractive proposition for obtaining geotechnical data in the offshore industry. Ideally in situ tests should stress a large enough mass of soil to give representative results, and the results should be capable of being analysed by a rigorous method with a sound theoretical basis.

Dr. Anderson is Senior Lecturer and Dr. Pyrah is Lecturer in the Department of Civil and Structural Engineering, The University, Mappin Street, Sheffield, S1 3JD, U.K.

Pressuremeters, particularly those with larger diameters, may be the most appropriate devices for offshore in situ testing and over the last few years a number of different types of pressuremeter have had trials offshore. In a review of in situ tests and their applications in offshore design Briaud and Meyer [1] have referred to the Push in Pressuremeter [2], the Wireline Expansometer [3], the Pressiometre Autoforeur Marin [4] and the Pressio-penetrometer [5]. In addition the Cambridge Self Boring Pressuremeter [6] has been used in offshore deposits, and Hughes et al [7] have described pressuremeter tests in which a similar device has been mud jetted into the seabed in the Arctic.

All these devices have different insertion techniques which will cause varying amounts of disturbance to the soil prior to testing. Also different testing techniques are used. For example, Hughes et al [7] have used both a stress increment technique and a constant rate of radial strain technique to expand the pressuremeter; Brucy et al [8] and Bisson et al [9] have used a constant rate of volume strain technique and Ligier [3] has used a constant volume increment technique.

When analysing the results of a pressuremeter test in a cohesive soil, it is assumed that the test is carried out under idealized undrained conditions. However a non uniform stress condition is set up in the soil due to the application of pressure and the resultant pore pressure gradient will result in local consolidation around the pressuremeter. The test results are also influenced by creep of the soil skeleton, which is not only time-dependent but also stress level dependent. Thus, the way in which a pressuremeter test is carried out may significantly affect the results and hence the derived 'undrained strength' and deformation modulus.

LABORATORY-SCALE STUDIES

Experimental Investigation

A major problem in carrying out field studies to examine the effects of using different test techniques is that the influence of natural soil variation is unknown. This problem may be overcome by carrying out laboratory simulation of the in situ test in clay beds prepared with known stress history, and Anderson et al [10] have carried out such a study on three clays with different consolidation and creep properties. Initially tests were carried out in a test cell sufficiently large to be considered a semi-infinite soil mass when compared to the size of the expanding cavity. A numerical analysis of the problem, and comparison with test results obtained by expanding a thick hollow cylinder under plane strain conditions, showed that the undrained pressuremeter test could be simulated with a high degree of accuracy by using a thick hollow cylinder.

Series of tests were carried out by expanding a 25mm diameter cylindrical cavity in a 150mm diameter cylindrical specimen of clay using different incremental pressure techniques. Fig. 1 and 2 show

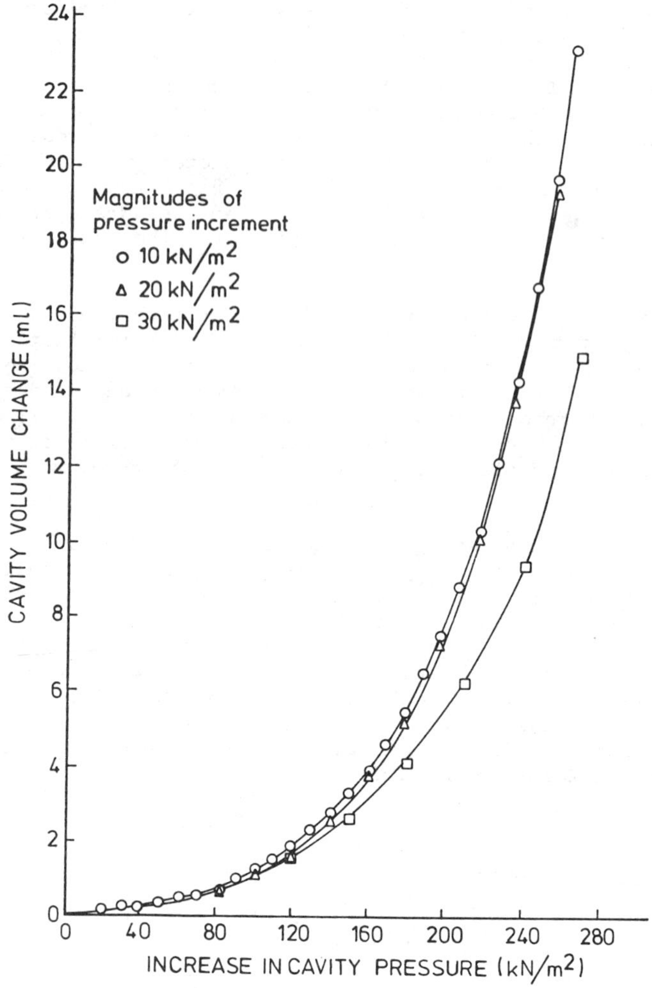

FIG. 1--Expansion curves for tests carried out using different pressure increments

the effects of varying magnitudes of pressure increment (holding time - 1 minute) and varying holding times (pressure increment - 10 kN/m^2) on the expansion curves for tests carried out on normally consolidated pottery clay (LL = 40% PL = 23%). These clearly demonstrate that varying test technique affects the volume change-pressure relationship. These curves have been analysed using the Palmer [11] method to derive the stress-strain curves which are shown in Fig. 3 and 4 and the effects of varying test technique are again apparent. For pottery clay deduced values of the undrained shear strength ranged from 102% to 118% of the value obtained from triaxial tests on samples prepared with the same stress history. Modulus values ranged from 74% to 113% of the triaxial value depending on the expansion technique used.

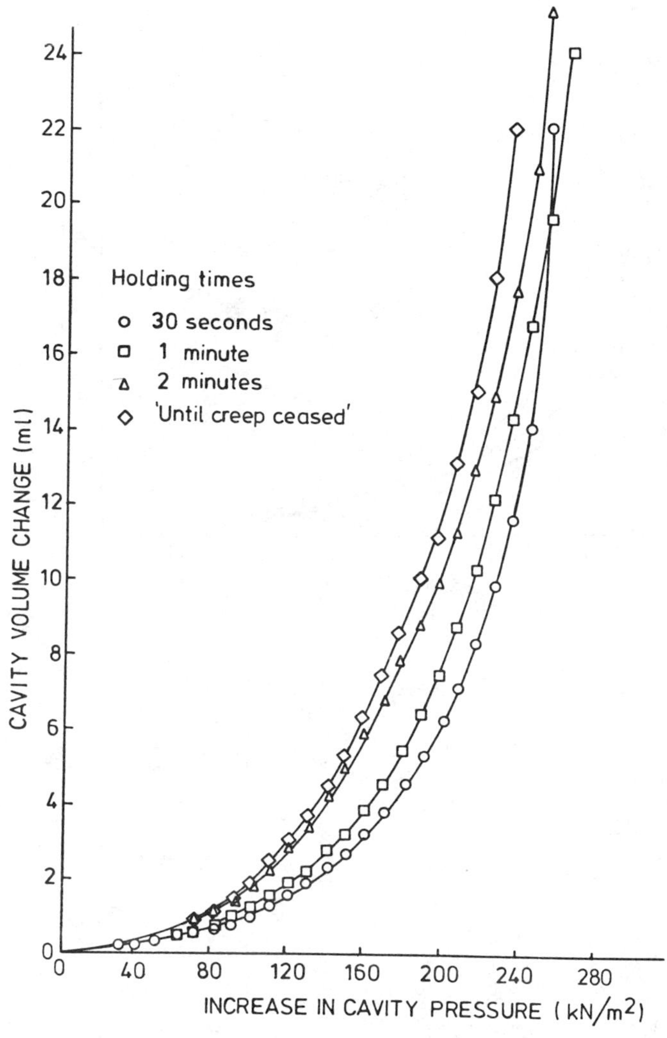

FIG. 2--Expansion curves for tests carried out using different holding times

Pore water pressures were measured at the expanding cavity wall and in the soil mass and these gave an indication of local consolidation. However, it was not possible to isolate the effects of consolidation and creep experimentally and an accompanying numerical study of the problem was carried out. This confirmed the trends of the laboratory studies.

Numerical Modelling

The numerical analysis was carried out using an extension of the

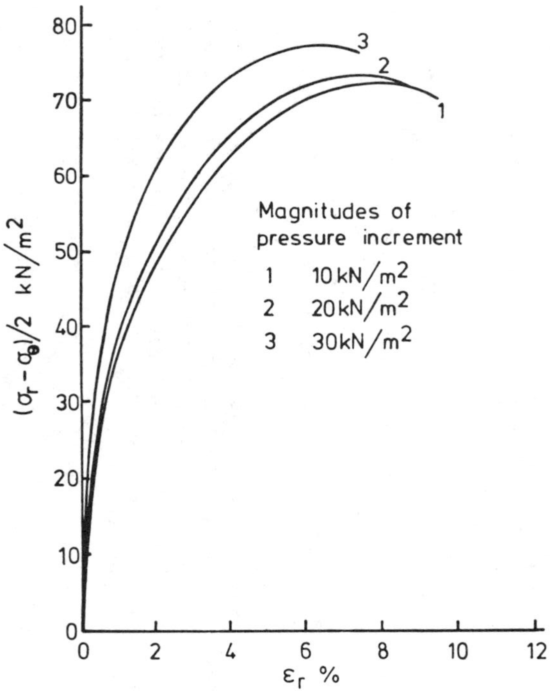

FIG. 3--Derived stress-strain curves for tests using different pressure increments

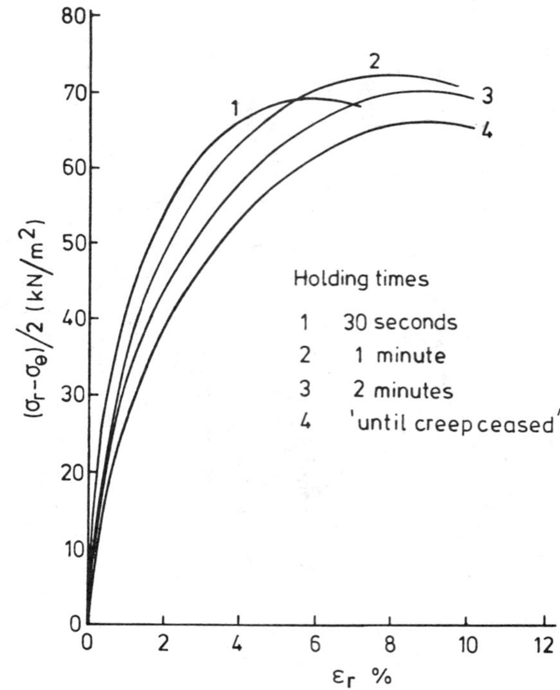

FIG. 4--Derived stress-strain curves for tests using different holding times

finite element computer program CAMFE, developed by Carter [12], and details of this modification have been given by Pyrah et al [13]. The program analyses the plane strain expansion of a cylindrical cavity using the modified Cam clay model [14] with a Biot type consolidation and a model of deviatoric creep proposed by Singh and Mitchell [15].

In this analysis it was possible to consider the prefectly undrained situation, the effects of creep or consolidation on their own and the combined effects of creep and consolidation as shown in Fig. 5. For the pottery clay tested, it demonstrated that if local consolidation only was considered, i.e. no creep, the effect of increasing the duration of a pressuremeter test, either by increasing the time for which each pressure increment was held or decreasing the magnitude of the pressure increment with a constant holding time, was to increase the deduced strength and strain to failure. When creep only was considered in the analysis the deduced strength was less than the undrained value. When both consolidation and creep were considered, the creep effects were found to be more significant than the consolidation effects.

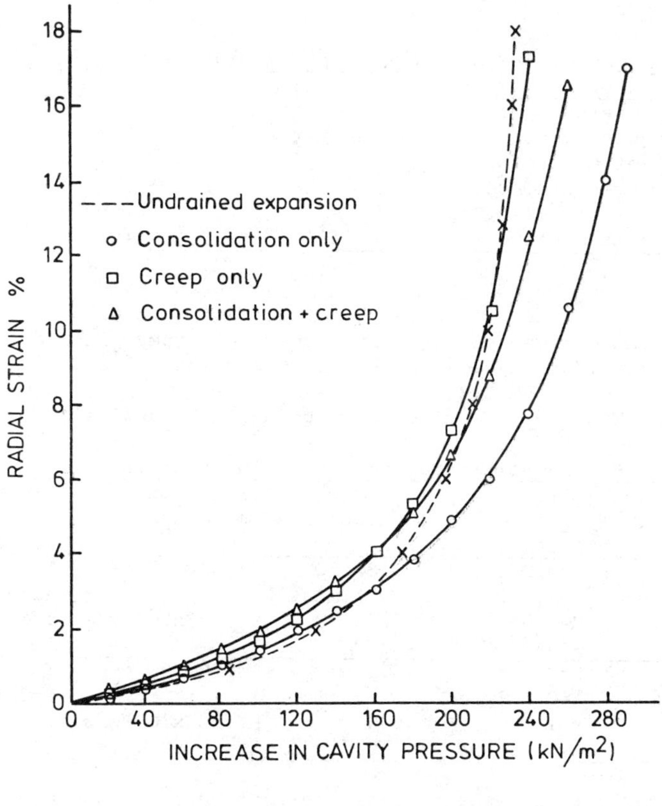

FIG. 5--Numerical predictions of the effects of consolidation and creep on the expansion curve (pressure increment 20 kN/m^2, holding time 1 minute)

The creep and consolidation properties of the other two clays tested, Kaolin (LL = 72%, PL = 36%) and Barnsley Clay (LL = 42%, PL = 23%) differed considerably from those of pottery clay and so their influences on the expansion curves and derived parameters were different. However, for all three materials the numerical predictions of the effects of varying test technique agreed closely with the actual test results.

NUMERICAL SIMULATION OF FULL-SCALE PRESSUREMETER TESTS

Since local consolidation and creep will affect pressuremeter test results, any factors which influence these phenomena will also affect the results. Local consolidation is dependent on the radial pore pressure distribution and the length of the drainage path, and therefore the initial diameter of the pressuremeter may have a significant effect on the deduced properties of the soil. For pressuremeters used offshore initial diameters vary as shown in Table 1.

TABLE 1--Pressuremeter probe diameters

Type	Diameter, mm
Wire line expansometer [3]	65
Cambridge self boring pressuremeter [6]	75
Pressiometre autoforeur [9]	76
Push in pressuremeter [2]	78
Pressio-penetrometer [5]	89
Pressiometre Autoforeur Marin [4]	160

In order to examine the possible effects of initial pressuremeter diameter on the derived parameters a numerical study, incorporating creep and consolidation effects, has been carried out using the modified version of CAMFE. Published soil parameters for San Francisco Bay Mud [16], Table 2, were used in the analysis.

TABLE 2--Soil parameters used in analysis

Soil type	Critical state parameters	Creep parameters	Other parameters	Initial Stress condition
Remoulded San Francisco Bay Mud	$K=0.04342$ $\lambda=0.3257$ $M=1.541$ $e_{cs}=2.075$	$A=8.80\times10^{-4}$ %/min $\bar{\alpha}=5.4$ $m=0.73$	$G=6800$ kN/m^2 $k=1.33\times10^{-9}$m/s	Isotropic effective stresses 100 kN/m^2
Pottery Clay	$K=0.025$ $\lambda=0.101$ $M=1.0$ $e_{cs}=1.265$	$A=5.25\times10^{-3}$ %/min $\bar{\alpha}=5.4$ $m=1.0$	$G=6300$ kN/m^2 $k=.338\times10^{-9}$m/s	Isotropic effective stresses 200 kN/m^2

The numerical analyses reported earlier were for the modelling of thick hollow cylinders of clay with an outer to inner diameter ratio of 6:1. For the present simulation of pressuremeter tests the soil was assumed to extend a distance of 50 times the initial cavity radius and pressuremeters ranging in size from 25 to 150mm diameter were studied. For most of the 'tests' an incremental pressure testing technique was adopted and the effects of different holding times examined. The pressure increment used was 10 kN/m^2 and the results using 30 seconds, 1 minute and 2 minutes holding times were compared with those from undrained 'tests'.

Effects of Consolidation

The effects of consolidation for different holding times can be seen in Fig. 6; for a given pressure the radial strain initially increases as the holding time increases but as the soil close to the cavity gains in strength the trend is reversed. The effect on the stress-strain curves, derived from the pressuremeter curves using Palmer's [11] method, is clearly seen in Fig. 7; the longer the holding time the greater the derived 'undrained' shear strength. The effect on the derived shear modulus is a reduction in its value.

Whilst the effects of soil consolidation around the pressuremeter are most pronounced for the smallest size considered, the trends are the same whatever the size of the cavity. The significance of consolidation, for different diameters and holding times, on the derived 'undrained' shear strength can be seen in Fig. 8. For a particular soil the derived strength is clearly dependent upon the holding time and the initial size of the pressuremeter. The extent of the overprediction is obviously related to the permeability of the soil.

Combined Effect of Consolidation and Creep

The above investigation was repeated but this time including creep in the analysis. Expansion curves were plotted and, as noted in previous analyses [13], the effect of creep on its own is to increase the strain for a given pressure, as with consolidation, but there is no increase in the limit pressure. This is reflected in the derived stress-strain curves which show decreases in strength and shear modulus compared to the true undrained values.

The results for the analyses including both creep and consolidation are summarised in Fig. 9. For an initial pressuremeter diameter of 25mm the increase in apparent strength due to consolidation is significant, whilst for pressuremeter diameters of 100mm or more the effects of creep are more significant as the derived strength is less than the undrained value. For intermediate diameters the relative significance of consolidation and creep is dependent upon the holding time.

Consolidation Effects for Incremental and Constant Rate of Radial Strain 'Tests'

As noted in the introduction, a number of different pressuremeter expansion techniques have been used for field tests. The numerical simulations described above are all based on an incremental pressure technique. The work is being extended to examine other test methods, such as a constant rate of radial strain technique. An indication

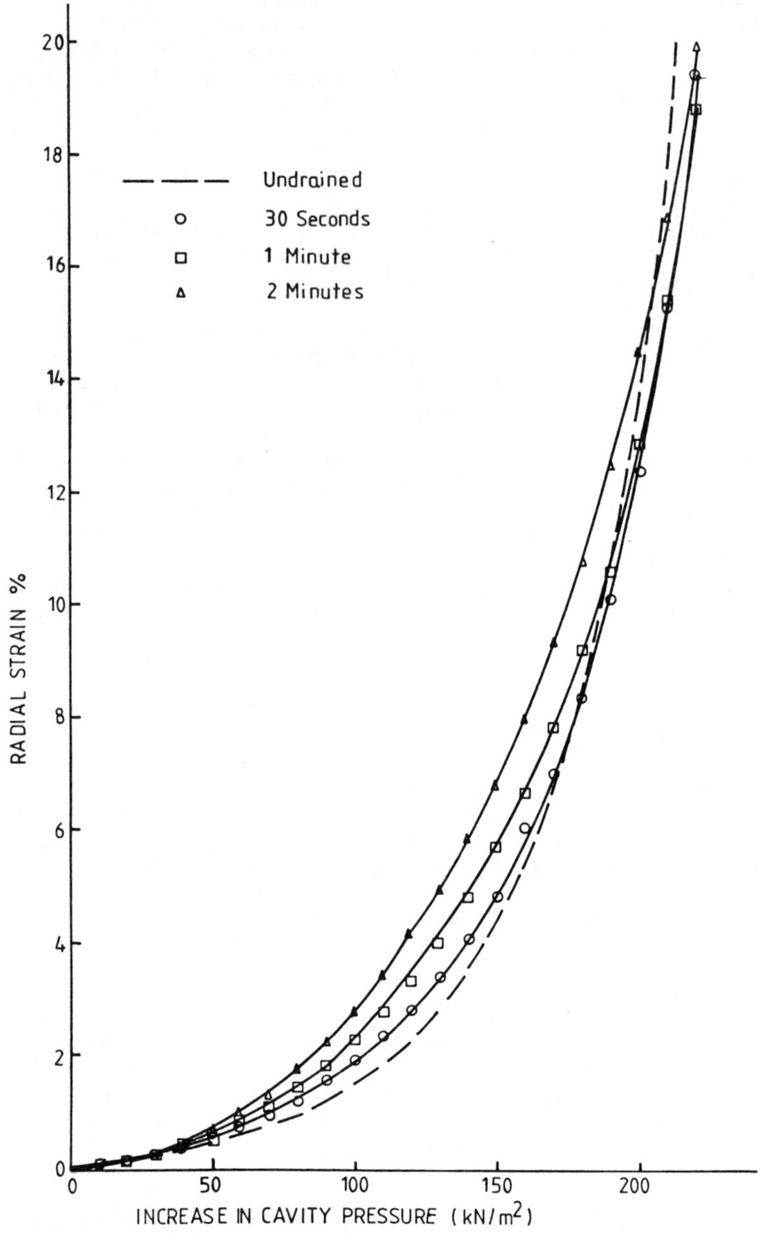

FIG. 6--Effects of consolidation on expansion curve for different holding times (initial diameter 25mm)

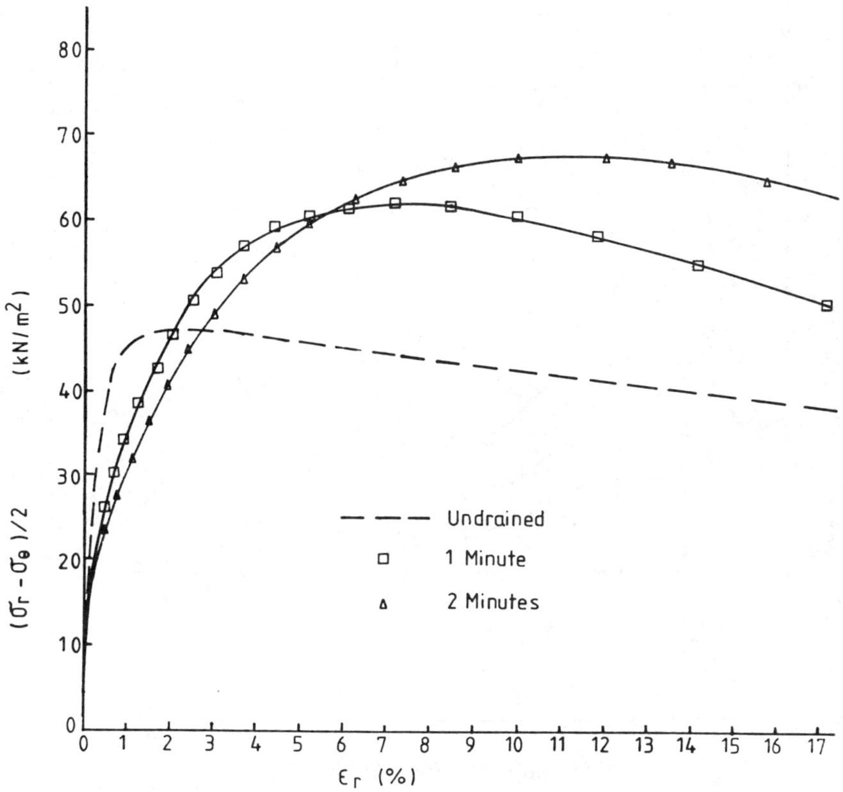

FIG. 7--Effect of consolidation on derived stress-strain curve for different holding times (initial diameter 25mm)

of the influence of test method can be obtained by considering the results from two 'tests' of similar duration but which follow different loading sequences. (For this initial study creep is not considered in the analysis.)

Two analyses were carried out to simulate pressuremeter tests (75mm dia.) in San Francisco Bay Mud; both 'tests' took approximately 40 minutes to reach 10% radial strain but one was stress-controlled (pressure increment 10 kN/m^2, holding time 2 minutes) whilst the other was strain-controlled (constant rate of radial strain 0.25%/min) 10% radial strain was reached in 38 minutes and 40 minutes respectively.

The derived stress-strain curves, together with the undrained curve, are shown in Fig. 10; the curves are significantly different.

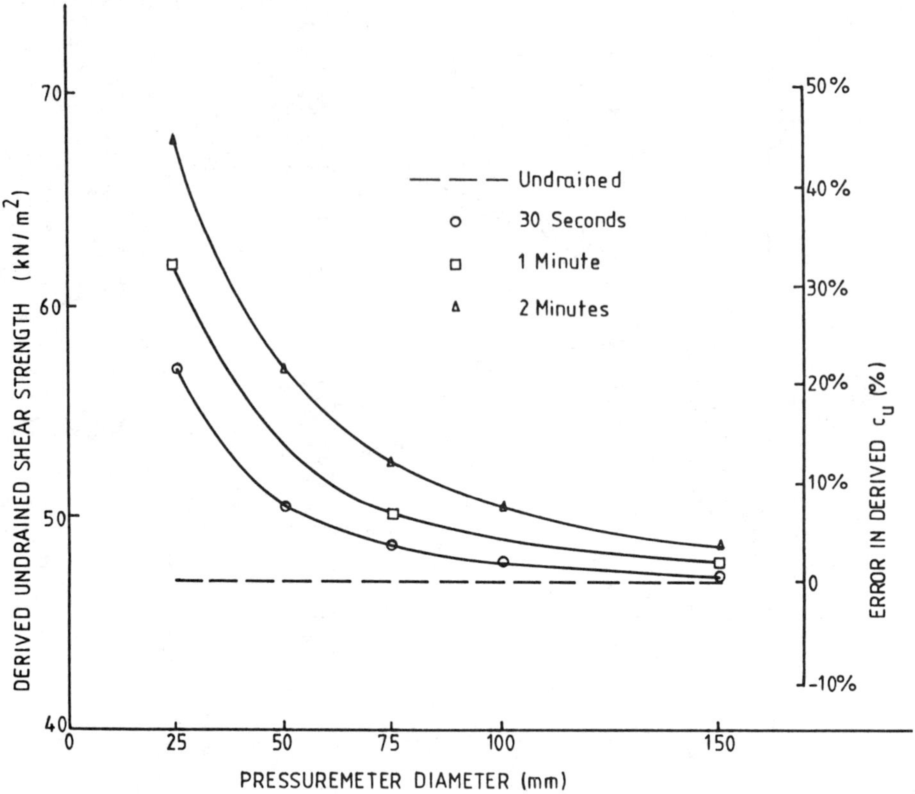

FIG. 8--Effect of consolidation on derived 'undrained' shear strength for different holding times and different initial diameters

DISCUSSION

The experimental and numerical results illustrate that the technique adopted in carrying out a pressuremeter test can significantly affect the results. The effects of consolidation and creep on their own are clear but their relative importance in practice, where both are acting concurrently, is more difficult to assess.

Considering consolidation only, the analyses confirm what might be expected intuitively. The amount by which the results are affected is dependent upon the test technique (e.g. duration and loading sequence), the size of the expanding cavity and, of course, the consolidation characteristics of the material.

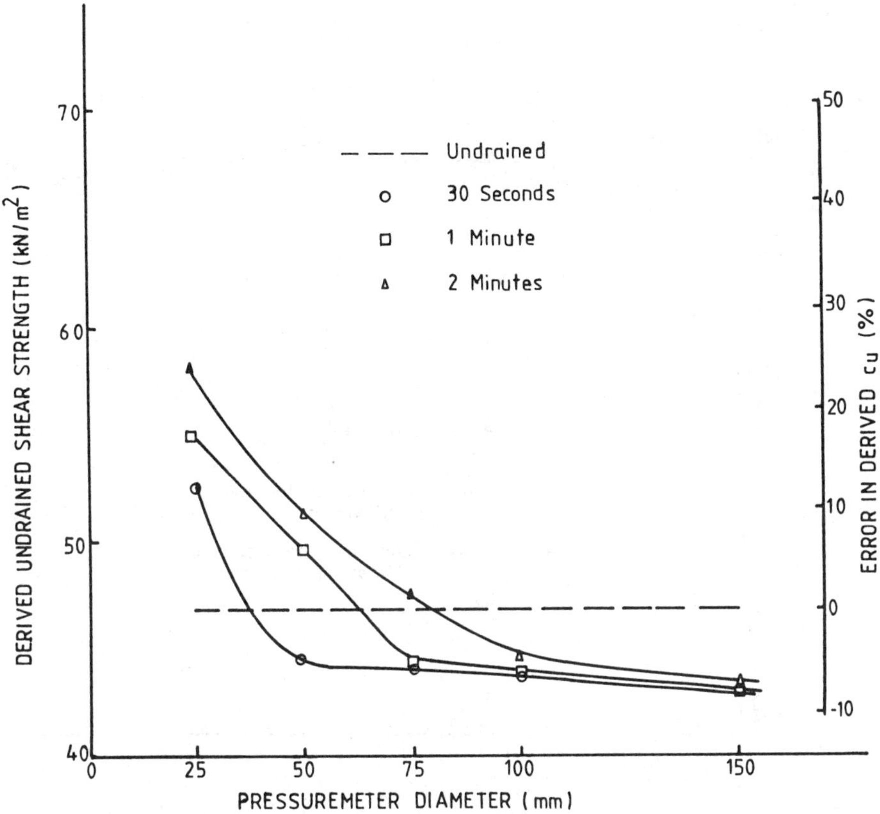

FIG. 9--Combined effect of consolidation and creep on derived 'undrained' shear strength for different holding times and different initial diameters

Considering creep only, the analyses show that the test technique and the creep characteristics of the soil clearly affect the results but the initial size of the pressuremeter is much less significant.

When the combined effects of creep and consolidation are considered their relative significance is dependent upon test technique, pressuremeter size and soil characteristics. For San Francisco Bay Mud consolidation is more significant than creep during the expansion of a 25mm diameter cavity whereas for Pottery Clay the influence of consolidation is balanced by the effects of creep and no increase in derived strength is apparent as the duration of the test increases. An examination of the parameters for the two clays (Table 2) shows that S.F. Bay Mud is more permeable than Pottery Clay whilst the latter has a higher value for the creep parameter A.

Although the parameters used in the analysis are for remoulded, rather than undistrubed, S.F. Bay Mud the high 'undrained' shear strength deduced from SBP tests in this deposit by Clough & Denby [17] may be attributable to some local consolidation during the tests.

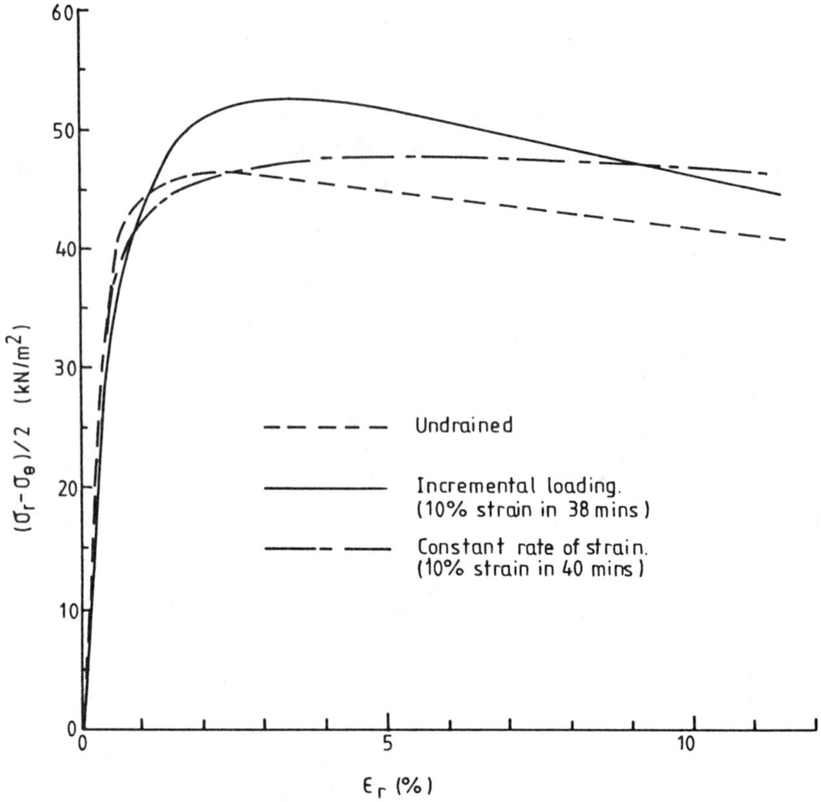

FIG. 10--Effect of different expansion techniques on the derived stress-strain curve (dia. 75mm, consolidation only)

CONCLUSIONS

The numerical simulation of full-scale in situ tests in San Francisco Bay Mud confirms that the test technique and the initial diameter of the pressuremeter may significantly affect the test results. In this deposit, for commonly used pressuremeter diameters (≃ 75mm), an error of up to 10% in the deduced undrained strength would be expected, its magnitude depending on the expansion technique used in the test. A single standardized test technique may be inappropriate because the relative effects of consolidation and creep will be different for different soils. Depending on the values of these soil parameters the discrepancy between undrained and pressuremeter deduced 'undrained' strength may be greater or smaller than those found in this study. Also the influence of consolidation and creep should be borne in mind when pressuremeter results are used to obtain undrained stress-strain relationships or for any other purpose, eg. p-y curves for the design of laterally loaded piles.

REFERENCES

[1] Briaud, J.-L. and Myer, B., "In Situ Tests and Their Application in Offshore Design", Proceedings of the Conference on Geotechnical Practice in Offshore Engineering, ASCE, University of Texas at Austin, April 1983, pp. 244-286.

[2] Reid, W. M., St. John, H. D., Fyffe, S. and Rigden, W. J., "The Push-In Pressuremeter", Proceedings of the Symposium on the Pressuremeter and its Marine Applications, Editions Technip, Paris, 1982, pp. 247-261.

[3] Leigier, A., "The Wire-Line Expansometer", Proceedings of the Symposium on the Pressuremeter and its Marine Applications, Editions Technip, Paris, 1982, pp. 263-274.

[4] Brucy, F., Fay, J. B. and Marignier, J., "Offshore Self-Boring Pressuremeter : the LPC/IFP Probe", Proceedings of the Symposium on the Pressuremeter and its Marine Applications, Editions Technip, Paris, 1982, pp. 127-142.

[5] Jezequel, J. F., Laury, J. L. and Perrier, M., "The LPC-TLM Pressio-Penetrometer", Proceedings of the Symposium on the Pressuremeter and its Marine Applications, Editions Technip, Paris, 1982, pp. 275-287.

[6] Wroth, C. P., "British Experience with the Self-Boring Pressuremeter", Proceedings of the Symposium on the Pressuremeter and its Marine Applications, Editions Technip, Paris, 1982, pp. 143-164.

[7] Hughes, J. M. O., Jeffries, M. G., and Morris, D. L., "Self-Bored Pressuremeter Testing in the Arctic Offshore", Proceedings of the Offshore Technology Conference, Houston, Vol. 1, 1984, pp. 255-264.

[8] Brucy, F., Fay, J. B. and Le Tirant, P., "Three Years' Experience with the Offshore Self Boring Pressuremeter 'PAM'", Proceedings of the Offshore Technology Conference, Houston, Vol. 1, 1984, pp. 265-274.

[9] Bisson, D., Bougnol, J. and Bourges, F., "Self-Boring Pressuremeter for Shallow Deep Water", Proceedings of the Symposium on the Pressuremeter and its Marine Applications, Editions Technip, Paris, 1982, pp. 289-323.

[10] Anderson, W. F., Pyrah, I. C. and Haji-Ali, F., "Pressuremeter Testing of Normally Consolidated Clays - the Effects of Varying Test Technique", Proceedings 20th Regional Meeting of the Geological Society, University of Surrey, Vol. 1, 1984, pp. 21-32.

[11] Palmer, A. C., "Undrained Plane Strain Expansion of a Cylindrical Cavity in Clay : a Simple Interpretation of the Pressuremeter Test", Geotechnique, Vol. 22, 1972, pp. 451-457.

[12] Carter, J. P., "CAMFE, a Computer Program for the Analysis of a Cylindrical Cavity Expansion in Soil", Cambridge University Report CUED-SOILS-TR52, 1978.

[13] Pyrah, I. C., Anderson, W. F. and Haji Ali, F., "The Interpretation of Pressuremeter Tests : Time Effects for Fine-Grained Soils", Proceedings 5th International Conference on Numerical Methods in Geomechanics, Nagoya, Vol. 3, 1985, pp. 1629-1636.

[14] Carter, J. P., Randolph, M. F. and Wroth, C. P., "Stress and Pore Pressure Changes in Clay During and After the Expansion of a Cylindrical Cavity", International Journal of Numerical and Analytical Methods in Geomechanics, Vol. 3, 1979, pp. 305-322.

[15] Singh, A. and Mitchell, J. K., "General Stress-Strain-Time Function for Soils", Journal of the Soil Mechanics and Foundation Engineering Division, Proceedings of the American Society of Civil Engineers, Vol. 94, No. SM1, 1968, pp. 21-46.

[16] Kavazanjian, E. and Mitchell, J. K., "Time-Dependent Deformation Behaviour of Clays", Journal of the Geotechnical Engineering Division, Proceedings of the American Society of Civil Engineers, Vol. 106, No. GT6, 1980, pp. 611-630.

[17] Clough, G.W. and Denby, G. M., "Self-Boring Pressuremeter Study of San Francisco Bay Mud", Journal of the Geotechnical Engineering Division, Proceedings of the American Society of Civil Engineers, Vol. 106, No. GT1, 1980, pp. 45-63.

Dennes T. Bergado and Md. Abdul Khaleque

CORRELATIONS OF LLT PRESSUREMETER, VANE, AND DUTCH CONE TESTS IN BANGKOK MARINE CLAY, THAILAND

REFERENCE: Bergado, D.T., and Khaleque, M.A., "Correlation of LLT Pressuremeter, Vane, and Dutch Cone Tests in Bangkok Marine Clay, Thailand", The Pressuremeter and Its Marine Applications: Second International Symposium, ASTM STP 950, J.L. Briaud and J.M.E. Audibert, Eds., American Society of Testing Materials, 1986.

ABSTRACT: Three types of in-situ tests were performed in Bangkok marine clay, namely: Pressuremeter test, Field Vane test, and Dutch Cone test. The Pressuremeter test using the OYO LLT monocell probe with wash boring technique yielded the horizontal earth pressure at rest (P_o), yield pressure (P_y), limit pressure (p_ℓ), undrained Young's modulus (E_u), and undrained strength (S_{up}). The Field Vane test gave the undrained strength (S_{uv}) while the Dutch Cone test yielded the cone resistance (q_c) and local friction (f_s). The tests were carried out in 5 sites in the campus of the Asian Institute of Technology (AIT) wherein the 3 uppermost subsoil layers consist of about 2 m thick weathered marine clay, about 8 m very soft marine clay and about 6 m thick stiff clay. In each site, each in-situ tests were done in 5 locations. The averaged in-situ soil parameters obtained were correlated. The correlations obtained have been confirmed, compared, and discussed in relation with past results.

KEYWORDS: pressuremeter test, vane shear test, Dutch cone test, marine clay, shear strength, penetration.

Since 1956, when the first pressuremeter was introduced [1], it has been widely used particularly in France. In Japan, the Oyo Corporation produced its first pressuremeter which was called Lateral Load Tester(LLT) in 1966. The LLT pressuremeter is based on the same principle as the Menard pressuremeter. The only difference between the two devices is that LLT is a one cell device while the latter has 3 cells. The purpose of this study is to find out the suitability of the LLT pressuremeter test in the Bangkok marine clay and to establish correlations between the soil parameters obtained from LLT pressuremeter tests and those obtained from the vane shear tests and Dutch cone tests.

Dr. D.T. Bergado is currently Assistant Professor of Geotechnical Engineering at the Asian Institute of Technology (AIT), P.O. Box 2754, Bangkok, Thailand; Engr. M.A. Khaleque recently obtained his M.Eng. degree from AIT.

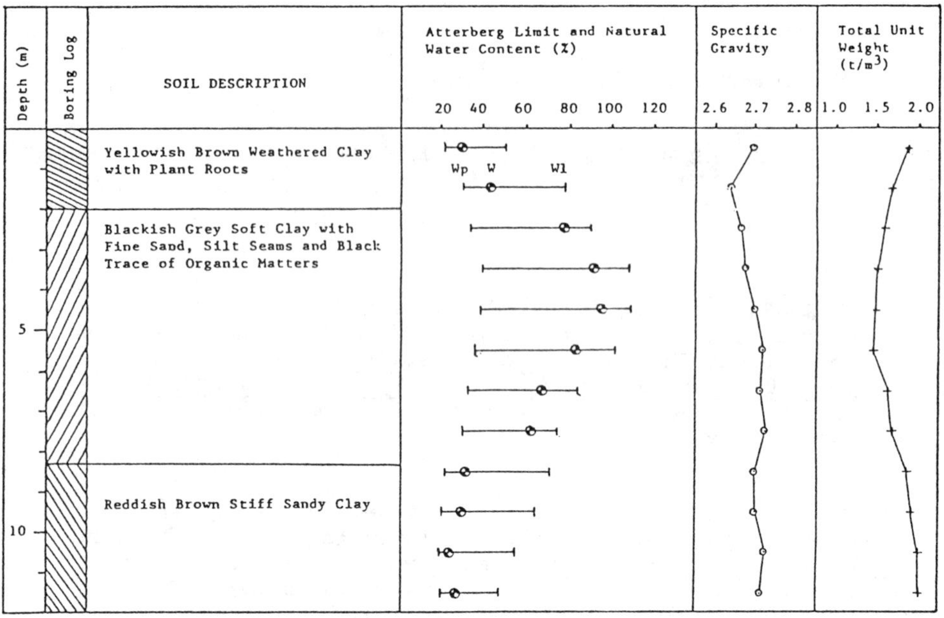

FIG. 1--Soil profile and general soil properties at the site.

A program of in-situ testing was performed at 4 sites in AIT campus, namely: behind the ET Building, near football field, near Korea House, and near New Zealand housing area. In each site, each of the 3 in-situ tests were done in 5 locations of about 10 m apart. The subsoil profile consists of about 2 m thick weathered marine clay, underlain by about 8 m of soft marine clay, and about 6 m of stiff clay as shown in Fig. 1.

THE OYO PRESSUREMETER AND METHOD OF TESTING

Unlike the Menard tricell pressuremeter, the Oyo LLT pressuremeter replaces the whole test segment in the borehole by an equivalent monocell cylindrical probe and its displacement is measured. Using the theory of elasticity, it was concluded [2] that the active length of the probe has little influence on the measured deformation as long as the deformation is measured at the midheight of the probe. The advantage of the monocell probe is the larger contact area between the measuring cell and the soil. Therefore, the measured parameters would be more reliable [3,4]. The LLT pressuremeter probe used in this study was model 4165 type M with cell diameter of 70 mm and length of 60 cm. A detailed diagram of the instrument is given in Fig. 2.

Three different methods were tried in advancing the borehole in the pressuremeter test [5], namely: hand augering method, thin-walled tube method, and wash boring method. It was found that the thin-walled tube method yielded the best result for the soft clay and that the hand augering method yielded poor results [5]. However, in a subsequent work [6], the wash boring method produced good results especially with

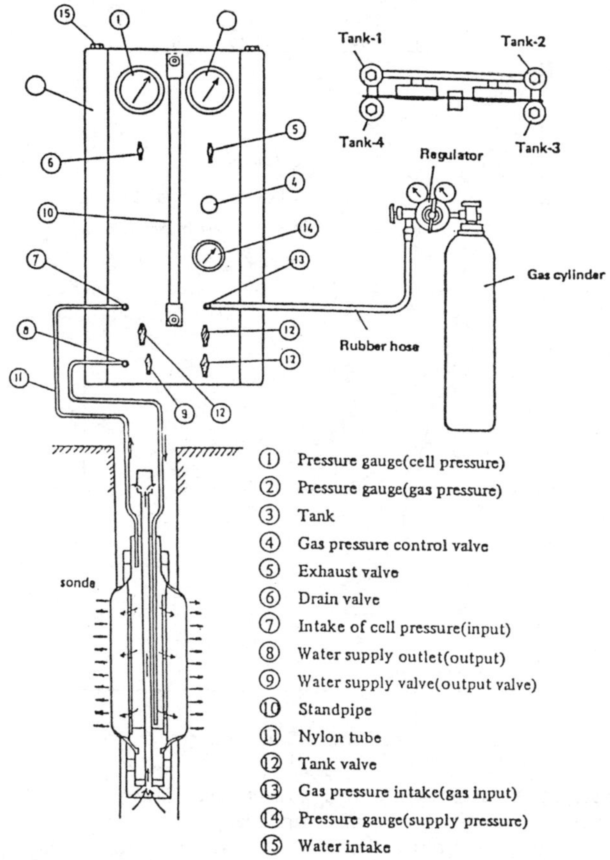

FIG. 2--Schematic diagram of LLT pressuremeter.

respect to the value of the deformation modulus, E_p, even in the soft clay layer. Using the wash boring method, the time required for advancing the hole is considerably reduced since the borehole can be kept open by the drilling mud pressure. In this method, the hole is drilled by a core barrel with wings to minimize the disturbance of the borehole wall during the circulation of the drilling mud.

SOIL PARAMETERS OBTAINED FROM PRESSUREMETER TESTS

Results from a pressuremeter test are plotted as effective pressure (P_e) versus radius (r) of the probe and as effective pressure versus deformation velocity (ΔH) curves. The effective pressure is the cell pressure corrected for rubber reaction and hydrostatic pressure. The deformation velocity is the ratio of deformation to time during the time interval of 30 sec. to 120 sec. after loading at every loading stage. This is the deformation that goes on with the lapse of time at a particular pressure due to the viscoelastic behavior of the ground. The plot directly yields 3 parameters, namely: horizontal earth pressure at

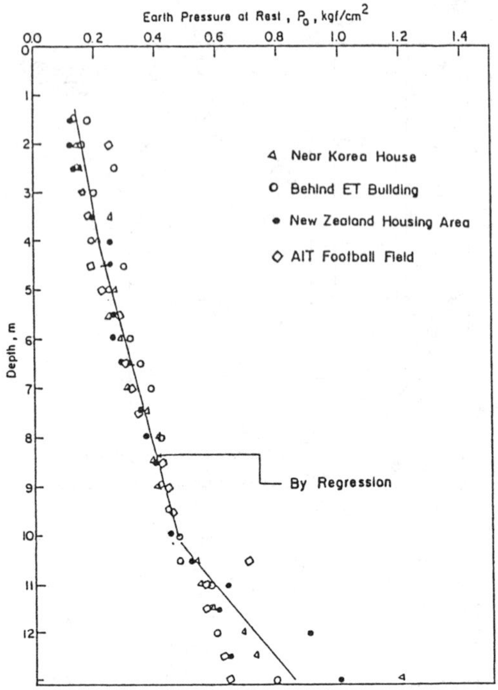

FIG. 3-- Variation of horizontal earth pressure with depth.

rest (P_o) yield pressure (P_y), and limit pressure (P_ℓ). Utilizing the theories developed for cavity expansion [7], the undrained strengths (S_{up}) and undrained Young's modulus (E_u) can be obtained.

The horizontal pressure at rest (P_o) corresponds to the pressure at which the probe gets into contact with the borehole wall and keeps the balance with the static earth pressure in the ground. In the P_e vs r curve, the starting point of the pseudo-elastic zone represents P_o. A plot of P_o with depth for all boreholes is given in Fig. 3. Each point in this figure is the average at that depth in each site. Starting from a depth of 1.5 m, the P_o values increased with depth with 3 different rate of increments corresponding to each distinguished layers in the subsoil. Similar results were obtained by subsequent investigator [8].

The pressure corresponding to a point at the end of the linear portion of the P_e-r curve is called the yield pressure (P_y). At this point, the soil moves from elastic to plastic state. A plot of the averaged values of P_y with depth is given in Fig. 4. It can be seen that the P_y values are approximately the same in the soft clay zone while the values increased with depth in the stiff clay zone. In the weathered zone, the values slightly decreased with depth. Similar values and trends were obtained in a subsequent investigation [8].

The pressure at which the cavity has doubled its volume is defined as the limit pressure (P_ℓ). Sometimes the test cannot be carried out to

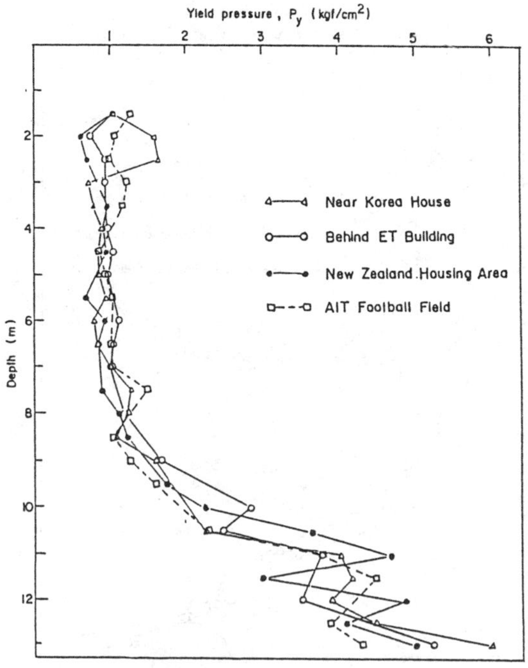

FIG. 4--Variation of yield pressure with depth.

obtain limit pressure due to limited water reservoir and due to the risk of damaging or bursting the tubing and the probe. In this case the pressuremeter curve has to be extrapolated [7] to estimate the limit pressure. Figure 5 shows the averaged P_ℓ values from 4 sites as a function of depth. The P_ℓ values are found to follow the same trend as P_y values with depth but at twice as much in numerical value. These values were later confirmed [8].

In this paper, the calculation method [7] of obtaining the pressuremeter undrained strength (S_{up}) was utilized. In this method, the undrained shear strength of the cohesive soil is calculated based on the ideal elasto-plastic behavior as follows:

$$S_{up} = \frac{P_\ell - P_o}{1 + \ln\left[\frac{Eu}{2S_{up}\,(1+\nu)}\right]} \qquad (1)$$

The terms in Eq. 1 have been previously defined. The Poisson's ratio (ν) is taken as 0.50. Many investigators [5, 8, 9) have found that this method always yielded reasonable values of undrained strengths closer to the values obtained by other field tests. Other methods resulted in much higher values [10, 11]. The undrained shear strengths and obtained using Eq. 1 are plotted in Fig. 6 as averaged values with depth.

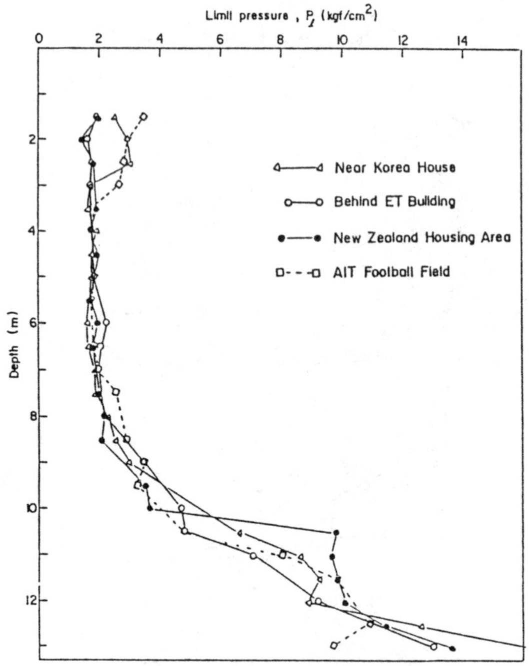

FIG. 5-- Variation of limit pressure with depth.

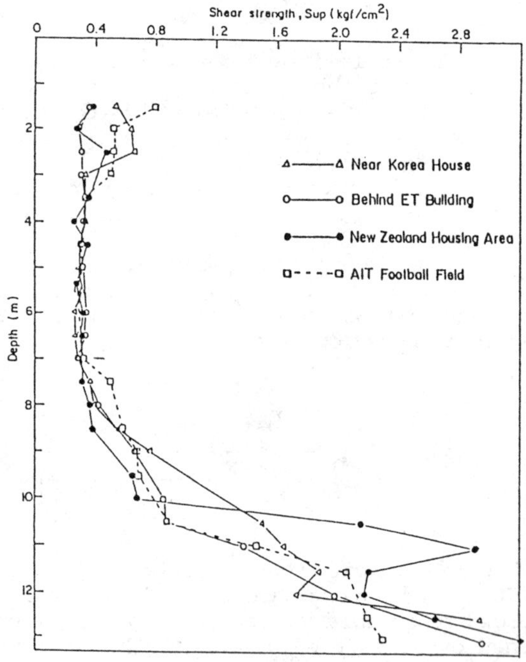

FIG. 6--Variation of pressuremeter undrained strength with depth.

The undrained Young's modulus (E_u) has been calculated as follows [7]:

$$E_u = \frac{2\Delta P}{(\Delta V/V_o)} (1+\nu) \quad (2)$$

where

ΔP = P_2-P_1 (linear part of the pressuremeter curve)

V_o = volume corresponding to P_1

V = volume corresponding to P_2

ΔV = $V - V_o$

The calculated values of E_u are plotted in Fig. 7. The values are scattered since this parameter is quite sensitive to the disturbance of the borehole wall [12, 13]. The disturbance has been shown by x-ray photographic methods [14] and is due to the advancement of the borehole, a temporary lack of confinement in the sides of the borehole, and finally due to insertion of the probe into the borehole. The disturbance affected the thin zone neighboring the wall of the borehole [13]. In the soft clay, E_u varies ranging from 20 kg/cm^2 to 60 kg/cm^2. These values were later confirmed [8]. The Poisson's ratio (ν) was taken as 0.50 corresponding to undrained incompressible case.

THE FIELD VANE TEST

A Geonor vane apparatus with blade dimensions of 130 mm high and 65 mm wide was used. In performing the test, an assembly of the apparatus with the vane blade inside a protective shoe was pressed vertically down into the ground 50 cm above the testing depth. The vane was then pushed out of the shoe and rotated at a strain rate of 6 degrees per minute up to the maximum torque. Measurements of the vane strengths were made at 0.5 m interval. Results of the vane shear strength measurements from the four sites in AIT Campus are plotted as average values with depth in Fig. 8. It can be seen that in the soft clay zone, the strengths are almost constant ranging from 0.25 kg/cm^2 to 0.40 kg/cm^2.

THE DUTCH CONE TEST

The Dutch cone penetrometer test of 2 tons capacity (2000 kg) and a friction sleeve was used. The cone and rod assembly were first pressed down, then the inner rod protruding from the casing was pushed down to the testing depth. Readings were taken at every 20 cm at a rate of penetration of 2 cm/sec. The average test results carried out in four sites are plotted in Fig. 9 and Fig. 10 for local friction and cone resistance, respectively. Both data sets showed similar trend with depth. In the weathered zone, both values, decreased with depth while in the soft clay zone, both values are almost constant.

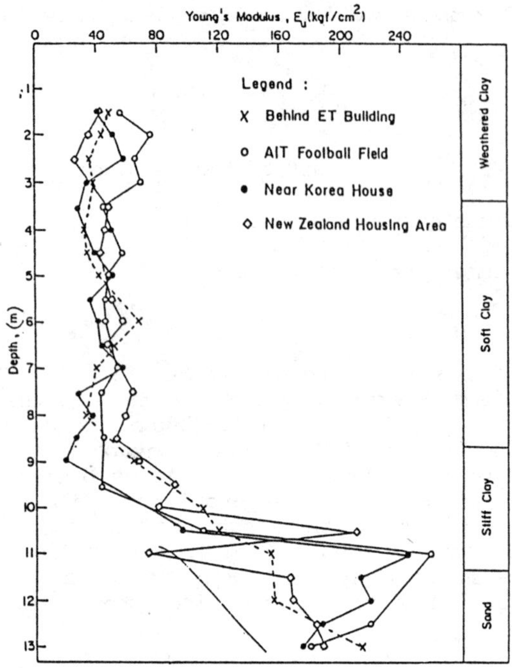

FIG. 7--Variation of Young's modulus with depth.

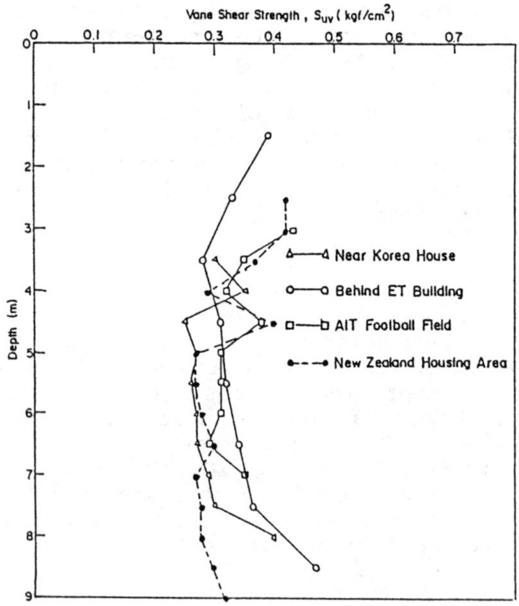

FIG. 8--Variation of vane shear strength with depth.

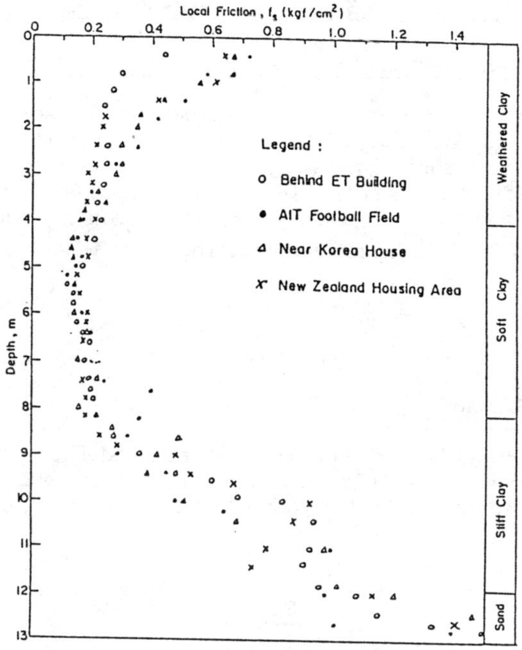

FIG. 9--Variation of local friction with depth.

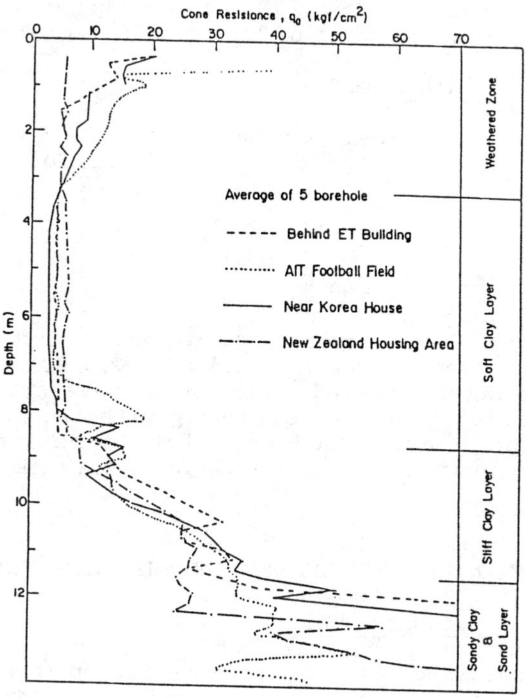

FIG. 10--Variation of cone resistance with depth.

RELATIONSHIP BETWEEN THE VANE AND THE PRESSUREMETER

Correlation of Undrained Shear Strengths

By regression analysis, the S_{up} values in the soft clay layer were found to be 10% to 25% higher than the S_{uv} values [15]. Similar values were obtained by other investigators in London clay [16] and in different sites in France [17]. Other investigators obtained S_{up} values up to 40% higher than S_{uv} values [18, 19]. The average relationship between S_{up} and S_{uv} in Bangkok soft marine clay is as follows:

$$S_{up} = 1.20\ Suv \quad (3)$$

Correlation of S_{uv} and E_u

Undrained Young's modulus (E_u) determined from pressuremeter test using Eq. 2 is correlated with the undrained vane shear strength (S_{uv}) for both weathered and soft Bangkok clay and is given by the following relationship:

$$E_u = 145\ S_{uv} \quad (4)$$

The values obtained in this study compare favorably with those obtained by other investigators [5, 20]. The modulus value of other investigation [21] were found to be too large for Bangkok clay [20].

Correlation of S_{uv} with P_y and P_ℓ

The correlation between vane shear strength (S_{uv}) with the yield pressure (P_y) and limit pressure (P_ℓ) yielded the following relationships:

$$P_y = 3.15\ S_{uv} \quad (5)$$

$$P_\ell = 5.90\ S_{uv} \quad (6)$$

In the case of the limit pressure (P_ℓ), the above correlation was found to be in good agreement with other works [5, 22, 23]. In the case of the yield pressure (Py), the main interest in correlating with the vane shear strength lies in its relationship with the preconsolidation pressure of the clay. Good correlation between Py and the preconsolidation pressure from consolidation tests were obtained [8, 22, 24].

RELATIONSHIP BETWEEN THE PRESSUREMETER AND THE DUTCH CONE

Correlation of S_{up} with q_c and f_s

The cone resistance (q_c) and local friction (f_s) have yielded a good correlation with the pressuremeter undrained strength (S_{up}). By regression analysis, the following relationships were found:

$$q_c = 16.4\ S_{up} \quad \text{(Soft clay)} \quad (7)$$

$$q_c = 19.4\ S_{up} \quad \text{(Stiff clay)} \quad (8)$$

$$f_s = 0.61\ S_{up} \quad \text{(Soft clay)} \quad (9)$$

$$f_s = 0.53\ S_{up} \quad \text{(Stiff clay)} \quad (10)$$

These cone factors agree well with the results of past investigators [25, 26] using undrained strengths converted to pressuremeter undrained strengths using Eq. 3.

Correlations of q_c and E_u

The undrained Young's modulus (E_u) from pressuremeter test was correlated with the cone resistance (q_c) from Dutch cone test for weathered and soft marine clays. The following relationships were obtained:

$$E_u = 8.48\ q_c \quad \text{(weathered clay)} \quad (11)$$

$$E_u = 8.85\ q_c \quad \text{(soft clay)} \quad (12)$$

For the stiff clay, similar relationship was obtained. By regression, the following equation has been found:

$$E_u = 4.5\ q_c \quad \text{(stiff clay)} \quad (13)$$

A value of E_u/q_c of 4.5 was obtained in a subsequent study [27].

Correlation of q_c with P_y and P_ℓ

Correlations were made for establishing the relationship between the cone resistance (q_c) with the yield pressure (P_y) and the limit pressure (P_ℓ), respectively. By regression analysis, the following relationships were plotted for both soft and stiff clay.

$$q_c = 7.4\ P_y \quad (14)$$

$$q_c = 3.36\ P_\ell \quad (15)$$

In an earlier work [5], a ratio of q_c/P_ℓ was found to be 3.5 for Bangkok marine clay. In a recent study [24], a ratio of q_c/P_y was found to be 14 and a ratio of q_c/P_ℓ was obtained as 3.24 in the stiff clay layer. Once again, in the case of Py, the main interest lies in its good relationship with preconsolidation pressure of the clay from consolidation test as stated earlier.

APPLICATIONS OF LLT PRESSUREMETER

Application of LLT pressuremeter have been illustrated by numerous investigators [4, 6, 8, 13, 24]. In Bangkok soft marine clay, the LLT pressuremeter was used to predict the settlement and ultimate load of vertically loaded piles, deformation of laterally loaded piles and settlement of embankments.

The settlement and ultimate vertical load of single piles predicted based on pressuremeter method [28] were compared with the observed results of full scale load tests [29, 30, 31]. It was observed that in the case of the ultimate vertical load of single piles, the calculated values from Menard's method [28] overestimated the observed values by a factor of 1.5 to 2.0 [6]. Methods developed by Gambin [32] and Cassan [33] have been used for predicting the settlement of vertically loaded single piles and compared to the observed values. It was observed that Gambin's method considerably overestimated while the corresponding results from Cassan's method were in good agreement with the observed values [6]. The discrepancy of the prediction by Menard's and Gambin's method may be thought to be caused by the empirical nature of these proposed methods.

The results of full scale tests on horizontally-loaded wooden piles in AIT campus [34] were used for comparison with values predicted from LLT pressuremeter tests [35]. It was observed that the method proposed by Yoshida and Yoshinaka yielded good agreement with the observed results. The calculated settlements of AIT embankment from pressuremeter results as proposed by Menard [36] was compared with the observed results obtained earlier [37]. It was found that the predicted settlement agreed with the observed value [6].

It should be emphasized that the main objective of this paper as the title indicates is to correlate the parameters obtained using the LLT pressuremeter test with other conventional in-situ tests such as the vane and Dutch cone tests which are commonly used in the Bangkok area. To that end, this study has achieved its objective. It is now possible to obtain the corresponding soil parameters from either the pressuremeter test, vane test, and Dutch cone test based on the correlations established in this study which have been compared and confirmed with past results.

CONCLUSIONS

The pressuremeter, vane, and Dutch cone field tests were performed in the weathered, soft, and stiff layers of Bangkok marine clay. The pressuremeter test yielded the following soil parameters, namely: the undrained shear strength (S_{up}), undrained Young's modulus (E_u), in-situ horizontal pressure (P_o), yield pressure (P_y), and the limit pressure (P_ℓ). The parameters obtained from the vane shear test is the undrained shear strength (S_{uv}) while the parameters from the Dutch cone test are the cone resistance (q_c) and the local friction (f_s). The following correlations were derived from the soil parameters obtained from the pressuremeter, the vane, and the Dutch cone tests.

<u>For the soft marine clay (2 m to 9 m depth):</u>

$$S_{uv} = S_{up}/1.20 = E_u/145$$

$$S_{uv} = P_y/3.15 = P_\ell/5.90$$

$$S_{up} = q_c/16.4 = f_s/0.61$$

$$q_c = E_u/8.85$$

For the weathered clay (0 m to 2 m depth)

$$q_c = E_u/8.48$$

For the stiff clay (9 m to 15 m depth)

$$S_{up} = q_c/19.4 = f_s/0.53$$

$$q_c = E_u/4.5$$

Furthermore, the results from LLT pressuremeter were applied to predict the settlement and ultimate capacity of single piles, the deformation of laterally-loaded piles, and settlement of embankments. In general, the predictions agreed well with the observed values based on full scale test.

ACKNOWLEDGEMENTS

The authors acknowledge with appreciation the help of Mrs. Vatinee Chern, Mrs. Uraiwan Singchinsuk and Miss Nualchan Sangthongsttit in the many typings of the manuscript.

REFERENCES

[1] Menard,L., "An Apparatus for Measuring the Strength of Soils In Place," M.Sc. Thesis, University of Illinois, 1957.

[2] Livneh, M., Gellert,M. and Uzan, S., "Determination of the Modulus of Soil by the Pressuremeter Test, Theoretical Background", Journal of Materials, Vol. 6, No. 2, 1971, pp. 348-355.

[3] Yoshii, R., "Comparative Trial of Lateral Load Tester", Report for Short Course of Engineering Geology, Asian Institute of Technology, Bangkok, Thailand, 1979.

[4] Tanaka, S., "Applications of the Lateral Load Tester in Engineering Geology", Report for Oyo Corporation for Short Course of Engineering Geology, Asian Institute of Technology, Bangkok, Thailand, 1979.

[5] Huang, W.F., "LLT Pressuremeter Test in Bangkok Clay", M.Eng. Thesis No. GT-79-11, Asian Institute of Technology, Bangkok, Thailand, 1980.

[6] Surya, I., "Application of the LLT Pressuremeter Test to Soil Engineering Problems in Bangkok Clay", M.Eng. Thesis No. GT-80-11, Asian Institute of Technology, Bangkok, Thailand, 1981.

[7] Gibson, R.E. and Anderson, W.F., "In-situ Measurement of Soil Properties with the Pressuremeter", Civil Engineering and Public Review, London, Vol. 56, May, 1961, pp. 615-620.

[8] Chang, C.C., "Correlation of Soil Parameters from Pressuremeter Test with Index Properties and Compressibility Characteristics of Bangkok Clay", M.Eng. Thesis No. GT-84-7, Asian Institute of Technology, Bangkok, Thailand, 1985.

[9] Wroth, C.P. and Windle, D., "Analysis of the Pressuremeter Test Allowing for Volume Change", Geotechnique, Vol. 25, No. 3, 1975, pp. 598-604.

[10] Palmer, A.C., "Undrained Plane-Strain Expansion of a Cylindrical Cavity in Clay: A Simple Interpretation of the Pressuremeter Test", Geotechnique, Vol. 22, No. 3, 1972, pp. 451-457.

[11] Ladanyi, B., "In-situ Determination of Undrained Stress-Strain Behaviour of Sensitive Clays with the Pressuremeter", Canadian Geotechnical Journal, Vol. 9, No. 3, 1972, pp. 313-319.

[12] Baguelin, F., Jezequel, J.F., Mee, L.E. and Le Mehaute, A., "Expansion of a Cylindrical Probe in Cohesive Soils", Journal of Soil Mechanics and Foundation Engineering Division, ASCE, SM11, 1972, pp. 1129-1141.

[13] Mori, H.,"Study on the Properties of Soils in the Northern Coast of Tokyo Bay Using a Self-Boring Pressuremeter", Soils and Foundations, Vol. 21, No. 3, Sept., 1981, pp. 83-98.

[14] Wroth, C.P. and Hughes, J.M.O., "An Instrument for the In-Situ Measurement of the Properties of Soft Clays", Proceedings of the 8th International Conference on Soil Mechanics and Foundation Engineering, Moscow, USSR, Vol. 1.2, 1973.

[15] Khaleque, M.A., "Correlation of Pressuremeter, Vane, and Dutch Cone Tests on Bangkok Clay", M.Eng. Thesis No. GT-83-20, Asian Institute of Technology, Bangkok, Thailand, 1984.

[16] Hughes,J.M.O., Wroth, C.P. and Pender, M.J., "A Comparison of the Results of Special Pressuremeter Tests on Deposit of Soft Clay at Convey Island", Proceedings Second Australia-New Zealand Conference on Geomechanics, Brisbane, July, 1975, pp. 292-296.

[17] Amar, S., Baguelin, F., Jezequel, J.F. and Le Mehaute, A., "In-Situ Shear Resistance of Clay", Proceedings of the Conference on In-situ Measurement of Soil Properties, ASCE, Raleigh, North Carolina, Vol. 1, June, 1975, pp. 22-44.

[18] Marsland, A. and Randolph, M.F., "Comparisons of the Results from Pressuremeter Tests and Large In-situ Plate Tests in London Clay', Geotechnique, Vol. 27, No. 2, 1979, pp. 217-243.

[19] Eden, W.J. and Law, K.T., "Comparison of Undrained Shear Strength Results Obtained by Different Test Methods in Soft Clays", Canadian Geotechnical Journal, Vol. 17, No. 3, 1980, pp. 369-381.

[20] Tsai, C.Y., "A Monograph on the Engineering Properties of Bangkok Subsoil", M.Eng. Thesis No. GT-82-18, Asian Institute of Technology, Bangkok, Thailand, 1982.

[21] Bjerrum, L., "Observed Versus Computed Settlements of Structures on Clay and Sand", Lecture presented at the Massachusetts Institute of Technology, Cambridge, Massachusetts, U.S.A., 1964.

[22] Lukas, R.G. and De Bussy, B.L., "Pressuremeter and Laboratory Test Correlations for Clays", Journal of the Geotechnical Engineering Division, ASCE, Vol. 102, No. GT9, 1976, pp. 945-962.

[23] Calhoon, M.L., "Pressuremeter Field Testing of Soils", Civil Engineering, ASCE, Vol. 39, No. 7, July 1970, pp. 71-74.

[24] Mori, H. and Tajima, S., "The Application of Pressuremeter Method to the Design of Deep Foundation, Soils and Foundations, Feb., 1964, p. 23.

[25] Wirojanagud, P., "Interpretation of Dutch Cone Test and Vane Shear Test in Soft Clays", M.Eng. Thesis No. 709, Asian Institute of Institute of Technology, Bangkok, Thailand, 1974.

[26] Pham, T.N., "Application of Dutch Cone Tests in Bangkok Area", M. Eng. Thesis No. 377, Asian Institute of Technology, Bangkok, Thailand, 1972.

[27] Ruj, N., "Correlation of Pressuremeter Test, Dutch Cone Test, and Standard Penetration Test in First Stiff Clay and First Sand Layers in Bangkok Clay", M.Eng. Thesis No. GT-84-30, Asian Institute of Technology, Bangkok, Thailand, 1985.

[28] Menard, L., "Calcul de la Force Portante des Fondations sur la Base des Resultats des Essais Pressiometrique", Sols-Soils, No. 5, 1963, pp. 9-24 and No. 6, 1963, pp. 9-27.

[29] Brand, E.W., Muktabhant, C. and Taechathummarak, A., "Load Tests on Small Foundations in Soft Clay", Proceedings of the Specialty Conference on the Performance of Earth and Earth-Supported Structures, ASCE, Purdue University, Vol. 1, June, 1972.

[30] Juttuniratsai, S., "Ultimate Bearing Capacity of Foundations of Short Piles in Soft Bangkok Clay", M. Eng. Thesis No. 914, Asian Institute of Technology, Bangkok, Thailand, 1976.

[31] Firoz, M.K., "Vertical Load Tests on a Single Instrumented Pile", Special Study Report No. SSPR-GT-80-9, Asian Institute of Technology, Bangkok, Thailand, 1980.

[32] Gambin, M., "Calcul du Tassement d'une Fondation Profonde en Fonction des Resultats Pressiometriques", Sols-Soils, No. 7, 1963, pp. 11-23.

[33] Cassan, M., "Le Tassement des Pieux - Syntheses des Recherches Recentes et Essais Comparatifs", Sols-Soils, No. 18/19/20, 1966, pp. 43-54.

[34] Tungboonterm, P., "Full Scale Load Tests on Single Piles and Pile Groups", M. Eng. Thesis No. 1289, Asian Institute of Technology, Bangkok, Thailand, 1978.

[35] Yoshida, I. and Yoshinaka, R., "A Method to Estimate Modulus of Horizontal Subgrade Reaction for a Pile", Soils and Foundations, Vol. 12, No. 3, 1972, pp. 3-17.

[36] Menard, L., "Rules for the Calculation of Bearing Capacity and Foundation Settlement Based on Pressuremeter Tests, Proceedings of the 6th International Conference on Soil Mechanics and Foundation Engineering, Montreal, Vol. II, 1965, pp. 295-299.

[37] Mukul, M.M.R., "Settlement Characteristics of Test Embankments on Soft Clay Foundation With and Without Sand Drains", M. Eng. Thesis, No. GT-79-15, Asian Institute of Technology, Bangkok, Thailand, 1980.

Foundation Design Using Pressuremeter Data

Jean-Pierre Bécue, Françoise Brucy and Pierre Le Tirant

PROPOSED METHOD FOR APPLICATION OF PRESSUREMETER TEST RESULTS TO DESIGNING OF OFFSHORE FOUNDATIONS

REFERENCE: Bécue, J.-P., Brucy, F. and Le Tirant P., "Proposed Method for Application of Pressuremeter Test Results to Designing of Offshore Foundations", The Pressuremeter and Its Marine Applications: Second International Symposium, J.-L. Briaud and J. M. E. Audibert, Eds, American Society for Testing and Materials, 1986.

ABSTRACT: Generalized procedures for soil classification and routine analysis of offshore pile and shallow foundations, directly using self-boring pressuremeter (SBP) test results, are yielded with detailed charts, thus enabling straight application to dimensioning.
The proposed geotechnical classification is based on:
- the soil identification coefficient β
- and the net pressure p^*_{20}
Recommended practice for determination of unit skin friction and end bearing capacity of pipe piles account for phenomena such as critical depth effect and possible pile plugging.
Procedures for generation of lateral load-deflection curves are based on definite analogy between the pressuremeter test and a laterally-loaded pile.
Similarly, the ultimate bearing capacity of shallow foundations may be estimated straightly using SBP test results with empirical coefficients depending on the embedment depth and corrective factors accounting for load eccentricity and inclination and for foundation base shape.
For heterogeneous or multilayered soil profiles procedures are given to derive equivalent parameter values.

KEYWORDS: soil classification, skin friction, end bearing, lateral load-deflection curves, shallow foundations, self-boring pressuremeter.

INTRODUCTION

The increasing use of the pressuremeter method in offshore soil investigation calls for the provision of recommendations for application of

Mr Becue is project manager at GEODIA offshore geotechnical consultants, 16 rue Médéric, 75017 Paris, France; Mrs Brucy and Mr Le Tirant are respectively research engineer and head of Marine Geotechnics Department, Institut Français du Pétrole, B.P. 311, 92506 Rueil Malmaison, France.

the test results to the design of offshore foundations. This is the purpose of the present paper which deals with semi empirical methods for routine design of piled and shallow foundations, based on the present state-of-the-art.

Available onshore design methods [1, 2] have been modified so as to suit the typical offshore conditions, namely the range of values of the soil characteristics likely to be encountered. In this regard, as tight as possible a fit has been tried with the still limited but growing data base relative to offshore sites or soils representative of marine conditions.

This work is part of a general effort undertaken at Institut Français du Pétrole for promoting the marine application of the pressuremeter [3, 4].

PRESSUREMETER TEST DATA AND SOIL CLASSIFICATION

Pressuremeter Curves

The basic principle of the pressuremeter test is the expansion of a cylindrical cavity in the ground, using specially developed pressuremeter probes and standard procedures.

The raw data - i.e. the pressure against the wall of the cavity and the corresponding expansion measured and recorded during the inflation of the cell - are corrected for pressure and volume losses, membrane inertia and hydrostatic pressure and then used to plot pressuremeter curves.

Fig. 1 shows a typical corrected pressure-relative variation in volume ($p - \frac{\Delta v}{V_o}$) curve where V_o is the initial volume of the cavity for SBP tests.

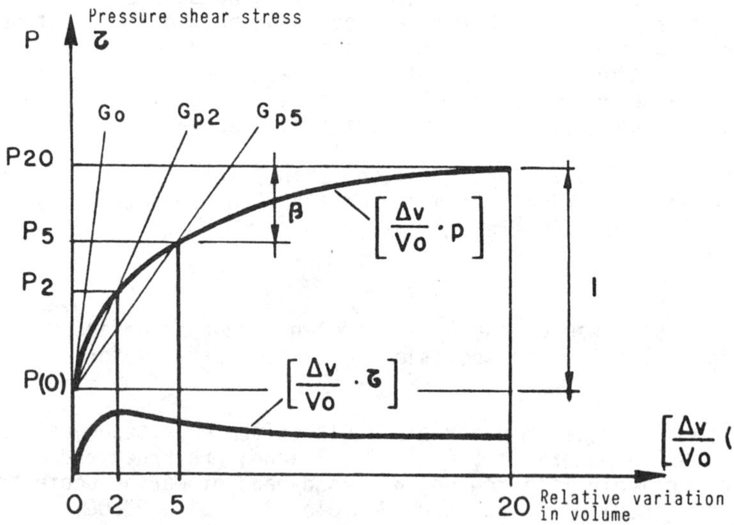

FIG. 1--Typical self-boring pressuremeter (SBP) curve and derived shear stress curve

Total Horizontal Pressures

The initial total horizontal pressure is conventionally defined as the pressure corresponding to zero dilation strain and noted $p_{(o)}$.

In the case of SBP tests the initial total horizontal pressure obtained as per standard procedures may constitute an approximate estimate of the at rest total horizontal pressure (p_o) which is generally suitable for introduction in routine foundation design methods.

Pressures characterizing the soil response for various volume dilatation ($\frac{\Delta v}{V_o}$) of the pressuremeter are read directly from the corrected pressuremeter curves (Fig. 1), namely:

- p_2 = pressure at 2% strain
- p_5 = pressure at 5% strain
- p_{20} = pressure at 20% strain

The net pressures (p*) are derived according to the following relationship:

$$p^* = p - p_o \tag{1}$$

Soil Moduli

Shear moduli (G) are calculated from the pressure variation during a relatively small strain change on the inflation curve - or during a reload loop -, using the classical equation for the radial expansion of a cylindrical cavity in a continuous elastic medium:

$$G = \frac{\Delta p}{\frac{\Delta v}{v}} \tag{2}$$

where Δp is the pressure variation over a volume strain range of $\frac{\Delta v}{v}$.

For SBP tests (Fig. 1) the following shear moduli may be defined:

- G_o initial tangent shear modulus
- $G_{p_2} = \frac{p_2^*}{0.02}$ secant modulus at 2% strain
- $G_{p_5} = \frac{p_5^*}{0.05}$ secant modulus at 5% strain

Soil Identification Coefficient

In the case of SBP tests the shape of the corrected pressuremeter curve appears to be related to the nature of the soil [5]. As a result, it may be characterized by the soil identification coefficient β (Fig. 1) given by Equation 3:

$$\beta = \frac{p_{20}^* - p_5^*}{p_{20}^*} \tag{3}$$

Shear Stress Curves

There definitely exists a relationship between the soil response during the pressuremeter test and its shear stress behavior. Many attempts have been made to derive the shear stress curve of soil from SBP data and several methods have been proposed by various authors [6 to 15].

A typical shear stress curve $(\frac{\Delta v}{V_o} \cdot \tau)$ is shown in Fig. 1.

Soil Classification

Main marine soils can be classified as proposed in Table 1 [1, 5], using the two following parameters obtained from SBP tests:

- the soil identification coefficient (β),
- the net pressure at 20% strain (p^*_{20}).

For geotechnical engineering purpose a distinction is made between:

- purely siliceous soils,
- and carbonate and calcareous sediments for which the total carbonate content may be variable.

TABLE 1--Indicative soil classification.

Type of soil	β, %	p^*_{20} , kPa
. Siliceous:		
- soft clays, soft silty clays	< 35	< 600
- medium stiff clays and silty clays	< 40	600-1200
- stiff clays, stiff silty clays	< 40	>1200
- loose silts	30-45	< 600
- medium dense silts	35-50	600-1200
- dense silts	35-50	>1200
- loose sands	40-55	< 700
- medium dense to dense sands	45-60	700-2000
- very dense sands	> 55	>2000
. Calcareous:		
- sands	30-40	

A tentative diagram for soil classification is shown in Fig. 2.

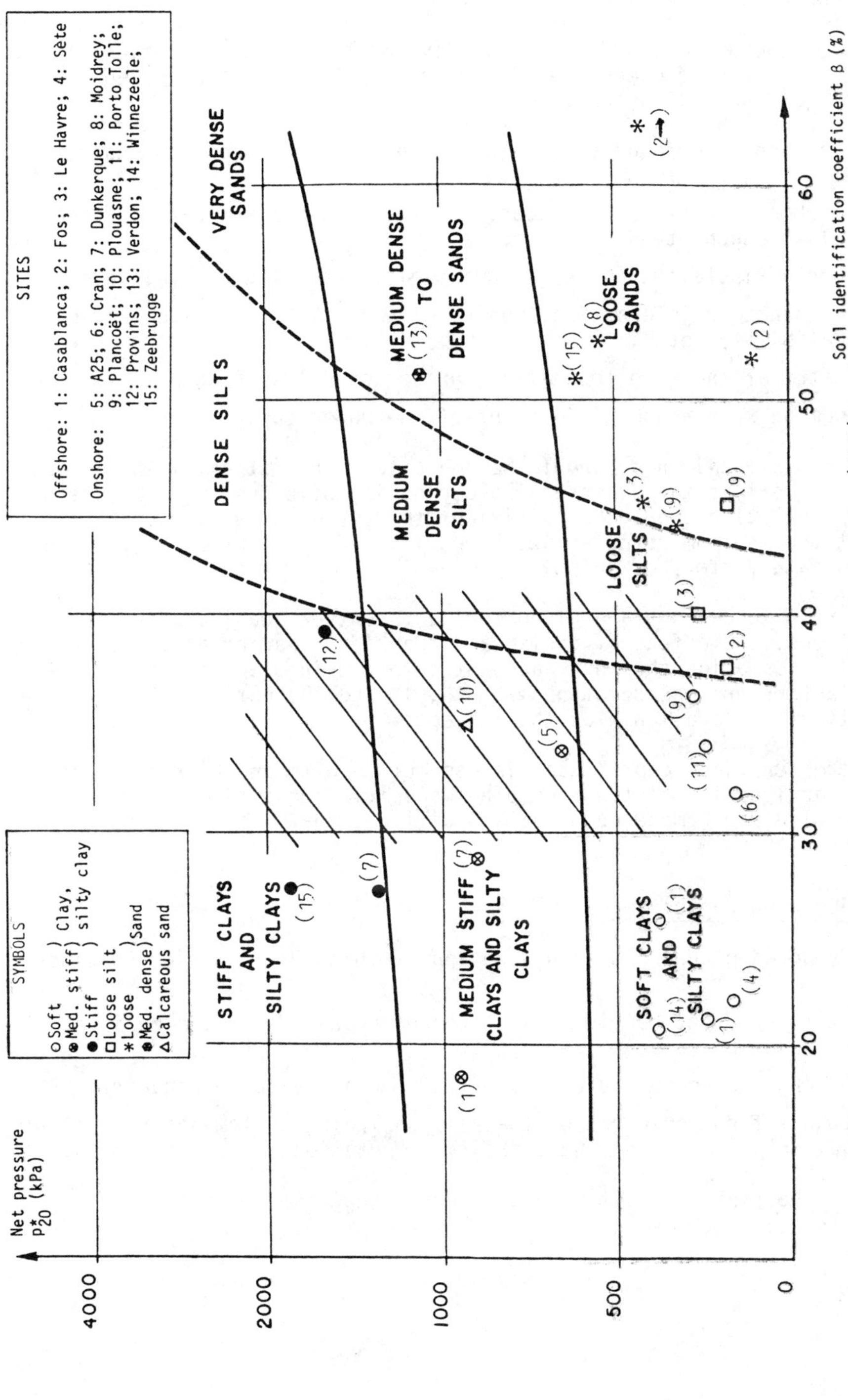

FIG. 2--Soil classification based on pressuremeter test data (SBP) (Hachures correspond to calcareous sands).

PILE FOUNDATIONS

The proposed procedures are mainly devoted to the determination of the design capacity of open-ended driven pipe pile foundations used to support offshore structures.

In designing pile foundations, the following items should be considered:

- pile characteristics: diameter, length, flexibility, wall thickness, steel strength, tip type, etc.;
- method of installation, e.g.: driving stages, plug removal, etc.;
- pile-structure interaction: restraint conditions, load-deflection compatibility, etc.;
- geometry of the pile foundation: number of piles, spacing, etc.;
- phenomena such as possible scour of the upper soil layers.

The pipe wall thickness should be adequate to resist axial and lateral loads as well as the stresses induced during pile driving. The design pile penetration should be sufficient to develop adequate capacity to resist the maximum computed axial bearing and pull-out capacity with appropriate factors of safety.

These recommendations are given for permanent or quasi static loading conditions. The effect of cyclic and transient loading on the behavior of pile foundation should be accounted for, using appropriate reduction factors or considerations [5, 16, 17 and 18] until further recommendations are available on the subject.

The term "critical depth" used in connection with end bearing capacity, lateral soil resistance or other specific property means the depth below which, in homogeneous soil conditions, that property remains constant [19].

Axial Bearing Capacity

Proposed procedures should be applied with usual design considerations relevant to:

- limitation of end bearing resistance of open-ended piles by internal shaft friction capacity;
- pile length or compressibility effects on actual pile capacity;
- allowance for the weight of the pile-soil plug system and hydrostatic uplift in pile loading and capacity computations.

End bearing: The ultimate base resistance should be computed using Equation 4:

$$q_u = q_o + k\, p^*_{20} \qquad (4)$$

where

q_u = ultimate bearing capacity,

q_o = total vertical stress of the overburden at pile tip level,

p^*_{20} = net pressure at 20% volume strain,

k = bearing capacity factor.

Generally speaking, the values of k depend on:

- the nature of the soil,
- the net pressure (p^*_{20}),
- the pile embedment depth.

For homogeneous soil conditions, the k values applicable to the critical depth and underneath are given in Fig. 3 as a function of the soil identification coefficient (β) and for various values of the net pressure (p^*_{20}).

As the length of the piles used for foundation of offshore structures is usually larger than the critical depth, only these values are usually required.

However, where pile penetration is less than the critical depth, appropriate reduction factors should be applied to the k limit values.

In case of heterogeneous soil conditions, equivalent net pressure $(p^*_{20})_e$ and equivalent soil identification coefficient β_e have to be used in Equation 4 in lieu of p^*_{20} and β, respectively. Since the zone of influence of the end of vertically loaded piles is assumed to extend from 1.5 diameter above to 1.5 diameter below the base of the foundation, these equivalent values should be computed as the geometrical mean of the p^*_{20} and β values within this zone, respectively. They are obtained through Equations 5 and 6, using $(p^*_{20})_i$ and β_i values measured at regular intervals:

$$(p^*_{20})_e = \sqrt[n]{(p^*_{20})_1 \times \ldots \times (p^*_{20})_i \times \ldots \times (p^*_{20})_n} \qquad (5)$$

and

$$\beta_e = \sqrt[n]{\beta_1 \times \ldots \times \beta_i \times \ldots \times \beta_n} \qquad (6)$$

This procedure must be used with caution as soon as the individual p^*_{20} values differ by 40% or more from $(p^*_{20})_e$ [20].

Skin friction: Skin friction or adhesion (f) along the pile wall depends on the extent of radial displacement that has affected the soil around the pile during the installation phase. This displacement is greater for plugged or closed-ended piles than for unplugged piles. Consequently, for analysis of a vertically loaded pile, distinction should be made between partial displacement piles and full displacement piles.

Respective f values are given in Figs. 4 and 5 for estimating the skin friction or adhesion in siliceous soils. Ultimate skin friction values measured during static pile load tests [17,21] are compared with those predicted by the present method in Fig. 6.

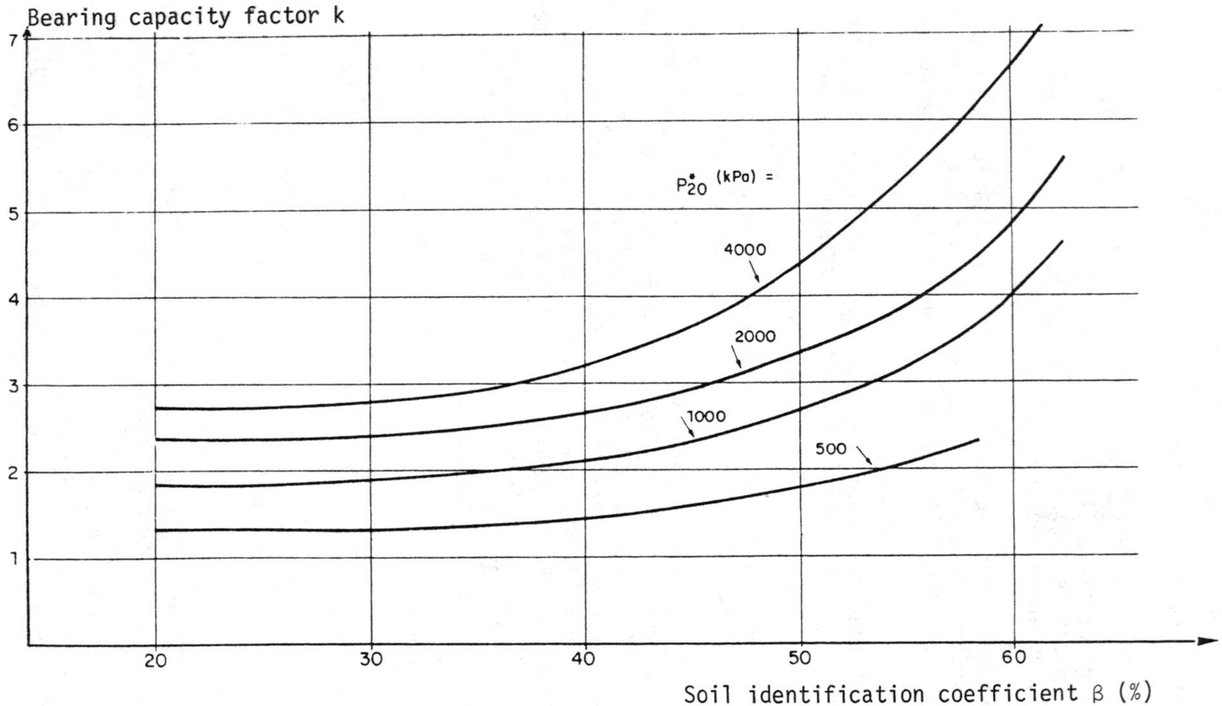

FIG. 3--Bearing capacity factor for determination of end bearing capacity of piles (applicable below the critical embedment depth).

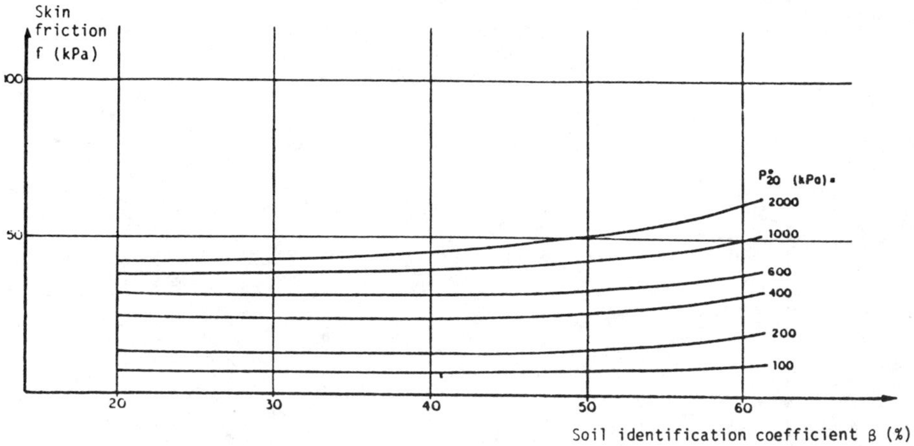

FIG. 4--Skin friction for layers through which pile is unplugged during driving operation.

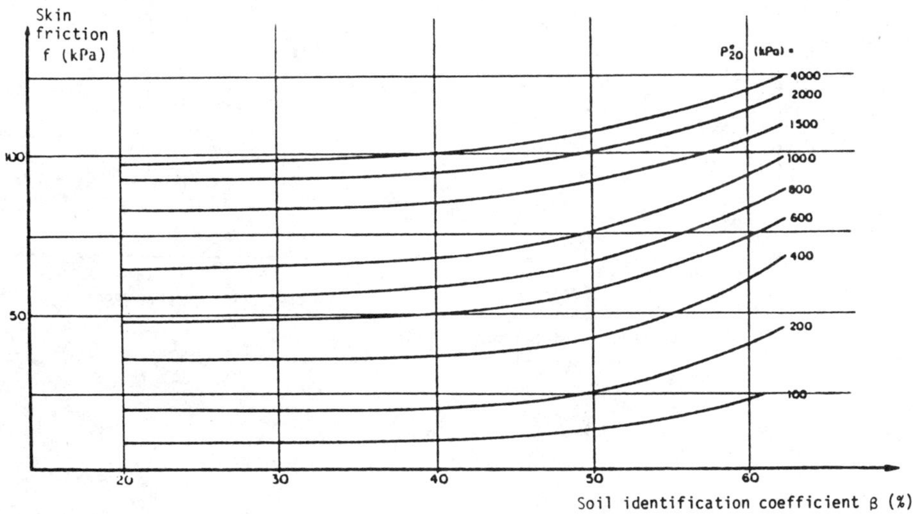

FIG. 5--Skin friction for layers through which pile is plugged during driving operation.

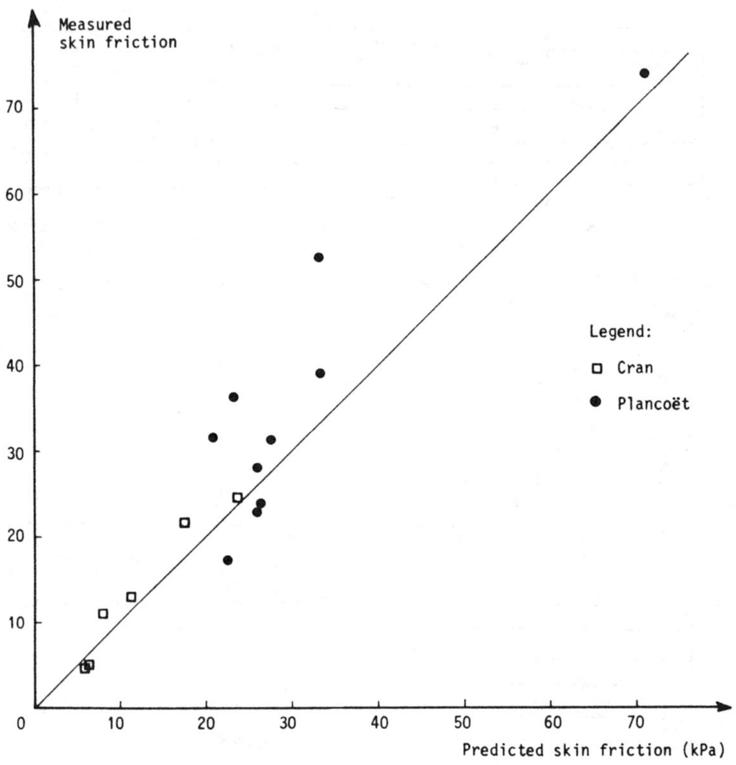

FIG. 6--Measured vs predicted skin friction.

Nota: Experimental evidence has shown that skin friction of piles driven through calcareous soils could be significantly lower than anticipated from usual methods applicable to siliceous soils. Actually skin friction may be closely dependent on parameters such as texture, compressibility, brittleness and degree of cementation [22].

Axial Pull-out Capacity

The net ultimate pile pull-out capacity may be equal to the total skin friction, which may be estimated by the same procedures as for axial bearing capacity.

Laterally-loaded Pile Capacity And Behavior

Background: The pile foundation should be designed to sustain lateral loads with appropriate safety considerations regarding:

- on the one hand, the stress ratio induced in the pile,
- on the other hand, the pile deflection at the mudline.

The non-linear subgrade reaction methods commonly used to predict the complete load-deflection curve for the pile call for the construction of load-deflection (p_{pil} - y) curves or load-relative deflection ($p_{pil} - \frac{y}{R}$) curves, where: p_{pil} = lateral soil resistance,
y = soil deflection,
R = pile radius.

Based on definite analogy of loading between the pressuremeter test and a laterally loaded pile, procedures for directly generating ($p_{pil} - \frac{y}{R}$) curves from pressuremeter curves are recommended, thus enhancing the reliability of pile behavior analysis [23, 24]. These procedures emphasize the generation of ($p_{pil} - \frac{y}{R}$) curves for the surface soil layers, whose behavior is significant in pile design. The critical depth for the pressuremeter probe is very shallow and always substantially less than the critical depth for lateral loading behavior for the pile. So, it is necessary to apply a reduction factor to the lateral bearing capacity of the layers encountered between the seabed surface and the critical depth of the pile.

Load-relative deflection ($p_{pil} - \frac{y}{R}$) curves: SBP test data are used to derive these curves for the application to static loading conditions. The abscissa of the curves is the radial strain ($\frac{y}{R}$) which is given by Equation 7:

$$\frac{y}{R} = \frac{1}{2}\frac{\Delta v}{V_o} \qquad (7)$$

where

y = pile deflection,
R = pile radius,
$\frac{\Delta v}{V_o}$ = volume strain at the desired level.

The ordinate of the $(p_{pil} - \frac{y}{R})$ curve is the lateral soil resistance (p_{pil}) which is given by:

$$p_{pil} = \eta p^* \tag{8}$$

where

p^* = net pressure corresponding to $\frac{\Delta v}{V_o}$ at the same level,
η = lateral resistance factor.

The lateral resistance factor is given in Fig. 7 as a function of the soil identification coefficient (β) and the relative depth ($\frac{z}{R}$ where z is depth below mudline).

The η values, between the mudline and the critical depth for lateral loading, are given by a set of curves corresponding to various values of relative depth $\frac{z}{R}$ (Fig. 7). Below the critical depth, the values of the upper curve are applicable.

ULTIMATE BEARING CAPACITY OF SHALLOW FOUNDATIONS

The following section deals with the estimation of the ultimate bearing capacity of shallow foundations such as jack-up rigs, gravity base platforms, gravity anchors and any kind of structures simply setting on the sea bottom.

The allowable bearing capacity of the foundation is then computed by dividing the ultimate capacity by a safety factor which depends on the type of structures and applicable standards.

General Formulation

The ultimate gross vertical load that a footing can support is expressed:

$$q_u = q_o + \lambda_e \cdot \lambda_s \cdot \lambda_i \cdot \Psi \cdot p^*_{20} \tag{9}$$

where

q_u = ultimate bearing capacity for the effective foundation area,
q_o = total vertical stress of the overburden at the level of the foundation,
λ_e = eccentricity factor,
λ_s = shape factor,
λ_i = load inclination factor,
Ψ = coefficient of bearing capacity for strip footings.

Eccentricity factors (λ_e) and effective foundation area

In the case of a strip footing (Fig. 8), the eccentricity factor is:

$$\lambda_e = \frac{B'}{B} = 1 - 2\frac{e}{B} \tag{10}$$

where

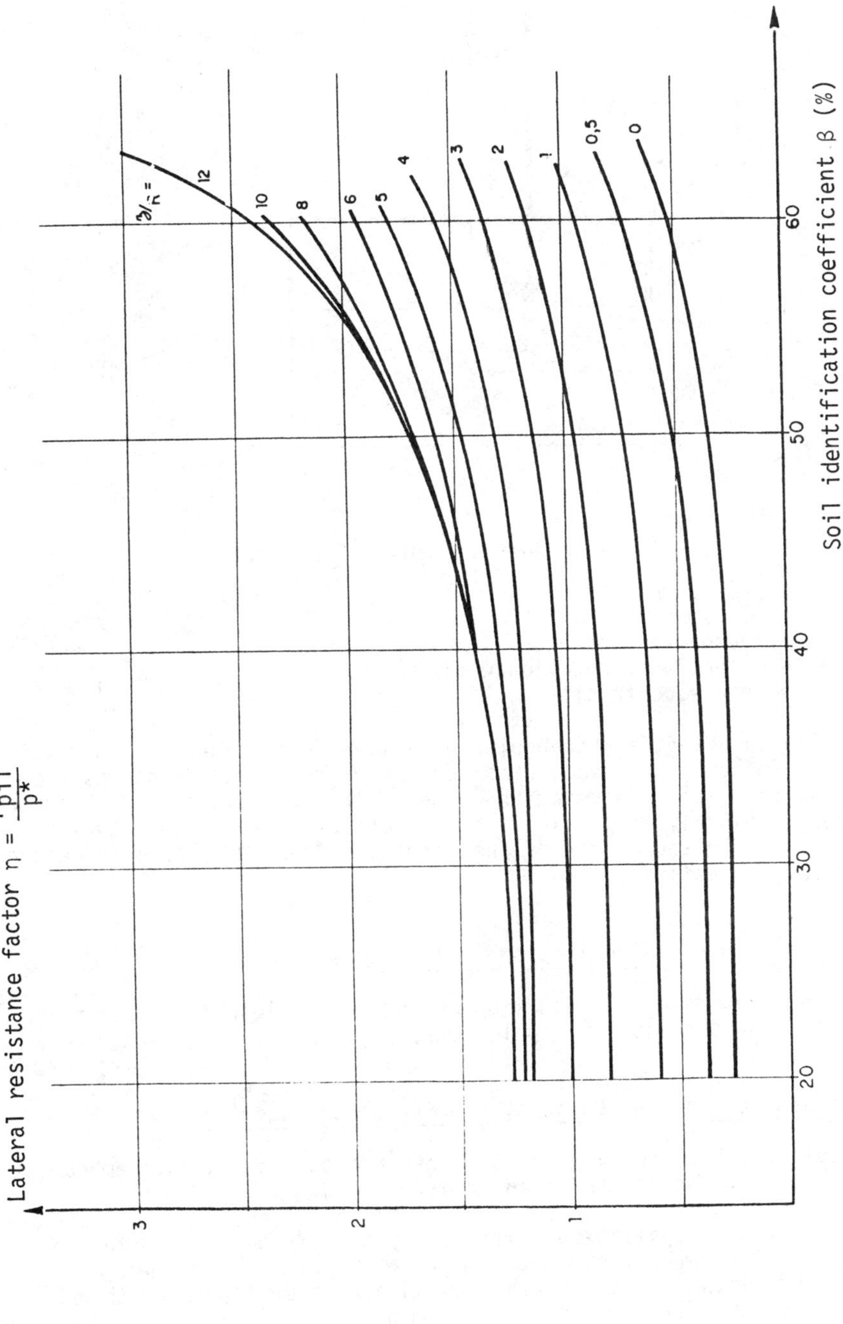

FIG. 7--Lateral resistance factor.

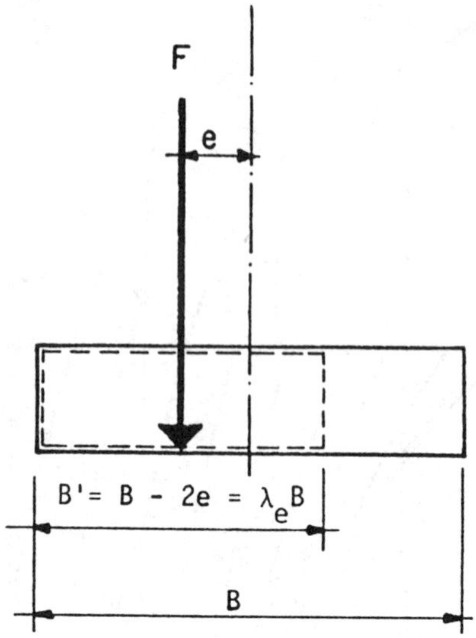

FIG. 8--Eccentric load on footing.

B = foundation width,
B' = effective foundation width,
e = load eccentricity.

Generally speaking, a rectangular effective foundation area is determined as follows: the geometrical centre of this area coincides with the resultant load centre at the base of the foundation and its contour follows as closely as possible the nearest contour of the actual base area. The width and length of the effective foundation area are noted B' and L' respectively.

Shape (λ_s) and Load Inclination (λ_i) Factors

Values such as given for usual embedment depth, load inclination and foundation shapes [16] are applicable.

Coefficient of Bearing Capacity for Strip Footings (Ψ)

The coefficient of bearing capacity depends on the relative embedment depth $x = D/B'$ and is expressed as $\Psi(x)$ [16].

Typical offshore shallow foundations may usually be surface footings with a relative embedment depth nearly zero, or jack-up type foundations with a relative embedment depth generally lower than 3. So Ψ values are given for x ranging from 0 to x_1 (> 3).

Homogeneous soil conditions: The coefficient of bearing capacity $\Psi(x)$ may be taken as follows:

- for $x = 0$ and $x = 1$, values of $\Psi(0)$ and $\Psi(1)$ respectively are given in Fig. 9, as a function of β and p^*_{20};
- for $0 < x < 1$, values of $\Psi(x)$ may be assumed to vary linearly between $\Psi(0)$ and $\Psi(1)$:

$$\Psi(x) = \Psi(0) + x \quad \Psi(1) - \Psi(0) \tag{11}$$

- for $1 < x < x_1$, $\Psi(x)$ may be obtained as follows:

$$\Psi(x) = \Psi(1) + a \log x \tag{12}$$

where

a = coefficient of variation with embedment depth, given in Fig. 10,

x_1 = maximum relative embedment depth to which Equation 12 is applicable (given in Fig. 10).

Nota: x_1 should not be confused with relative critical depth.

Heterogeneous soil conditions: The following procedures may be used to derive:

- an equivalent net pressure $(p^*_{20})_e$,
- an equivalent soil identification coefficient β_e,
- and an effective depth of embedment D_e,

to be used in computations in lieu of p^*_{20}, β and D, respectively.

The zone of influence of a foundation is assumed to extend from 1.5 B' above the base - or from ground surface level where embedment depth is lower than 1.5 B' - to 1.5 B' below the base of the foundation. Thus, the equivalent net pressure $(p^*_{20})_e$ (respectively the equivalent soil identification coefficient β_e) may be computed, using Equations 5 and 6, respectively, with the same precaution as for pile end bearing capacity [20].

The effective depth of embedment may be calculated using the following summation over the actual embedment depth D:

$$D_e = \sum_{i=1}^{n} \Delta z_i \frac{(p^*_{20})_i}{(p^*_{20})_e} \tag{13}$$

with $D = \sum_{i=1}^{n} \Delta z_i$

where

$(p^*_{20})_i$ = net pressure p^*_{20} at the level i,

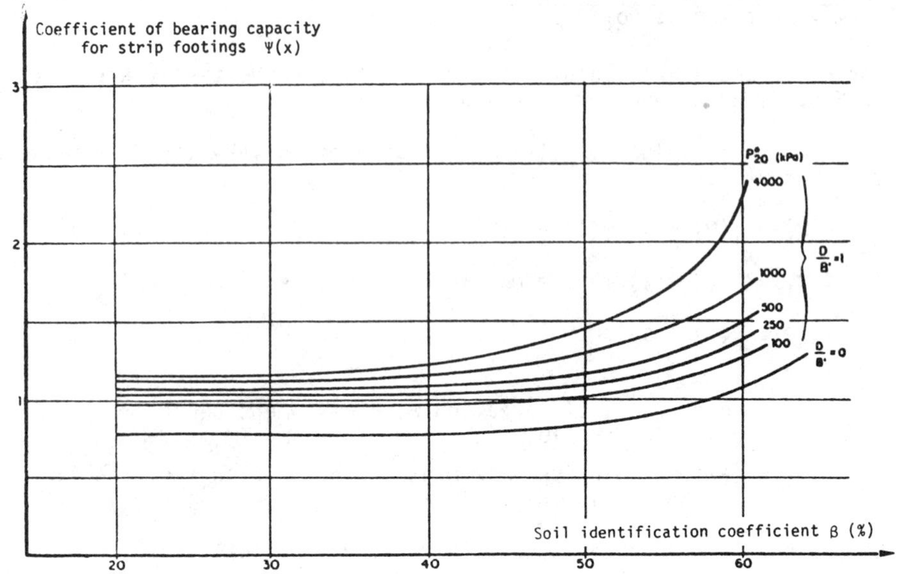

FIG. 9--Coefficient of bearing capacity for strip footings with embedment depth $0 < x\ (\frac{D}{B'}) < 1$.

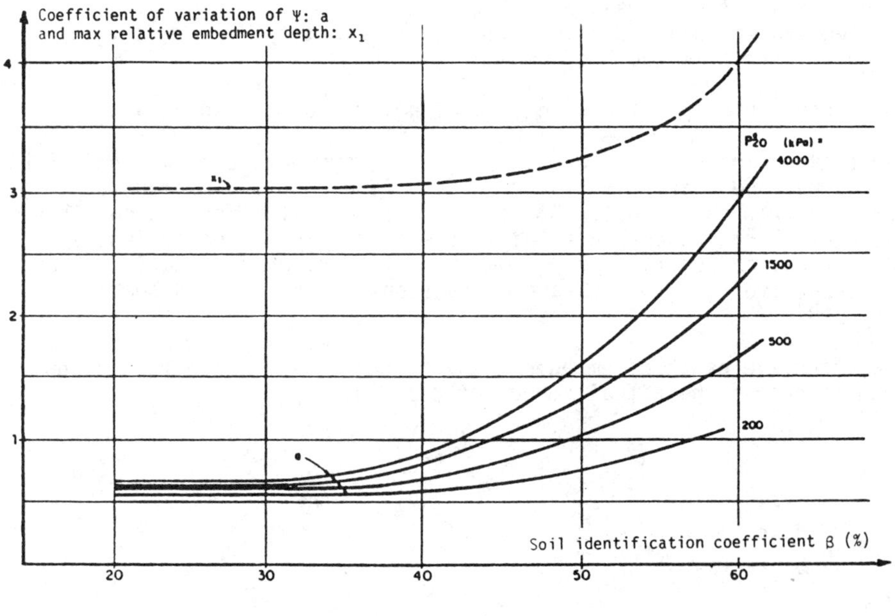

FIG. 10--Coefficient of variation of Ψ for embedment depth $\frac{D}{B'}$ higher than 1 and lower than x_1.

Δz_i = thickness of the corresponding elementary layer,

$(p^*_{20})_e$ = equivalent net pressure.

CONCLUSIONS

The presentation used in this paper makes it possible to proceed easily to sensitivity analyses of the geotechnical parameters, which may be of concern for foundation engineering.

Further research and experience will undoubtedly enable the proposed soil classification and foundation design procedures to be improved, namely through the growing of the data base.

Although other pressuremeter test results may be introduced into the design procedures through appropriate correlations with the SBP parameters [16], emphasis is borne on the increasing interest of SBP tests in offshore geotechnics.

Namely, the SBP tests offer very interesting possibilities such as the determination of in-situ stress state or the investigation of stress-strain behavior, that should result in significant foundation engineering advances.

REFERENCES

[1] BAGUELIN, F., JEZEQUEL, J.F. and LE MEHAUTE, A., "Le pressiomètre autoforeur et le calcul des fondations", Comptes-rendus du 7ème Congrès Européen de Mécanique des Sols et des Travaux de Fondations Brighton, England, Vol. 2, Sept. 1979, pp. 185-190.

[2] BUSTAMANTE, M. et GIANESELLI, L., "Prévision de la capacité portante des pieux isolés sous charge verticale. Règles pressiométriques", Bulletin de liaison du Laboratoire des Ponts et Chaussées, No. 113, Mai-Juin 1983, pp. 83-108.

[3] LE TIRANT, P., FAŸ, J.B. and BRUCY F., "The offshore self-boring pressuremeter (PAM), its application at great water depth", Deep Offshore Technology, Malta, Vol. 2, 1983, pp. 45-62.

[4] BRUCY, F., FAŸ, J.B. and LE TIRANT, P., "Three years' experience with the offshroe self-boring pressuremeter PAM", Offshore Technology Conference, Houston, Texas, May 1984, Paper 4677.

[5] BAGUELIN, F., "Rules of foundations design using self boring pressuremeter test results", in Symposium on the pressuremeter and its marine applications, Ed. Technip, Paris, 1982.

[6] ARNOLD, M., "An empirical evaluation of pressuremeter test data", Canadian Geotechnical Journal, Vol. 18, No. 3, 1981, pp. 455-459.

[7] BAGUELIN, F., JEZEQUEL, J.F., LE MEE, E. and LE MEHAUTE, A., "Expansion of cylindrical probes in cohesive soils", Journal of Soil Mechanics and Foundation Division, A.S.C.E., Vol. 98, No. SM 11, Nov. 1972, pp. 1129-1142.

[8] DENBY, G.M. and CLOUGH, G.W., "Self-boring pressuremeter tests in clay", Journal of the Geotechnical Division, A.S.C.E., Vol. 106, No. GT 12, 1980, pp. 1369-1387.

[9] GIBSON, R.E. and ANDERSON, W.F., "In situ measurement of soil properties with the pressuremeter", Civil Engineering and Public Works Review, Vol. 56, No. 658, 1961, pp. 615-618.

[10] JAMIOLKOWSKI, M., LANCELLOTTA, R., PASQUALINI, E., MARCHETTI, S. and NOVA, R., "Design parameters for soft clay", European Conference on Soil Mechanics and Foundation Engineering, 7, Brighton, Vol. 5, 1981, pp. 25-57.

[11] LADANYI, B., "In situ determination of undrained stress-strain behavior of sensitive clays with the pressuremeter", Canadian Geotechnical Journal, Vol. 9, No. 3, 1972, pp. 313-319.

[12] LADD, C.C., FOOT, R., ISHIHARA, K., SCHLOSSER, F. and POULOS, H.G., "Stress-deformation and strength characteristics. State-of-the-art report", in "International Conference on Soil Mechanics and Foundation Engineering, 9, Tokyo, Vol. 2, 1977, pp. 421-494.

[13] PALMER, A.C., "Undrained plane-strain expansion of a cylindrical cavity in clay ; a simple interpretation of the pressuremeter test", Geotechnique, Vol. 22, No. 3, 1972, pp. 451-457.

[14] PREVOST, J.H. and HOEG, K., "Analysis of pressuremeter in strain-softening soil", Journal of the Geotechnical Engineering Division, A.S.C.E., Vol. 101, No. GT 8, 1975, pp. 717-732.

[15] WINDLE, D. and WROTH, C.P., "In situ measurement of the properties of stiff clays", in International Conference on Soil Mechanics and Foundation Engineering, 9, Tokyo, Vol. 1, 1977, pp. 347-352.

[16] BAGUELIN, F., JEZEQUEL, J.F. and SHIELDS, D., Ed., The pressuremeter and foundation engineering, Trans Tech Publications, Germany, 1978.

[17] PUECH, A., "Basic data for the design of tension piles in silty soils", Third International Conference on the Behaviour of Off-shore Structure, M.I.T., Cambridge, Massachusetts, Paper F4, 1982.

[18] PUECH, A., BRUCY, F. and MA, E., "Tension pile design using self-boring pressuremeter test results", in Symposium on the pressuremeter and its marine applications, Ed. Technip, Paris, 1982.

[19] BRIAUD, J.L. and SHIELDS, D.H., "Pressuremeter tests at very shallow depth", Journal of the Geotechnical Engineering Division, ASCE, Vol. 105, No. GT 8, 1981.

[20] MENARD, L., "Calcul de la force portante des fondations sur la base des résultats des essais pressiométriques", Sols-Soils, Vol. 5, June 1963, pp. 9-28.

[21] PUECH, A., BRUCY, F. and MA, E., "Design of tension piles using self-boring pressuremeter test results", in Symposium on the pressuremeter and its marine applications, Ed. Technip, Paris, 1982.

[22] NAUROY, J.F. and LE TIRANT, P., "Driven piles and drilled and grouted piles in calcareous sand ", Offshore Technology Conference, Houston, Texas, May 1985, Paper 4850.

[23] BRIAUD, J.L., SMITH, T. and MEYER, B., "Design of laterally loaded piles using pressuremeter test results", in Symposium on the Pressuremeter and its Marine Applications, Ed. Technip, Paris, 1982.

[24] BRIAUD, J.L., SMITH, T. and MEYER, B., "Using the pressuremeter curve to design laterally loaded piles", Offshore Technology Conference, Houston, Texas, May 1983, Paper 4501.

Jean-Louis Briaud

PRESSUREMETER AND DEEP FOUNDATION DESIGN

REFERENCE: Briaud, J.-L., "Pressuremeter and Deep Foundation Design," The Pressuremeter and Its Marine Applications: Second International Symposium, ASTM STP 950, J. L. Briaud and J.M.E. Audibert, Eds., American Society for Testing Materials, 1986.

ABSTRACT: The rules of design for vertically and horizontally loaded piles on the basis of pressuremeter test results are reviewed. The pressuremeter test considered is the one where the pressuremeter probe is inserted in a prebored hole. The rules are then evaluated on the basis of a series of well-documented vertical and horizontal pile load tests. The accuracy of the predictions is shown and recommendations are made for future use.

KEYWORDS: pressuremeter, clay, sand, piles, piers, drilled shafts, ultimate capacity, settlement.

INTRODUCTION

The pressuremeter can be used for the design of piles subjected to vertical and horizontal loads. Rules of design for vertical loads were developed in the early sixties by Menard and his co-workers. These rules were adjusted in the mid seventies by the Laboratoire des Ponts et Chaussees (LPC). Newer rules were also proposed in the early eighties by the LPC. Presently (1985) these three sets of existing rules are being consolidated by the LPC into a unified set of rules.

The development of rules for the design of laterally loaded piles using pressuremeter data has been lagging that of vertically loaded piles and appears to be at the stage of development where many different design methods are being proposed. There is a need for a unification of methods.

In a first part, load test data is used to evaluate the precision of the various design rules for vertical loads. A similar

Dr. Briaud is an Associate Professor in the Civil Engineering Department at Texas A&M University, College Station, Texas, 77843.

approach is taken in a second part for lateral loads. All the rules make use of preboring pressuremeter test results.

VERTICAL LOADS: METHODS FOR ULTIMATE POINT CAPACITY

Over the last 25 years, four methods have been proposed. These methods are very similar in concept and differ mainly because the load test data base has been increasing over the years and has allowed the adjustment of the design rules. The basic concept of the methods is to obtain the pile ultimate point resistance, q_L, from the pressuremeter net limit pressure, p^*_L, by use of a bearing capacity factor, k, and to obtain the pile ultimate friction resistance, f_L, from correlations with p^*_L. The net limit pressure is obtained from the pressuremeter limit pressure, p_L, as:

$$p^*_L = p_L - p_0 \quad (1)$$

where p_0 is the at rest total horizontal stress at the testing depth.

The pile ultimate point resistance q_L is calculated as follows:

$$q_L = k\, p^*_{Le} + q_o \quad (2)$$

where q_0 is the at rest total vertical stress at the depth of the pile point, and p^*_{Le} is the equivalent net limit pressure near the point. The parameter p^*_{Le} is obtained from:

$$p^*_{Le} = \sqrt[n]{p^*_{L1} \times p^*_{L2} \times \ldots \times p^*_{Ln}} \quad (3)$$

where p^*_{L1}, --, and p^*_{Ln} are the net limit pressures obtained from tests performed within the + 1.5B/- 1.5B zone near the pile point with B being the pile diameter. The equivalent depth of embedment H_e is obtained from:

$$H_e = \sum_{1}^{n} \frac{\Delta z_i\, p^*_{Li}}{p^*_{Le}} \quad (4)$$

where p^*_{Li} are the net limit pressures obtained from tests performed between the ground surface and the pile point, ΔZ_i are the thicknesses of the elementary layers corresponding to the pressuremeter tests. This definition of H_e allows to take into consideration layers within the depth of embedment which are stronger or weaker than the layer on which the pile point is resting.

Method 1 was proposed in 1963 [1, 2] by Menard who recommended the use of Figure 1 for obtaining the bearing capacity factor, k. The load test data base that Menard used to propose the design curves was made of 8 plate tests and 6 short piles at one site in sand and silt.

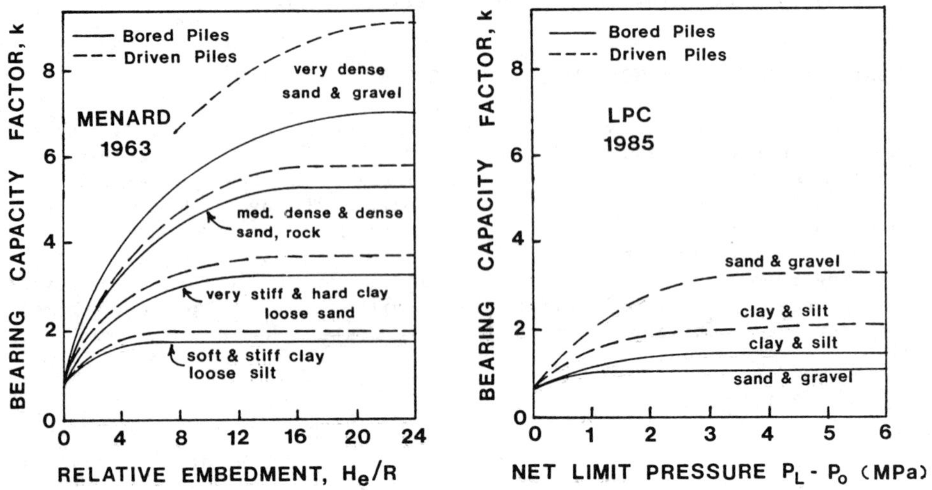

Figure 1. Menard 1963 Rules versus LPC 1985 Rules for Point Resistance k Values; Simplified Version

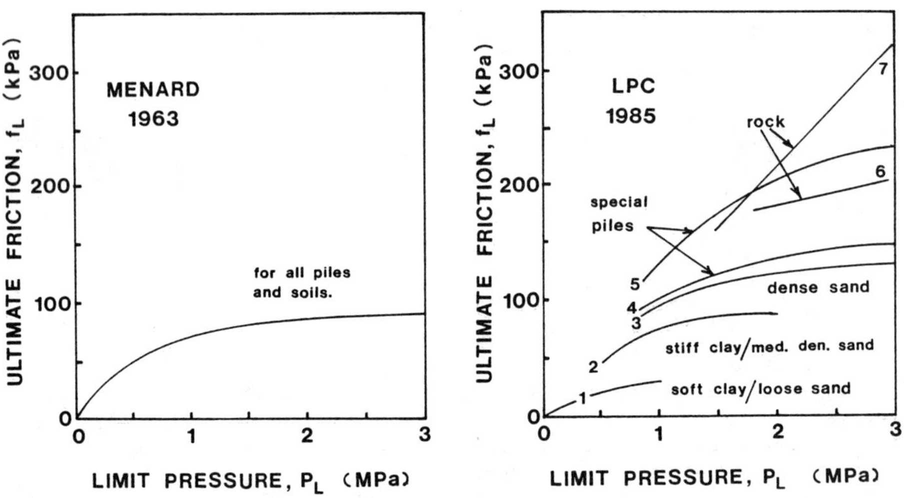

Figure 2. Menard 1963 Rules versus LPC 1985 Rules for Friction Values; Simplified Version

Although the data base was limited, Menard was careful to include in the proposed rules the most fundamental aspects of pile foundation behavior and supplement his data base by comparison to other soil tests, especially the cone penetrometer test. Even though the k values have been adjusted over the years in light of additional load test data, Menard's original concepts, namely equation 2 and the shape of the design curves for k, were sound enough to withstand over 20 years of experience.

Method 2 was proposed in 1978 [3] by Baguelin, Jezequel and Shields. The main difference with Menard's rules was that the soil categories were simplified and for each soil category several curves were proposed for various net limit pressures. The effort then was to make the rules more precise but the overall magnitude of the k values did not change.

Method 3 was proposed in 1981 [4] by Bustamante and Gianeselli. It is based on 45 load tests on instrumented piles at 23 different sites. The soil categories have been changed slightly but the major difference is that the k values are about one half the values recommended in methods 1 and 2.

Method 4 was developed in 1982 [5] because of the problems arising from the existence of three different methods; it is proposed by the Laboratoire des Ponts et Chaussees (LPC) in France (Figure 1). The LPC has assembled all the data used for methods 1, 2 and 3 plus data available since method 3. In method 4 the soil categories have been simplified. the effective relative depth of embedment, H_e/B, has disappeared from the abscissa axis on the k-chart and has been replaced by a net limit pressure axis. This implies that the chart applies to piles having sufficient H_e/B and that, since k increases with p^*_L, q_L increases faster than p^*_L. But the major difference with method 3 is the fact that the k values have decreased further by about 25% on the average.

VERTICAL LOADS: METHODS FOR ULTIMATE FRICTION CAPACITY

If the rules for point resistance have grown towards simplification, the rules for friction resistance grew from simple to very detailed. The basic approach is to correlate the ultimate unit friction, f_L, on the pile to the net limit pressure, p^*_L. The aforementioned four methods proposed various f_L versus p^*_L design curves.

Method 1 was proposed in 1963 [1, 2] by Menard who recommended the use of Figure 2 for obtaining the value of f_L for all piles. The limited extent of Menard's data base is probably the reason for the fact that a single design curve applies to all soils and all types of piles.

Method 2 was proposed in 1978 [3] by Baguelin, Jezequel and Shields. The main difference with Menard's rules is that Menard's single curve is replaced by four curves to be used depending on soil type, pile installation and pile material. The values of f_L for method 2 bracket Menard's single curve.

Method 3 was proposed in 1981 [4] by Bustamante and Gianeselli. It is based on 64 load tests on instrumented piles at 40 sites. Seven different f_L-$p*_L$ design curves are proposed. The proper f_L-$p*_L$ design curve is chosen on the basis of soil type, pile material and pile installation technique. More pile installation techniques (e.g.: injected piles) and more soil types (e.g: rock) are included. The details of the recommendations show clearly the drastic influence of the pile placement technique and the quality of execution on the f_L friction values.

Method 4 [5] developed by the Laboratoire des Ponts et Chaussees in France evolved from method 3. Seven f_L-$p*_L$ design curves are used (Figure 2) and the recommended f_L values are somewhat lower than the ones of method 3. A first table is entered to identify the pile material and the pile placement technique. With this, a second table is entered to determine which curve to use considering the soil type. The proper f_L value is then obtained using that curve and the measured $p*_L$ values. Method 4 is also described in details in reference 6; the official French document will be available soon. A microcomputer program has been written to perform predictions according to Method 4 [7].

Method 5 is based on the results of selfboring presuremeter tests [8,9]; it was also developed by the Laboratoire des Ponts et Chaussees and is parallel to the preboring pressuremeter method. It makes use of the net pressure at 20% relative volumetric expansion, $p*_{20}$, to obtain the pile unit point resistance and unit skin friction.

ULTIMATE VERTICAL LOAD: PMT VS. SPT/S_U APPROACH

Equation 2 can be compared with the undrained shear strength plasticity approach for clays:

$$q_L = N_c \, S_U + \gamma D \tag{5}$$

The term N_cS_U compares directly with k $p*_{Le}$ in Equation 2. The factor N_C is taken as 9 and in clay, method 4 recommends k values which range from 1 to 2. This leads to a value of $p*_L$ ranging from 4.5 S_U to 9 S_U. This is well within the range of existing correlations between $p*_L$ and S_U [3, 10, 11].

Equation 2 can also be compared with the friction angle-plasticity approach for sands:

$$q_L = \frac{1}{2} \gamma B N_\gamma + \gamma D N_q \tag{6}$$

The term 1/2 γBN_γ is usually negligible compared to γDN_q. The term γDN_q is usually limited by using a critical depth approach for D [12] or by using an upper limit for q_L of 10,000 to 15,000 kPa (100 to 150 tsf). There is growing controversy about the critical depth concept [13,14] but the q_L limits still stand since higher q_L values have rarely been measured in pile load tests. Figure 2 shows that for very dense sands which may have net limit pressures of 5000 kPa (50 tsf) the point resistance q_L would be about 15,000 kPa (150 tsf) which is consistent with the aforementioned limits.

If the penetration of the pile is associated with the expansion of a sphere, the bearing capacity factor is the ratio of the sphere net limit pressure q^*_L over the cylinder (PMT) net limit pressure p^*_L:

$$k_{clay} = \frac{q^*_L}{q^*_L} = \frac{\frac{4}{3} S_U \left(1 + Ln \frac{G}{S_U}\right)}{S_U \left(1 + Ln \frac{G}{S_U}\right)} = \frac{4}{3} \tag{7}$$

$$k_{sand} = \frac{q_L}{p_L} = \frac{p_o \frac{3(1+\sin\phi)\cos\phi}{(3-\sin\phi)\sin\phi}\left[\frac{G(3-\sin\phi)}{3 p_o \sin\phi}\right]^{0.66(1-K_a)}}{p_o (1+\sin\phi)\left(\frac{G}{p_o \sin\phi}\right)^{0.5(1-Ka)}} \tag{8}$$

where $q^*_L = q_L - p_o$, q_L is the sphere limit pressure, p_o the total stress at rest at the pile point, S_U the undrained shear strength, ϕ the friction angle, G the shear modulus, and K_a the coefficient of active earth pressure.

The expressions for q^*_L and p^*_L for the clay assume a Tresca yield criterion while the expressions for q_L and p_L for the sand assume a Mohr Coulomb yield criterion with no volume change. Equation 7 leads to a k value for clay of 1.33 while Equation 8, for reasonable values of the parameters, varies between 3 and 4. These theoretical k values are within the range of proposed values on Figure 2.

One factor which influences q_L is the compressibility of the soil; N_c, N_γ and N_q factors which depend on a compressibility index were proposed by Vesic [15]. This was done in an effort to correct for the shortcomings of the rigid-plastic solution. This important factor is incorporated directly into p^*_{Le} since the compressibility of the soil affects the pressuremeter limit pressure. Indeed the shear modulus is a factor in the theoretical expression of of p_L for both clay and sand (Equations 7 and 8).

The friction values recommended in method 4 (Figure 2) are somewhat higher than the friction values from other methods. The correlation of f_L to p^*_L is validated to some extent by the fact that p^*_L is a function of the undrained shear strength S_U (Equation 7). Obtaining f_L from p^*_L however, does not appear to be superior to obtaining f_L from other test results, such as sample strength or SPT blow count, except that: 1. the data base for the PMT method 4 includes a wide variety of insertion techniques as well as soil types, 2. the method can be used for piles in soils where samples cannot be recovered (sands), or where laboratory strength is doubtful (fissured clays), 3. it can be used in rock because PMT tests are possible in rock. In soft rock, friction values of up to

1000 kPa (10 tsf) are allowed.

VERTICAL LOAD TEST DATA BASE

Several comparisons have been made between measured and predicted ultimate loads using one or more of the above methods [7,16, 17,18,19,20,21,22]. In the following, predictions according to method 4, which was developed from a pile load test data base collected in France, are compared to the results of 51 pile load tests performed in the United States over the last 5 years. The 51 pile load tests are listed in Table 1.

At Lock & Dam No. 26 Replacement Site on the Mississippi River near St. Louis, 28 vertical load tests were performed on 22 H piles and 6 pipe piles by the Corps of Engineers, District of St. Louis [22,23]. Of the 28 tests, 16 were tension tests and 12 compression tests. Five of the 12 compression tests were loaded to 400 tons without any sign of plunging failure; three of the 16 tension tests were loaded to 150 tons without any sign of pullout failure. Therefore 20 load tests were used for this data base. Piles 1 to 20 are listed on Table 1 with pertinent information. The soil at the site is sand and gravel with 5% passing sieve No. 200 and grain size up to 4 cm (1.57 in.) as reported from 7.5 cm (3 in.) diameter SPT tests. Limestone exists at 19.8 m (65 ft) on the average. At depth, boulders up to 30 cm (1 ft) in size were identified from cut off trench excavations. The SPT blow count averaged 10 blows per 30 cm (1 ft) at the surface to 60 blows per 30 cm (1 ft) at 18.3 m (60 ft). The electric cone penetrometer point resistance average 7500 kPa (75 tsf) at the surface to 22500 kPa (225 tsf) at 18.3 m (60 ft) depth. Sleeve friction and piezo-cone measurements were also made. The pressuremeter net limit pressure varied from 500 kPa (5 tsf) at the surface to 3500 kPa (35 tsf) at 18.3 m (60 ft).

The Mississippi Highway Department has performed and collected 100 pile load test results over the last ten years. At 8 of the pile load test sites, pressuremeter tests were performed; this allowed comparison of predicted and measured behavior for 15 piles (piles 21 to 35 on Table 1) including square concrete, bored and H piles all tested in compression. The soils in general were stiff clays and medium dense sands with pressuremeter net limit pressures varying from 300 kPa (3 tsf) to 2500 kPa (25 tsf) [7,24].

At the construction site of Lock & Dam 2 on the Red River near Alexandria, Louisiana, pressuremeter tests were performed next to a few of the many piles which were load tested by the Corps of Engineers, Vicksburg District [17] . This led to the comparison between measured and predicted response for 8 piles (piles 36 to 43 on Table 1) including 5 H piles and 3 concrete piles. At the H piles site, the soil is made of a 7.93 m (26 ft) thick stiff clay layer over a 12.81 m (42 ft) thick layer of sand, underlain by hard clay. At the concrete piles site the clay layer is 21.65 m (71 ft) thick, and the sand layer is 11.0 m (36 ft) thick. The stiff clay has the following average properties: undrained shear strength from unconsolidated undrained triaxial tests, 1.2 tsf, CPT point resistance 3500 kPa (35 tsf), and PMT limit pressure 800 kPa (8 tsf). The sand layer has the following average properties: SPT blow count of 100 blows per 30 cm (1 ft), CPT point resistance 16000 kPa (160 tsf) and

TABLE 1 - Vertical pile load test data base

Pile ID No.	Site	Pile Type	Pile Length (m)	Pile Diam. (m)	Load Test	Soil	Ult. Load Predicted (tons)	Measured Load at B/10 (tons)
1	L&D 26	HP 14x73	16.5	0.36	Tens	Sand/Grav	304	120**
2	L&D 26	HP 14x73	16.5	0.36	Comp	Sand/Grav	356	356**
3	L&D 26	HP 14x73	16.5	0.36	Tens	Sand/Grav	304	65
4	L&D 26	HP 14x73	16.2	0.36	Comp	Sand/Grav	350	430**
5	L&D 26	HP 14x73	16.2	0.36	Comp	Sand/Grav	380	320**
6	L&D 26	HP 14x73	16.8	0.36	Tens	Sand/Grav	313	130**
7	L&D 26	HP 14x73	18.0	0.36	Tens	Sand/Grav	338	180*
8	L&D 26	HP 14x73	21.1	0.36	Tens	Sand/Grav	405	150
9	L&D 26	HP 14x73	17.7	0.36	Tens	Sand/Grav	331	100**
10	L&D 26	HP 14x73	18.0	0.36	Comp	Sand/Grav	391	225
11	L&D 26	HP 14x73	12.2	0.36	Tens	Sand/Grav	223	50
12	L&D 26	Pipe	14.2	0.30	Comp	Sand/Grav	230	130**
13	L&D 26	Pipe	11.0	0.30	Tens	Sand/Grav	111	61
14	L&D 26	Pipe	14.4	0.36	Comp	Sand/Grav	284	125**
15	L&D 26	Pipe	11.1	0.36	Tens	Sand/Grav	132	68
16	L&D 26	Pipe	14.6	0.41	Comp	Sand/Grav	343	180**
17	L&D 26	Pipe	11.1	0.41	Tens	Sand/Grav	151	100
18	L&D 26	HP 14x73	11.9	0.36	Tens	Sand/Grav	185	130*
19	L&D 26	HP 14x73	11.3	0.36	Tens	Sand/Grav	172	90
20	L&D 26	HP 14x73	11.3	0.36	Tens	Sand/Grav	172	95
21	Miss.	Bored	12.5	0.36	Comp	Clay	93	124
22	Miss.	Sq. Conc	6.3	0.36	Comp	Clay	66	90
23	Miss.	Sq. Conc	7.6	0.46	Comp	Clay	176	120
24	Miss.	Sq. Conc	10.4	0.46	Comp	Clay	255	200*
25	Miss.	Sq. Conc	7.0	0.36	Comp	Sand	152	55
26	Miss.	Sq. Conc	7.1	0.36	Comp	Clay/Sand	75	100**
27	Miss.	Sq. Conc	8.6	0.36	Comp	Clay/Sand	93	67
28	Miss.	Sq. Conc	10.2	0.36	Comp	Clay/Sand	115	108
29	Miss.	Bored	10.6	0.36	Comp	Clay	138	130
30	Miss.	HP 12x53	19.1	0.30	Comp	Sand	113	154**
31	Miss.	Sq. Conc	16.0	0.36	Comp	Sand	243	120
32	Miss.	HP 12x53	12.4	0.30	Comp	Sand	58	58
33	Miss.	HP 12x53	8.6	0.30	Comp	Sand	37	35
34	Miss.	Sq. Conc	14.0	0.41	Comp	Clay/Sand	184	117
35	Miss.	Sq. Conc	16.4	0.41	Comp	Clay/Sand	283	143
36	L&D 2	HP 14x89	16.5	0.36	Comp	Clay/Sand	293	200**
37	L&D 2	HP 14x89	18.0	0.36	Tens	Clay/Sand	284	85
38	L&D 2	HP 14x89	19.5	0.36	Comp	Clay/Sand	366	250**
39	L&D 2	Sq. Conc	16.2	0.41	Comp	Clay/Sand	213	220
40	L&D 2	Sq. Conc	19.7	0.41	Comp	Clay/Sand	266	230
41	L&D 2	Sq. Conc	22.7	0.41	Comp	Clay/Sand	325	325*
42	L&D 2	HP 14x73	15.9	0.36	Tens	Clay/Sand	240	100
43	L&D 2	HP 14x73	18.3	0.36	Tens	Clay/Sand	287	60
44	LADWP	Bored	2.9	0.66	Tens	Clay	19	55
45	LADWP	Bored	2.9	0.64	Tens	Clay	18	55
46	LADWP	Bored	4.4	0.66	Tens	Clay	46	85
47	LADWP	Bored	2.9	0.71	Tens	Sand	79	125*
48	LADWP	Bored	2.2	0.75	Tens	Sand	61	75
49	Lackland	Bored	10.5	0.46	Comp	Clay	121	170*
50	U of Hous	Driven	13.1	0.27	Comp	Clay	55	80
51	U of Hous	Bored	2.36	2.41	Comp	Clay	252	336

* Load test curve extended to $B/10 + Q_m L/AE$ with reasonable confidence.

** Slope at $B/10 + Q_m L/AE$ does not indicate plunging failure; time ultimate load most probably higher than load listed in table by at least 20%.

PMT limit pressure, 2500 kPa (25 tsf). The hard clay has a PMT limit pressure of 3000 kPa (30 tsf).

The Los Angeles Department of Water and Power performed 5 tension tests on short drilledshafts for a major power line project [16]. Two sites were tested, one on stiff clay (p^*_L = 500 kPa or 5 tsf), one on medium dense sand (p^*_L = 2000 kPa or 20 tsf). The shafts were about 0.62 m (2 ft) in diameter and 3 to 5 m (10 to 15 ft) long.

At Lackland Air Force Base a 0.46 m (1.5 ft) diameter and 10.5 m (34.5 ft) long drilled shaft was load tested [18]. A gap existed under the base and therefore the ultimate load represents only the friction load developed in the compression test. The soil is a stiff clay with net limit pressures increasing from 500 kPa (5 tsf) at the surface to over 2000 kPa (20 tsf) at depth.

At the University of Houston, two research piles were tested; one pipe pile and one large diameter drilled shaft [25,26,27]. The drilled shaft was built in such a way that zero friction would exist on the shaft and the point resistance would be directly measured. The soil at the site is a stiff clay with net limit pressures increasing with depth from 200 kPa (2 tsf) at the surface to 1000 kPa (10 tsf) at 7.6 m (25 ft) depth.

VERTICAL LOAD: PREDICTED VS. MEASURED ULTIMATE LOAD

The load settlement curve for each load test was used to determine the measured ultimate load, Q_m. The load Q_m was defined as the load reached for a settlement equal to one tenth of the diameter or equivalent diameter of the pile plus the elastic compression of the pile under Q_m ($D/10 + Q_mL/AE$). This load was not always reached during the load tests collected; only the piles where Q_m or a plunging load was reached (88% of the data) or where Q_m could be obtained with reasonable confidence by extension of the load test curve were used. Note in Table 1 that, especially for piles in sand, at the ($D/10 + Q_L/AE$) settlement the tangent to the load-settlement curve did not indicate plunging failure. It is estimated that in those cases the true ultimate load was larger than the "D/10" ultimate load listed in Table 1 by at least 20% [20].

The predictions were performed automatically by using an IBM-PC computer program written for that purpose. For the H piles the ultimate pile capacity was calculated first by using the steel-soil contact areas for point and shaft resistances, Q_{U1}, and second by using the enclosing rectangular areas for point and shaft resistances, Q_{U2}. For these H piles the predicted ultimate load was considered to be the lowest of the two loads Q_{U1} and Q_{U2}. In all cases Q_{U2} was smaller than Q_{U1}. Note that Q_{U2} corresponds to a plugged condition where a soil plug exists between the flanges of the H pile.

Figure 3 shows predicted versus measured ultimate loads for compression tests. On the average the method overpredicts the measured loads by 20%. However, as was mentioned earlier, for a significant number of piles (Table 1) no plunging failure was reached at a

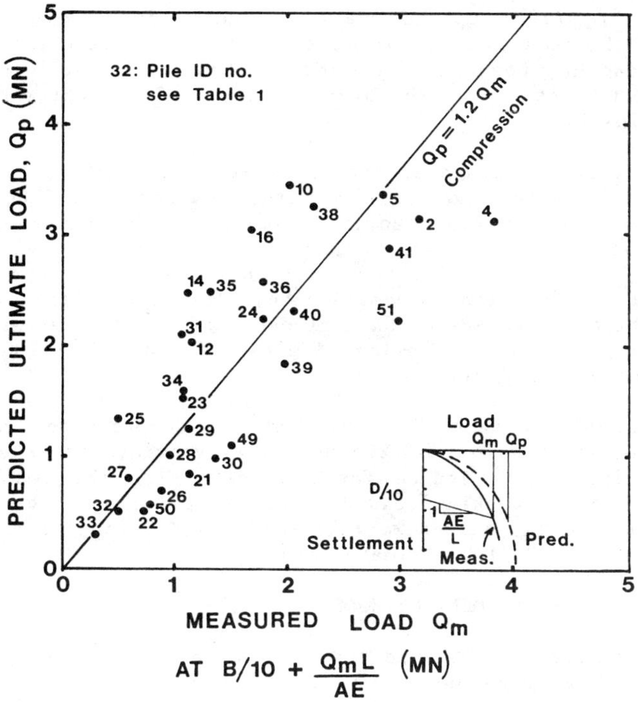

Figure 3. Predicted vs Measured Ultimate Loads for Compression Tests

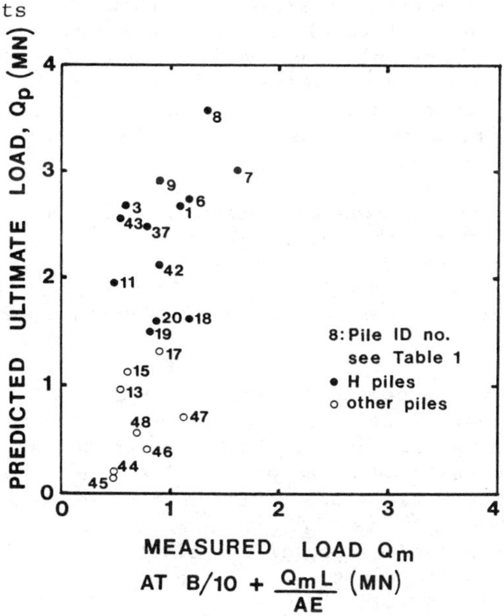

Figure 4. Predicted vs Measured Ultimate Leads for Tension Tests

settlement of $B/10 + Q_mL/AE$ and the true ultimate load for those piles was estimated to be at least 20% larger. Therefore if the true ultimate load had been measured and plotted on Figure 3 instead of the "B/10" ultimate load, the overprediction of the PMT method would have been lower.

Figure 4 shows predicted versus measured ultimate load for tension tests. On the average the method overpredicts the measured loads by a factor of 2.5. However almost all the tension tests in the data base were performed on H piles and there is evidence [20, 22] showing that for H piles the friction in tension is approximately one half of the friction in compression; this may explain in part the overprediction of the PMT method. Indeed, if the H piles are ignored on Figure 4, the few other piles (3 pipe piles and 5 bored piles) do not show the overprediction trend.

Inspection of Figure 3 shows that for compression loading the use of a factor of safety of 2.5 may be appropriate to determine the design working load. Indeed for this data base this would ensure that at the design working load none of the piles would be loaded to failure and that the design working load would be on the average one half the measured "D/10" ultimate load.

VERTICAL LOAD: SETTLEMENT METHODS

The q-w curve is the load transfer curve at the tip of the pile, where q is the average point pressure developed for a pile tip movement w. An f-w curve is a load transfer curve along the shaft of the pile where f is the pile soil friction developed for a pile shaft movement w. Three methods exist to obtain the f-w and q-w curves from pressuremeter test results. All three methods model the f-w and q-w curves by elastic-plastic models (Figure 5) where the ultimate values f_L and q_L are obtained from one of the methods for ultimate loads described in the previous sections. The three methods make different recommendations for the slope of the elastic part of the f-w and q-w curves.

Method 1 was proposed by Gambin [28] in 1979. This method is based on elasticity with some simplifications. The recommended slope for the q-w curve is 2E/R where E is the pressuremeter modulus and R is the pile radius. The recommended slope for the f-w curve is E/CR where C is a factor dependent upon the length to diameter ratio of the pile.

Method 2 was proposed by Frank and Zhao [29] in 1982. This method is semi-empirical in that it uses the results of elastic theory empirically corrected to better match the experimental data of full scale load tests on bored piles. The elastic part of the curves is bilinear. The first slope of the q-w curve is 5.5 E/R; the first slope of the f-w curve is $\alpha E/R$ where α is 2.5 times the pile radius in meters. The second slopes of the elastic part of the curves are 5 times softer than the first one and the change in slope occurs at one half of the ultimate value.

Method 3 was proposed by Baguelin, Frank and Jezequel [30] in 1982. This method recommends a slope for the f-w curve only. It is

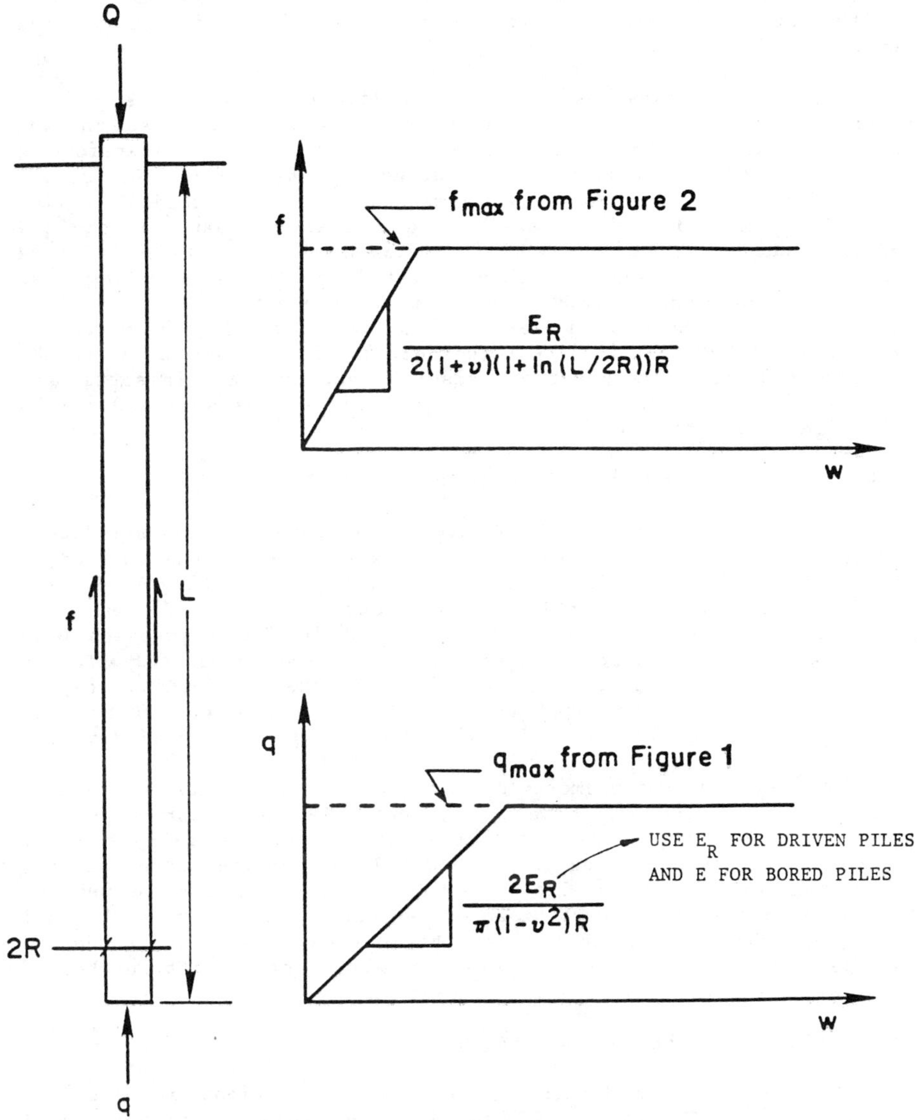

Figure 5. Load Transfer Curves from Pressuremeter Data.

based on the radial integration of the shear curve obtained from the selfboring pressuremeter curve and was verified by finite element work. The recommended slope of the f-w curve is $E_{SB}/2R(1+\nu)(1+Ln(L/2R))$ where E_{SB} is the initial tangent modulus from a selfboring pressuremeter test, ν is Poisson's ratio, L is the embedded pile length.

Method 4 makes use of the selfboring modulus E_{SB}; E_{SB} correlates favorably with the modulus ER obtained from an unload-reload cycle in a preboring pressuremeter test [3]. Therefore a pseudo method 4 consists of replacing E_{SB} by E_R in the recommended slope for the f-w curve in method 4. This method is called method 5. Method 5 also takes from elasticity a slope for the q-w curve equal to $2E_R/(1-\nu^2)R$ for driven piles and $2E/(1-\nu^2)R$ for drilled shafts where E is the first load pressuremeter modulus. This slope corresponds to the equation for the settlement of a rigid plate at the surface of an elastic half space and therefore makes the reasonable assumption [31] that the decrease in settlement due to the embedment of the pile point is offset by the increase in settlement at the pile point due to the shear stresses in the soil induced by the pile shaft friction (Figure 5).

VERTICAL LOAD: PREDICTED VS. MEASURED SETTLEMENT

Method 5 above was used to develop the complete load settlement curve for the pile. Then the factor of safety of 2.5 was applied to the predicted ultimate load. This gave the design working load. The predicted and measured settlements were obtained at the design working load; they are plotted on Figure 6. Only the compression pile load testd are considered because of the problems associated with the tension tests on H piles. As can be seen the predictions compare favorably with the measured values. The settlement magnitude of these single piles is very small since it averages 2.5 mm (0.1 in.).

No specific pressuremeter method exists to calculate the settlement of pile groups; however, potentially two approaches are possible. The first one would consist of using a PILGP1 type of approach [32] by combining the method 5 load transfer curves and the pressuremeter reload modulus as elastic modulus. The second one would be to use the concept of an equivalent footing at a depth equal to 2/3 of the pile length [33] and treat it as a shallow foundation settlement problem [10, 34]. These two methods have not been checked against case histories.

HORIZONTAL LOAD: METHODS AVAILABLE

One of the most direct and obvious applications of pressuremeter test results is the design of piles subjected to horizontal loads. Indeed some analogy exists between the cylindrical expansion of the PMT probe and the horizontal movement of a pile segment. This analogy is not a direct analogy and steps are necessary to go from the pressuremeter expansion curve to the horizontal soil resistance curve for the pile. It is in these steps that the various existing methods differ.

When a pile is loaded laterally there are several components to

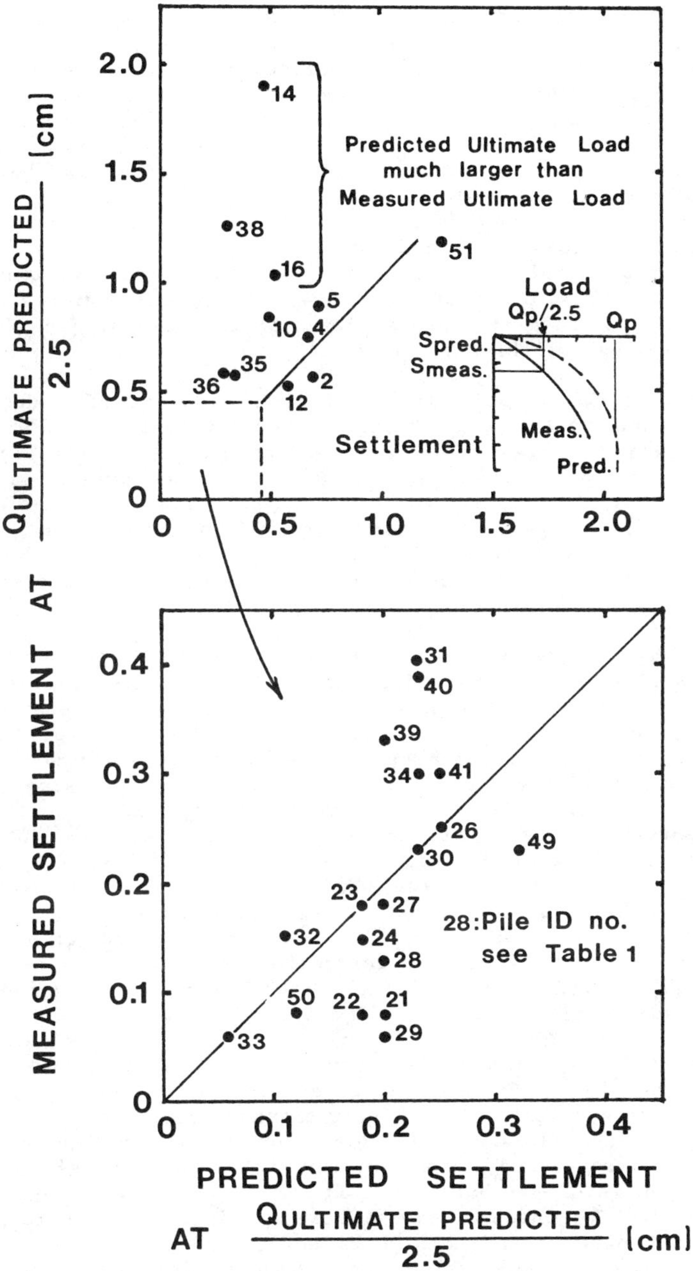

Figure.6 Predicted vs Measured Settlement at 1/2.5 the Predicted Ultimate Load

the soil resistance (Figure 7): The front resistance due to normal stresses, σ_r, the shaft friction resistance due to shear stresses, $\tau_{r\theta}$ and τ_{rz}. the base friction resistance due to shear stresses $\tau_{z\theta}$ and τ_{zr}. and the base moment resistance due to normal stresses, σ_z. Except for very short stubby piles (D/B < 3) the major components of soil resistance are due to σ_r and $\tau_{r\theta}$. At working loads the contribution due to the $\tau_{r\theta}$ effect may be as much as 50% of the total resistance [35]. At any depth, z, the resultant of the above soil resistances is the P-y curve where P is the resultant soil resistance in force per unit length of pile, and y is the horizontal displacement of the pile. The horizontal load versus horizontal deflection of the pile at the ground surface is then obtained by finite difference solution of the governing differential equation [36] using the P-y curves.

Close to the ground surface there is a zone of reduced soil resistance due to the lack of vertical confinement. This phenomenon affects both the pressuremeter expansion and the laterally loaded pile down to a critical depth, D_{pm}, for the pressuremeter and D_{pi} for the pile. Below the critical depth the deformation process is no longer influenced directly by the ground surface. Within the critical depth the soil resistance decreases from the "deep" resistance at the critical depth to a minimum resistance at the ground surface.

Method 1 [37,38] was developed originally by Menard, Bourdon and Gambin in 1969 and then expanded upon by Gambin in 1979. This method uses the results of preboring pressuremeter tests. It considers that the P-y curve is bilinear elastic-plastic. The slope of the first linear portion of the curve is obtained from Menard's equation for the settlement of a strip footing [39] and the second slope is half the first slope. The soil ultimate resistance, p_{ult}, is given by the pressuremeter limit pressure. The critical depth, D_{pi}, varies from 3B to 8B depending on the soil type with B being the pile diameter. A reduction factor which varies from 1 at D_{pi} to 0.5 at the surface is applied to p_{ult} within the critical depth, D_{pi}. The author's experience is that this method gives deflections larger than the measured deflections often by a factor of two. A hand calculation version of this method is presented by Baguelin, Jezequel and Shields [3] for small displacements.

Method 2 [40,41] was developed by GAI Consultants in 1980 for the rigid drilled shafts used as power line tower foundations. This method uses the results of preboring pressuremeter tests. It has the advantage of including all the components shown on Figure 7. The σ_r and $\tau_{r\theta}$ resistances are combined into one lateral resistance model which is a parabola cut off at p_{ult} obtained by Hansen's theory [42]. The three other resistance models are elastic-plastic. The initial part of all models is correlated to the pressuremeter modulus. The ultimate values are obtained from the cohesion and friction angle of the soil. The critical depth effect is incorporated through Hansen's theory. GAI's experience with this method is that when the measured deflection is 2.5 cm (1 in.) the predicted deflection is between 1.2 cm (0.5 in.) and 3.7 cm (1.5 in.).

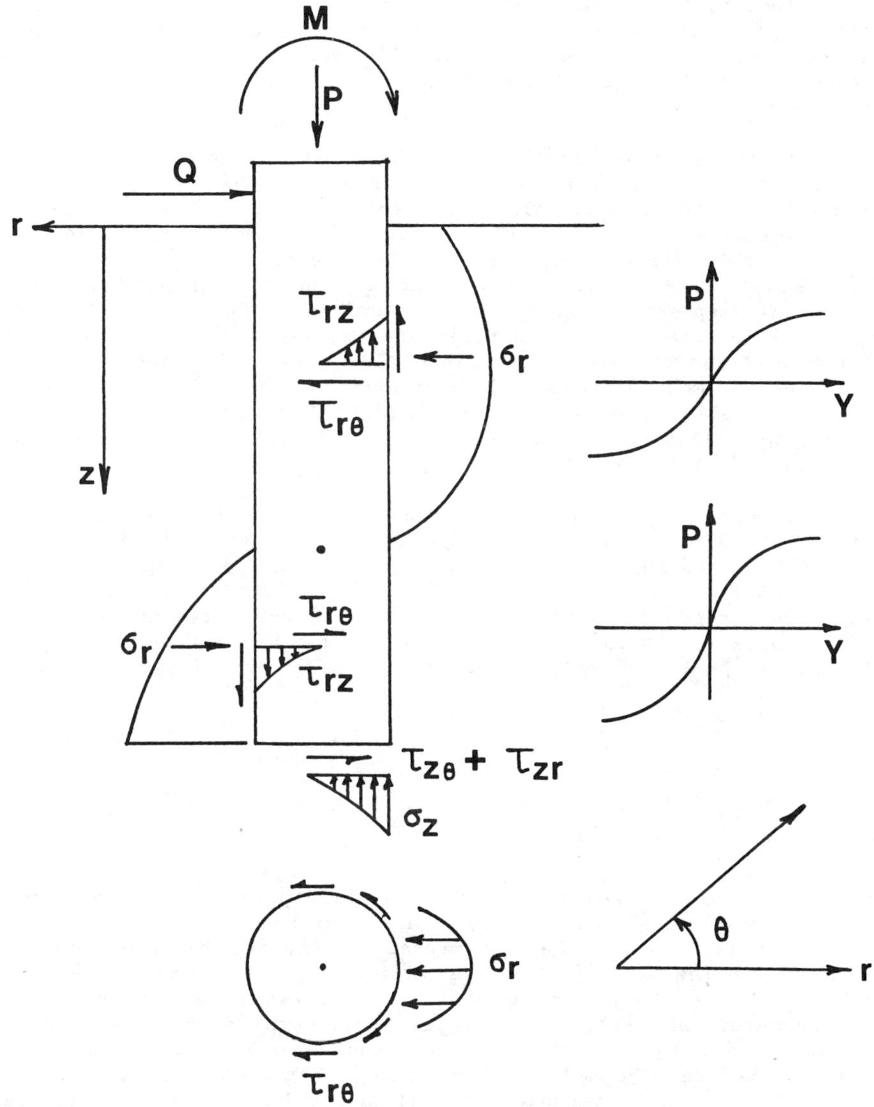

Figure 7 - Components of Soil Resistance (after Reference 40)

Method 3 [43] was developed by Dunand in 1981. This method is based on the results of preboring pressuremeter tests. An elastic-plastic model is used for the frontal reaction. The slope of the elastic portion is obtained from the pressuremeter modulus and elasticity theory, while the ultimate value is considered to be the limit pressure from the pressuremeter. A friction model is also proposed and the critical depth approach is the same as in method 1. This method was developed for the pile foundations of electric power lines in France.

Method 4 [44] was developed by the OYO Corporation; it uses the results of preboring pressuremeter tests. It is based on the closed form solution to the governing differential equation for the problem of an infinitely long laterally loaded pile embedded in a uniform soil with a constant modulus of subgrade reaction k. The solution which gives the relationship between the horizontal load and the horizontal displacement at the ground surface is then modified with empirical factors to match experimental data including nonlinearity of the soil and scale effects. The value of k is obtained from the pressuremeter modulus value. This method has the advantage of being a hand calculation method. A small data base of model piles and steel pipe piles is presented and shows that the predicted and measured data coincide perfectly!

Method 5 [3,8,9] was first proposed by Baguelin, Jezequel and Shields in 1978; it was presented with more details in 1982 [8] and was incorporated in a yet more complete form into a design manual by the French Petroleum Institute in 1983 [9]. It uses the results of selfboring pressuremeter tests (SBPMT). The P-y curve at a depth z for the pile is obtained point by point from the pressuremeter expansion curve ($p^* - 1/2\ V/V_o$) at the same depth z as follows:

$$P = \eta p^* B \tag{9}$$

$$y = \frac{1}{2}\frac{\Delta V}{V_o} R \tag{10}$$

where P is the soil resistance on the pile expressed as a force per unit length of pile, η is the lateral resistance factor varying from 0.33 to 3 [9], p^* is the net pressure $p-p_o$ in the pressuremeter curve, p is the pressure on the wall of the soil cavity in the pressuremeter test, p_o is the horizontal pressure at the start of the pressuremeter test, y is the pile horizontal displacement, R is the pile radius, V_o is the initial pressuremeter probe volume, and ΔV is the volume injected into the probe. The factor η incorporates the critical depth phenomenons and acknowledges the difference between the pile P-y curve and the pressuremeter expansion curve. Comparisons between SBPMT expansion curves and pile P-y curves are shown by Baguelin, Jezequel and Shields [3].

Method 6 [6,45,46] was developed by Briaud, Smith and Meyer in 1981. It uses the results of preboring pressuremeter tests and considers that the P-y curve is made of a front resistance Q-y curve and a friction resistance F-y curve [35]. The Q-y curve is obtained point by point from the pressuremeter curve (first load curve for bored piles and reload curve for driven piles) as follows:

$$Q = SQ \times p^* \times B \quad (11)$$

$$y = \frac{\Delta R}{R_o} \times R \quad (12)$$

where Q is the frontal soil resistance on the pile (force/unit length of pile), SQ is a shape factor from elasticity, p* is the net pressure as in method 5, B or 2R is the pile diameter, y is the pile horizontal displacement, and ΔR is the increase in soil cavity radius from the initial cavity radius R_o.

The F-y curve is obtained point by point from the reload pressuremeter curve by the subtangent method [3] as follows:

$$F = SF \times B \times X\ (1+X)\ \frac{\Delta p^*}{\Delta X} \quad (13)$$

$$y = \frac{\Delta R}{R_o} \times R \quad (14)$$

where F is the friction soil resistance on the pile (force/unit length of pile), SF is a shape factor from elasticity, and X is $\Delta V/V_o$ with V_o being the volume of the pressuremeter probe at the beginning of the reload curve and ΔV the volume injected in the probe from that point on. The critical depths for the pile and for the pressuremeter are calculated and within the critical depths; reduction factor profiles are recommended. The precision of this method is evaluated on the basis of 27 pile load tests in the following section. A microcomputer program to reduce pressuremeter data and produce the P-y curves according to this method has been written [6]. A simplified version of this method based on a subgrade modulus approach has been proposed and can be used in a first crude approximation:

$$k = \frac{P}{y} = E_o + E_r \text{ for bored piles} \quad (15)$$

$$k = \frac{P}{y} = 2E_r \qquad \text{for driven piles} \quad (16)$$

These k values together with the closed form solution for an infinitely long laterally loaded pile in a homogeneous soil (3) lead to predictions which correspond to measurements at a deflection approximately equal to 2% of the pile diameter. Method 6 is described in detail in reference 6.

Method 7 [47,48] was developed by Fugro B.V. in 1983. It uses the results of push-in or full-displacement pressuremeter test results. Pressuremeter test derived parameters, namely shear modulus G, undrained shear strength S_U, friction angles ϕ', and at rest pressure coefficient K_o, are input in a pseudo three dimensional finite element code. The soil is modelled as a linear-elastic, perfectly-plastic material. This method has the advantage of modelling the soil as a continuum and of being potentially extendable to pile groups. Comparisons between predicted and measured response of 4 piles show encouraging results [48].

Method 8 stemmed from work done by Hughes in the late seventies [49] and was later refined at the University of British Columbia [50,51]. It uses the results of driven pressuremeter tests for driven piles, and preboring or selfboring pressuremeter tests for bored piles. The P-y curve at a depth z for the pile is obtained point by point from the pressuremeter expansion curve at the same depth z by using equations 9 and 10. The value of in equation 9 is taken as 2 for cohesive soils and 1.5 in cohesionless soils. The critical depth phenomenon is handled as in method 6 for the pressuremeter. For the pile, a unique value of critical depth equal to 4 pile diameters is proposed and reduction factors are recommended within that critical depth. Comparisons between predicted and measured pile response for 4 case histories show good performance of the method.

Method 9 [19] is in use at Woodward-Clyde Consultants for piles in clay. It is a version of the Matlock and Reese method [52,53] which makes use of preboring pressuremeter test results. The Matlock and Reese method requires the undrained shear strength, S_u, and the strain at 50% of the peak stress ε_{50} in order to construct the various P-y curves. Davidson and Bodine [19] use S_u obtained from preboring pressuremeter tests by the Gibson/Anderson method [54], $\varepsilon_{50} = S_u/E_R$ for deflection less than 1.2 cm (0.5 in.) where E_R is the reload pressuremeter modulus, and $\varepsilon_{50} = S_u/E$ for deflections of 2.5 cm (1 in.) or larger where E is the first load pressuremeter modulus.

HORIZONTAL LOAD: PREDICTIONS VS. LOAD TEST RESULTS

Predicted responses by Methods 1, 2, 4 and 6 have been compared to the results of one lateral load test [55]. In this section, a 27 pile load test data base which has been collected over the last five years is used in conjunction with method 6. The piles listed in Table 2 show a wide range of pile types and pile insertion techniques including bored piles, pipe piles, H piles, concrete piles. The pile lengths vary from 3 m (10 ft) to 25 m (82 ft) and the diameters from 0.27 m (0.88 ft) to 1.37 m (4.5 ft). The soils include sand, silt and clay as well as layered profiles.

For each pile, preboring pressuremeter tests were performed next to the pile, predictions according to Method 6 were performed leading to a predicted horizontal load-horizontal deflection curve at the pile top. On the same graph the measured curve obtained during the load test was plotted (Figure 8). The site names followed by an asterisk in Table 2 represent cases where predictions were sent to the sponsor prior to receiving the load test results. In order to compare predicted with measured behavior, the loads obtained at a value of the horizontal deflection equal to 10% (ultimate) and 2% (small movements) of the pile diameter were compared (Fig. 8 and 9). Figs. 8 and 9 show that the method predicts the measured behavior

TABLE 2. - Horizontal pile load test data base

I.D. .	Site	Pile Type	Pile Embedded Length (m)	Pile Diam. (m)	Soil	Reference
1	Sabine	Pipe	12.2	0.32	Clay	56, 46
2	Mustang Island	Pipe	21.0	0.61	Sand	57, 46
3	Lake Austin	Pipe	12.2	0.32	Clay	58, 46
4	Houston	Bored	13.0	0.76	Clay	59, 46
5	Texas A&M 1977	Bored	6.1	0.91	Clay	60, 46
6	Texas A&M 1978	Bored	4.6	0.76	Clay	61, 46
7	Texas A&M 1979	Bored	4.6	0.76	Clay	62, 46
8	Univ. of Houston*	H	11.8	0.27	Clay	63, 72
9	Univ. of Houston*	Pipe	11.4	1.22	Clay	63, 72
0	L&D 26 (1983)	HP14x73	20.4	0.36	Sand	64
1	L&D 26 (1983)	HP14x73	20.4	0.36	Sand	64
2	L&D 26 (1978)	H	15.2	0.36	Sand	65, 46
3	L&D 26 (1978)	Pipe	15.2	0.36	Sand	65, 46
4	Virginia	Bored	3.5	1.37	Clay	40, 46
5	Carolina	Bored	4.5	1.37	Sand	40, 46
6	Iowa	Bored	4.6	1.37	Clay	40, 46
7	LADWP Delta*	Bored	3.0	0.61	Clay	16
8	LADWP Caliente*	Bored	3.0	0.74	Sand	16
9	LADWP Alamo*	Bored	3.0	0.65	Clay	16
0	Baytown	Bored	11.9	0.61	Clay	66, 46
1	Lackland*	Bored	10.5	0.46	Clay	18
2	La Baule 1	Rd. Conc.	6.0	0.61	Sand/Clay	68, 67
3	La Baule 2	Rd. Conc.	6.0	0.61	Sand/Clay	68, 67
4	Plancoet	Caisson	4.4	0.95	Silt	71,67,47
5	Plancoet	H	6.1	0.36	Silt	71,67,47
6	Cubzac	Pipe	24.7	0.91	Clay	69, 67
7	Provins	Pipe	23.0	0.93	Silt/Peat	70, 67

.d test results unknown at time of predictions

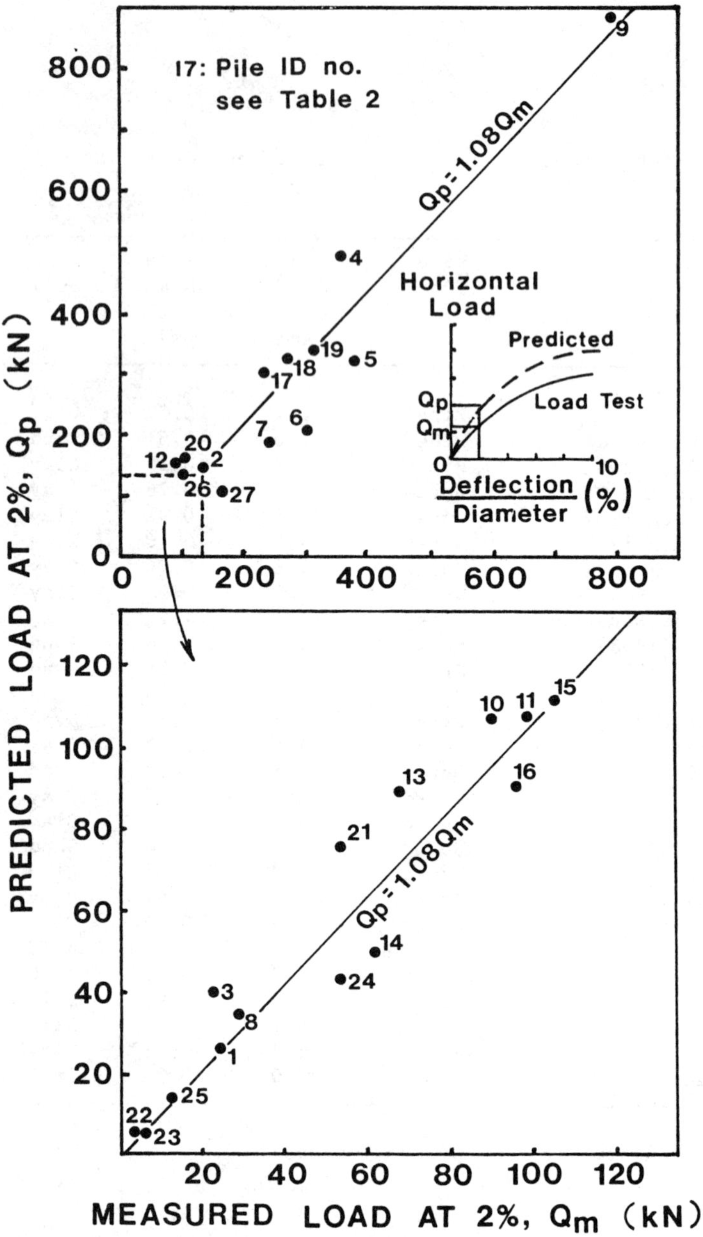

Figure 8. Predicted vs Measured Loads for Method 6 at a Groundline Deflection Equal to 2% of the Pile Diameter

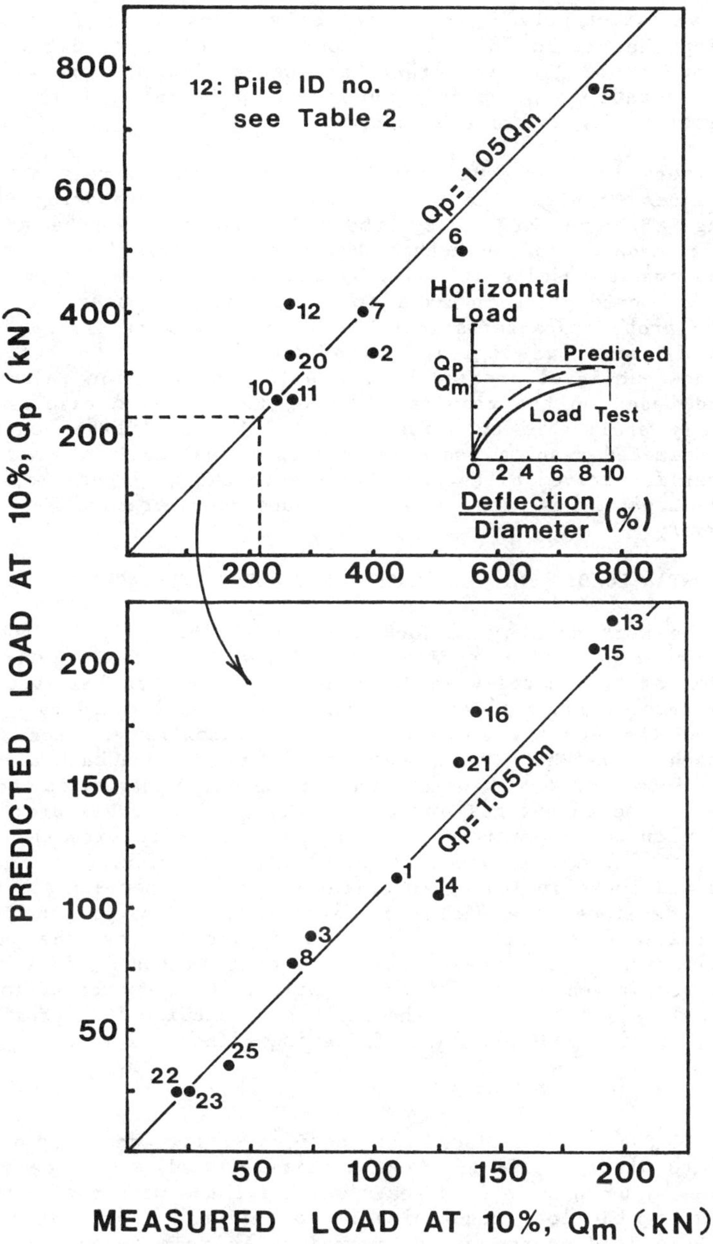

Figure 9. Predicted vs Measured Horizontal Loads for Method 5 at a Groundline Deflection Equal to 10% of the Pile Diameter

very satisfactorily. Note that only piles 1 to 7 were used to develop the method initially. Comparison of Figs. 3 and 4 on one hand and 8 and 9 on the other hand shows that the pressuremeter is more accurate at predicting horizontal pile behavior than it is at predicting vertical pile behavior.

There is a growing trend in using various types of insertion techniques to place the pressuremeter probe: preboring PMT, self-boring PMT, cone PMT, shelby tube PMT. The following recommendations can be made. For a method based on the results of selfboring pressuremeter results it would be logical to use an oversize cutting shoe for bored piles in order to create the horizontal stress relief on the probe characteristic of bored pile construction. For driven piles it would be logical to inflate the probe to the estimated displacement level created by the pile and wait for relaxation. A method based on the results of cone pressuremeter results would be directly applicable to driven piles. A method based on preboring pressuremeter results would be directly applicable to bored piles; however for driven piles preinflation to the horizontal displacement level created by the pile and subsequent relaxation would be necessary.

HORIZONTAL LOAD: PMT VS. CURRENT P-Y CURVE APPROACH

The work done by Matlock and Reese [56,57,58,59] from 1955 to 1970 has represented a major step forward in the solution to the problem of horizontally loaded piles. The author believes that the pressuremeter can represent an additional step forward by allowing to improve the quality and range of applicability of the P-y curve approach. The PMT P-y curve approach has the drawback of requiring to perform PMT tests which are not routine tests in the United States. The advantages over the current P-y curves are: 1. the PMT P-y curve is obtained point by point in situ with the pressuremeter, 2. the pressuremeter test can be performed in almost all soils and rocks including in calcareous soils, permafrost, peat, and rock; therefore, the method can always be used, 3. the method of installation of the pile can be duplicated by the method of installation of the pressuremeter such as prebored holes for bored piles and driven probes for driven piles, 4. the type of loading can be easily simulated during the PMT test including long term sustained loads, cyclic loads and rate of loading effects.

CYCLIC AND SUSTAINED LOAD

All the aforementioned methods for vertical and horizontal loads apply to static loading. More specifically they were developed and checked by using pile load tests which for the most part consisted in increasing the load monotonically to failure. The time to failure for those load tests varied approximately from 1 hour to 20 hours. Offshore, cyclic and rapid loading are particularly important while onshore, sustained long term loads are of concern. The application of the pressuremeter to predict the behavior of piles under cyclic, rapid and sustained loads holds promise. Indeed, the PMT test is well suited to performing unload-reload cycles or to holding the pressure constant while recording the increase in strain versus time. Research in this area of special PMT testing is still in its infancy, however.

Soil behavior under sustained long-term loading can potentially be handled by performing creep PMT tests where the pressure is brought to the proper stress level and maintained while recording injected volume versus time. Soil behavior under rapid loading can potentially be handled by performing tests more rapidly than the standard test. These types of test allow to obtain time dependent properties; the following models are proposed for vertical loads [73,74,75]:

$$E_{(t)} = E_o \left(\frac{t}{t_o}\right)^{-n} \tag{17}$$

$$p_L^*(t) = p_{Lo}^* \left(\frac{t}{t_o}\right)^{-n} \tag{18}$$

where E_o and $E(t)$ are the PMT moduli measured by increasing the pressure from the horizontal pressure at rest, p_o, to the end of the elastic phase in a time t_o and t, respectively, where p^*_{Lo} and $p^*_L(t)$ are the net limit pressure measured by increasing the pressure from p_o to the limit pressure in a time t_o and t, respectively. For horizontal sustained loads the following is proposed:

$$y_{(t)} = y_o \left(\frac{t}{t_o}\right)^{n} \tag{19}$$

where, for a given P of the P-y curve, $y(t)$ and (y_o) are the horizontal displacements reached in a time t and t_o, respectively. Measured values of n in clays vary from 0.02 to 0.10 and average 0.06 [73,74]; these n values were obtained from shear strength tests and pile load tests for values of t varying from a few seconds to 100 hours. Similar values of n were obtained from pressuremeter tests [76].

Soil behavior under cyclic loads can potentially be handled by performing cyclic PMT tests where the pressure is cycled between appropriate bounds. Such tests have been performed up to 100 cycles [27,77,72] and 10000 cycles [78]. Application of cyclic pressuremeter data to piles subjected to cyclic vertical loads has been discussed [79]. Also, a cyclic load threshold, above which displacements accelerate more rapidly, has been measured in cyclic rod shear tests [80] and cyclic pressuremeter tests [77]. The applicability of cyclic PMT test results to piles subjected to cyclic horizontal loads has been shown [63,72]. The following model is proposed:

$$y_N = y_1 (N)^a \tag{20}$$

where, for a given P of the P-y curve, y_1 and y_N are the horizontal displacements at the top of the first and of the Nth cycle, respectively. Note that the point on the curve where the load or pressure is first decreased carries the cycle number 1. The parameter a varies from 0.02 to 0.25 averaging 0.09 in clays and from 0.01 to 0.15 averaging 0.05 in sands; these a values were obtained from the study of 34 full scale piles subjected to cyclic horizontal loads [72]. This model is appropriate for slow cyclic loading up to 1000 cycles but predicts y_N values larger than the measured values for larger numbers of cycles [72,78].

CONCLUSIONS

Rules of design for vertically and horizontally loaded piles on the basis of pressuremeter test results are reviewed. Predicted and measured behaviors are compared by using a 51 vertical pile load test data base and a 27 horizontal pile load test data base.

For ultimate vertical loads, methods were developed in 1963, 1978, 1981. Recently the Laboratoire des Ponts et Chaussees has proposed a method intended to regroup and supercede the previous methods. Comparison of predicted and measured ultimate loads, Q_p and Q_m, respectively, using that latest method on 51 piles shows that the ratio Q_p/Q_m averages 1.20 for compression tests (Figure 3). A factor of safety of 2.5 is recommended for use with this method. For tension tests on H piles the predicted ultimate load is much higher than the measured ultimate load with an average Q_p/Q_m ratio of 2.5 (Figure 4).

For settlement under vertical loading, 3 methods were developed from 1979 to 1982. A modified version of the 1982 method by Baguelin, Frank and Jezequel is evaluated using the 29 compression tests for the data base. It shows that at predicted working loads (factor of safety = 2.5) the predicted settlement compares well with the measured settlement (Figure 6) except for 14% of the piles where the ultimate load is significantly overpredicted.

For horizontal loads, 9 methods were developed, most of them in the last decade. All methods appear to perform very well when comparing predicted and measured behavior; this is possibly due to the fact that there is sufficient analogy between the pressuremeter expansion curve and the P-y curve to start with. One method is used together with a 27 pile load test data base in order to evaluate its precision. Predicted and measured horizontal loads are compared at two displacement levels namely 2% and 10% of the pile diameter (Figs. 8 and 9). In both cases predicted and measured loads compare very well.

Comparison of the performance and prediction of the methods for vertical and horizontal loads indicates that the pressuremeter performs better in the prediction of horizontally loaded piles than in the prediction of vertically loaded piles. For vertically loaded piles it performs better for compression loading than for tension loading.

In the future, there is a need for: 1. standardizing the pressuremeter equipment, test and insertion procedure, 2. comparing pressuremeter data from preboring PMT, selfboring PMT, cone PMT, and shelby tube PMT, 3. continuing to document the performance of the design methods, 4. regrouping the 9 methods for horizontally loaded piles, 5. expanding the method(s) for horizontally loaded piles to cyclic loading, rapid loading, long term sustained loading and unusual soils, 6. more generally obtaining cyclic and creep soil properties from cyclic and creep PMT tests as well as studying unusual soils by PMT tests, and 7. studying the applicability of the PMT to the problem of horizontally loaded cantilever retaining walls.

REFERENCES

1. Menard. L., "Calcul de la Force Portante des Fondations sur la Base des Resultats des Essais Pressiometriques," Sols-Soils, No. 5, June 1963.
2. Menard. L., "Calcul de la Force Portante des Fondations sur la Base des Resultats des Essais Pressiometriques, Seconde partie: Resultats Experimentaux et Conclusions," Sols-Soils, No. 6, September 1963.
3. Bagueline, F., Jezequel, J.F. and Shields, D.H., The Pressuremeter and Foundation Engineering, Transtech Publications, Rockport, Mass., 1978.
4. Bustamante, M. and Gianeselli, L., "Prevision de la Capacite Portante des Pieux Isoles, sous Charge Verticale: Regles Pressiometriques et Penetrometriques," Bull. Liaison Labo. P. et Ch. 113, May-June 1981.
5. Bustamante, M. and Gianeselli L., "Prevision de la Capacite Portante des pieux par la Methode Pressiometrique (Version 82)," Compte rendu de recherche F.A.E.R. 1.05.02.2, Laboratoire Central des Ponts et Chaussees, en communication privee.
6. Briaud, J.L., Tucker, L.M., and Olsen, R.S., "Pressuremeter and Foundation Design," Volume 1, Short Course Notes, Civil Engineering, Texas A&M University, 1985.
7. Briaud, J.L., Anderson, J.S., Tucker, L.M., and Coyle, H.M., "Measured and Predicted Response for 100 Piles," Research Report 4981-1, Civil Engineering, Texas A&M University, July 1985.
8. Baguelin, F., "Rules of Foundation Design Using Selfboring Pressuremeter Test Results," Symposium on the Pressuremeter and Its Marine Applications," Editions Technip, Paris, 1982.
9. Institut Francais du Petrole, Recommended Practice for Application of Pressuremeter Test Results to Designing of Offshore Foundation," Ref. 31763, Paris, December 1983.
10. Briaud, J.L., Tand, K.E., and Funegard, E.G., "Pressuremeter and Shallow Foundations on Stiff Clay," Proceedings of the Session on Spread Foundations, Transportation Research Board Meeting, Washington, January 1986.
11. Briaud, J.L., Noubani, A., and Tucker, L.M., "Correlations Between Pressuremeter Data and Other Soil Parameters," Research Report, Civil Engineering Department, Texas A&M University, 1985.
12. Meyerhof, G.G., "Bearing Capacity and Settlement of Pile Foundations," ASCE Geotechnical Engineering Journal, Vol. 102, GT3, 1976.
13. Coyle, H.M. and Castello, R., "New Design Correlations for Piles in Sand," ASCE Geotechnical Engineering Journal, Vol. 107, GT7, 1981.
14. Kulhawy, F.H., "Limiting Tip and Side Resistance: Fact or Fallacy?" Analysis and Design of Pile Foundations, ASCE National Convention, San Francisco, October 1984.
15. Vesic, A.S., "Expansion of Cavities in Infinite Soil Mass,"

ASCE, Journal of Soil Mechanics and Foundation Engineering, Vol. 98, SM3, 1972.

16. Briaud, J.L., Pacal, A.J., and Shively, A.W., "Powerline Foundation Design Using the Pressuremeter," International Conference on Case Histories in Geotechnical Engineering, St. Louis, 1984.

17. Briaud Engineers, "Pressuremeter Test and Pile Capacity at Lock and Dam 2, Red River, Alexandria, Louisiana, Report to Corps of Engineers, Vicksburg District, June 1984.

18. Briaud Engineers, "Lackland Air Force Base Vertical and Lateral Drilled Shaft Load Test: Pressuremeter Testing," Report to the U.S. Army Waterways Experiment Station, September 1982.

19. Davidson, R.R. and Bodine, D.G., "Analysis and Verification of Louisiana Pile Foundation Design Based on Pressuremeter Results," The Pressuremeter and Its Marine Applications: Second International Symposium, ASTM, STP 950, 1986.

20. Briaud, J.L., Tucker, L.M., Lytton, R.L., and Coyle, H.M., "Behavior of Piles in Cohesionless Soils," Research Reort 4777, Civil Engineering, Texas A&M University, September 1983.

21. Hansbo, S., "Prediction of Load vs. settlement Curves for Friction Piles," Contribution to Symposium on the Pressuremeter and Its Marine Applications, Editions Technip, Paris, 1982.

22. Briaud, J.L., Huff, L.G., Tucker, L.M., and Coyle, H.M., "Evaluation of In Situ Test Methods for Vertically Loaded H Piles at Lock and Dam 26 Replacement Site," Research Report 4690-1, Civil Engineering, Texas A&M University, July 1984.

23. Briaud, J.L., Lawson, W. and Tucker, L.M., "Wave Equation for Piles in Gravel," Research Report 4690-2, Civil Engineering, Texas A&M University, September 1984.

24. Briaud, J.L., Perdomo, D., Tucker, L.M., and Coyle, L.M., "Statistical and Risk Analysis for the Measured and Predicted Response of 100 Piles," Research Report 4981-2, Civil Engineering Department, Texas A&M University, October 1985.

25. O'Neill, M.W., Hawkins, R.A., and Mahar, L.T., "Field Study of Pile Group Action," Research Report No. FHWA/RD-81, Civil Engineering, University of Houston, October 1980.

26. O'Neill, M.W. and Sheikh, S.A., "Geotechnical Behavior of Underreams in Pleistocene Clay," Drilled Piers and Caissons II, ASCE Convention, May 1985.

27. Briaud, J.L. and Riner, K.B., "A Study of Cyclic Pressuremeter Testing for Offshore Applications," Research Report 4839, Civil Engineering, Texas A&M University, 1984.

28. Gambin, M., "Calcul du Tassement d'une Fondation Profonde en Fonction des Resultats Pressiometriques," Sols-Soils, No. 30/31, 1979.

29. Frank, R. and Zhao, S.R., "Estimation par les Parametres Pressiometriques de l'Enfoncement Sous Charge Axiale de Pieux Fores dans les Sols Fins," Bull. Liaison Labo. P. et Ch. 119, May-June 1982.

30. Baguelin, F., Frank, R. and Jezequel, J.F., "Parameters for Friction Piles in Marine Soils," Second Int. Conf. in Numerical Methods for Offshore Piling, Austin, April 1982.

31. Randolph, M.F. and Wroth, C.P., "Analysis of Deformation of Vertically Loaded Piles," ASCE Geotechnical Engineering Journal, Vol. 104, GT12, December 1978.

32. O'Neill, M.W. Hawkins, R.A. and Mahar, L.J., "Field Study of

Pile Group Action: Appendix A, User's Guide, Program PILGP1," Report No. FHWA/RD-81/003, Federal Highway Administration, March 1981.

33. Terzaghi, K. and Peck, R.B., Soil Mechanics in Engineering Practice, John Wiley & Sons, New York, p. 550, 1967.
34. Briaud, J.L. and Tucker, L.M., "Pressuremeter and Shallow Foundations on Sand," Settlement of Shallow Foundations on Sand, ASCE Convention, Seattle, April 1986.
35. Briaud, J.L., Smith, T.D. and Meyer, B.J., "Pressuremeter Gives Elementary Model for Laterally Loaded Piles," Int. Symp. on In Situ Testing of Soil and Rock, Paris, May 1983.
36. Reese, L.C., "Laterally Loaded Piles: Program Documentation," ASCE Journal of the Geotechnical Engineering Division, Vol. 103, GT4, April 1977.
37. Menard, L., Bourdon, G. and Gambin, M., "Methode Generale de Calcul d'un Rideau ou Pieu Sollicite Horizontalement en Fonction des Resultats Pressiometriques," Sols-Soils, No. 20/23, 1969.
38. Gambin, M., "Calculations of Foundations Subjected to Horizontal Forces Using Pressuremeter Data," Sols-Soils, No. 30/31, 1979.
39. Menard, L., "The Menard Pressuremeter: Interpretation and Application of the Pressuremeter Test Results to Foundation Design," Sols-Soils, No. 26, 1975.
40. GAI Consultants, Inc., "Laterally Loaded Drilled Pier Research: Volume 1: Design Methodology, Volume 2: Research Documentation," Research Report EPRI EL-2197, January 1982.
41. DiGioa, A., Davidson, H.L. and Donovan, T.D., "Laterally Loaded Drilled Piers: A Design Model," Drilled Piers and Caissons, ASCE Convention, St. Louis, October 1981.
42. Hansen, B.J., "The Ultimate Resistance of Rigid Piles Against Transversal Forces," The Danish Geotechnical Institute Bulletin No. 12, 1961.
43. Dunand, M., "Etude Experimentale du Comportement des Fondations Soumises au Renversement," These de Docteur Ingenieur, Institut de Mecanique de Grenoble, France, 1981.
44. Suyama, K., Imai, T. and Ohya, S., "Development of LLT Pressuremeter and Its Application in Prediction of Pile Behavior under Horizontal Load," Symposium on the Pressuremeter and Its Marine Applications, Editions Technip, Paris, 1982.
45. Briaud, J.L., Smith, T.D., and Tucker, L.M., "A Pressuremeter Method for Laterally Loaded Piles," Int. Conf. of Soil Mechanics and Foundation Engineering, Vol. 3, p. 1353, San Francisco, 1985.
46. Smith, T.D., "Pressuremeter Design Method for Single Piles Subjected to Static Lateral Load," Ph.D. dissertation, Civil Engineering Department, Texas A&M University, College Station, Texas, August 1983.
47. Kay, S., Kolk, H. J., and Hooydonk, W.R., "Site Specific Design of Laterally Loaded Piles," ASCE Speicalty Conference on Geotechnical Practice in Offshore Engineering, Austin, 1983.
48. Kay, S., Grifiths, V. and Kolk, H.J., "Application of Pressuremeter Testing to Assess Lateral Pile Response in Clays," The Pressuremeter and Its Marine Applications: Second International Symposium, ASTM STP 950, 1986.
49. Hughes, J.M.O., Goldsmith, P.R. and Fendall, H.D.W., "Predicted and Measured Behavior of Laterally Loaded Piles for the Westgate

Freeway Bridge," Victoria Geomechanics Society, Australia, August 1979.

50. Robertson, P.K., Hughes, J.M.O., Campanella, R.G. and Sy, A., Design of Laterally Loaded Displacement Piles Using a Driven Pressuremeter," Laterally Loaded Deep Foundations, ASTM STP 835, Kansas City, June 1982.

51. Robertson, P.K., Hughes, J.M.O., Campanella, R.G., Brown, P., and KcKeown, S., "Design of Laterally Loaded Piles Using the Pressuremeter," The Pressuremeter and Its Marine Applications: Second International Symposium, ASTM STP 950, 1986.

52. Matlock, H., "Correlations for Design of Laterally Loaded Piles in Soft Clay," Offshore Technology Conference, Paper 1204, Houston, 1970.

53. Reese, L.C., Cox, W.R., and Koop, F., "Analysis of Laterally Loaded Piles in Sand," Offshore Technology Conference, Paper 2080, Houston, 1974.

54. Gibson, R.E. and Anderson, W.F., "In Situ Measurement of Soil Properties with the Pressuremeter," Civil Engineering and Public Works Review, London, 56 No. 658, May 1961.

55. Briaud, J.L., Smith, T.D., and Meyer, B.J., "Laterally Loaded Piles and the Pressuremeter: Comparison of Existing Methods," Laterally Loaded Deep Foundations ASTM STP 835, 1983.

56. Matlock, H. and Tucker, R.L., "Lateral Load Tests of an Instrumented Pile at Sabine, Texas," A Report to Shell Development Company, Houston, September 1961.

57. Reese, L.C., Cox, W.R. and Koop, F.D., "Analysis of Laterally Loaded Pile Tests in Sand at Mustang Island," A Report to Shell Development Company, Houston, November 1967.

58. Matlock, H., Ripperger, E.A. and Fitzgibbon, D.P., "Static and Cyclic Lateral Loading of an Instrumented Pile," A Report to Shell Oil Company, Houston, July 1956.

59. Welch, R.C. and Reese, L.C., "Lateral Load Behavior of Drilled Shafts," Research Report 89-10, Center for Highway Research, The University of Texas at Austin, May 1972.

60. Kasch, V.R., Coyle, H.M., Bartoskewitz, R.E. and Sarver, W.G., "Lateral Load Test of a Drilled Shaft in Clay," Report TTI 211-1, Texas Transportation Institute, Texas A&M University, November 1977.

61. Holloway, G.L., Coyle, H.M., Bartoskewitz, R.E. and Sarver, W.G., "Field Test and Preliminary Design Method for Laterally Loaded Drilled Shafts in Clay," Report TTI 211-2, Texas Transportation Institute, Texas A&M University, September 1978.

62. Bierschwale, M.W., Coyle, H.M., and Bartoskewitz, R.E., "Field Tests and New Design Procedure for Laterally Loaded Drilled Shafts in Clay," Report TTI 211-3F, Texas Transportation Institute, Texas A&M University, January 1981.

63. O'Neill, M.W. and Dunnavant, T.W., "A Study of the Effect of Scale, Velocity, and Cyclic Degradability on Laterally Loaded Singel Piles in Overconsolidated Clay," Research Report No. UHCE 84-7, Civil Engineering, University of Houston, October 1984.

64. Briaud, J.L., Braswell, T.E. and Tucker L.M., "Case History of Two Laterally Loaded Piles at Lock & Dam 26 Replacement Site," Research Report 4690-2, Civil Engineering, Texas A&M University, June 1984.

65. Woodward-Clyde Consultants, "Foundation Investigation and Test

Program, Lock & Dam No. 26, Mississippi, Alton, Illinois," Report to the St. Louis Distric, U.S. Army Corps of Engineers, 1979.

66. Spencer, J. Buchanan and Associates, Inc., "Lateral Pile Load Tests at Exxon Baytown Olefins Plant," A report to Exxon Chemical Americas, Bryan, Texas, January 1982.

67. Laboratoire des Ponts et Chaussees, "Internal Documents on Lateral Pile Load Tests," Paris, 1982.

68. Le Mauff, H. and Peignaud, M., "Sollicitations Horizontales de Dispositifs Legers d'Amarrage," Bull. Liaison Labo. P. et Ch. No. 64, March-April, 1973.

69. Ambrosino, R., Bru, J.P. and Ledoux, J.L., "Comportement d'un Pieu Sollicite Horizontalement," Bull. Liaison Labo. P. et Ch. No. 67, September/October 1973.

70. Bigot, G., Bourges, F. and Frank, R., "Etude Experimentale d'un Pieu Soumis aux Poussees Laterales des Sols," Revue Francaise de Geotechnique, No. 18, February 1982.

71. Baguelin, F. and Jezequel, J.F., "Etude Experimentale du Comportement de Pieux Sollicites Horizontalement," Bull. Liaison Labo. P. et Ch. No. 62, November/December 1972.

72. Briaud, J.L., Makarim, C.A., Little, R. and Tucker, L.M., "Development of a Pressuremeter Method for Predicting the Behavior of Single Piles in Clay Subjected to Cyclic Lateral Loads," Research Report 5112, Civil Engineering, Texas A&M University, November 1985.

73. Briaud, J.L. and Garland, E., "Influence of Loading Rate on Axially Loded Piles in Clay," Research Report 4695, Civil Engineering, Texas A&M University, March 1984.

74. Briaud, J.L. and Garland, E., "Loading Rate Method for Piles Response in Clay," ASCE Geotechnical Engineering Journal, Vol. III, No. 3, March 1985.

75. Riggins, M., "The Viscoelastic Characteristics of Marine Sediments in Large Scale Simple Shear," Ph.D. dissertation, Civil Engineering, Texas A&M University, 1981.

76. Briaud, J.L., Terry, T.A., "Rate Effect for Vertical and Horizontal Pile Response," 3rd International Conference on Numerical Methods in Offshore Piling, Nantes, May 1986.

77. Briaud, J.L., Riner, K.B. and Ohya, S., "Cyclic Pressuremeter Tests for Cyclic Lateral Loads," Offshore Technology Conference, Paper 4678, Houston, 1984.

78. Jezequel, J.F. and LeMehaute, A., "Cyclic Tests with the Selfboring Pressuremeter," Symposium on the Pressuremeter and Its Marine Applications, Editions Technip, Paris, 1982.

79. Puech, A., Brucy, F. and Ma, E., "Tension Pile Design Using Selfboring Prerssuremeter Test Results," Symposium on the Pressuremeter and Its Marine Applications, Editions Technip, Paris, 1982.

80. Briaud, J.L. and Felio, G.Y., "Influence of Cyclic Loading on Axially Loaded Piles in Clay," Research Report 4980, Civil Engineering, Texas A&M University, April 1985.

R. G. Lukas

SETTLEMENT PREDICTION USING THE PRESSUREMETER

REFERENCE: Lukas, R.G., "Settlement Prediction Using the Pressuremeter," The Pressuremeter and its Marine Applications: Second International Symposium, ASTM STP-950, J.L. Briaud and J.M.E. Audibert, Eds. American Society for Testing and Materials, 1986.

ABSTRACT: At seven projects sites where a complete geotechnical investigation had previously been completed, settlement readings of the foundations were obtained after the structures were completed. The measured settlements were compared with the settlements predicted using the results of pressuremeter tests and the pressuremeter method of predicting settlements. A good correlation was observed between settlements predicted for isolated load points such as for spread footings or drilled piers and a poorer correlation was obtained for large loaded areas consisting of a raft foundation. At all seven project sites, the subsoils consisted of deposits where settlement is not time dependent after loading.

KEY WORDS: Pressuremeter, settlement prediction, spread footing foundations, drilled pier foundations, raft foundations.

R.G. Lukas is a Senior Principal Engineer, STS Consultants, Ltd. Northbrook, Illinois 60062

The prediction of settlement that will occur within soil deposits under new loading is an important consideration in geotechnical design. This is especially true for settlement sensitive structures such as buildings where the settlement prediction often dictates the type of foundation that is used. If the settlement of a shallow foundation is predicted to be larger than the structure can tolerate, a mat or deep foundation will probably be required. Furthermore, the bearing level for the deep foundation is often governed by the compressibility of the various potential bearing strata.

In cohesive deposits that are fully saturated where the new loading exceeds the preconsolidation pressure, reasonably accurate settlement predictions can be obtained by using consolidation tests and the Terzaghi method of settlement prediction or modifications suggested by others [1, 2, 3, 4]. In these deposits, settlement occurs over a period of time as the excess pore pressure dissipates. However, there are many deposits where settlement under loading is not primarily time dependent and it is in these deposits that the pressuremeter has been found to be a useful tool for evaluating the compressibility of the soil. These deposits include;

1.) Granular soil deposits of gravel, sand, and coarse silt, either natural or fill.

2.) Manmade fills such as building rubble, broken concrete, or crushed rock.

3.) Highly overconsolidated cohesive deposits such as glacial tills where the new loading will not exceed the preconsolidation pressure.

4.) Weathered and decomposed rock including some residual soils.

5.) Waste fill deposits such as decomposed land fill refuse or fly ash.

6.) Partially saturated fill deposits including cohesive fills.

In these deposits, the case histories presented in this paper or previously published papers [5, 6, 7] indicate that the pressuremeter can be used to determine the compressibility from which reasonably accurate settlement predictions can be made.

PRESSUREMETER METHOD OF SETTLEMENT PREDICTION

Four soil parameters are generally measured with each pressuremeter test; the limit pressure, the creep pressure, the pressuremeter modulus, and the reload modulus. The limit pressure is generally used for calculating the ultimate bearing capacity of soil and will not be discussed further. However, the other three parameters are used either directly or indirectly in settlement prediction.

The general formula for predicting settlement of isolated foundations is a follows;

$$w = \frac{1.33}{3E} p (\lambda_1 R)^{\alpha} + \frac{\alpha}{4.5E} p \lambda_2 R \qquad (1)$$

where:

w = settlement
E = pressuremeter modulus
E^+ = reload modulus
p = pressure transmitted to the soil deposit
R = half width of loaded area
λ_1, λ_2 = empirical factors to account for shape of loaded area
α = rheological soil parameter

This formula was developed by the Menard Organization [8]. In their publication they list typical values for α and λ_1, λ_2. The pressuremeter modulus as determined from actual in-situ tests is used directly in this formula. The ratio of pressuremeter modulus to reload modulus has been suggested [9] as an indicator of the rheological factor, α, for the soil so it is useful to run a cycled pressuremeter test to determine the reload modulus. The creep pressure has been found by Lukas [5] as an indicator of preconsolidation pressure in cohesive soils. Since the pressuremeter modulus is accurate only in the range of loading from the in-situ soil pressure up to the creep pressure, this implies that the settlement prediction using equation 1 is accurate for soils up to the preconsolidation pressure which is the settlement due to reloading in this deposit.

The Menard Organization [8] has also suggested an abbreviated formula for a large loaded area such as a raft foundation of a structure or an embankment founded on a relatively thin layer of compressible soil.

$$w = \sum_{0}^{z} \frac{\alpha(z)\beta(F)p(z)}{E_{(z)}} dz \qquad (2)$$

where:

w = settlement
$p_{(z)}$ = the increase in vertical pressure at depth z
$E_{(z)}$ = the pressuremeter modulus at depth z
$\alpha_{(z)}$ = the rheologic factor corresponding to the soil layer at depth z
$\beta_{(F)}$ = coefficient depending on the safety factor for the foundation.

Modification of these formulas are available for special conditions of closely spaced load points, basement excavations, and compressible layers of limited thickness.

SETTLEMENT PREDICTION CASE HISTORIES

The accuracy of the settlement predictions using pressuremeter data is illustrated by comparing the predicted settlements with the measured settlements for seven (7) projects. On each of these projects, the structural loading information was known, settlement readings were obtained during and shortly after construction to compare against the predictions, and the soils loaded by the new construction fall into one of the category of the six soil deposits previously described where settlement is not time dependent. The projects were also selected to illustrate the settlement predictions for a variety of foundation types including spread footings, drilled pier foundations, and raft foundations.

Spread Footing Case Histories

Three projects where spread footings were used to support a building, are illustrated on the figures labeled Case History 1 to Case History 3. The soil profile at each site is shown on these figures along with structural loading information, the settlement prediction, and the measured settlement.

Case History 1 consists of a twelve story building supported on spread footing foundations which bear upon an overconsolidated glacial till deposit. The upper portion of the till is desiccated and of slightly higher strength and lower compressibility than the underlying till. The footing foundations are spaced sufficiently far apart so that the footings can be considered to act independent of each other. The settlement prediction based upon equation 1 was calculated to be 22 mm (0.85 inches). The measured settlement was found to be 12.5 mm (0.5 inches) which is in reasonable agreement with the predictions. For comparison purposes, 76 mm diameter shelby-tube specimens were taken and consolidation tests performed upon various samples obtained. Based upon consolidation theory, the settlement prediction was 33 mm (1.5 inches) which was three times the measured value.

Case History 2 consists of a two-story building supported on spread footings which are situated within a former landfill that was densified by means of dynamic compaction. The deposit can be described as decomposed refuse, cinders, rubble, and miscellaneous materials. The dynamic compaction densified the upper portion of this deposit more than the lower portions so the pressuremeter modulus decreases with increasing depth. Based upon pressuremeter tests and equation 1, the settlement was predicted to be 12.5 mm (0.5 inches) at the end of construction. An additional 12.5 mm (0.5 inches) of settlement was arbitrarily predicted to account for the potential long-term decomposition of the landfill. Settlement readings were obtained at 20 to 30 footing locations at the end of construction and the settlement was found to range from 6 mm to 18 mm (0.25 to 0.7 inches) which agrees very well with the pressuremeter prediction.

Case History 3 describes a seven-story apartment building supported on spread footings. The subsoil deposits consist of a densified rubble fill overlying a natural silty sand or sandy silt soil deposit. The rubble fill was densified by dynamic compaction and the pressuremeter modulus in the rubble was therefore higher than the sands or silts underlying the rubble. Using equation 1, the predicted settlement was 9 mm (0.36 inches). Since the footings for this building were situated relatively close together there was some overlap of stress below footing level so a modification of equation 1 to account for the overlap effect was also used to predict the settlement. According to this procedure, the settlement is calculated in two parts. First, there is the settlement due to an average pressure increase due to the total building weight divided by the area of the building. Second, the settlement is calculated from the pressure beneath each footing where the induced pressure is the footing pressure less the average building pressure. Using this modification of equation 1, the settlement was predicted to be 18 mm (0.7 inches). Since the footings were closely spaced, another method of predicting settlement would be to assume that the footings constituted a raft foundation. Using equation 2 resulted in a settlement prediction of 41 mm (1.61 inches). The measured settlement was found to be 16 mm (0.63 inches) which agrees very close with equation 1 taking into account stress overlap from closely spaced footings.

Deep Foundation Case Histories

Case History 4 consists of a seventy five story apartment building supported on drilled piers extending to a very dense silty clay or sandy clay deposit commonly referred to as hardpan. The column spacing is sufficiently far apart that each drilled pier acts as an individual foundation. Equation 1 was used to predict the settlement for both the heavily loaded interior piers as well as the lighter loaded exterior piers. The predicted settlements were adjusted for elastic compression of the drilled pier shaft and the stress relief due to the deep basement excavation. The elastic compression of the shaft was computed assuming the full column load is applied over the length of the shaft. The settlement reduction from the basement excavation area was computed both by stress relief causing rebound in the bearing stratum and by a reduction in bearing pressure at the base of the drilled pier by an amount equal to the stress relief. Both methods resulted in a similar and minor settlement reduction. For both the interior and exterior piers, the predicted settlement is very close to the mean between the range of measured settlements.

Case History 5 consists of a forty seven story office building supported on drilled piers which bear upon a very dense silt, sand or gravel formation with extremely high pressuremeter modulus values. The drilled piers are located sufficiently far apart so that there is no stress overlap from adjacent foundations. The settlement predicted from equation 1 calculated to be 23 mm (0.92 inches). Adding 9.7 mm (0.38 inches) for elastic compression in the concrete shaft and reducing the settlement by 2 mm (0.08 inches) for the basement excavation results in a net settlement prediction of 26 mm (1.02 inches). This agrees closely with the measured settlement which ranges from 21 to 25 mm (0.82 to 1.0 inches).

Raft Foundations Case Histories

Case History 6 consists of a twenty six story office building supported on a mat foundation embedded within a silty sand with the pressuremeter modulus increasing with increasing depth. The average contact pressure at the base of the mat from structural loading was 244 kP_a (2.55 tsf) but the stress relief from the basement soil amounted to 144 kP_a (1.50 tsf) which resulted in a net pressure applied to the soil of 96 kP_a (1.05 tsf). The net pressure was used in the settlement prediction although there is some heave followed by rebound due to excavation and reloading that is not accounted for when using net pressure. Using equation 2 which was developed for a raft foundation, the settlement prediction was 26 mm (1.02 inches). The measured settlement was found to be 16.5 mm (0.65 inches) or only about half of the predicted value. The reasons for the poor agreement could be due to a number of factors. It is possible that the full structural loading is not applied since a portion of the loads are live loads. Thus, the effective dead load pressure could be significantly less. Equation 2 is based on elastic compression of a short member but soils do not behave elastically and the use of the pressuremeter modulus divided by α apparently does not convert the pressuremeter modulus into the modulus of elasticity that is required in this formula.

A better correlation between predicted and measured settlement for Case Histories 6 and 7 is obtained when using the second half of equation 1.

This portion of equation 1 computes the spherical compression which is the settlement which occurs directly beneath the loaded area. When the thickness of the compressible deposit is less than the half width of the mat, this value is used for R in the second half of equation 1. In effect the raft is considered as an isolated load point on a compressible deposit of limited thickness where the deviatoric settlement is neglected because of the increased stiffness of the layers underlying the soils within the spherical zone of compression.

Case History 7 consists of a fifteen story office building supported on a mat foundation. The subsoil profile is geologically classified as a glacial till deposit that contains layers of relatively dense sand within and below the till. The pressure at the base of the mat from the structural loading was 134 kP_a (1.40 tsf) and the stress relief from basement excavation was 58 kP_a (0.61 tsf) resulting in a net soil pressure of 75 kP_a (0.79 tsf). Using equation 2, the settlement prediction was 81 mm (3.2 inches) and the measured settlement was only 15.2 mm (0.6 inches) which is only about 20% of the predicted amount. The settlement prediction using the second half of equation 1 was 28 mm (1.1 inches) which was still larger than the measured settlement. For comparison purposes consolidation tests were performed upon 76 mm shelby tubes from various layers within the deposit. The settlement prediction from consolidation theory was 63.5 mm (2.5 inches). For this site, none of the settlement predictions agreed well with the measured settlement.

CONCLUSIONS

For soil deposits where settlement is primarily not time dependent, the pressuremeter settlement predictions agreed reasonably well with observed values of settlement. The best correlation between predicted and observed settlement occurs for isolated load points such as spread footing or drilled pier foundations. A less accurate correlation occurs for large loaded areas which in this study included the raft foundations. Since the pressuremeter testing was undertaken with the same device and same person performing the tests at all sites, the poorer correlation between predicted and observed settlement for the raft foundations is not attributed to the field pressuremeter tests, but rather to inaccuracies associated with the use of equation 2. A better prediction of settlement for a raft foundation was obtained by using the second half of equation 1.

At two of the project sites cohesive soils formed the bearing stratum and consolidation tests were performed. The predicted settlement from the consolidation tests was about double the prediction from the pressuremeter tests and about 3 to 4 times greater then the measured settlement.

REFERENCES

[1] Crawford, C.B., Sutherland, J.G., "The Empress Hotel, Victoria British Columbia, Sixty Five Years of Foundation Settlement," Canadian Geotechnical Journal, Volume 8, November 1, February 1971.

[2] Horn, H.M., Lambe, T.W., "Settlement of Buildings on the MIT Campus," Journal of the Soil Mechanics and Foundations Division, ASCE, Volume 90, Number SM5, September 1964.

[3] Moore, P.J., Spencer, G.K., "Settlement of a Building on Deep Compressible Soil," Journal of the Soil Mechanics and Foundations Division, ASCE, Volume 95, Number SM3, May 1969.

[4] Peck, R.B., Uyanik, M.E., "Observed and Computed Settlements of Structures in Chicago," University of Illinois Experiment Station Bulletin No. 429, March 1955.

[5] Lukas, R.G., Leclerc de Bussy, B., "Pressuremeter and Laboratory Test Correlations for Clays," Journal of the Geotechnical Engineering Division, ASCE, Volume 102, GT9, September 1976

[6] Martin, R.E., "Estimating Foundation Settlements in Residual Soils," Journal of the Geotechnical Engineering Division, ASCE, Volume 103 , GT3, March 1977.

[7] Steinberg, S.B., Lukas, R.G., "Densifying a Landfill for Commercial Development," International Conference on Case Histories in Geotechnical Engineering, Rolla, Missouri, May 1984.

[8] Menard, L., "The Menard Pressuremeter - Interpretation and Application of Pressuremeter Test Results," Sols No. 26, 1975.

[9] Baguelin, F., Jezequel, J.F., Shields, D.H., "The Pressuremeter and Foundation Engineering," Trans Tech Publications, Clausthal Germany, 1979, p. 155

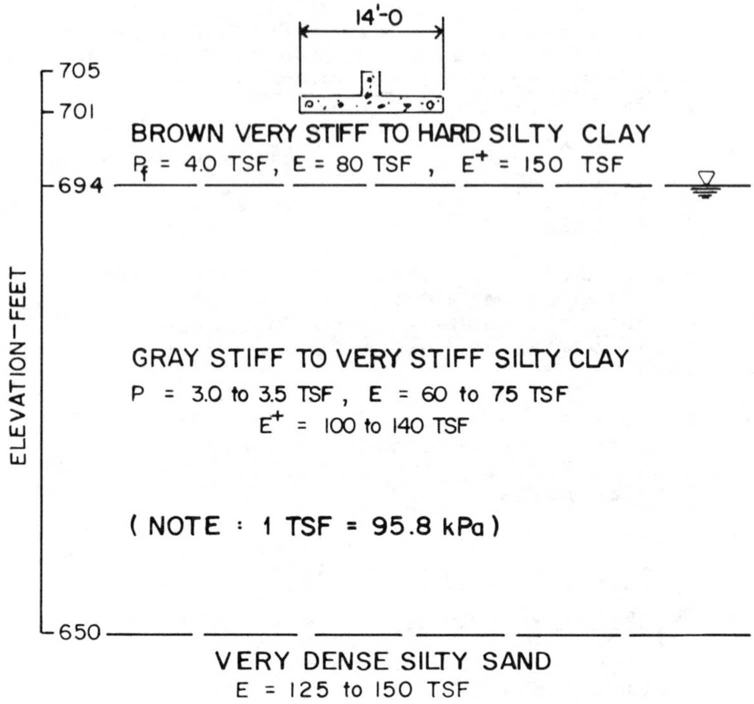

Loading Information:

Spread Footing Foundations - No Basement
Loaded Area 190' by 260'
Column Loads 1000 kips
Soil Pressure 5000 psf
Bay Spacing 28 feet

Settlement Predictions:

From Equation 1,	0.85 inches
Consolidation Tests & Theory,	1.50 inches
Measured Settlement,	0.50 inches

Note: 1 inch = 25.4 mm, 1 ft = .305 m, 1 kip = 4448 N

CASE HISTORY 1 - 12 Story Building on Spread Footings

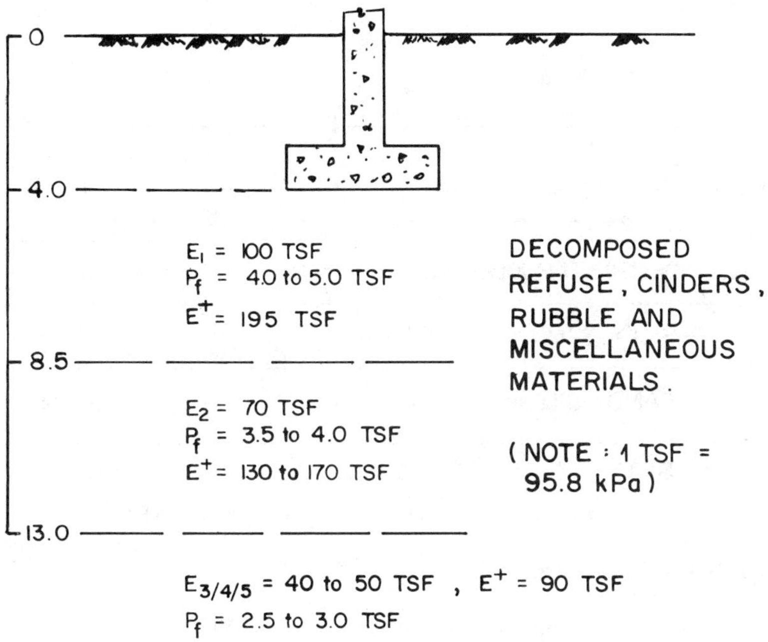

Loading Information:

Spread Footing Foundation - No Basement
Column Loads 250 to 300 kips
Soil Pressure 3000 psf
Footing Size 9 foot square

Settlement:

Calculated by Equation 1,	0.50 inches
Measured at End of Construction,	0.25 to 0.7 inches

Note: 1 inch = 25.4 mm, 1 ft = .305 m, 1 kip = 4448 N

CASE HISTORY 2 - 2 Story Building on Spread Footings

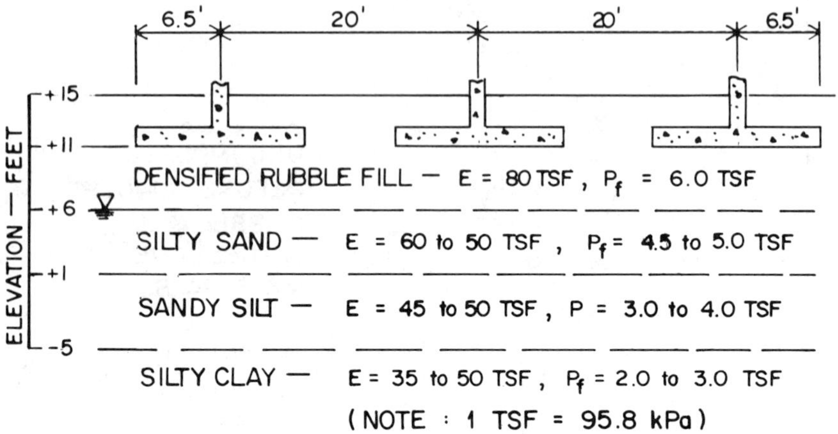

Loading Information:

Spread Footing Foundations - No Basement
Column Loads 500 kips
Soil Pressure 3000 psf
Footing Size 13 foot square

Settlement Predictions:

From Equation 1,	0.36 inches
Equation 1 with Overlap Effects,	0.70 inches
From Equation 2,	1.61 inches
Measured Settlement,	0.63 inches

Note: 1 inch = 25.4 mm, 1 ft = .305 m, 1 kip = 4448 N

CASE HISTORY 3 - 7 Story Apartment Building on Spread Footings

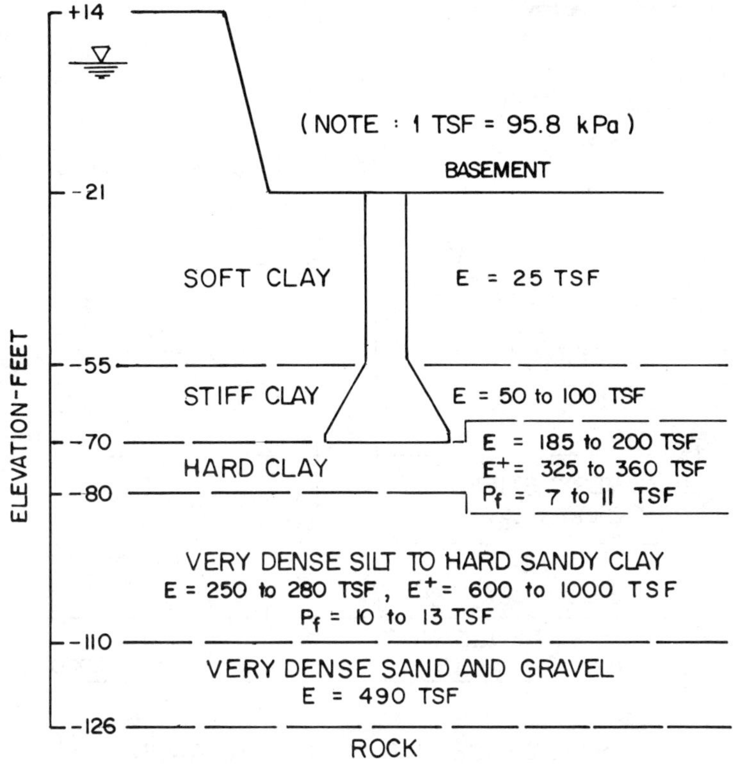

LOADING INFORMATION AND SETTLEMENTS

	Interior Pier	Exterior Pier
Bell Size	27 feet	25 feet
Soil Pressure	13.5 tsf	10 tsf
Predicted Settlement Equation 1	1.9 inches	1.3 inches
Measured Settlement	1.6 - 2.1 inches	1.0 to 1.6 inches

Note: 1 inch = 25.4 mm, 1 ft = .305 m

CASE HISTORY 4 - 75 Story Apartment Building on Drilled Piers

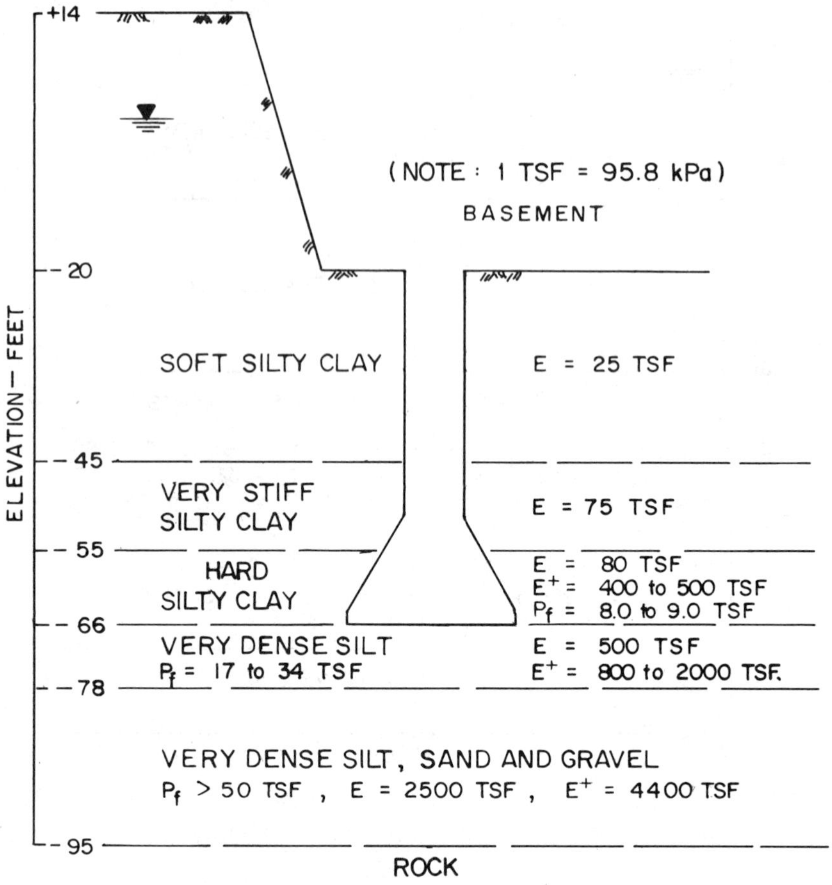

Loading Information:

Column Load,	16,000 kips
Bell Size,	24 feet
Soil Pressure,	20 tsf

Settlement Prediction:

From Equation 1,	0.92 inches
Elastic Compression in Shaft,	0.38 inches
Reduction in Settlement from Excavation,	0.08 inches
Total Anticipated Movement,	1.22 inches
Measured Settlement,	0.82 to 1.0 inches

Note: 1 inch = 25.4 mm, 1 ft = .304 m, 1 kip = 4448 N

CASE HISTORY 5 - 47 Story Office Building on Drilled Piers

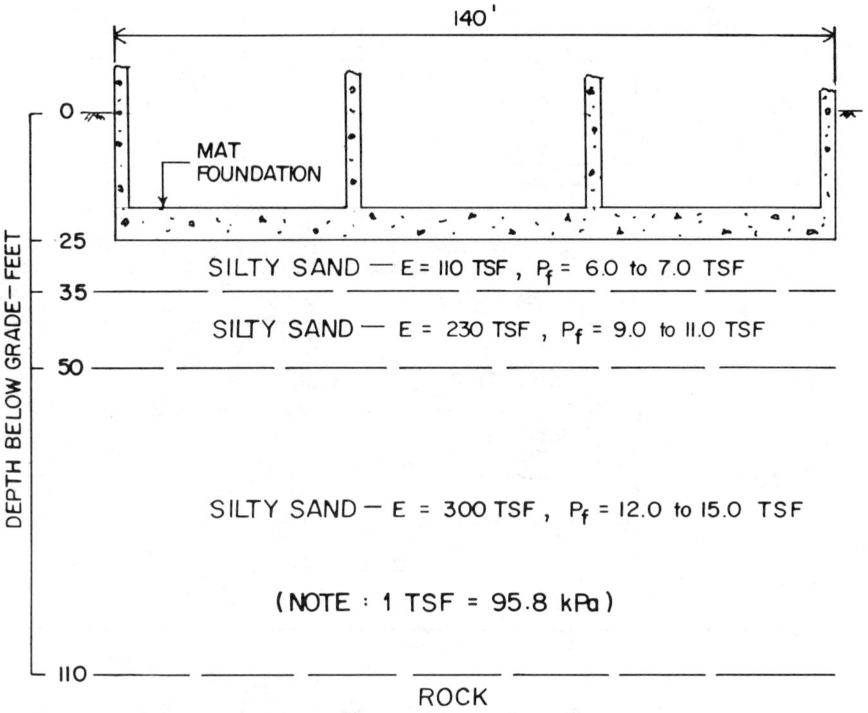

Loading Details:

Size of Loaded Area,	140 ft by 140 ft
Soil Pressure at Base of Mat,	2.55 tsf
Stress Relief from Basement,	1.50 tsf
Net Pressure Applied to Soil,	1.05 tsf

Settlement Predictions:

Using Second Half of Equation 1,	0.55 inches
Using Equation 2,	1.02 inches
Measured Settlement,	0.63 inches

Note: 1 nich = 25.4 mm, 1 ft = .304 m

CASE HISTORY 6 - 26 Story Office Building Supported on a Mat Foundation

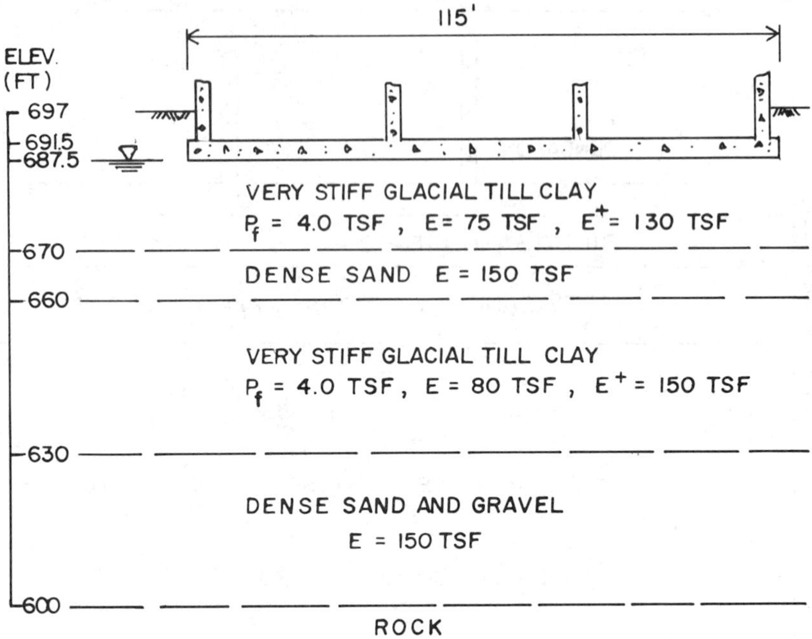

(NOTE : 1 TSF = 95.8 kPa)

Loading Details:

Size of Loaded Area,	115 ft by 220 ft
Soil Pressure at Base of Mat,	1.40 tsf
Stress Relief from Basement,	0.61 tsf
Net Pressure Applied to Soil,	0.79 tsf

Settlement Predictions:

Using Second Half of Equation 1,	1.1 inches
Using Equation 2,	3.2 inches
From Consolidation Tests,	2.5 inches
Measured Settlement,	0.6 inches

Note: 1 inch = 25.4 mm. 1 ft = .305 m

CASE HISTORY 7 - 15 Story Office Building
Supported on a Mat Foundation

Richard R. Davidson [(1)] and Daniel G. Bodine [(2)]

ANALYSIS AND VERIFICATION OF LOUISIANA PILE FOUNDATION DESIGN BASED ON PRESSUREMETER RESULTS

REFERENCE: Davidson, R. R. and Bodine, D. G., "Analysis and Verification of Louisiana Pile Foundation Design Based on Pressuremeter Results", The Pressuremeter and Its Marine Applications: Second International Symposium, ASTM STP 950, J.-L. Briaud and J.M.E. Audibert, Eds., American Society for Testing and Materials, 1986.

ABSTRACT: The Menard pressuremeter was utilized as one element of a comprehensive soil property characterization of medium to stiff, plastic, Pleistocene clays on which the pile foundations for a proposed coal-fired power plant would be supported. Pressuremeter testing was supplemented by drilling, sampling and laboratory testing. Various design methods were utilized to predict the ultimate capacity and load-deformation behavior of the pile foundations under axial and lateral loads. Analytical predictions were verified by a program of full-scale instrumented pile load testing on various types of piles including hel-cor corrugated straight-shaft, monotube, step-taper, precast concrete and pipe piles. Five axial tension and compression load tests and one lateral load test were conducted. The field and laboratory testing program confirmed that the pressuremeter provided accurate properties of the clay and that reliable predictions of axial and lateral pile capacity and deformation can be made using these properties.

KEYWORDS: in situ testing, pressuremeter, pile foundation, clay, pile load test

In southern Louisiana, conventional geotechnical engineering practice for design of foundations for heavily loaded structures dictates that axial pile capacity is determined from empirical correlations with undrained shear strength derived from laboratory unconfined compression tests on undisturbed Shelby tube samples. [1] These empirical correlations have been developed from an abundant

(1) Senior Project Engineer, Woodward-Clyde Consultants, 4582 S. Ulster Street Parkway, Denver, Colorado 80237

(2) Supervisor, Geotechnical Division, Sargent & Lundy Engineers, 55 E. Monroe Street, Chicago, Illinois 60603

data base of local full-scale prototype pile load testing. [2] This conventional method can lead to overly conservative pile design in areas with overconsolidated clays that have fissures and slickensides, because unconfined compressive strengths are governed by these planes of weakness which in situ may have a far less significant impact on mobilized pile-soil adhesion. To deal with this problem some engineers conduct laboratory strength testing on remolded samples which have destroyed the in situ soil structure, or the laboratory strength testing results are disregarded and piles are designed based on past experience with pile load tests made in similar soil conditions. In other cases, more sophisticated laboratory triaxial testing is conducted under the in situ stress regime to provide a more reliable soil property measurement, so long as the soil with its discontinuity remains intact during the sampling and test preparation process. Where sufficient uncertainty remains, full-scale pile load tests are made to verify foundation design.

A new approach to this problem was implemented for the proposed Wilton Steam Electric Station situated on the east bank of the Mississippi River, approximately 35 miles south of Baton Rouge, Louisiana, near Donaldsonville. Wilton Station is to include a power block of two 700 MW coal fired units. For this important facility, a comprehensive program of field investigation, laboratory testing and prototype foundation testing was implemented to significantly reduce foundation design uncertainty and effect a resulting economy in foundation construction. This program included conventional drilling and sampling, Menard pressuremeter testing, a suite of routine and sophisticated laboratory testing, and a pile load testing program. Each element provided a complementary contribution to the geotechnical data base from which the foundation was designed.

Menard pressuremeter testing was conducted to provide an in situ measurement of the engineering properties of the soil mass including its heterogenieties and discontinuties. Careful hole preparation minimized disturbance to the borehole walls and maintained the in situ soil structure intact. The pressuremeter was used to measure in situ stress, stress history, undrained elastic modulus and undrained shear strength, from which design parameters and predictions of foundation performance were made.

FIELD INVESTIGATION

Twenty-two soil borings were drilled in the main power block area to depths of 30(100) to 60(200) m(ft). Each boring was advanced with rotary wash drilling using bentonite stabilizing fluid. Undisturbed Shelby tube samples were taken on 1.5-m (5-ft) increments in the boring. Each sample was classified in the field, tested with a pocket penetrometer, and sealed for laboratory testing.

Sixty-one pressuremeter tests were performed in five borings using the Menard GAm pressuremeter system manufactured by Roctest. This conventional pressuremeter system has a tri-cellular probe with its measuring cell filled with water and contiguous guard cells filled with gas. The system sustains constant stress increments during which the probe volume change measurements are made on a sight tube in the instrument.

Pressuremeter tests were performed every 3 m (10 ft) of depth in holes prepared by pushing a 76-mm (3-in.) diameter Shelby tube 85 cm (33.5 in.), removing the sampler, and then immediately inserting the probe into the hole created during sampling. This procedure not only minimized disturbance to the borehole walls but provided undisturbed samples for laboratory testing. Undrained, stress-controlled pressuremeter tests were performed utilizing a standard one minute stress increment duration with an unload-reload cycle midway through the test. Further information on the pressuremeter apparatus and test procedure is presented by Baguelin et al. [3]

LABORATORY TESTING

A comprehensive suite of geotechical laboratory tests was made on the large number of undisturbed samples obtained during the field investigation to provide a reliable characterization of soil properties for design and to evaluate the spatial variability of the data across the site. Index testing included natural water content tests, Atterberg limits determinations, seive and hydrometer analyses, and measurements of unit weight and specific gravity. One-dimensional consolidation tests were conducted in the oedometer to measure stress history and compressibility characteristics. Undrained shear strength was measured by unconfined (UC), unconsolidated-undrained (UU) triaxial and consolidated-undrained (CIU) triaxial compression tests. Pore pressure measurements were made during CIU triaxial testing at confining stresses selected to be representative of in situ lateral stresses measured by the pressuremeter, to determine effective stress shear strength parameters.

SOIL PROPERTY CHARACTERIZATION

Soil properties have been characterized from the classification of materials encountered during drilling, from laboratory testing results, and from pressuremeter testing results.

Geology and Stratigraphy. The Wilton site is situated in the Mississippi River delta floodplain just north of the Mississippi River some 8 miles east-southeast of Donaldsonville, Louisiana. Located entirely on the natural levee of the river, the site is underlain by a variable thickness, 1.5(5) to 17(55) m(ft) of Holocene (Recent) age clays, silty clays, and clayey silts resting on 46(150) to 58(190) m(ft) of stiffer Pleistocene-age strata of similar composition. Below the clays, at a depth of about 52(170) to 61(200) m(ft), is a thick water-bearing Pleistocene sand and gravel which likely extends to a depth of 122(400) to 152(500) m(ft). [4]

The top stratum consists of an upper natural levee deposit of silty clay (CL) which thins rapidly from a maximum thickness at the river of 6 m (20 ft) to approximately 2 m (5 ft) inland. This deposit was created as the result of overbank flooding of the modern river. The remainder of the top stratum is higher plasticity clays (CH) deposited in marsh, swamp or brackish environments.

The contact between the top stratum of recent sediments and Pleistocene age clays is distinguished by oxidation color change, reduction in water content and overconsolidation, all induced by dessication of the exposed portions of this layer during Pleistocene times. The Pleistocene deposit is generally clay (CH) with interbeds of silts, fine sands and organics. Below a depth of 52(170) to 61(200) m(ft) the lower Pleistocene deposit consists of alternating thick sheets of very dense sand/gravel and hard clay. [4] A generalized soil profile, including the results of index property testing, is shown in Fig. 1.

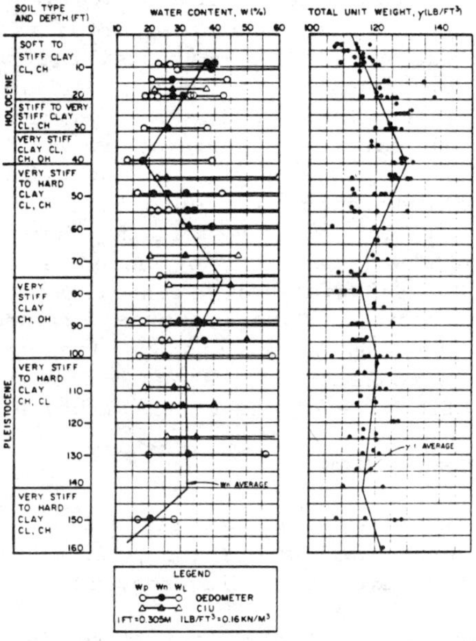

FIG. 1--Generalized Subsurface Profile and Index Properties.

In Situ Stresses. The in situ horizontal effective stress was determined from the configuration of the initial portion of the pressuremeter test curve using the procedure presented by Davidson. [5] The in situ horizontal stress P_o is interpreted as the pressure which initiates the linear elastic portion of the pressuremeter curve shown in Fig. 2. If the level of hole disturbance has been sufficiently mininized, as was the case at Wilton, reliable horizontal stress measurements can be made. However, if the borehole walls are disturbed or the hole size is too large, horizontal stress measurements can be grossly overestimated. This procedure is the least reliable and most difficult to implement in soils which are easily disturbed during hole preparation.

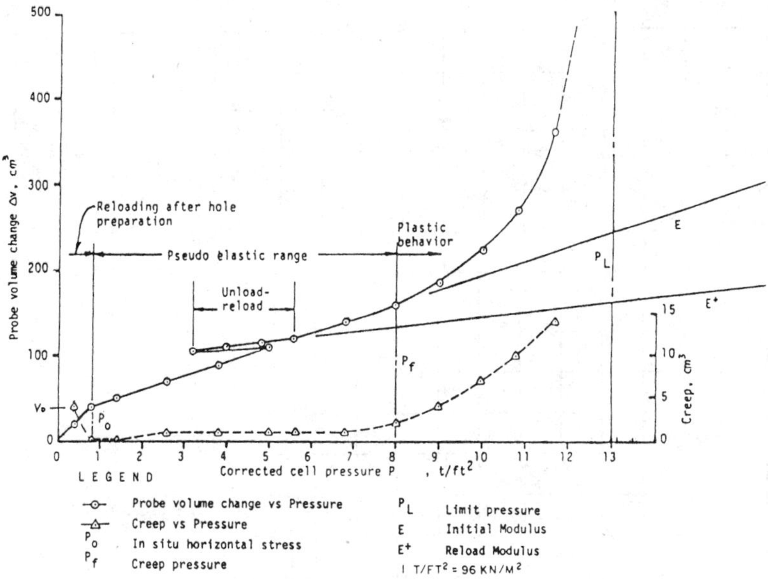

FIG. 2--Typical Pressuremeter Test Plot.

In medium to stiff clays, an independent check of the interpreted horizontal stress can be made by utilizing a relationship proposed by Marsland and Randolph [6] that initial plastic yield will occur once the shear strength is mobilized. The initiation of pressuremeter yield corresponds to the creep pressure P_f, so:

$$P_f = P_o + S_u$$

An interpretation of the pressure P_f is shown on Fig. 2. The undrained shear strength (S_u) is determined as discussed later in this section. Values of P_o are iterated until the Marsland and Randolph relationship is satisfied.

Values of in situ horizontal effective stress are plotted on Figure 3. Also plotted on this figure are lines of horizontal effective stress which correspond to specific values of the coefficient of earth pressure at rest K_o, which relates the horizontal effective stress σ_h' to vertical effective stress σ_v' by:

$$K_o = \frac{\sigma_h'}{\sigma_v'}$$

Brooker and Ireland [7] presented data on stiff clays which suggested that a relationship between K_o, overconsolidation ratio (OCR), and angle of internal friction (ϕ') exists for each type of clay:

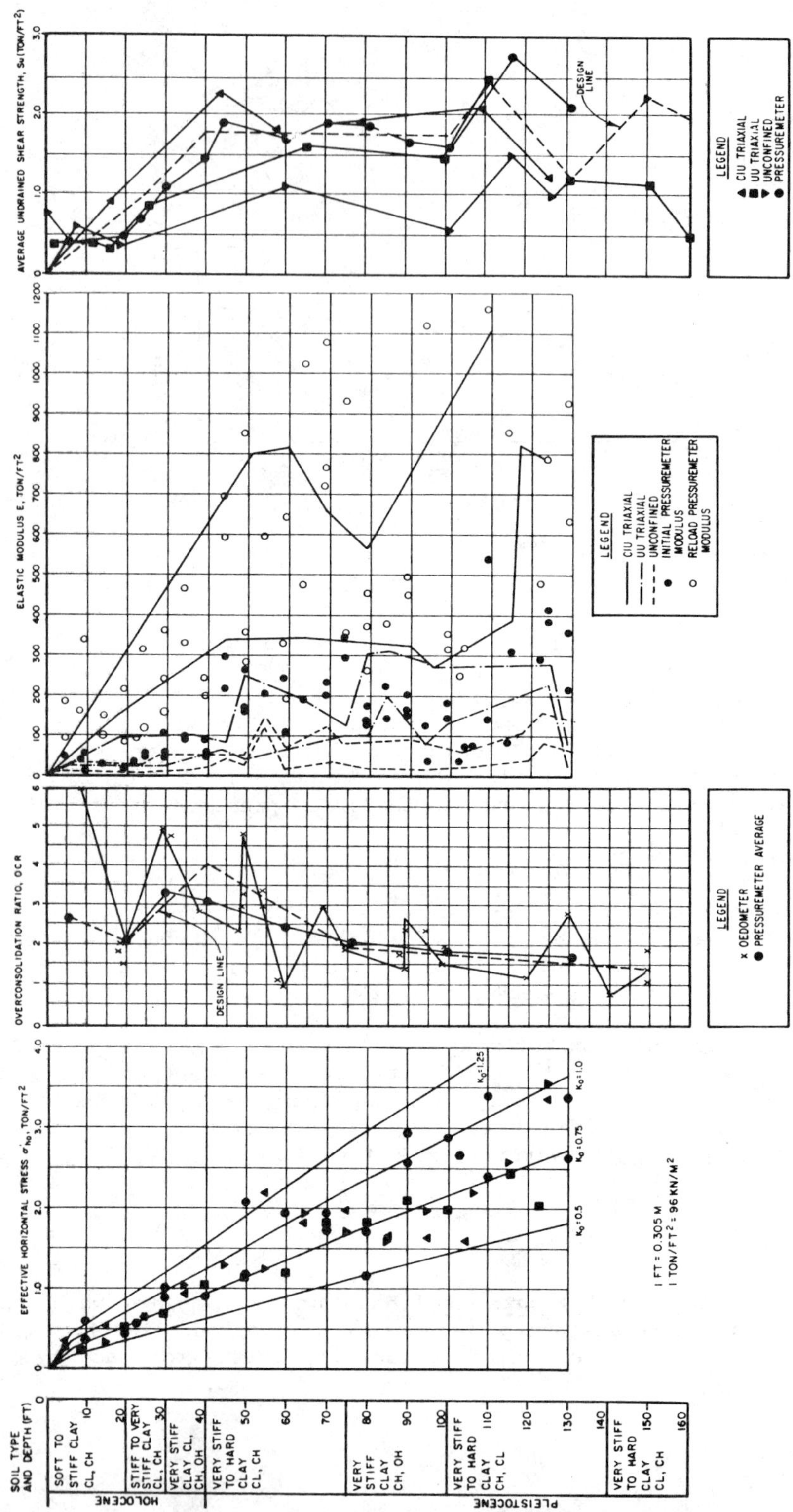

Figure 3--Soil Property Characterization

$$K_o = (1 - \sin\phi')\ OCR^m$$

where $m = 0.58\ (PI)^{-0.12}$

PI = plasticity index of the soil.

Using laboratory test results to obtain values of OCR, PI and ϕ', a range of K_o from 0.6 to 1.1 would be expected for the clays encountered at this site. The data presented in Figure 3 generally falls within this range.

Stress History. A horizontal overconsolidation ratio (OCR) can be interpreted from the pressuremeter curve as:

$$OCR = \frac{P_f - u}{P_o - u} = \frac{\sigma_h'\ max}{\sigma_h'}$$

where u = ambient pore pressure

$\sigma'_{h\ max}$ = maximum past horizontal effective stress.

This procedure assumes that a break in the linear elastic slope or the initiation of yield in the pressuremeter test expansion curve corresponds to a maximum past horizontal stress, analogous to the maximum past vertical stress measured in a 1-D consolidation test. This procedure breaks down in a normally consolidated soil, but works well for lightly to heavily overconsolidated clays, silts and sands. This method is sensitive to the reliability of the in situ lateral stress interpretation. Pressuremeter OCR values should always be checked against laboratory oedometer results, stress history from geologic information and the Brooker and Ireland [7] relationship.

Values of pressuremeter horizontal OCR are plotted on Figure 3. Comparisons with interpretations of laboratory 1-D consolidation test results show excellent agreement below a depth of 4.5 m (15 ft). Above 4.5 m (15 ft), laboratory vertical OCR values of 6 to 9 significantly exceed pressuremeter horizontal OCR values of 2 to 3. Since this high OCR is due to dessication, perhaps this process has a more severe influence on the vertical stress regime than the horizontal stresses.

Elastic Modulus. An elastic modulus E can be calculated from linear portions of the pressuremeter test curve using cylindrical cavity expansion theory, see Figure 2.

$$E = 2(1+u)\ V_o \frac{\Delta P}{\Delta V}$$

$$V_o = V_i + \Delta V_o$$

where u = Poisson's ratio

ΔV_o = volume change corresponding to P_o

$\Delta P/\Delta V$ = linear slope of test curve

V_i = initial volume of probe.

Two moduli are typically calculated, the initial pressuremeter modulus (E) corresponding to the pseudo-elastic portion of the test curve, and the reload pressuremeter modulus (E+) corresponding to the reload following unloading at the initiation of yield. Initial and reload pressuremeter moduli are plotted on Fig. 3 along with initial tangent moduli from UU and CIU triaxial tests at comparable confining stresses, and elastic moduli from unconfined compression tests.

Pressuremeter test results on many other projects with stiff clay located throughout the United States have indicated that the initial pressuremeter modulus is not the initial tangent modulus of the soil but a modulus which is influenced by the level of stress relief or unloading of the soil during hole preparation. This modulus typically equals or exceeds the initial tangent modulus from unconfined compression (UC) or unconsolidated-undrained triaxial (UU) tests on fissured clay. However, compared to good quality laboratory tests such as isotropically or anisotropically-consolidated (CIU, CAU) triaxial testing, the initial pressuremeter modulus falls somewhere between the secant modulus at 50 percent of the peak stress and secant modulus at failure. These trends are apparently in agreement with the data from the Wilton site.

Our broad base of experience has also indicated that the pressuremeter reload modulus is comparable to an initial tangent modulus measured in a "good quality" consolidated-undrained triaxial test under appropriate confining stress conditions. Following a similar reasoning to that advocated in the normalized stress parameter approach, unloading at the initiation of yield in effect erases the soil's memory of hole preparation induced disturbance. The reload response follows a low strain elastic response comparable to recompression in a 1-D consolidation test, which is insensitive to moderate levels of hole disturbance. This interesting discovery was first made in 1979 on a project in Seattle [8] where comparisons of Menard pressuremeter tests with self boring pressuremeter, CIU triaxial and plane strain triaxial tests revealed that the pressuremeter reload modulus was approximately equal to the "true" initial tangent modulus of the soil. This "true" modulus was back-calculated with finite element modeling from measured performance of a nearby tied-back excavation wall. This observation was verified on projects throughout the United States by comparing the pressuremeter reload modulus to the initial tangent modulus from CIU or CAU triaxial tests. Interpretation of initial tangent modulus values from CIU triaxial tests on Wilton samples agree reasonably well with pressuremeter reload moduli. However, the significant scatter in this sensitive parameter should be noted.

Undrained Shear Strength. The undrained shear strength (S_u) was determined from the failure portion of the pressuremeter test curve using the Gibson and Anderson [9] procedure:

$$P_L - P_o = S_u\left(1+\log_e\left\{\frac{E}{2\,S_u(1+\mu)}\right\}\right)$$

If the factor in parenthesis is defined as β then:

$$P_L = P_o + \beta S_u$$

where P_L = Pressuremeter limit pressure.

The term β resembles the bearing capacity factor (N_c) of a long footing. The $E/2S_u$ term is known to decrease with increasing OCR, and for many clays, is in the range of from 100 to 600. Therefore, if $\mu = 0.5$, β is in the range of 5.2 to 7.0, with the upper part of the range corresponding to normally to slightly overconsolidated soils.

Undrained shear strength values interpreted from the pressuremeter test results are presented in Fig. 3. Our experience in Pleistocene clays in Houston [10] indicated that the pressuremeter derived shear strength typically exceeds the undrained shear strength from UC and UU laboratory tests. However, samples without slickensides tested in a consolidated undrained triaxial test typically yield similar undrained shear strength results to the pressuremeter derived strengths. This experience was borne out at Wilton as the pressuremeter derived shear strengths typically exceeded UU or UC shear strengths, and the CIU shear strength from slickensided samples. However, for samples without slickensides, the CIU triaxial shear strength compared favorably with the pressuremeter derived shear strength. In some cases, the UU shear strengths fell within the range of pressuremeter strengths, but typically near the lower bound.

Drained Shear Strength. Drained shear strength parameters were determined from the effective stress response during CIU triaxial compression with pore pressure measurements. Drained shear strength envelopes have been plotted in Fig. 4, including several slickensided samples which have low friction angles and zero cohesion.

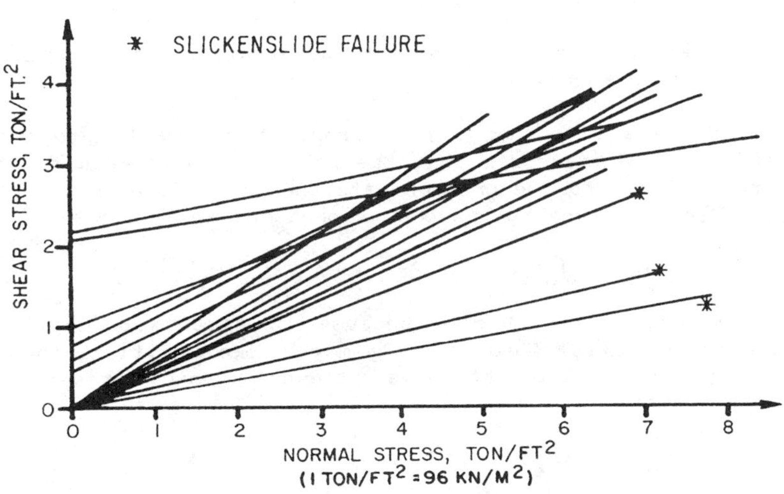

FIG. 4--Drained shear strength

PILE DESIGN

Ultimate Axial Capacity. The ultimate axial capacity in compression of individual piles was calculated assuming that the pile developed its capacity from side friction and end bearing. For a design load of 1,112 KN (125 tons) and a factor of safety of 2, a 32-m (105-ft) long pile was selected. Four design methods were utilized; three for undrained pile capacity including alpha, lambda and pressuremeter, and one for drained pile capacity, beta.

The alpha method, first introduced by Tomlinson [11] and updated by many others, assumes that the pile-soil adhesion for a driven pile in clay is proportional to the undrained shear strength (S_u). For design, an average adhesion/shear strength (α) relationship was selected, see McClelland [1], using the design undrained shear strength profile. The ultimate axial pile capacity (Pu) was then calculated by integrating along the length of the pile (L):

$$Pu = \int_o^L C \quad S_u \alpha \, dz + A_b \, S_u \, N_c$$

where
- C = pile perimeter
- A_b = area of pile base or step taper sections bearing on soil
- N_c = bearing capacity factor = 8.5.

The lambda method [1] was developed originally for pipe piles in clays to improve the empirical correlation with a number of pile load tests:

$$Pu = \lambda \, (\sigma'm + 2 \, \overline{S_u}) \, A_s + A_b \, S_u \, N_c$$

where
- $\sigma'm$ = mean effective vertical stress along length of pile
- $\overline{S_u}$ = mean undrained shear strength
- A_s = pile surface area
- λ = dimensionless coefficient.

The λ coefficient is a function of pile penetration and is presented by Vijayvergiya and Focht. [12]

The pressuremeter method [3], introduced by the Menard organization in the 1960's assumes that the mobilized pile skin friction (f_L) is proportional to the pressuremeter net limit pressure ($P_L{}^* = P_L - Po$). This relationship has been reproduced in Fig. 5

$$P_u = \int_o^L C \quad f_L \quad dz \; + \; A_b \quad S_u \quad N_c$$

The beta method was checked because in some overconsolidated clays, the drained axial load capacity may be more critical than the undrained capacity. The effective stress approach presented by Vesic [13] includes:

$$P_u = \int_o^L C \, \sigma_v' \beta \, dz \; + \; A_b \; \sigma_{vb}' \; N_q$$

where σ_v' = vertical effective stress at any depth Z including base b

β = $K_s \tan \phi'a$

K_s = coefficient of lateral stress

ϕ'_a = drained pile-soil friction angle

N_q = bearing capacity factor = 32.

For stiff clay, Burland [14] suggests that:

$$Ks = Ko$$
$$Ko = (1-\sin \phi')\sqrt{OCR}.$$

For driven piles in clay, ϕ_a' can be taken as ϕ', so

$$\beta = (1-\sin \phi')\sqrt{OCR} \tan \phi'.$$

Axial pile capacities were calculated using these four methods for the five types of piles load tested. For the tapered portion of the Monotube pile, a multiplication factor of 2.1 was used to compute the frictional resistance. This empirical factor compensates for additional pile capacity mobilized from increased adhesion and bearing along the tapered portion of the pile, and has been based on review of many pile load test results by the pile manufacturer.

The ultimate axial capacity of the pile in tension was assumed to be approximately equal to the frictional capacity in compression. Results of the calculated pile capacities are presented in Table 1.

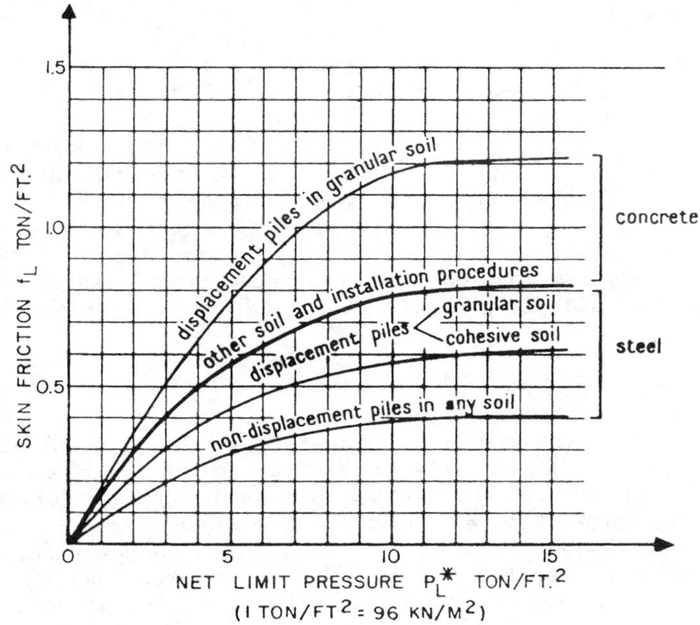

FIG. 5--Pressuremeter Derived Pile Skin Friction [3]

TABLE 1-- Ultimate Axial Pile Load Capacity

ULTIMATE AXIAL LOAD CAPACITY, tons (1 ton = 8.9 KN)

PILE TYPE	PREDICTION								MEASUREMENT	
	ALPHA		LAMBDA		BETA		PRESSUREMETER		PILE LOAD TEST	
	Friction	Total	Friction	Total	Friction	Total	Friction	Total	Compression	Tension
Steel Pipe	201	217	279	294	223	319	210	226	230	230
Hel-cor	231	252	320	341	258	383	242	263	270	230
Step Taper	215	239	326	350	222	356	220	244	270	290
Precast Concrete	256	276	354	375	285	407	315	335	310	150*
Monotube	256	262	395	401	270	302	299	305	330	300

* Failed Splice

Lateral Load Capacity. The lateral load-deflection response of the Monotube pile was predicted using the p-y curve approach developed by Reese. [15] Critical soil parameters required to develop p-y curves include the undrained shear strength (S_u) and strain at 50 percent of the peak stress (ϵ_{50}). Both parameters were developed directly from pressuremeter test results. Determination of the undrained shear strength was discussed previously. The ϵ_{50} parameter was determined by the following:

o For small expected lateral deflections, say less than 0.5 in.

$$\epsilon_{50} = \frac{S_u}{E+}$$

o For larger expected lateral deflections, say 1.0 in. or larger

$$\epsilon_{50} = \frac{S_u}{E}$$

where E + = pressuremeter reload modulus
E = pressuremeter initial modulus.

This approach, which is based on our experience with piles and piers in various soils, takes advantage of the similarity of the mode of loading provided by the pressuremeter test to the lateral loading of a pile. It also attempts to account for the nonlinear relationship between stiffness and pile deflection. Predicted lateral load-deflection curves obtained using Reese's empirical recommendations and the pressuremeter data are presented in Fig. 8.

PILE LOAD TESTS

Five axial compression and five axial tension load tests were conducted on 36 cm (14 in.) diameter steel pipe piles, 41 cm (16 in.) diameter Armco Hel-cor corrugated straight shaft piles, 36 cm (14 in.) square precast, prestressed concrete piles, 36 cm (14 in.) Monotube tapered piles and Raymond Step Taper piles (Section Nos. 18-12), Fig. 6. A lateral load test was conducted on a 36 cm (14 in.) Monotube pile. All piles were preaugered to a depth of 9(30) to 12(40) m(ft) and driven to a total depth of 32 m (105 ft). Each pile was allowed to cure after concreting at least one month before loading. The loading tests were conducted in accordance with ASTM D 1143-81, D 3689-83 and D 3966-81 for compression, tension and lateral loading, respectively.

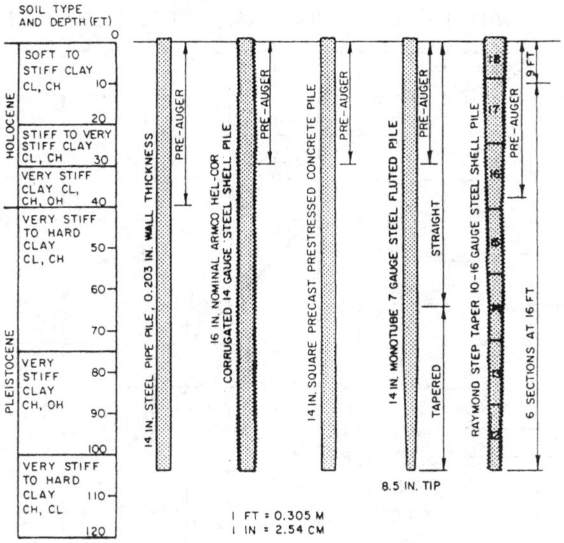

FIG. 6--Load Test Piles

Axial pile load test results are presented in Fig. 7 and the lateral load test result is shown in Fig. 8. Ultimate axial load capacities interpreted from these curves are presented in Table 1, along with the analytical predictions. Lateral load-deflection predictions are shown with the measured test data in Fig. 8.

Several load test anomalies were observed. The low tension capacity of the precast concrete pile was the result of failure of the splice. The disproportionally larger level of displacement of the Hel-cor pile than the other piles, under tension and compression loadings, was probably caused by pile installation problems. A new mandrel was used to drive the piles and great difficulty was encountered in removing it after driving. The effort put into removing the mandrel dislodged the pile slightly. Also the mandrel was only 30 m (100 ft) long which required a 2-2.5 m (7-8 ft) deep excavation to drive the pile. This hole was backfilled prior to testing.

Comparison of the predicted and measured ultimate pile load capacities in Table 2 indicates that the pressuremeter method provided predictions generally within 10 percent of the measured capacity. The alpha method generally underpredicted the measured pile capacity by approximately 10 percent while the lambda method overpredicted the measured pile capacity by approximately 25 percent. The beta method overpredicted the total pile capacity in compression while generally predicting the frictional component within 10 percent. The high end bearing prediction resulted from assuming fully drained conditions which were not allowed to develop during the quick pile load testing. The underpredictions for the Monotube pile generally indiate that the empirical factor of 2.1 for the tapered portion of the pile was slightly conservative.

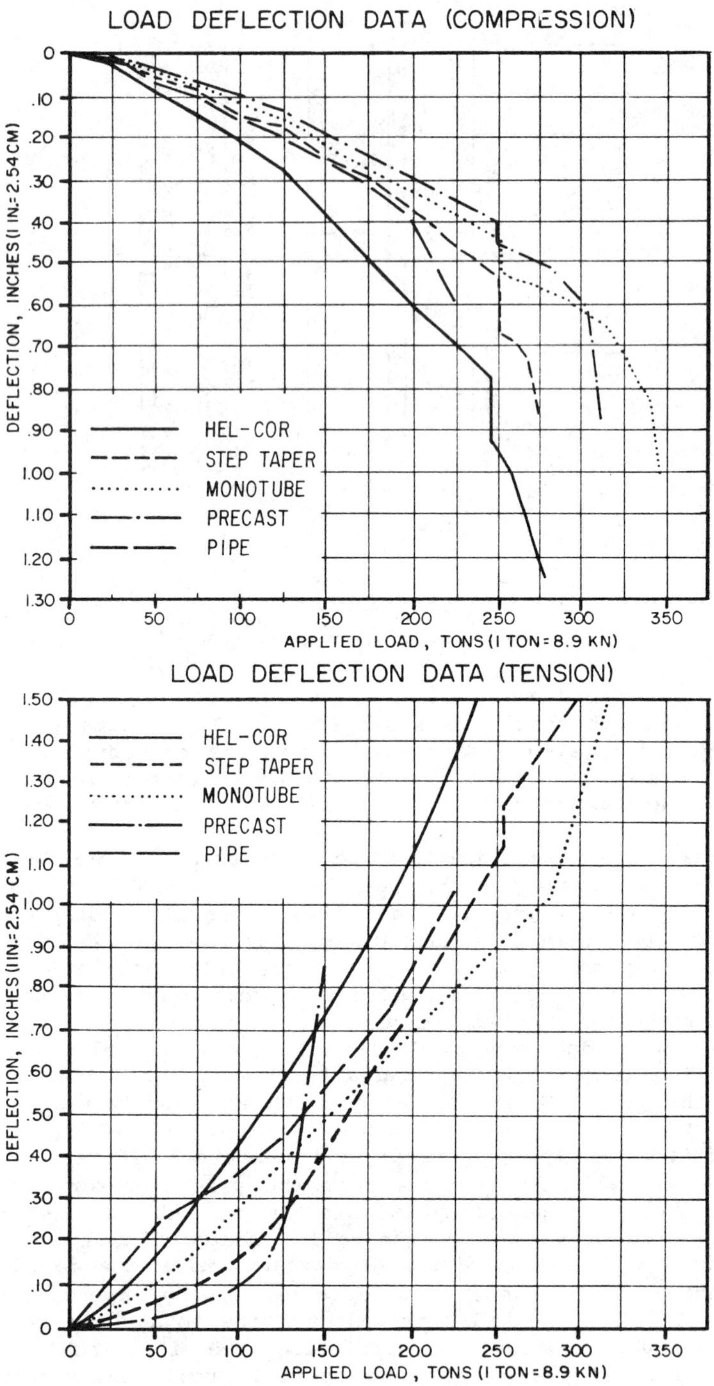

FIG. 7--Results of Axial Pile Load Tests

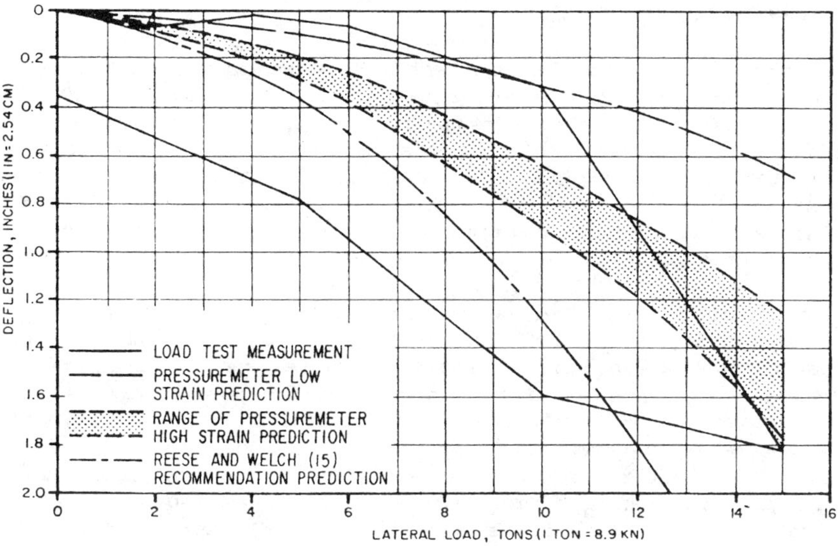

FIG. 8--Lateral Load Test Result

TABLE 2--Predicted versus Measured Utlimate Pile Load Capacity (P_u)

PILE TYPE	RATIO = P_u (Predicted)/P_u (Measured)			
	COMPRESSION TESTS			
	ALPHA	LAMBDA	BETA	PRESSUREMETER
Steel Pipe	0.96	1.31	1.42	0.98
Hel-cor	0.93	1.26	1.42	0.97
Step Taper	0.88	1.30	1.32	0.90
Precast Concrete	0.89	1.14	1.31	1.08
Monotube	0.79	1.22	0.92	0.92
	TENSION TESTS			
Steel Pipe	0.87	1.21	0.97	0.91
Hel-cor	1.00	1.39	1.12	1.05
Step Taper	0.74	1.12	0.77	0.76
Monotube	0.85	1.32	0.90	1.00

Lateral load-deflection predictions using the pressuremeter reload modulus compare well at small strains (deflections less than 1.0 cm, 0.5 in.) while predictions using the range of pressuremeter initial modulus slightly underpredict deflections near the ultimate load. Predictions using Reese's [15] empirical recommendations significantly over-predict the initial portion of the load deflection curve and overpredict deflection at the ultimate load.

CONCLUSIONS

With proper hole preparation technique, the pressuremeter is an effective soil property characterization tool in soils which are difficult to sample and maintain their in situ structure intact. In the stiff, slickensided clays at the Wilton site, the results of pressuremeter and CIU triaxial testing verified that the soils have higher strength and modulus than would be indicated from a conventional geotechnical investigation. The pressuremeter also provided realistic measurements of in situ stress and stress history.

Piles in clay can be designed based on the results of pressuremeter tests using both empirical correlations and soil property characterization. The soil property characterization approach has the advantage that conventional design techniques are utilized and soil properties are determined by theoretical means. Furthermore, the pressuremeter derived soil properties can be checked against soil properties determined more conventionally in the laboratory. The soil property characterization approach can also be used to check the applicability of the empirical correlation approach for the specific conditions at the site.

At Wilton, both the pressuremeter empirical and soil property characterization approaches provided realistic predictions of ultimate axial load capacity and lateral load-deflection performance for five different types of piles. Conventional methods would have provided more conservative pile designs. On large projects, such as Wilton where several thousand piles will be driven, a substantial savings to the owner will occur as a result of using more realistic foundation designs.

Another advantage of the pressuremeter approach is that realistic geotechnical properties and pile performance predictions can be made very quickly (within a day of completing the field testing) and economically (compared to the cost of a comprehensive sophisticated laboratory testing program). Pressuremeter testing can provide a large volume of data encompassing the spatial variability of various geotechnical conditions across the site. The pressuremeter approach does not replace the need for laboratory testing, rather it complements and enables the laboratory testing to be optimized to specific sets of conditions identified by field measurements.

ACKNOWLEDGEMENTS

Wilton Steam Electric Station will be owned by Louisiana Power and Light Company who gave their permission to publish the results of the geotechnical investigation and test program. Sargent and Lundy has designed the power block facilities for Middle South Services. Geotechnical drilling, sampling and testing services were provided to Sargent and Lundy by Woodward-Clyde Consultants. Pile load testing was conducted by Boh Brothers Construction Company.

REFERENCES

[1] McClelland, B., "Design of Deep Penetration Piles for Ocean Structures," Journal of Geotechnical Engineering Division, ASCE, Vol. 100, No. G77, 1974, pp 705-747

[2] Darragh, R.D. and Bell, N.A. "Load Tests on Long Bearing Piles," ASTM, STP 444, 41-67, 1969, pp 41-67.

[3] Baguelin, F., Jezequel, J.F., and Shields, D. The Pressuremeter and Foundation Engineering, Series in Rock and Soil Mechanics, 214, Trans Tech Publications, Germany, 1978

[4] Cardwell, G. T., and Rollo, J. R., "Interim Report on Ground Water Between Baton Rouge and New Orleans, Louisiana," Louisiana Geological Survey and Louisiana Department of Public Works, Water Resources Pamphlet 9, 1960.

[5] Davidson, R. R. "Interpretation of the Pressuremeter Test" Woodward-Clyde Consultants Techical Memorandum No. 104-78 December 1979.

[6] Marsland, A. and Randolph, M. E. "Comparison of the Results from Pressuremeter Tests and Large In Situ Plate Tests in London Clay" Geotechnique, Vol. 27, No. 2, 1977, pp 217-243.

[7] Brooker, E.W., and Ireland, H.O. "Earth Pressures at Rest Related to Stress History," Canadian Geotechnical Journal, Vol. 2, No. 1, February 1965.

[8] Denby, G.M., Aboim Costa, C, Clough, G.W. and Davidson, R.R. "Laboratory and Pressuremeter Tests on a Stiff Clay," Proceedings of the Tenth International Conference on Soil Mechanics and Foundation Engineering, Stockholm, Sweden, July 1981.

[9] Gibson, R.E. and Anderson, W.F. "In-Situ Measurement of Soil Properties with the Pressuremeter" Civil Engineering and Public Works Review, Vol. 56, No. 658, London, May 1961, pp 615-618.

[10] Merritt, B.K., Davidson, R.R. and Baker, G.L. "Menard Pressuremeter Experience in Beaumont Clay" Spring Texas Section ASCE Mtg. Austin, TX, March, 1979.

[11] Tomlinson, M.J., "The Adhesion of Piles Driven in Clay Soils," Proceedings of the Fourth International Conference on Soil Mechanics and Foundation Engineering, Vol. 2, 1957, pp 66-71.

[12] Vijayvergiya, V.N. and Focht, J.A., Jr. "A New Way to Predict the Capacity of Piles in Clay," 4th Annual Offshore Technology Conference, Houston, Vol. 2, 1972, pp 865-874.

[13] Vesic, A.S., "Design of Pile Foundations," NCHRP Synthesis No. 42, Transportation Research Board, 1977.

[14] Burland, J.D. "Shaft Friction of Piles in Clay - A Simple Fundemental Approach, Ground Engineering, Vol. 6, No. 3, May 1973, pp 30-42.

[15] Reese, L.C. and Welch, R.C. "Lateral Loading of Deep Foundations in Stiff Clay," Journal of the Geotechnical Engineering Division, ASCE, Vol. 103 No. GT4, pp 287-305, April 1977, pp 287-305.

Pressuremeter and Laterally Loaded Piles

Peter K. Robertson, John M.O. Hughes, Richard G. Campanella, Peter Brown and Shawn McKeown

DESIGN OF LATERALLY LOADED PILES USING THE PRESSUREMETER

REFERENCE: Robertson, P.K., Hughes, J.M.O., Campanella, R.G., Brown, P. and McKeown, S., "Design of Laterally Loaded Piles using the Pressuremeter", The Pressuremeter and its Marine Applications: Second International Symposium, ASTM STP 950, J.-L. Briaud and J.M.E. Audibert, Eds., American Society for Testing and Materials, 1986.

ABSTRACT: The pressuremeter is ideally suitable for determining the in-situ non-linear parameters for the design of laterally loaded piles. A method for designing laterally loaded piles using presuremeter data is presented. For driven displacement piles, which are commonly used offshore, the pressuremeter can be pushed into the soil in a similar full-displacement manner. Several case histories are presented to illustrate the proposed method.

KEY WORDS: pressuremeter, piles, lateral loading.

The non-linear subgrade reaction method is widely used for the design of laterally-loaded piles. This method replaces the soil reaction with a series of independent springs. The non-linear behavior of the soil springs is represented by P-y curves, which relate soil reaction and pile deflection at points along the pile length. Most of the existing methods for obtaining P-y curves are highly empirical. Often little account is taken of the method of pile installation and the influence that this may have on the soil behaviour. The pressuremeter, however, offers the potential to measure the soil reaction in-situ, under similar loading conditions.

Drs. Robertson and Campanella and Mr. P.T. Brown are NSERC University Research Fellow, Professor and student, respectively, in the Department of Civil Engineering, 2324 Main Mall, The University of British Columbia, Vancouver, B.C., Canada, V6T 1W5; Dr. Hughes is a consultant at 3009 St. Andrews St., N. Vancouver, B.C., Canada, V7N 1Z5; Mr. S. McKoewn is an engineer with Golder Associates (Western Canada) Ltd., 7017 Farrell Road S.E., Calgary, Alberta, Canada, T2H OT3.

Several methods have been proposed for the design of laterally loaded piles using pressuremeter data (Briaud et al. [1], Baguelin et al. [2], Imai [3], Robertson et al. [4], Baguelin [5]). Most of these methods make use of preboring pressuremeter results, using a Ménard type pressuremeter, and cannot model the disturbance caused from a driven pile since the pressuremeters are placed in a prebored hole. It is possible to install the pressuremeter in a manner which models the disturbance caused during pile installation. For driven displacement piles, the pressuremeter can be pushed into the soil in a full-displacement manner. For cast-in-place or bored piles, a pre-bored or self-bored pressuremeter test can model the disturbance during pile installation. The method by Robertson et al. [4,6] uses the results from a pressuremeter pushed into the soil to model the installation of a driven displacement pile.

During the pressuremeter test, the soil deforms in a simple radial direction, whereas the displacements in the soil surrounding a laterally loaded pile are more complex as the soil moves radially away from the front face of the pile and inwards towards the back face [7], as shown on Fig. 1. However, it is reasonable to expect that the soil in the center region of the front face of the pile would deform in a similar manner to that about a pressuremeter (see Fig. 1). Therefore, it is reasonable to suppose that the geometric form of the pressure expansion curve obtained from the pressuremeter would be similar to the load displacement (P-y) curve for the soil acting on the front face of the pile, provided the pressuremeter was installed to model the soil disturbance during pile installation.

It is interesting to note from Fig. 1 that when a circular pile is loaded laterally, virtually all the soil displacements are radially away or toward the pile. There appears to be very little slip along the side of the pile to generate lateral friction.

Hughes et al. [8] suggested that the pressuremeter curves should be increased by some factor (α) to give the correct P-y curves for the pile. This multiplying factor is a factor to account for the fact that laterally loaded piles have limiting soil reactions that are higher than those for radially expanding pressuremeters. Hughes et al. [8] and Robertson et al. [4] suggested that the multiplying factors are as follows:

$$\alpha = 2 \quad \text{for cohesive soils} \tag{1}$$

$$\alpha = 1.5 \quad \text{for cohesionless soils} \tag{2}$$

These multiplying factors were also confirmed by Byrne and Atukorala [9] using finite element analyses. Further research and field evaluation, however, is still required to refine these numbers for many different soil types.

The multiplying factors (α) are applied to the pressure component of the pressure expansion curves, as shown in Fig. 2.

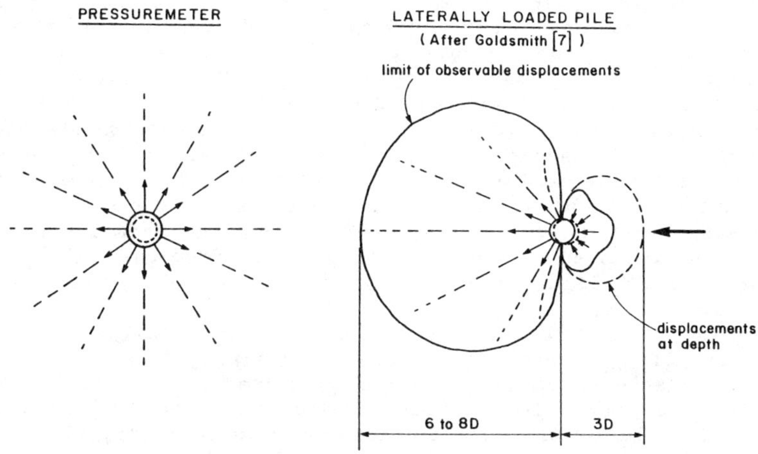

Fig. 1. Displacements in Soil around Radially Expanding Pressuremeter and Laterally Loaded Pile.

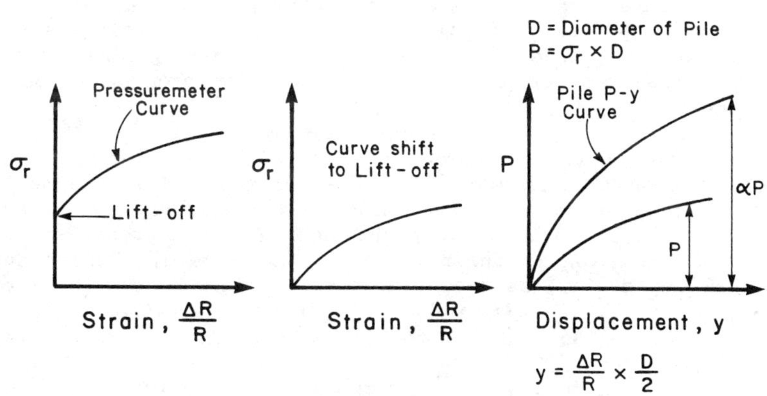

Fig. 2. Schematic Representation of Development of Pile P-y Curves from Pressuremeter Curves.

The lateral stresses acting against the sides of the pile will be influenced by the pile driving and subsequent waiting or set-up period. The pile is in equilibrium with the lateral stresses, therefore, with the application of a lateral load, the pile will immediately start to move. However, when a pressuremeter is installed, the pressure inside the probe is increased from zero until the membrane is in equilibrium with the soil and expansion starts. By installing the pressuremeter in a similar manner, the pressure expansion curve will have a lift-off pressure approximately equivalent to the initial lateral stress around the pile. Therefore, to obtain the subsequent P-y curve for lateral loading, the pressure axis for the P-y curve is moved to the lift-off pressure from the pressuremeter curve, as shown on Figure 2.

In fine grained soils, such as silty sands, silts and clays, significant pore pressures can be generated during driving of a pile. These pore pressures usually dissipate before any lateral loads are applied to the pile. If the pressuremeter is to be used as a model for the pile, the excess pore pressures generated during the installation of the pressuremeter must also be allowed to dissipate before expanding the membrane.

The time required for excess pore pressures to dissipate around a cylindrical probe is proportional to the radius of the probe squared. Therefore, the time required for a significant portion of the pore pressures to dissipate around a pressuremeter depends on the diameter of the instrument. Most pressuremeter probes are in the order of 75 mm in diameter. Some pressuremeters are as small as 44 mm in diameter. Therefore, the time required for significant pore pressure dissipation around most pressuremeters is usually less than about 30 minutes for most silty soils. Longer times are required for low permeability clays. Smaller times for dissipation are required for smaller diameter probes. If piezo-cone penetration testing is available, the time required for dissipation of excess pore pressures around the pressuremeter can be estimated from a dissipation test using the cone.

The approach given above can be applied to pressuremeter tests performed at some depth remote from the surface. Near the surface, the displacements of both the laterally loaded pile and the expanding pressuremeter can be influenced by the surface. The pressuremeter is subject to a reduction in the mobilized resistance at shallow depth. The critical depths (z_c) as recommended by Baguelin et al. [2] are:

$$z_c = 15\ D_{PMT} \quad \text{for cohesive soils} \tag{3}$$

$$z_c = 30\ D_{PMT} \quad \text{for cohesionless soils} \tag{4}$$

where D_{PMT} = diameter of pressuremeter.

Therefore, for 75 mm diameter pressuremeters, the pressure values of the pressuremeter test results require some increase due to surface effects to a depth of between 1.1 m to 2.2 m. The suggested reduction factors given by Briaud et al. [1] are shown in Fig. 3. The pressuremeter curve is then corrected by using:

$$P_{corr} = \frac{P}{\beta} \qquad (5)$$

where P_{corr} = corrected pressure.
β = reduction in mobilized pressuremeter pressure at all strains.

The correction is only applied to pressuremeter test results that may be influenced by the close proximity of the ground surface. The depth of the pressuremeter should be based on the position of the center of the membrane.

The displacements of soil around a laterally loaded pile are also influenced by the ground surface. The critical depth (D_c), to which displacements are influenced, depend on the pile load, diameter and stiffness. Briaud et al. [1] proposed a relative rigidity factor, RR, given by:

$$RR = \frac{1}{B} \sqrt[4]{\frac{EI}{P_L}} \qquad (6)$$

where EI = pile flexural stiffness
B = pile diameter
P_L = net pressuremeter limit pressure.

The correlation between the ratio of critical depth to pile diameter and RR proposed by Briaud et al. [1] is shown on Fig. 4.

For most prebored pressuremeter tests, there is a clear definition of the net pressuremeter limit pressure, P_L. However, for self-bored or full-displacement pressuremeter tests in sandy soils, there is not always is a clear definition of P_L.

Robertson et al. [6] suggested using a general critical depth of four pile diameters (D_c = 4). However, the relationship proposed by Briaud et al. [1] in Fig. 4 is more logical and is recommended since it shows that for a high relative rigidity, (RR > 10), i.e. a stiff pile relative to the soil, the critical depth will be slightly larger than 4 and for flexible piles, the critical depth can be less than 4.

To account for the reduced soil reaction mobilized within the pile's critical depth, the multiplying factors are progressively reduced. Based on a review of experience in the literature, the writers recommend that the factors be reduced from a value of 2 at the pile critical depth to 0.67 at the ground surface for cohesive soils and from 1.5 down to 0 for cohesionless soils, as shown on Fig. 5.

The P-y curves required for the analysis are in units of force per unit length (P) and displacement (y), whereas the pressuremeter curves are in units of stress (σ_r) and circumferential strain ($\frac{\Delta R}{R}$), where R is the initial radius of the probe and ΔR is the change in radius. Thus, to convert the pressuremeter stress to force per unit

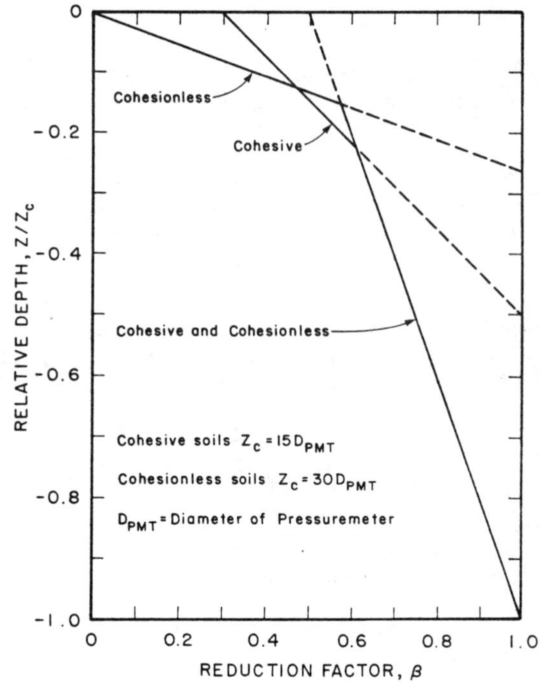

Fig. 3. Reduction Factors for Pressuremeter Test Results at Shallow Depth. (After Briaud et al. [1])

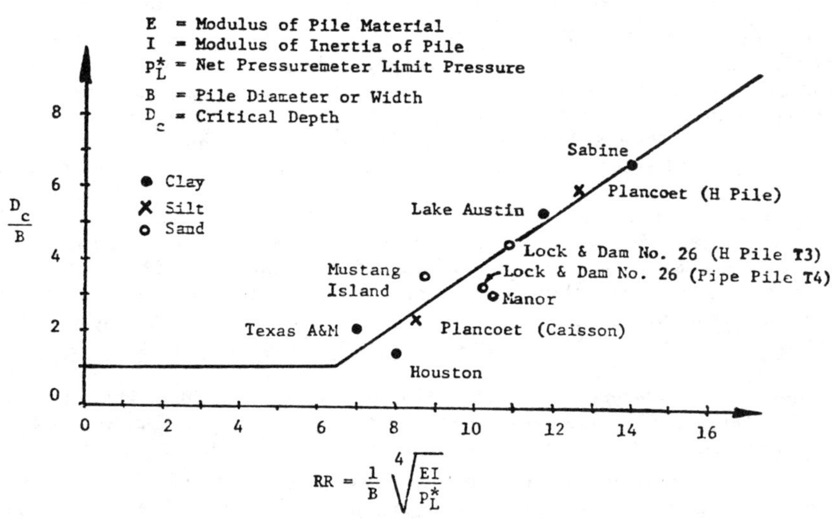

Fig. 4. Critical Depth of Laterally Loaded Pile as a Function of Relative Rigidity. (After Briaud et al. [1])

length, the stress data is multiplied by the pile width. The pressuremeter strain data is multiplied by the pile half-width to obtain the displacement (y), as shown on Fig. 2.

CASE HISTORIES

The writers have been involved in several case histories in recent years where the above method has been applied. The following sections will briefly present the results of these case histories.

Burnaby Mountain, B.C.

Full-displacement pressuremeter tests were performed to model the behaviour of four 300 mm square precast concrete piles driven to a depth of 8 m. The soil conditions consisted of 1.5 m of sand and gravel fill overlying 4.5 m of organic peat and silt. The piles were 6 m in length and driven end bearing onto a dense glacial till. Full details of the case history is given by Robertson et al. [4].

A summary of the calculated and measured load deflection curves at the ground surface are shown in Fig. 6.

The predicted deflection shows good agreement with the measured deflection for both test piles.

Annacis Crossing, B.C.

Full-displacement pressuremeter tests were performed to model the behaviour of a 915 mm diameter steel pipe pile driven open-ended to a maximum depth of 94 m. Full details of the case history is given by Robertson et al. [6]. A summary of a CPT profile showing pressuremeter test locations is given on Fig. 7. Also included in Fig. 7 is a summary of the multiplication factors (α) used in the calculation of the P-y curves.

A summary of the calculated and measured load deflection curves at the pile head and the pile deflection versus depth at a lateral load of 1100 kN are given in Figs. 8 and 9. The predicted deflection agrees remarkably well with the measured deflection and is approximately 20% larger at the pile head. It is worth noting that the prediction of lateral behavior was based solely on the pressuremeter and CPT results before the measured results were made available from the lateral load test. The predictions were based on the approach described above.

Calgary, Alberta

Self-bored pressuremeter tests were performed to model the behaviour of cast-in-place reinforced concrete piles. The piles

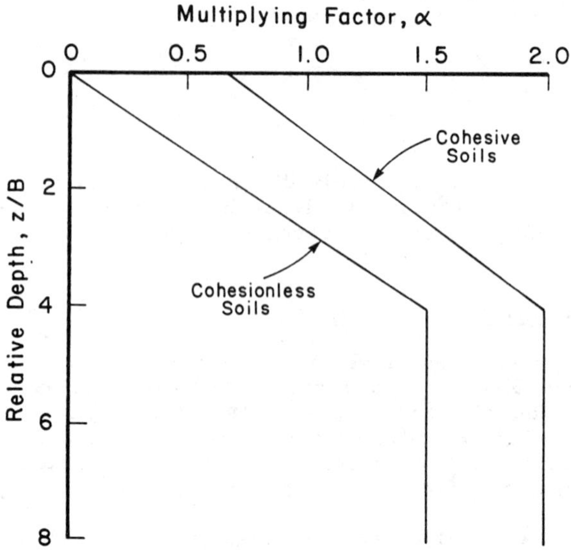

Fig. 5. Variation of Multiplying Factor with Relative Depth.

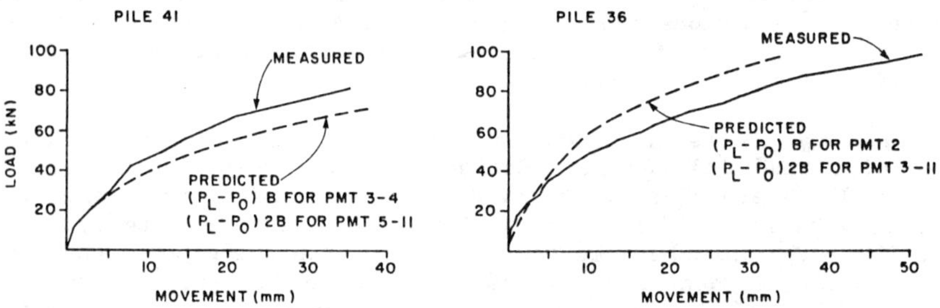

Fig. 6. Summary of Computed and Measured Load Deflection Curves at Ground Surface for Burnaby Mountain, B.C., Test Piles.

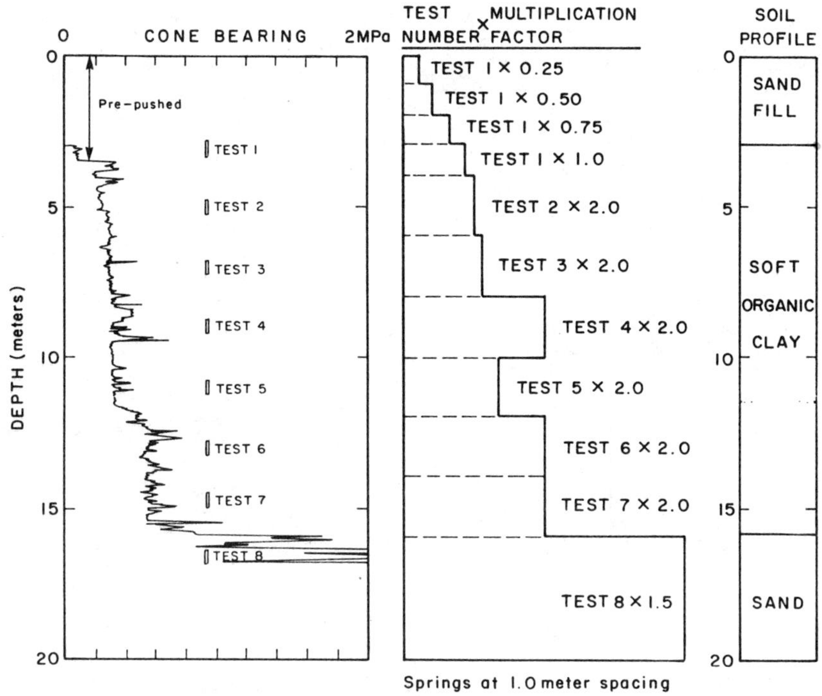

Fig. 7. Summary of CPT Profile Showing Pressuremeter Test Locations and Soil Spring Stiffness at Annacis Crossing, B.C., Site.

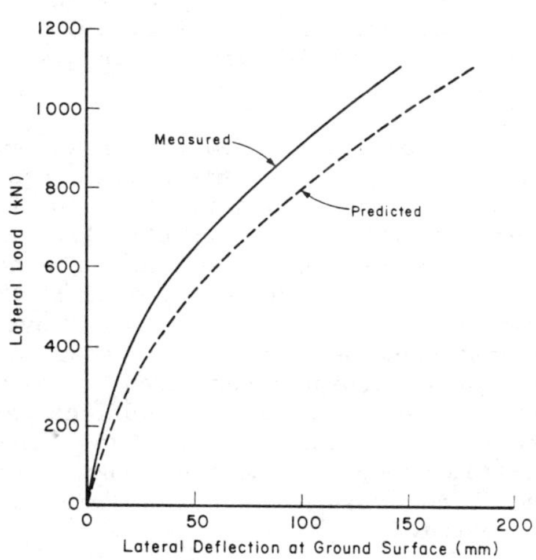

Fig. 8. Summary of Computed and Measured Load Deflection Curves at Ground Surface, for Annacis Crossing, Test Pile.

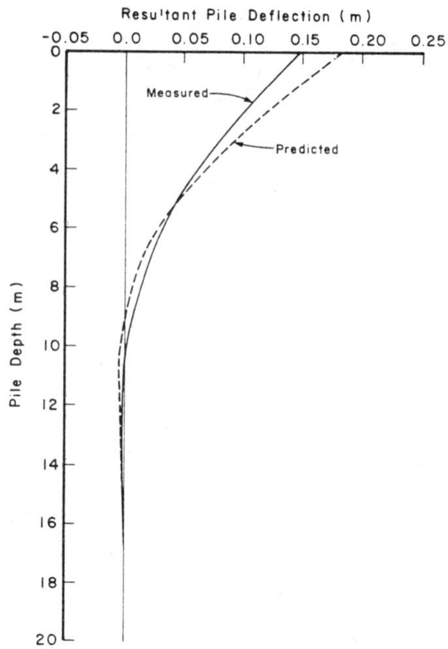

Fig. 9. Summary of Computed and Measured Lateral Pile Deflection versus Depth at a Lateral Load of 1100 kN for Annacis Crossing Test Pile.

varied in size and were 1.5 m, 1.2 m and 0.9 m in diameter, and installed to a depth of 15 m. A summary of a CPT profile showing pressuremeter test locations is given in Fig. 10. Also included in Fig. 10 is a summary of the multiplication factors (α) used in the calculation of the P-y curves.

The deflected shapes calculated from the pressuremeter tests for the free head piles showed good agreement with the observed pile deformations at the design loads. This is shown in Fig. 11 (a and b) for piles of diameter 1.5 and 0.9 m respectively. For the maximum test load, the computer analysis underestimated the total deflection (see Fig. 11a). The authors [10] considered this to be primarily due to the fact that the pile load test was carried out over a long period of time whereas the pressuremeter test was of short duration. Load-deflection-creep data was recorded for each pile and when creep deformations are taken out of the total pile deflections, the predicted and observed deflection are comparable. The authors [10] noted, however, that very little creep deformation occurred at less than the design load and that the pressuremeter model was considered appropriate for the piles tested.

It is worth noting that an evaluation of the creep behavior can be estimated from pressuremeter tests if the pressure during inflation is maintained constant and the rate of strain is recorded. These 'creep tests' can be performed at different stress levels.

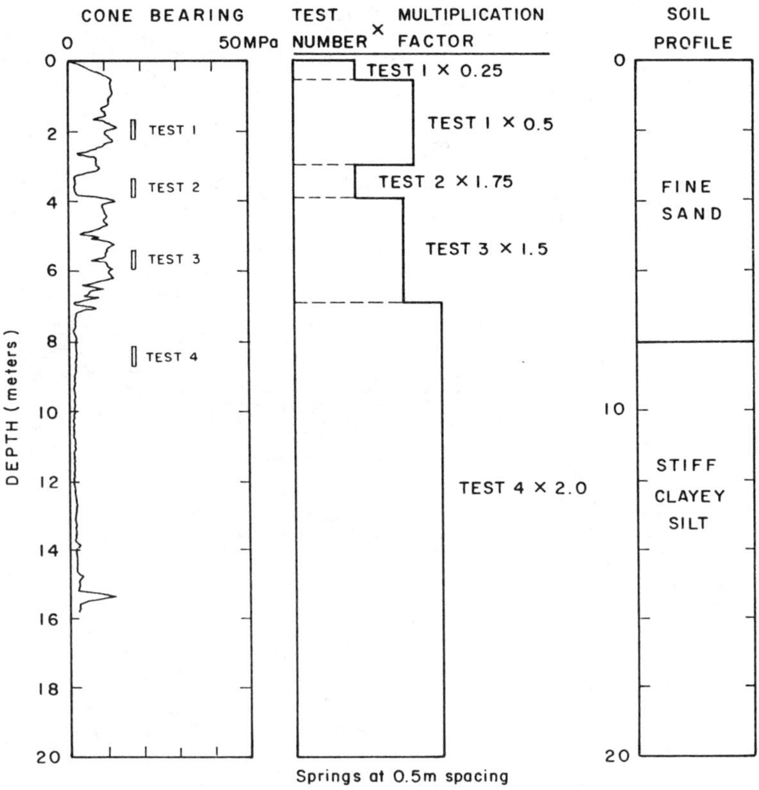

Fig. 10. Summary of CPT Profile Showing Pressuremeter Test Locations and Soil Spring Stiffness at Calgary, Alberta, Site.

Fraser River, B.C.

Full-displacement pressuremeter tests were performed to model the behaviour of a 324 mm diameter steel pipe pile driven open-ended to a depth of 18 m in a Fraser River sand deposit. The pile was installed at an incline of 3 vertical to 1 horizontal and loaded horizontally 600 mm above the ground surface. Full details of the case history are given by Brown [11]. A summary of a CPT profile showing pressuremeter test locations is given in Fig. 12. Also included in Fig. 12 is a summary of the multiplication factors (α) used in the calculation of the P-y curves.

A summary of the calculated and measured load deflection curves at the pile head are shown in Fig. 13. The horizontal lateral load was resolved perpendicular to the test pile for the analysis to account for the pile inclination.

It is interesting to note that significant creep deformation took place at large lateral loads during the lateral load test. If the creep deformations at the larger loads are taken out of the total pile deflections, the predicted and observed deflections are more compar-

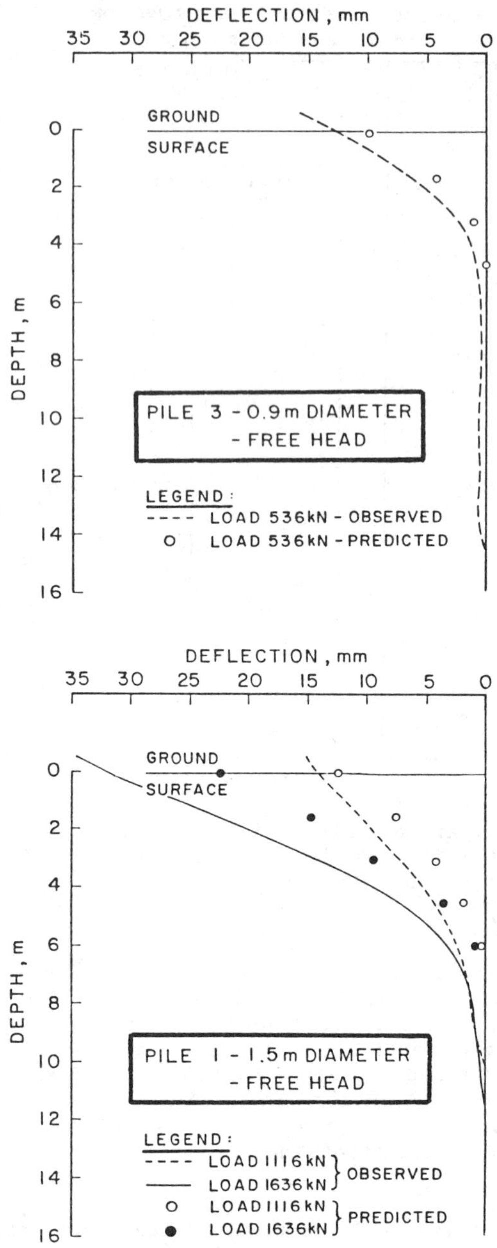

Fig. 11. Summary of Computed and Measured Lateral Pile Deflection versus Depth for 1.5 m and 0.9 m Diameter Piles at Calgary, Alberta, Site.

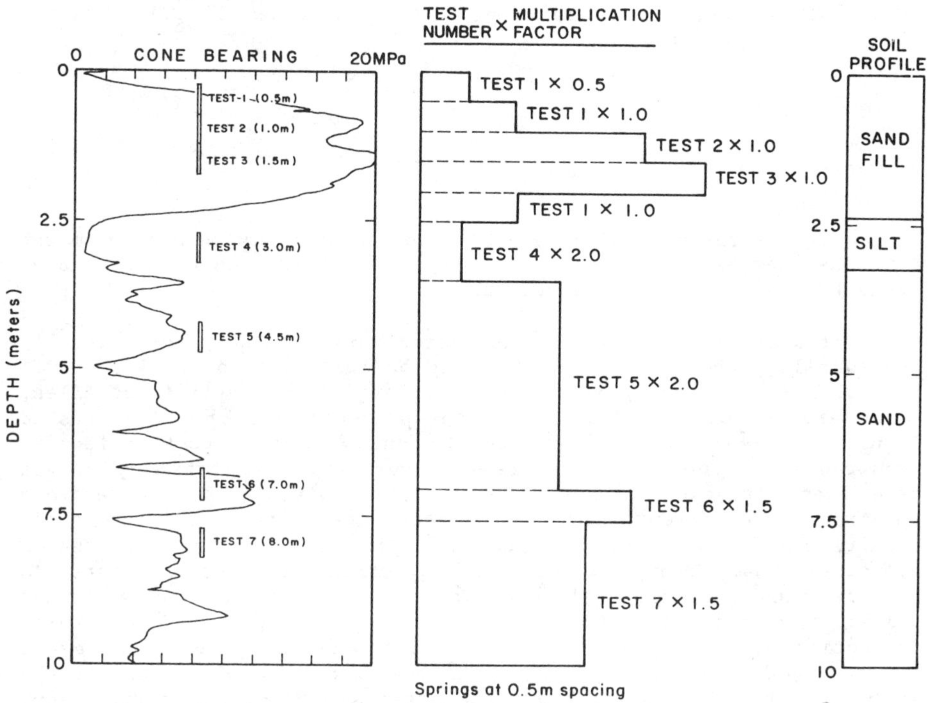

Fig. 12. Summary of CPT Profile Showing Pressuremeter Test Locations and Soil Spring Stiffness at Fraser River, B.C., Site.

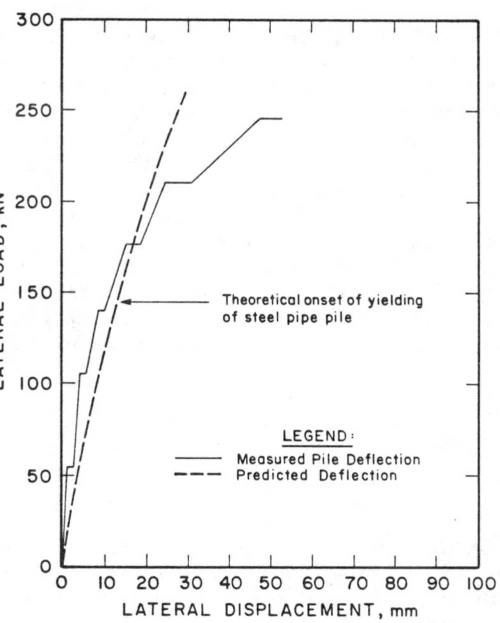

Fig. 13. Summary of Computed and Measured Lateral Load Deflection Curves for Fraser River Test Pile.

able. The effects of creep should be investigated if piles are to experience large lateral loads.

SUMMARY

A method for designing laterally loaded piles using pressuremeter data has been briefly summarized. Four case histories have also been presented to illustrate the method.

For a rational use of pressuremeter data for design of laterally loaded piles, the pressuremeter should be installed to model the soil disturbance during pile installation. For driven displacement piles, which are commonly used offshore, the pressuremeter should be pushed into the soil in a full-displacement manner. The correction factors required to adjust the pressuremeter curves to generate P-y curves have been discussed and preliminary values given. The corrections required for different soil types and the stress free surface effects require further evaluation. However, the case histories presented illustrate that the method and existing correction values suggested do provide excellent predictions of the behavior of laterally loaded piles. The proposed method was generally able to predict the lateral deflection at the ground surface and the overall deflected shape of the test piles to within 20% of the measured values. Further field evaluation is required to study the effects of long duration, creep, loads and cyclic loading. However, recent field work (Brown, 1985) has shown the potential of the pressuremeter to provide qualitative data on both creep and cyclic loading effects.

The ease of performing full-displacement pressuremeter tests offshore depends somewhat on the diameter of the probe since the pushing force to install the probe increases with increasing size. Research and development is presently underway at the University of British Columbia and Fugro, Holland to combine a 44 mm diameter (15 cm^2 base area) pressuremeter probe with an electric piezometer cone of the same diameter. This would enable the pressuremeter to be installed by currently available offshore CPT equipment. The CPT would also provide valuable data on the soil profile and assist in the optimum choice for pressuremeter test locations. The case histories presented have illustrated the usefulness of the combined CPT and pressuremeter. The CPT profile often provides confidence in the variation of pressuremeter test results [6] due to variable soil stratigraphy.

ACKNOWLEDGEMENTS

The assistance of the Natural Sciences and Engineering Research Council of Canada; Golder Associates, Calgary and the technical staff of the Civil Engineering Department, University of British Columbia, is very much appreciated.

REFERENCES

[1] Briaud, J.-L., Smith, T.D. and Meyer, B.J., 1983, "Laterally Loaded Piles and the Pressuremeter: Comparison of Existing Methods", ASTM Special Technical Publication STP 835 on the Design and Performance of Laterally Loaded Piles and Pile Groups, June.

[2] Baguelin, F., Jezequel, J.F. and Shields, D.H., 1978, The Pressuremeter and Foundation Engineering, Trans. Tech. Publications, Rockport, Mass.

[3] Imai, T., 1982, "The Estimation Method of Pile Behaviour Under Horizontal Load using LLT Test Results", Proceedings of the Symposium on The Pressuremeter and Its Marine Applications, Paris, 1982.

[4] Robertson, P.K., Hughes, J.M.O., Campanella, R.G. and Sy, A., 1983, "Design of Laterally Loaded Displacement Piles using a Driven Pressuremeter", ASTM STP 835, Design and Performance of Laterally Loaded Piles and Pile Groups, June, Kansas City, Mo.

[5] Baguelin, F., 1982, "Rules for the Structural Design of Foundations Based on the Selfboring Pressuremeter Test", Symposium on the Pressuremeter and its Marine Applications, Paris, April.

[6] Robertson, P.K. Campanella, R.G., Brown, P.T., Grof, I. and Hughes, J.M.O., 1985, "Design of Axially and Laterally Loaded Piles using In-situ Tests: A Case History", Canadian Geotechnical Journal, Vol. 22, No. 4.

[7] Goldsmith, P.P., 1979, "Aspects of Soil-Pile Interaction under Static Loads", Ph.D. Thesis, University of Auckland, New Zealand.

[8] Hughes, J.M.O., Goldsmith, P.R. and Fendall, H.D.W., 1979, "Predicted and Measured Behaviour of Laterally Loaded Piles for the Westgate Freeways Melbourne", a paper presented to the Victoria Geomechanics Society, Australia, August.

[9] Byrne, P.M. and Atukorala, U., 1983, "Prediction of P-y Curves from Pressuremeter Tests and Finite Element Analyses", University of B.C., Civil Eng. Dept., Soil Mechanics Series No. 66.

[10] Clark, J.I., McKeown, S., Lester, W.B. and Eibner, L.J., 1985, "The Lateral Load Capacity of Large Diameter Concrete Piles", Proceedings of the 38th Canadian Geotechnical Conference, Edmonton.

[11] Brown, P.T., 1985, "Predicting Laterally Loaded Pile Capacity Using the Pressuremeter", M.A.Sc. Thesis, University of British Columbia, Dept. of Civil Engineering.

Stephen Kay, D. Vaughan Griffiths, and Harry J. Kolk

APPLICATION OF PRESSUREMETER TESTING TO ASSESS LATERAL PILE RESPONSE IN CLAYS

REFERENCE: Kay, S., Griffiths, D.V., and Kolk, H.J., "Application of Pressuremeter Testing to assess Lateral Pile response in Clays," The Pressuremeter and Its Marine Applications: Second International Symposium, J.-L. Briaud and J.M.E. Audibert, Eds., American Society for Testing and Materials, 1986.

ABSTRACT: Typical lateral load magnitudes encountered on land and offshore are compared. The effect of pile deflection and stiffness response on jacket performance is discussed. The limitations of current design methods are noted. Alternative design procedures are given, using site specific pressuremeter load-deformation data and a quasi three-dimensional non-linear finite element method. Comparisons are made between the observed and computed behaviour of lateral load tests on small diameter piles in clay for which pressuremeter data are available. The problem of extrapolation to larger diameter single piles, and the behaviour of pile groups, is discussed.

KEYWORDS: soils, pressuremeter, piles, lateral load, finite elements

Messrs. Kay and Kolk are respectively Computing Manager and Manager of Engineering at Fugro B.V., P.O. Box 63, 2260 AB Leidschendam, Holland; Dr. Griffiths is Lecturer at the Simon Engineering Laboratories, University of Manchester, Manchester M13 9PL, United Kingdom.

INTRODUCTION

For single piles, the lateral pile response to static loads is normally determined using 'p-y' curves. These are non-linear force displacement relationships for the soil mass supporting a unit length of pile. These 'p-y' curves in design codes have been derived from back-analysis of laterally loaded piles and laboratory test data. The number of such case histories is small; consequently the number of soil types is also small - three in fact.

Problems therefore occur when different soil types are encountered. A secondary problem occurs in the extrapolation to the behaviour of piles which are considerably larger than used in the original lateral load tests. Such pile sizes are commonly encountered offshore.

Pile groups are generally required at each leg when the platform is situated in more than about 50 m water depth. The group response is normally obtained from the above p-y method plus elastic solutions to allow for interaction effects. Significant assumptions are generally made which lead to conservative deflection estimates.

An alternative method is discussed. This uses site specific data, derived from in-situ pressuremeter testing, plus a finite element method. Recommendations are given for the pressuremeter testing and analysis procedures.

The alternative method has been evaluated for 4 case studies in clay soils. These include the historic laterally loaded test pile sites at Lake Austin, Sabine River and Manor. The E, ν and Su data from pressuremeter tests of Smith [1] have been used.

CONVENTIONAL DESIGN PRACTICE

Beam-column Method

In this widely used model, the beam (i.e. pile) is supported by a set of springs which model the soil behaviour. The spring characteristics are given by non-linear 'p-y' curves. They represent the lateral soil resistance per unit length of the pile (p) against pile deflection (y).

If the 'p-y' curves can be accurately specified, then the results of deflected shape and bending moment profiles are acceptable.

Formulae and tables to construct the 'p-y' curves are given in various design codes. For offshore constructions, the most commonly used are the API code [2]. Therein, the curves are limited to three soil types: soft clay, stiff clay and (non carbonate) sand.

The 'p-y' criteria for soft clay were derived from lateral load tests performed on 12.75" (325 mm) dia. piles at Lake Austin, and at the mouth of the Sabine River [3]. The stiff clay criteria were obtained from lateral load tests on 6" (152 mm) and 24" (610 mm) diameter piles at Manor [4]. In addition a unified approach for both clay types has been proposed [5]. This was based on the above sites, plus additional testing made in San Francisco Bay. Similarly, the sand 'p-y' curves are based on tests at Mustang Island [6].

For clay, the above criteria are expressed in terms of the pile diameter, undrained shear strength, the axial strain half way to failure in a compression test, and empirical coefficients. For sands, the soil friction angle, empirical coefficients and the modulus of subgrade reaction, are required. Generally, the selection of the latter value is usually of concern for offshore structures.

The site specific data is thus obtained from laboratory tests plus empiricism.

Further problems arise when soils are encountered (such as soft rock, silts and carbonate sands) whose behaviour is different to those tested above.

Pressuremeter Method

More recently, methods have been proposed which develop the 'p-y' curves from the entire pressuremeter expansion curve. Briaud et al [7] show that the 'p-y' curves may be obtained from the addition of the frontal and side resistance components. Each of these is subject to a pile shape factor and a reduction factor to allow for the critical depth.

Robertson et al. [8] present a simpler method based on both full displacement and self boring pressuremeter curves. For the latter type, pressures are magnified by a factor 9/5, and the strains by half the pile diameter to obtain displacement. Again the critical depth is allowed for by reduction of the resultant pressure values.

Such pressuremeter based methods constitute an improvement, since the soils data is now obtained in-situ and is site specific. However some degree of empiricism remains in the analytical procedures.

OFFSHORE CONSIDERATIONS

Offshore Pile Loads

Offshore, Hult, the ultimate lateral capacity of the piles is never achieved due to their long length and fixed head condition. In order to provide a rough indication, Hult has been evaluated from the formula

$$H_{ult} = N_c * S_u * D_o * L$$

where N_c = bearing capacity factor = 9
S_u = undrained strength = 200 kN/m^2 for N.Sea clays
= reported values for test sites
D_o = outer pile diameter
L = pile length, assumed = 10 D_o

In addition, Randolph [9] has shown that the relevant measure of pile/soil stiffness is the value of Ep/G*
where Ep = pile effective Young's modulus
= $(EI)p/(\pi D_o{}^4/64)$
$G^* = G(1 + 3\nu/4)$
G = soil shear modulus
ν = soil Poisson's ratio

Table 1 compares typical load and stiffness ratios for offshore piles and the original laterally loaded piles on land.

TABLE 1--Typical Pile Data.

Location	Clay Type	Pile Diameter D_0 (m)	Max. Lateral Load H (MN)	Load Ratio H/Hult (%)	Pile/Soil Stiffness Ep/G*
Offshore piled platforms					
Southern N. Sea	stiff	1.00-1.25	0.5-1.0	2.8-4.4	1100-1400
Northern N. Sea	stiff	1.50-2.10	1.0-3.0	5.6-7.9	1400-1700
Land 'p-y' test sites					
Lake Austin	soft	0.323	0.10	57.0	32700
Sabine River	soft	0.323	0.08	56.8	41400
Manor	stiff	0.641	0.60	10.8	1250

The above table illustrates the following points:

(1) The significant increase in scale from onshore to offshore foundations. Design of onshore piles can be done with a relatively large confidence in their performance, since it is based on pile load tests on similar pile sizes installed in comparable soil conditions. Virtually no such data are available for offshore piles.

(2) The offshore piles mobilise a smaller percentage of their ultimate capacity despite their larger loads.

(3) The pile/soil stiffness ratio offshore differs little from the land test made in stiff clay.

Structural Performance

For shallow water depths, the effect of reasonable variations in horizontal soil stiffness on the jacket design is not great. Under horizontal loading, the smallest eigen frequency of the jacket/soil system is lower than the environmental (i.e. wave) frequency.

However, as structures move into deeper water, the platform and wave frequencies may coincide. Thus, in this case, better estimates are required of the foundation stiffness components both for a single pile and the interaction effects for the pile group.

PRESSUREMETER METHOD

The Finite Element Method requires as input real soil properties. These are obtained using the pressuremeter. This may be either of the push in (self boring) [10, 11] or the full displacement [12] type. In the pressuremeter test, at least 2 unload/ reload cycles should be performed prior to reaching the maximum volumetric strain. This is in order to check the variation of shear modulus G with stress level. Current Fugro practice is to cycle at the third points of the maximum strain range; however this procedure is soil formation dependent.

Table 2 shows how the soil parameters are determined for the push-in type of pressuremeter.

TABLE 2--Soil Parameters

Parameter	Obtained from PMT	Reference
Shear modulus G	unload/reload loops	Henderson et al. [10]
Undrained Shear Strength S_u	total pressure vs log volumetric strain	Gibson + Anderson [13]
Friction angle Ø'	log effective pressure vs log volumetric strain	Hughes et al. [14]
At rest pressure coefficient K_o	lift off pressure	Lacasse et al. [22]

By performing a series of tests at various depths in one or more boreholes, profiles of the above parameters versus depth are obtained. These are then used in the Finite Element analyses.

FINITE ELEMENT METHOD

The Finite Element Method was used for this case study. It offers the following advantages over the traditional beam column method:

(1) real soil properties are used.

(2) a soil model identical with that used for back-analysis of the pressuremeter data [10, 13, 14] is used, namely a linear elastic-perfectly plastic (Mohr-Coulomb) model [23]. In this model, parameters E and ν define the elastic soil behaviour, and c', Ø' and θ' (effective stress-sands) or S_u (total stress-clays) define the plastic work components.

(3) no additional value judgements are required; for example the reduction in ultimate lateral pressure at shallow depth, or the base shear stiffness for short piles, is automatically included. The analysis of laterally loaded retaining walls (both flexible [24] and stiff [25]) has demonstrated the capability of the Finite Element Method to obtain classical ultimate Rankine earth pressure distributions for the equivalent plane strain problem.

(4) the method may be also used for pile groups.

In this application, the soil is considered as a continuum initially attached to the pile shaft. A Fourier harmonic analysis technique is used to reduce the three dimensional problem to a two dimensional one. Each node has three degrees of freedom (r, z, θ). In the r-z plane the normal shape functions are used. In the θ plane a Fourier series is used since the material and geometry remain constant. Elastic analyses were first reported by Wilson [15]. Elasto-plastic formulations have been subsequently incorporated [16, 17, 18].

To handle the non-linearity, a constant stiffness matrix formulation was used, together with the viscoplastic method for redistribution of the excess stresses in the plastic zones. A no tension criterion [19] was incorporated. Eight noded rectangular (isoparametric) elements with two point gauss integration were used. The stresses were checked at 45 degree intervals. Five harmonics were used. Further program details are given by Griffiths [20]. Typical run-times of 200 seconds were required on a CDC 7600 computer system to obtain the presented results.

SITES SELECTED

Four sites were selected where both pile and pressuremeter tests had been performed. The latter were all conducted by Smith [1]. The pressuremeter testing was done either in an augered or a drilled borehole using a prototype TEXAM unit. The values of E and S_u presented by Smith were reused in the finite element analyses. Table 3 shows that the time lapse between the two types of tests was up to 25 years.

TABLE 3--Site History

Sites	Date of Lateral Pile Test	Date of Pressuremeter Test	Soil Type
Lake Austin	1956	1981	soft clay
Sabine River	1960	1982	soft clay
Manor	1967	1982	stiff clay
Texas A + M	1977	1981	stiff clay

Common to all sites were the following features:
(1) clay soils were tested
(2) the piles were free headed
(3) the initial static loading curve was analysed
(4) the reload modulus (E_R) was used in preference to the initial modulus (E_i)
(5) a fixed Poisson's ratio ν = 0.3 was used by Smith to derive the Young's modulus value E from the shear modulus G.

Lake Austin (Figure 1)

The soft clays and silts had been deposited this century behind the Lake Austin Dam, Texas. Due to exposure during drawdown, the upper soils are dessicated, jointed and fissured. During the pile test in 1956, vane testing gave shear strength values averaging 38 kN/m^2 with little variation with depth [3]. Unconfined triaxial tests were also done, giving shear strengths of about 24 kN/m^2 [1].

The soil profiles determined by Smith [1] from the pressuremeter testing in 1981 are shown on Fig. 1a. The shear strength values so derived (17 to 20 kN/m^2) are slightly lower than those previously reported.

Sabine River (Figure 2)

The site is located at the mouth of the Sabine River, which forms the Texas/Louisiana State boundary. The pile test was performed in 1960. This, plus the Austin test, formed the basis for the "soft clay" design method [3]. The soils are mainly soft slightly overconsolidated marine clays with a sand layer occuring from 5.0 to 6.1 m below ground level. Sand seams and partings are present throughout the soft clay. Vane shear strengths averaged 14.5 kN/m^2 in the significant upper zone. Unconfined tests showed a general strength increase with depth, ranging from 5 kN/m^2 near the mudline to 24 kN/m^2 at 9.1 m depth. The pile test was made in a 1.2 m deep pit flooded to 0.15 m depth.

Pressuremeter tests were made in 1982. The shear strength derived was again less than obtained previously by other tests. Smith [1] concluded that the soil profile was representative of that twenty years ago.

Manor (Figure 3)

The Manor site is located some 8 km North East of Austin, Texas. The surface soils are stiff overconsolidated clays of marine origin. They have a secondary slickensided and fissured structure. The shear strength increases rapidly with depth. The "stiff clay" design method [4] was based on results of testing 6.25" and 25" diameter pipe piles.

To simulate an offshore environment, a flooded pit 1.8 m deep was used. After excavation was completed, the pit was flooded with water some 5 months prior to pile installation. Another month passed before the piles were loaded. Borings made at various stages of the pit excavation indicated that the water content had increased and the shear strength had decreased in the upper zones.

The reported final shear strength profile from the test mudline gave an increase from 96 kN/m^2 at 0.3 m depth to 385 kN/m^2 at 3.6 m depth.

In 1982, pressuremeter testing was done. Due to the placement of about 1.6 m of fill and the limited depth to which testing could be done, only two tests could be made below the original pile mudline. Thus, at depth, the profiles used for the finite element analysis have been estimated on the originally reported shear strength profile shown on Fig. 3a, plus the ratio $E/S_u = 300$.

Texas A & M (Figure 4)

Lateral pile tests have been conducted on the campus of Texas A & M University in 1977, 1978 and 1979. All were on cast in-situ concrete piles. In this study, only the first was analysed. The clay is firm to stiff in the upper 1.5 m, and thereafter stiff to very stiff. Above the water table (4.5 - 5.5 m depth), the soil has been preconsolidated by dessication. The concrete pile failed structurally during the test [7].

In 1981 a pavement pressuremeter was used, giving the profiles shown on Fig. 4a. No fill had been placed and no water content alterations had occurred; thus, like the Sabine test site, a reasonable correlation was anticipated.

FINITE ELEMENT ANALYSES

Results

The profiles adopted for the analysis are shown on Figs. 1 to 4. The ratio E/s_u was around 300 (using $\nu = 0.30$).

The computed and observed load-deflection results are compared on the same Figures. As is obvious from the results, the stiffness has been overestimated, though not grossly, in all cases. However, the initial stiffnesses are reasonable, particularly for the Sabine and Texas A+M cases.

The cases studied illustrate the need to determine soil conditions at the time of the actual tests. The Lake Austin and Manor Sites have probably been influenced by dessication and flooding respectively. In addition, the latter portion of the Texas A+M result may be less stiff due to plasticity of the reinforced concrete pile occurring prior to failure.

Softened Zone Influence

As part of a sensitivity study, a softened zone with $E/2$ and $s_u/2$ immediately adjacent to the pile shaft for the Sabine and Texas A + M cases was adopted. The match was improved slightly, but the initial stiffness decreased marginally (Figs. 2b and 4b). The results were for an annulus 50 mm thick. Little alteration was obtained for 100 and 150 mm thicknesses. It was concluded from this that the initial pile stiffness is primarily influenced by the stiffness of the soil mass in the far field as opposed to the soil zone near to the pile.

Elastic Modulus Influence

The probable cause of the over-stiff response lies in in the fact that the non-linear load-displacement behaviour of the clay has been modelled in the analyses by two straight lines. A better approximation may have been by a hyperbola (Fig. 5). Time constraints precluded the insertion of this feature. The following procedure was therefore adopted. The Sabine River case was re-analysed with $E/s_u = 100$ to give a lower bound deflection response. The results (Fig. 2b) are encouragingly close to the measured pile response at the high load levels. It is concluded that an E/s_u ratio of 300 gives reasonably good estimates of initial stiffness, and that an E/s_u ratio of 100 should be used when soil strains in excess of, say, $\varepsilon_{max}/2$ are encountered.

LARGE DIAMETER PILES

Offshore, extremely large piles, up to 7.0 m diameter, are occasionally required to serve as monopile constructions or as seabed anchor points. It is generally believed that, for such piles, the displacements computed in accordance with the API code are an overestimate. However, there is very little published quantitative data on full scale large diameter pile tests to positively confirm the "diameter effect".

However, the present study indicates the probable reason for the diameter effect: The pile head stiffness is influenced greatly by the soil secant Young's modulus E_s. For example, in the beam column/p-y model for clays,

E_s is determined through a point in the triaxial stress-strain curve at one half of the maximum (peak) strain. In practice the ratio E_s/S_u is taken to be between 50 and 200, and typically 100 [3]. The four case studies have indicated that a better indication of initial pile head stiffness can be made using $E/S_u = 300$.

For a given load, the soil strains are inversely proportional to the pile diameter. Since such large piles operate at relatively low load levels, the strains are extremely small for the majority of the soil mass. At such low strain levels, a higher soil stiffness value is in operation.

OFFSHORE PILE GROUPS

For piled platforms in high water depths, groups of piles are required. The piles are driven at relatively close spacings (2 to 3 diameter spacing). The lateral stiffness of the group is then a function of the single pile response, which has been described above, plus effects due to interaction between the piles.

The finite element method may be used to also improve the accuracy of the interaction factors. Errors, due to incompatibility between the beam column and continuum analyses, used in the Poulos/p-y method [21], are eliminated, since both the single pile and pile group behaviour are obtained from the same finite element analysis.

Briefly, the finite element output (Figs. 6 to 8) is re-analysed to determine the additional displacement occurring within the soil mass surrounding the loaded pile. By taking the displacements components in the direction of loading, the interaction factors for perfectly flexible piles can be derived. A typical result is shown in contour form (Fig. 9). More complex methods, incorporating pile stiffness, have been described elsewhere [11].

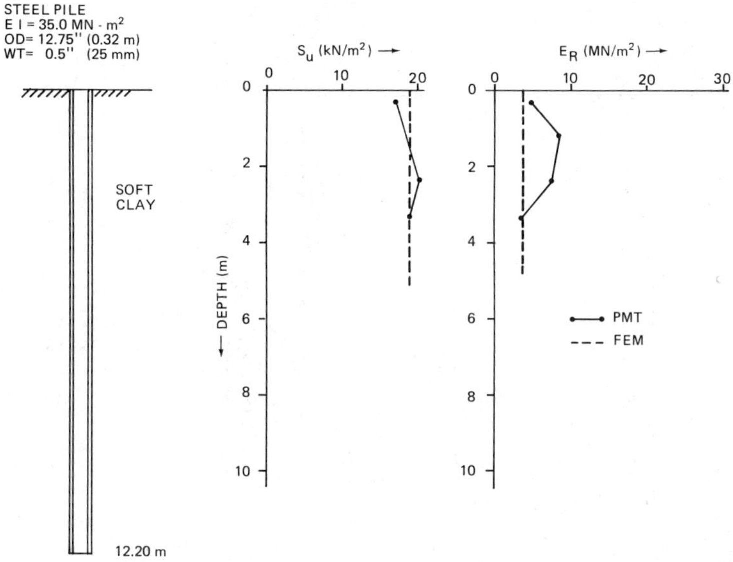

FIG. 1a Properties - Lake Austin Site

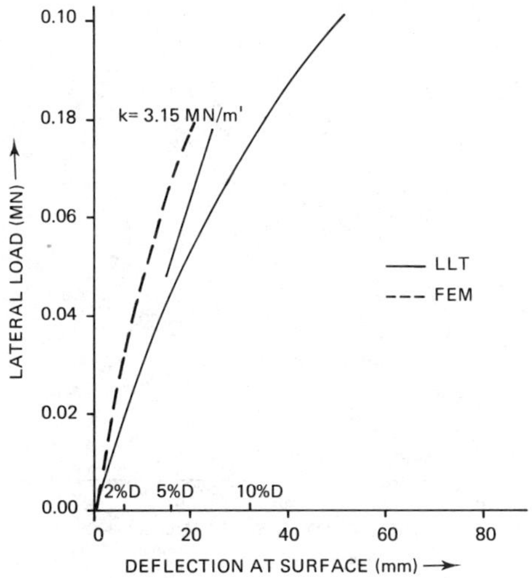

FIG. 1b Results - Lake Austin Site

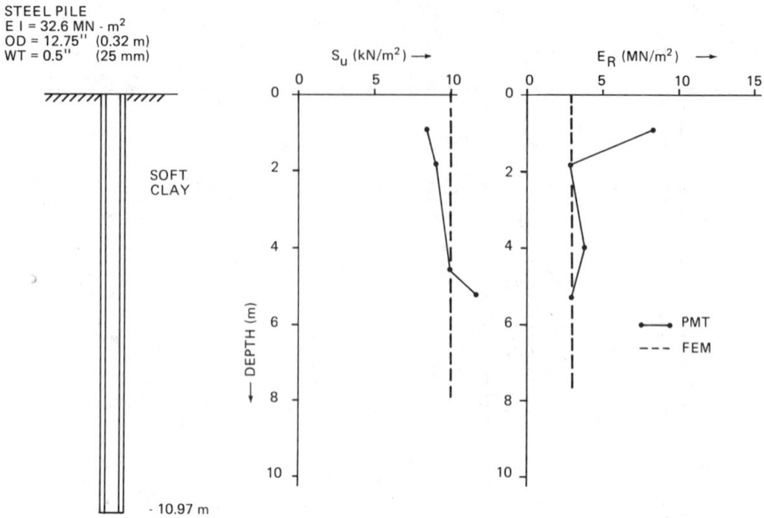

FIG. 2a Properties - Sabine Site

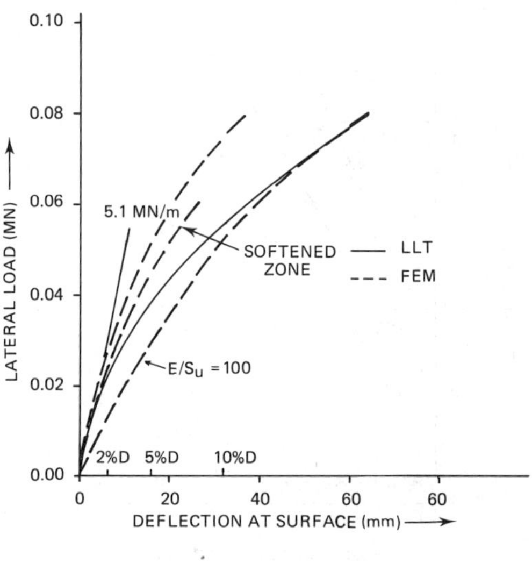

FIG. 2b Results - Sabine Site

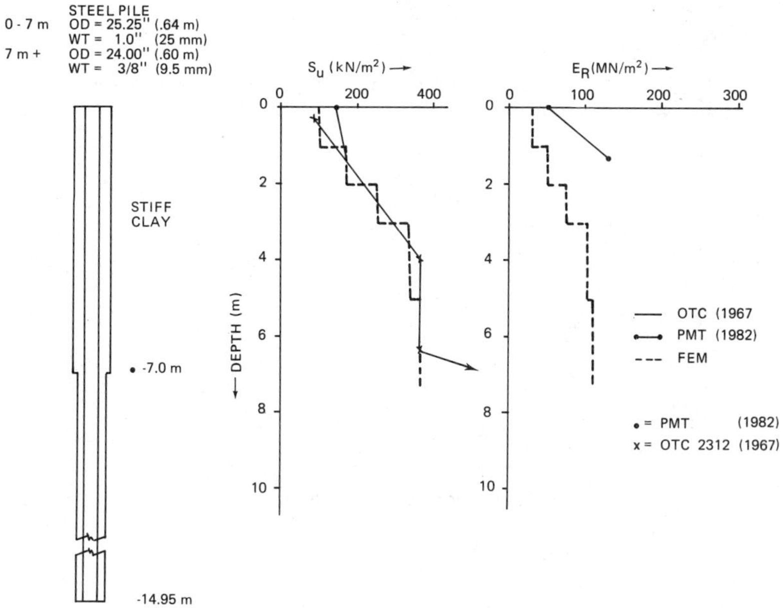

FIG. 3a Properties - Manor Site

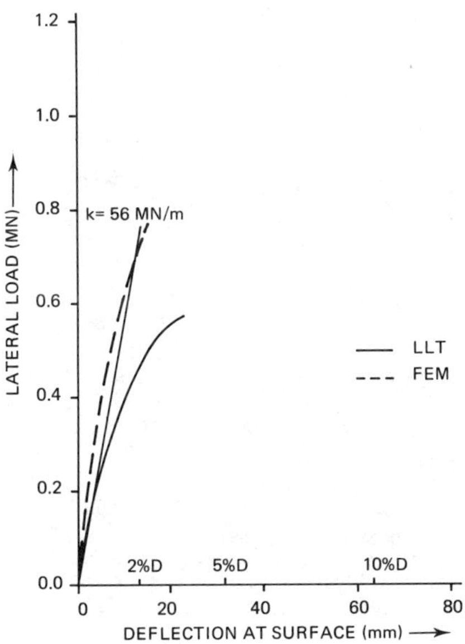

FIG. 3b Results - Manor Site 25"pile

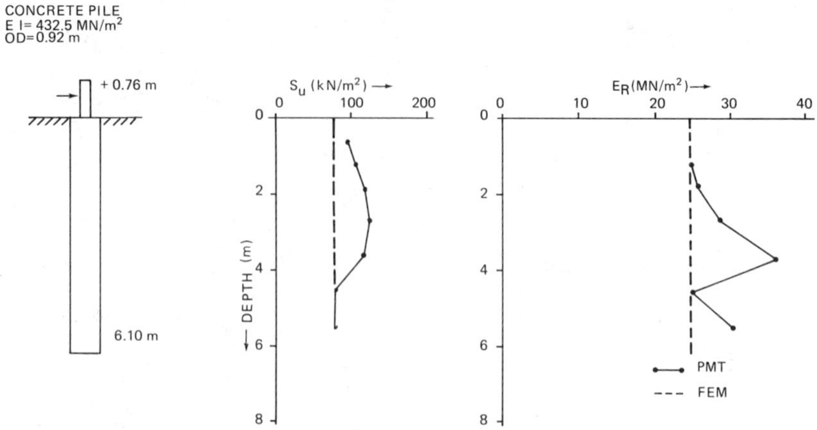

FIG. 4a Properties - Texas A&M Site

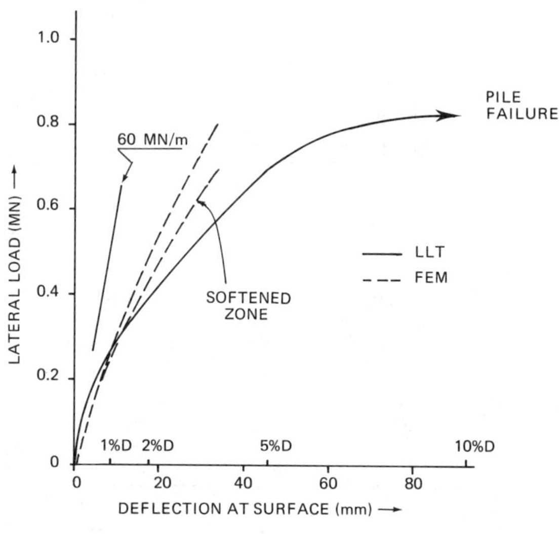

FIG. 4b Results - Texas A&M Site

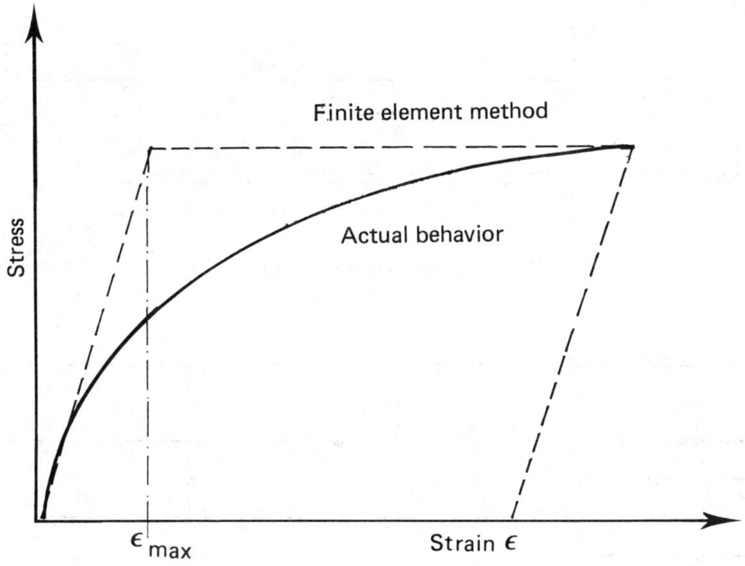

FIG. 5 Stress- strain curve

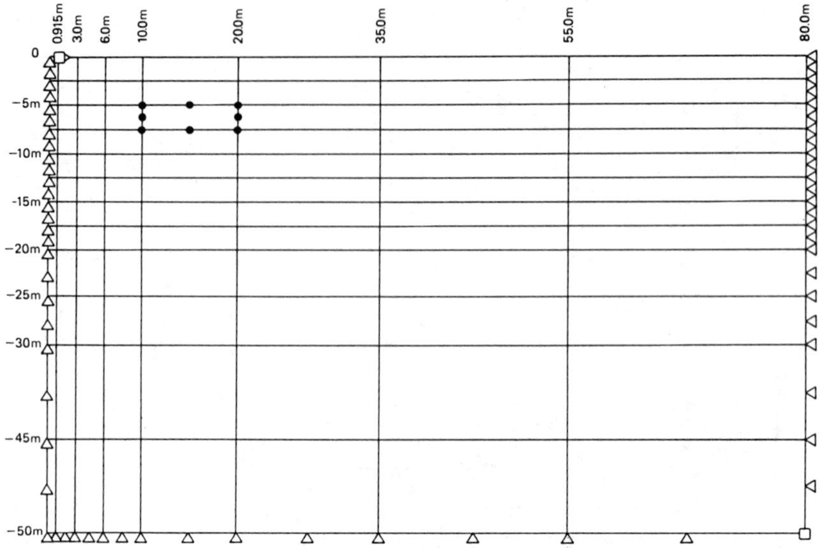

FIG. 6 Finite Element Mesh

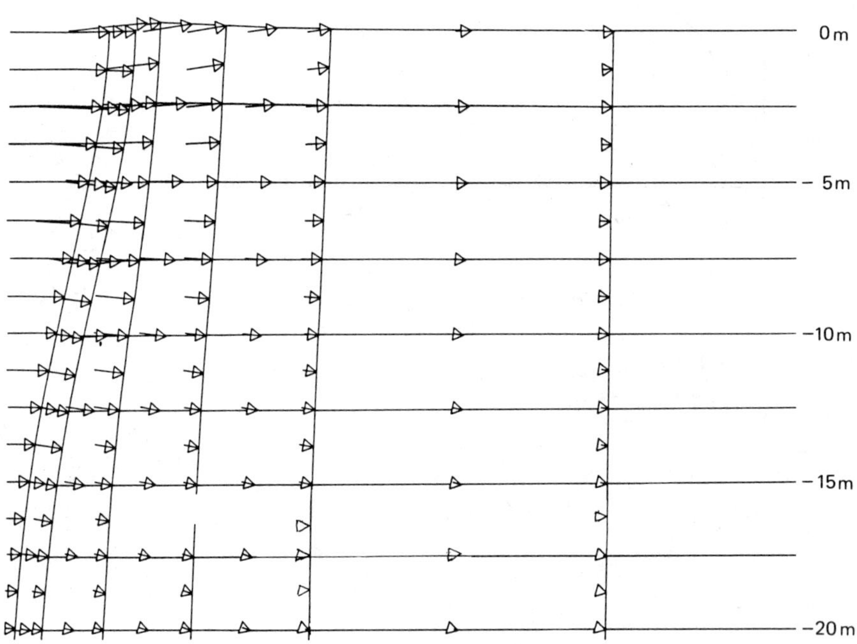

FIG. 7 Deformed Mesh and Displacement Vectors at Front of Pile

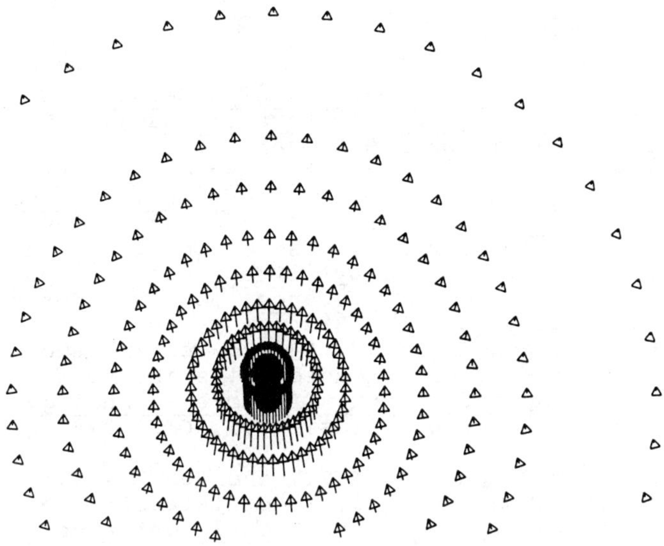

FIG. 8 Displacement Vectors at Ground Surface

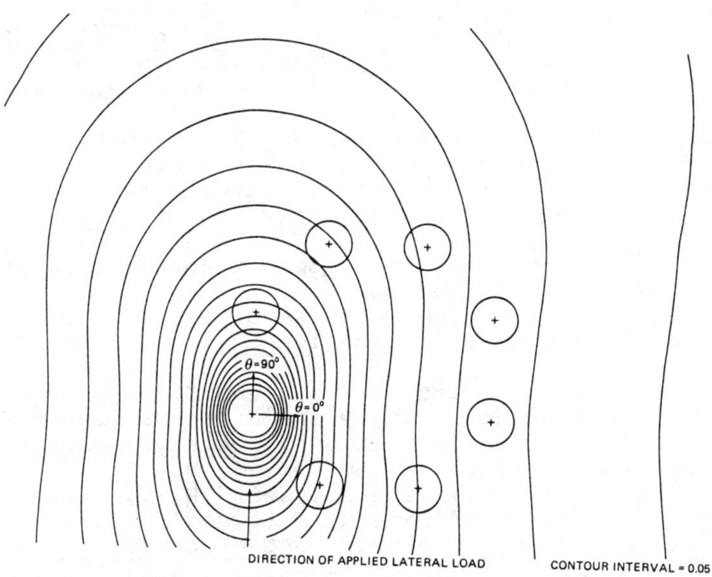

FIG. 9 Interaction Factors for Flexible Pile

REFERENCES

[1] Smith, T.D., Pressuremeter Design Method for Single Piles subjected to Static Lateral Load, Ph.D. thesis, 1983, Texas A&M University.

[2] American Petroleum Institute, Recommended Practice for Planning, Designing and Constructing Fixed Offshore Platforms, API RP2A, 16th ed., 1984.

[3] Matlock, H.M., "Correlations for Design of Laterally Loaded Piles in Soft Clay", Preprints, Second Annual Offshore Technology Conference, Houston, Texas, 1970, OTC 1204, pp. 578-588.

[4] Reese, L.C. and Co, W.R., "Field Testing and Analysis of Laterally Loaded Piles in Stiff Clay", Preprints, Seventh Annual Offshore Technology Conference, Houston, Texas, 1975, OTC 2313, pp. 671-690.

[5] Sullivan, W.R., Reece, L.C., and Fenske, C.W. "Unified Method for Analysis of Laterally Loaded Piles in Clay". 1st International Conference on Numerical Methods in Offshore Piling, ICE, London, 1980, pp. 135-146.

[6] Reece, L.C., Cox, W.R., and Koop, F.D., "Analysis of Laterally Loaded Piles in Sand". Preprints, Sixth Annual Offshore Technology Conference, Houston, Texas, 1974, OTC 2080.

[7] Briaud, J.-L., Smith, T.D. and Meyer, B., "Laterally Loaded Piles and the Pressuremeter: Comparison of Existing Methods.", Laterally Loaded Deep Foundations: Analysis and Performance, ASTM STP 835, J.A. Langer, E.T. Mosley, and C.D. Thompson, Eds., American Society for Testing and Materials 1984, pp. 97-111.

[8] Robertson, P.K., Hughes, J.M.O.M., Campanella, R.G. and Sy, A., "Design of Laterally Loaded Displacement Piles Using a Driven Pressuremeter", Laterally Loaded Deep Foundations: Analysis and Performance, ASTM STP 835, J.A. Langer, E.T. Mosley and C.D. Thompson, Eds., American Society for Testing and Materials, 1984, pp. 229-238.

[9] Randolph, M.F., "The Response of Flexible Piles to Lateral Loading", Géotechnique 31, 1981, No. 2, pp. 247-259.

[10] Henderson, G., Smith, P.P.K., and St.John, H.D. "The Development of the Push-in Pressuremeter for Offshore Site Investigation", Society for Underwater Technology Conference, London, 1979.

[11] Kay, S., Kolk, H.J. and Hooydonk, W.R., "Site Specific Design of Laterally Loaded Piles", Proceedings of Conference on Geotechnical Practice in Offshore Engineering, Austin, Texas, 1983, pp. 557-580.

[12] Withers, N.J., Schaap, L.H.J., and Dalton, J.C.P., "The Development of a Full Displacement Pressuremeter," The Pressuremeter and Its Marine Applications: Second International Symposium, J.-L. Briaud and J.M.E. Audibert, Eds., American Society for Testing and Materials, 1986.

[13] Gibson, R.E and Anderson, W.F., "In-situ measurements of Soil Properties with the Pressuremeter". Civil Engineering and Public Works Review, Vol. 56, No. 658, 1961, pp. 615-618.

[14] Hughes, J.M.O.M., Wroth, C.P., and Windle, D., "Pressuremeter Tests in Sands". Géotechnique Vol. 27, No. 4, 1977, pp. 455-477.

[15] Wilson, E.L., "Structural Analysis of Axisymmetric Solids", Journal American Institute of Aeronautics and Astronautics, Vol. 3, 1965, pp. 2249-2274.

[16] Meissner, H.E., "Laterally Loaded Pile in Cohesionless Soil", Numerical Methods in Geomechanics, ASCE, Vol. 3, 1976, pp. 1353-1365.

[17] Winniki, L.A., and Zienkeiwicz, O.C., "Plastic (or Visco-Plastic) Bodies Subjected to Non-symmetric Loading - Semi-analytical Finite Element Solution", International Journal Numerical Methods in Engineering, Vol. 4, 1979, pp. 1399-1412.

[18] Barton, Y.S. and Pande, G.N., "Laterally Loaded Piles in Sand: Centrifuge Tests and Finite Element Analyses", Numerical Models in Geomechanics, R. Dungar, G.N. Pande and J.A. Studer, Eds., Balkema, Rotterdam, 1982, pp. 749-758.

[19] Pandé, G.N., "Visco-plastic Algorithm in Numerical Modelling of Jointed Rock and Concrete Masses", Mechanics of Bi-Modulus Materials, American Society of Mechanical Engineers, AMD 33, 1979.

[20] Griffiths, D.V., HARMONY: Fourier Analysis Program, Report to Fugro B.V., 1984.

[21] Focht, J.A., and Koch, K.J., "Rational Analysis of the Lateral Performance of Pile Groups", Preprints, Fifth Offshore Technology Conference, Houston, Texas, 1973, OTC 1896.

[22] Lacasse, S., and Lunne, T., "In-situ horizontal Stresses from Pressuremeter Test", Symposium on the Pressuremeter and Its Marine Applications, Editions Technip, Paris, 1982, pp. 187-208.

[23] Smith, I.M., Programming the Finite Element Method with application to Geomechanics, John Wiley, London, 1982, pp. 155-158.

[24] Smith, I.M., and Boorman, S., "Analysis of Flexible Bulkheads in Sands", Proceedings Institution of Civil Engineers, London, part 2, 1974, pp. 413-416.

[25] Potts, D.M. and Fourie, A.B., "The behaviour of a propped retaining wall: results of a numerical experiment", Géotechnique, Vol. 34, no 3, 1984, pp. 383-404.

Trevor David Smith and Robert Slyh

SIDE FRICTION MOBILIZATION RATES FOR LATERALLY LOADED PILES FROM THE PRESSUREMETER

REFERENCE: Smith, T.-D., Slyh, R., "Side Friction Mobilization Rates for Laterally Loaded Piles from the Pressuremeter," The Pressuremeter and Its Marine Applications: Second International Symposium ASTM STP 950, J.-L. Briaud and J. M. E. Audibert, Eds., American Society for Testing and Materials, 1986.

ABSTRACT: Pile foundations in the offshore environment are frequently called upon to resist high horizontal load. Many pressuremeter methods have recently been developed to design piles, including techniques to formulate soil spring P-y curves directly from pressuremeter test results. Lateral soil resistance is now recognized to arise from two components: side shear and front resistance. The field measurement of mobilization rates and prediction of lateral side shear has never been made. This paper reports the attempts to measure lateral friction rates in sand and clay by employing torsion testing as a "yardstick" method. Pure lateral shears are mobilized and the effects of depth and installation method can be studied qualatively. It is suggested that since the soil shear modulus, G, is the controlling variable, the pressuremeter is a suitable insitu tool to duplicate pile installation disturbance and predict side shear mobilization rates.

KEYWORDS: piles, pressuremeter, side friction, torsion.

INTRODUCTION

All design approaches for laterally loaded piles currently make use of a predetermined lateral soil modulus. The modulus is either independant of deflection in hand, (non-dimensional), solutions, or is

Dr. Smith is an Assistant Professor of Civil Engineering at Portland State University, Portland, Oregon: Mr. Slyh is a research engineer with the U.S. Navy Civil Engineering laboratory, Port Hueneme, California.

a function of deflection in the non-linear P-y approach. It is well acepted that axially loaded piles derive their resistance from side friction and end bearing. The same two components are also mobilized for the laterally loaded pile, expressed in total by the modulus value, or P value of the P-y curve. Since friction is mobilized quickly, at working loads the side friction now mobilized in a horizontal direction may provide the majority of the resistance. Smith (1) illustrated for a circular rigid concrete shaft that with a factor of safety of 8 on ultimate load side shear contributed 84% of the soil reaction. Hence, it is instructive to consider the two components of P separately, as side shear F and front pressure, Q, such that:

$$P = F + Q \tag{1}$$

which is illustrated qualitatively in Figure 1.

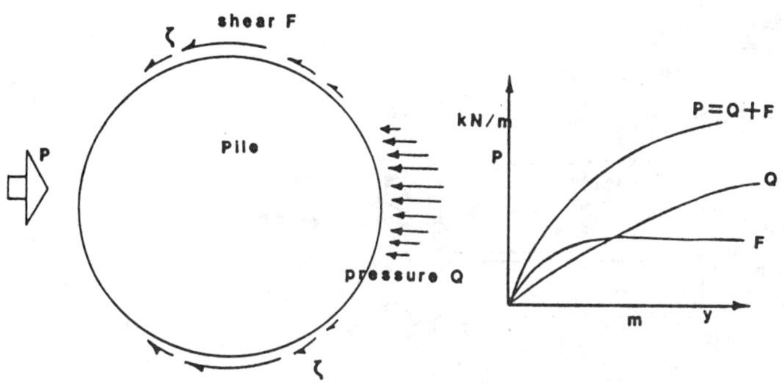

FIG. 1 - The Two Components of P

At least seven pressuremeter methods to design laterally loaded piles exist, (2), and it is accepted that estimates of pile movement at working load by these methods are probably conservative. The small strain resistance of the soil is underestimated by pressuremeter methods due to the lack of any shear model. Smith (3) illustrated from field measurements that the mobilization of the Q-y curve can be obtained from the prebored pressuremeter. To complete the prediction of total resistance P accurate assessment of shear F must also be made.

This paper describes the first stage of testing to measure lateral side friction mobilization rates in pure shear, relate the results to lateral load resistance mobilization, and assess the role of the pressuremeter. This is achieved through a model pile study applying torsion at the pile head to mobilize lateral side friction in pure shear and conduct pressuremeter tests and lateral load tests. Attempts are made to establish if any depth effects from a free surface are present (known to significantly affect P), and if pile movement to peak F is a function of diameter.

MODEL TEST SERIES

Tests were performed (4) on 19mm diameter solid steel rod and 50mm dia, 4.7mm wall thickness, steel pipe in compacted clay, loose and dense sand. A schematic of the testing frame and soil container is shown in Figure 2. The soil container has braced, rigid, side walls of plywood and has no surface restraint. The size of the box is consistent with previous model test studies and is designed to minimize boundary effects, Smith (3).

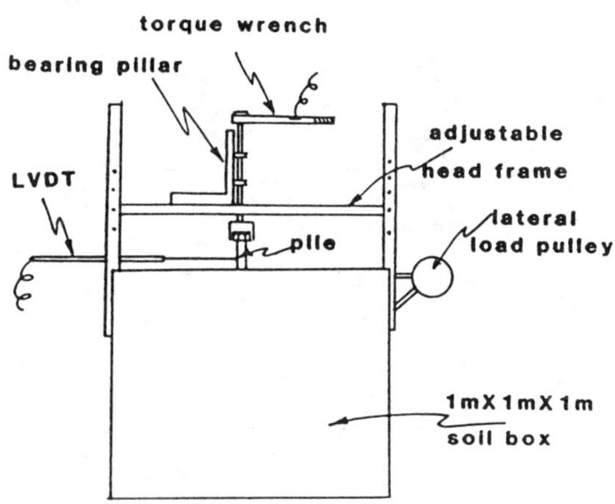

FIG. 2 - Schematic of Test Setup

Soil Placement and Torque Test Procedures

The clay was placed in 50mm layers, compacted with a hand tamp and sprayed with water prior to the next lift to achieve high saturation. The finished deposit showed a uniform vane shear strength of 25 kPa at a water content of 27%. Liquid and Plastic limits were 32% and 25% respectively. Torque testing in the clay was conducted by two procedures to examine installation disturbance and strain effects.

i) The pile was advanced by driving and a torque test immediately conducted to ultimate. Thereafter further advancement was made and again re-torqued to failure. This is termed Torque-Drive-Torque.

ii) After each torque test the pile was withdrawn, a 400mm diameter clay column was removed, recompacted and the pile driven to a new embedment and finally torqued to failure. This is termed Torque-Excavate-Torque; time between completion of recompaction and failure was approximately 20 minutes.

The sand comprised a fine uniform clean "Mason" Sand with 100% passing #8 sieve 50% passing the #30 and 0% passing #80 sieve. The sand was deposited by the same rain method to give a uniform loose unit weight of 14.5kN/m^3. The dense deposit was achieved by application of 2" air vibrating poker to give a unit weight of 17.5kN/m^3. Both loose and dense deposits were tested dry. Torque-Drive-Torque tests only were performed.

Torque was applied to the pile top by a strain gauge instrumented wrench calibrated to 100Nm. The vertical delivery shaft was supported in 2 bearings to remove the net horizontal load. Angular rotation was recorded from the linear movements of the pile circumference at the soil surface by a wire attached to an LVDT. Output from both the torque wrench and LVDT were fed via an analogue to digital converter through an Apple IIe and stored on floppy disks. In addition both channels were continously recorded by x-y plotter. This degree of sensitivity proved invaluable to record the pile rotation to reach peak torque. Piles were brought to residual failure in approximately 10 seconds.

Pressuremeter Test Procedure

Driven pressuremeter tests were conducted with a 32mm diameter 260mm long single cell probe inflated by a hydraulic system, the unit is similar in operation to the PENCEL unit manufactured by Roctest Inc. of Plattsburg, New York. Inflation took place at a constant rate of volumetric strain to give double the initial cavity size in 10 minutes. Both injected volume and mobilized pressure were recorded by the Apple IIe through the analogue to digitial converter.

A total of eight tests were conducted, three in clay, two in loose sand and three in dense sand. The pressuremeter was driven into position for testing in sand and clay. Approximate depth to test center line from the unrestrained ground surface in all tests was 650mm.

Results

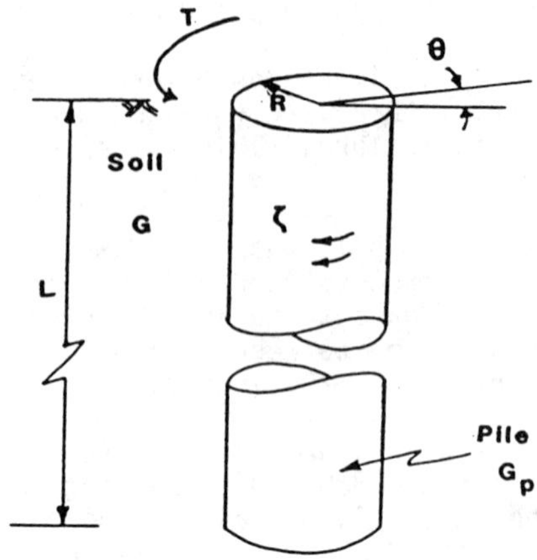

FIG. 3 - Notation for Torque Testing

With the notation given in Figure 3 Table 1 gives a summary of typical results with average adhesion, τ_f, calculated by Equation 2,

$$\tau_f = \frac{T_f}{2\pi L R^2} \qquad (2)$$

The incremetental shear, τ_i, is derived by considering only the increase in torque mobilized in the Torque-Drive-Torque series from the peak less the previous residual torque. A typical series of torque against rotation tests in clay and dense sand is shown in Figure 4 and 5 respectively.

TABLE 1--Summary of Torque Test Results

CLAY

TORQUE-DRIVE-TORQUE 19mm Pile

L/D	4	6	8	10
T_f	2.6	2.15	2.5	2.9
y_f	.051	0.1	0.127	.05
τ_f	59.0	32.4	28.5	27.0
τ_i		2.26	3.01	

19mm Pile (cont)

L/D	12	18	34
T_f	1.7	2.7	5.4
y_f	0.3	0.38	0.2
τ_f	14.6	13.8	14.4
τ_i	1.85	2.17	

50mm Pile

L/D	1	2	3	4	5
T_f	9.6	11.7	8.7	13.6	13.6
y_f	0.5	0.51	0.5	0.6	0.4
τ_f	46	28.4	14	13.2	8.2
τ_i	14.2		12.0	4.5	

TORQUE-EXCAVATE-TORQUE 50mm Pile

L/D	2	3.8	6	8	10
T_f	7.2	13.3	14.7	16.7	21.4
y_f	0.31	0.63	0.4	1.27	0.6
τ_f	17.5	16.7	11.8	10.1	10.4

50mm Pile

L/D	2	4	6	8	10
T_f	8.2	11.0	14.1	18.6	24.3
y_f	0.5	1.0	1.0	1.0	1.7
τ_f	20.0	13.2	11.3	11.3	11.7

SAND

TORQUE-DRIVE-TORQUE 50mm Pile Dense Sand

L/D	4	6	8	10
T_f	3.4	7.3	15.8	25.4
y_f	1.8	1.3	1.5	1.5
τ_f	4.1	5.9	9.6	12.3
K	K=5.0	4.82	5.84	6.0

Loose Sand 50mm Pile

L/D	13
T_f	7.3
y_f	1.3
τ_f	2.7
K	K=1.43

NOTATION

L/D = Length/Diameter
T_f = Failure Torque (N m)
y_f = Circumference movement to peak (mm)
τ_f = Failure shear over L (kPa)
τ_f = Incremental shear (kPa)(see text)
K = Ratio of horizontal to vertical stress

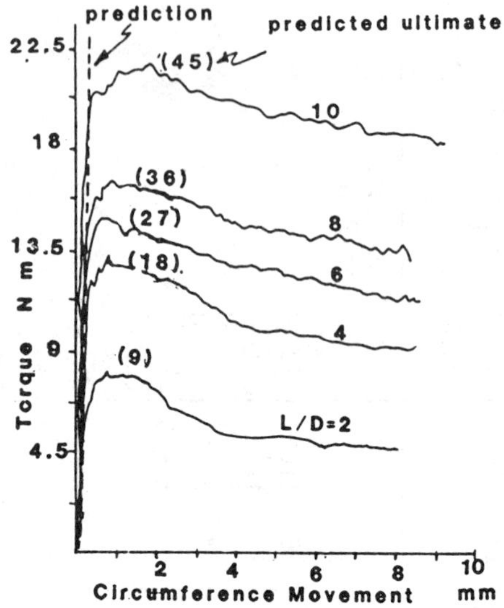

FIG. 4 - Torque vs. rotation in Clay (50mm Pile)

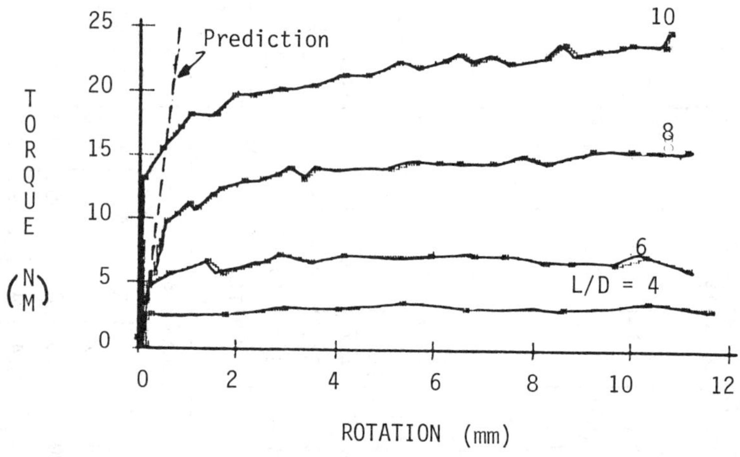

FIG. 5 Torque vs. Rotation in Dense Sand (50mm Pile)

A representative pair of pressuremeter tests results in sand and clay are given in Figure 6 and is uncorrected for any limited embedment depth effects.

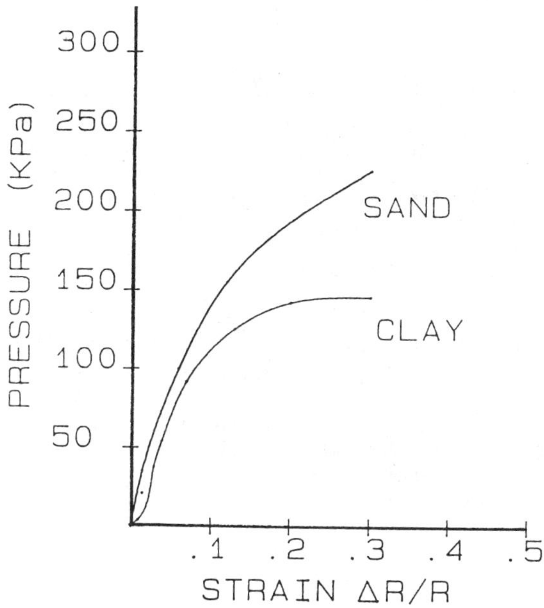

FIG. 6 Pressuremeter Test Results in Sand and Clay

ANALYSIS OF RESULTS

Theory

Analytical solutions for the pile torsion problem are available. Poulos (5) used an integral equation numerical method, and Randolph (6) presented more general closed form expressions for both elastic soil and elastic-plastic soil. Both reported the potential of torsion testing for prediction of vertical shear capacity of axially loaded piles. From the limited data available in the literature both solutions are in general agreement. Relative pile/soil flexibility considerations and length leads to a gradual reduction in the angle of twist down the pile. Hence piles may be classified in a similar manner to laterally loaded piles as rigid, seim-flexible and flexible. When considering the surface Torque versus Rotation response Randolph expressed torsional stiffness in the following way:

$$\frac{T}{\theta} = GR^3\left(\frac{16}{3} + 4\pi\frac{L}{R}\right) \quad (3)$$

for rigid piles, and

$$\frac{T}{\theta} = GR^3 4.44\left(\frac{G_p}{G}\right)^{1/2} \quad (4)$$

for perfectly flexible piles in uniform clays where length, L, exceeded a critical value L_c, given by

$$L_c = R\left(\frac{G_p}{G}\right) \quad (5)$$

By Randolph's solutions the 19mm diameter piles range from rigid to semi-flexible in both sand and clay. All 50mm diameter pile tests were rigid in behavior. It should be noted that the elastic response of the soil under pure shear is governed by the shear modulus G. However, real soil behavior is characterized by slip at peak adhesion, τ_f, and the shear attainable thereafter may reduce to a residual value in strain softening behavior.

Interpretation

The majority of pile tests used in the study used the rigid 50mm pile, therefore, the torque rotation measured is characteristic of the soil to pile shear response. The predictions given by Randolph from Equation 3, employing G values calculated from pressuremeter tests assuming elasticity in the psuedo linear range on the initial cycle, are shown dashed on Figure 4 and 5 for sand and clay respectively. Failure was assumed at peak miniature vane shear strengths in the clay. The excellent correspondence in the modulus range suggests the pressurmeter evaluation of G is valid for lateral shear response.

Figure 4 suggests that peak adhesion mobilized in the clay is less than the undrained shear strength and Table 1 indicates movement to failure is a function of both pile diameter and installation disturbance. This substantiates recent discussion which has questioned the use of a constant vertical displacement required to mobilize peak shear in both the Wave Equation and settlement predictions by load-transfer methods (7).

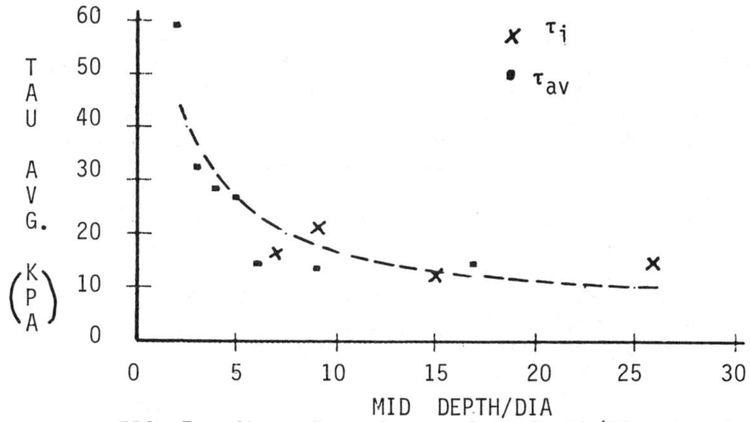

FIG. 7 - Shear Transfer against Depth/Diameter Ratio

Figure 7 shows the percentage of adhesion mobilized in the clay tests as a function of mid depth in the increment considered. Clearly a pronounced depth effect is indicated. The top 5 pile diameters exhibit shear transfer in excess of the known undrained shear strength prior to driving. A striking resemblance is found to the variation of axial shear reported in field load tests in the λ method, Vijayvergiya and Focht (8) and also reported by Aurora et al. (9).

Table 1 reports the backcalculated K value, the ratio of vertical to horizontal stress for the sand tests, which shows no trend with depth. If the mobilized shear stress is given by a friction relationship then at any depth Z.

$$\tau_f = K\gamma z \tan\delta \tag{6}$$

where γ = sand unit weight
δ = sand to steel friction angle
$= \frac{2}{3}\phi$

The average τ_f is assumed at mid depth i.e., $Z = \frac{L}{2}$ and by substitution into equation (6) and then equating eq's (2) and (6), then rearranging gives

$$K = \frac{T_f}{\pi L R^2 \gamma \tan\delta} \tag{7}$$

The K value approaches K_p for these driven piles, a phenomena reported by Coyle and Castello (10) from full scale load test results.

LATERAL LOAD STUDY

The comparison shown on Figures 4 and 5 suggest that in the elastic range of shear mobilization the pressuremeter prediction is good. To confirm that shear mobilized in pure torsion tests are valid as models appropriate under lateral load a series of piles were tested in clay and sand under lateral load. Lateral load tests were conducted free head by a main delivery wire passing over a pully and supporting increasing weights, pile deflection given by LVDT. Loads were applied in 0.2kN increments and held constant for 15 seconds. Predictions of the measured load-deflection response were made by the finite difference scheme COM622, Reese (11), with bielastic-plastic P-y curves. Based on the work of Smith (3) the front pressure, Q, contribution is found from the mobilized pressuremeter p* by

$$Q = p^* \times SQ \times D \tag{8}$$

where SQ = Shape factor (= 0.8 circular piles, = 1.0 square piles)

D = Pile Dia

p^* = Pressure mobilized in excess of at rest pressure

This occurs at equal radial strains, ε_θ, between pile and pressuremeter, (3), and hence pile movement y is given by

$$y = \varepsilon_\theta \times \frac{D}{2} \tag{9}$$

With the friction, τ, mobilization rate taken directly from the torsion test in the linear range then

$$F = \tau \times SF \times D$$

where SF = shape factor (=1.0 circular piles, = 2.0 square piles)

a P-y curve can be assembled using equation (1).

The comparisons between predicted and measured response from COM622 is shown in Figures 8 and 9 for clay and dense sand respectively.

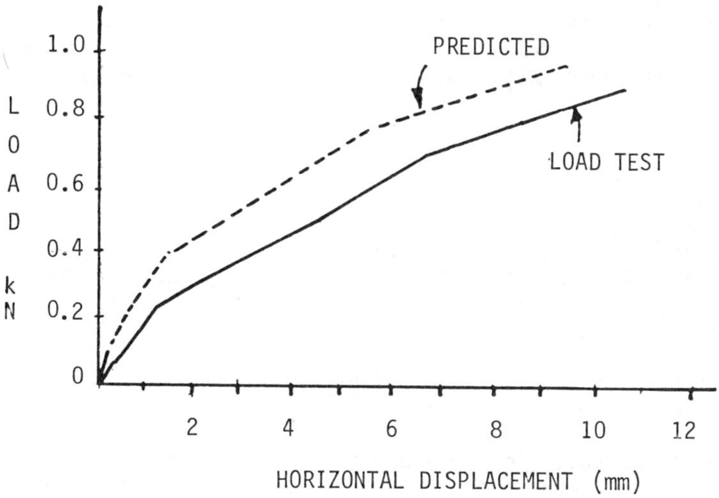

FIG. 8 - Lateral Load Test in Clay

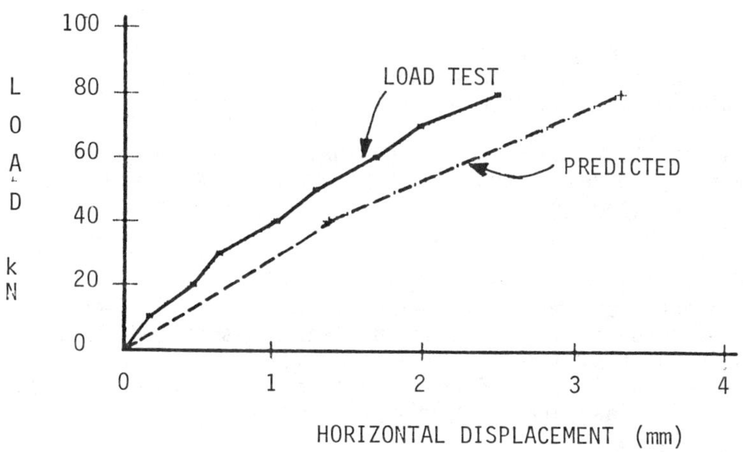

FIG. 9 - Lateral Load Test in Sand

Without any empiricism, both predictions are in good agreement with the measured response up to 12% and 50% radial movement in the clay and sand respectively. The stiffer prediction for the clay undoubtedly results from the neglect of critical depth effects. This would appear to substantiate that the friction mobilized in torsion testing is valid for use in lateral load calculation. The second phase of testing at Portland State is to formulate a method to predict non-linear lateral shear curves directly from pressuremeter test results.

CONCLUSIONS

Torsion and lateral load tests on model piles, together with published results of axial tests on full size piles, show that there is a rational link between the soil behavior in each test. From the results of this 1st stage study of lateral shear transfer the following tentative conclusion can be made:

1) Torsion testing provides directly the shear transfer soil response in laterally loaded piles to evaluate F-y curves.

2) The response of soils in torsion upto yield is dictated by the shear modulus, G. The pressuremeter is an ideal insitu tool to directly measure G. In the linear range of behavior the torsion test is well predicted by the pressuremeter.

3) From this limited study the amount of displacement to reach peak shear is a function of pile diameter and probably installation disturbance.

4) All of the most recent understood phenomena affecting vertical shear transfer appears to be present in lateral shear mobilization and can be studied by torsion testing.

5) Popular insitu tests such as the cone provide information directly for pile design under axial load only. A torsion test however may provide similar information for both axial and lateral designs.

As an interesting corollary to #5 it may be postulated that more specific soil properties would be obtained if the friction sleeve behind a cone is torqued to failure not pushed! This procedure would give the shear modulus by backcalcultion from the linear response portion, and allow calculation of the side shear contribution for both vertical and lateral load pile design.

With sound theoretically based design methods the role played by variables such as depth effects, soil/pile stiffness and embedment depth, can be better studied by torsion testing at the model level. In particular these torsion tests have shown that most full scale pile friction transfer phenomena can be represented without associated scale effects. A large range of full size pile/soil relationships may also be duplicated by torsion testing from fully rigid to fully flexible.

A further stage of testing currently underway at Portland State University to study lateral shear phenomena in frozen ground and cemmented sands and investigate the role played by pressurmeter testing. Of particular interest is the prediction of the full non-linear F-y curve from the shear stress strain curve given by the pressuremeter. At the conclusion of the research it is hoped to give a suggested procedure to construct F-y curves from pressuremeter tests that take account of installation effects and depth effects.

REFERENCES:

(1) Smith, T.D., and Ray, B., "Shear Mobilization on Laterally Loaded Shafts", to be published in the Geotechnical Aspects of Stiff and Hard Clays, ASCE Spring Convention, Seattle, April 7-11, 1986.

(2) Briaud, J.-L., Smith, T.D. and Meyer, B., "Laterally Loaded Piles and the Pressuremeter: Comparison of Existing Methods", ASTM STP 938, June 1983.

(3) Smith, T.D., "Pressuremeter Design Method for Single Piles Subjected to Static Lateral Load", Dissertation submitted to Graduate College, Texas A&M University, in partial fulfillment for Doctor of Philosophy, August 1983. Vol. I and II.

(4) Smith, T.D., "Lateral Shear Transfer on Model Piles", Research Report to Portland State University Graduate Office, Grant 90-050-5801 TSA, January 1986.

(5) Poulos, H.G., "Torsional Response of Piles", Journal of Geotechnical Engineering Division, ASCE, August 1981, pp. 1019-1035.

(6) Randolph, M.F., "Piles subjected to Torsion", Journal of Geotechnical Engineering Division, ASCE, August 1981, pp. 1095-1112.

(7) Coyle, H.M. and Reese, L., "Load Transfer of Axially Loaded Piles in Clay", Journal of Soil Mechanics and Foundations Division, ASCE, March 1966, p. 1-26.

(8) Vijayvergiya, V.N. and Focht, J.A., "A New Way to Predict Capacity of Piles in Clay", 4th Annual Offshore Technology Conference, OTC Texas, 1982.

(9) Aurora, R., Peterson, E.H. and O'Neill, M.W., "Model Studies of Long Piles in Clay", 13th Annual Offshore Technology Conference, OTC 4149, Texas, 1981.

(10) Coyle, M.M. and Castello, R., "New Design Correlations for Piles in Sand", Journal of Geotechnical Engineering Division, ASCE, July 1981, pp. 965-986.

(11) Reese, L., "Laterally Loaded Piles: Program Documentation," Journal of Geotechnical Engineering Division, ASCE, Vol. 103, April 1977, pp. 287-305.

Author Index

Subject Index